S0-AUF-343

PATTERNS OF MAMMALIAN REPRODUCTION

SECOND EDITION

PATTERNS OF MAMMALIAN REPRODUCTION

SECOND EDITION

By S. A. Asdell

Professor of Animal Physiology
New York State College of Agriculture
Cornell University

COMSTOCK PUBLISHING ASSOCIATES

A DIVISION OF

Cornell University Press

ITHACA, NEW YORK

CORNELL UNIVERSITY PRESS

First edition 1946

Second edition 1964

Library of Congress Catalog Card Number: 64–25162

PRINTED IN THE UNITED STATES OF AMERICA

BY KINGSPORT PRESS, INC.

PREFACE TO THE SECOND EDITION

IN this new edition each species has been treated separately in order to make the references readily available. Since the work has been done without the help of professional bibliographers, some errors may have crept in, although I have made every effort to avoid them. In dealing with the lesser-known species I have attempted to bring together reports from naturalists and from field expeditions as a contribution toward eventually determining the reproductive pattern for these species.

No attempt has been made in the new edition to bring geographical references into conformity with present political divisions. Names, now no longer current, of areas where research was originally done (such as the Belgian Congo, French Guinea, and the Cameroons) have deliberately been retained.

Critics of the first edition objected chiefly to the taxonomy. Several took exception to my identification of *Rattus norvegicus* with *R. rattus*. This error has been rectified, but no doubt new ones will be called to my attention, because whenever I have found a reasonable excuse to "lump" I have lumped. The general arrangement has been changed to conform as closely as possible with G. G. Simpson's classification. Probable errors have been changed to standard errors. The introductory chapter in the original edition has been omitted here in view of the many works on the physiology of reproduction now available.

It is a pleasure to acknowledge help from others. Mainly I am indebted to Biological Abstracts and Animal Breeding Abstracts. The bibliographies in the Journal of Mammalogy have also been helpful. For Japanese literature I had the help of Professor Y. Nishikawa and Dr. S. Mikami; for Russian, that of Mr. P. Kitzberg. The staffs of the Balfour Library, Cambridge, and The British Museum of Natural History, London, were helpful while I was on sabbatical leave in 1960. Dr. G. van Wagenen has checked the article on *Macaca mulata* and added to it from her vast experience with this species.

S. A. Asdell

Ithaca, New York
February 1964

PREFACE TO THE FIRST EDITION

DURING the past few years several good books have appeared dealing with reproduction in general and with certain phases of it in particular. The approach has been that of the physiologist who wishes to synthesize; the books have dealt with functions and with the relations of functions to structures. With the advance in knowledge it has become apparent that reproductive phenomena, while depending upon similar basic principles in all species, fit together in diverse ways to produce the very varied patterns which constitute species differences. Recent work has tended to emphasize the importance of the quantitative aspects of reproduction. In our laboratory we have found that this method of approach has given us the explanation of several peculiarities in the estrous cycle of the cow, and it has enabled us to make certain generalizations upon hormone thresholds and reproductive patterns that are useful. Accordingly, it seemed to me that a beneficial purpose would be served if the available information on mammalian reproduction were brought together, species by species, with particular stress, whenever possible, on the quantitative aspects. I hope that this compilation will help lay the foundations of a physiological zoology of reproduction, so well begun by Dr. F. H. A. Marshall. The work involved has been heavy, but when I broached the subject to Dr. Carl Hartman his reaction was so encouraging that I undertook it; I could hardly do otherwise with his own example in the field to inspire me.

I have attempted to pick out the most important work for each species—that which would give some idea of the species' quantitative reactions—and to present the histological and physiological foundations upon which further progress in the subject may be built. The information available has been pieced together in as logical an order as possible, with occasional comment in the nature of comparisons or notations of inconsistencies in the data. I have indicated in places, especially in the introductions to each Order, certain gaps in our knowledge, as suggestions to those who may have opportunities to carry on the work. Other problems will naturally suggest themselves if the abundance of knowledge regarding one species be compared with its dearth in another. I should like to point out here the comparative neglect from which the American fauna has suffered and to suggest that this is a field which might profitably be explored to the benefit of science and of the practical man, be he conservationist, pest controller, or animal breeder.

The literature is vast and widely scattered. I cannot fail to have made errors, but I hope that the sins of omission are more frequent than those of commission; the former do less harm in a work of this sort and can be more easily rectified. I trust that the book will be found sufficiently useful to warrant keeping it up to date. Accordingly, I shall be grateful if readers will draw my attention to papers containing material which should have been included....

In picking out my references I have made no attempt to settle questions of priority; my choice has always been dictated by circumstances of completeness, seeming reliability, or of availability. All who have worked in the field are, in a sense, coauthors, and share responsibility for the book.

It has not been easy for me to pick my way among the pitfalls of zoological nomenclature, and, doubtless, this part is open to a good deal of criticism. I have tried to adhere to the best of modern methods of classification and to avoid duplication of species....

Many statistical data have been brought together and recalculated, and I have taken considerable trouble to indicate the degree of variability of such data. The probable error has been used throughout, together with the range within which about 80 per cent of the material falls. The mode, also, has been reported wherever possible, since the practical man is mainly interested in what happens most frequently and in the range within which the majority of observations may be expected to fall. Averages are useful for comparing species but not for use within a species. The counts of the number of young in most wild rodents are embryo counts....

S. A. ASDELL

Ithaca, New York
January 1946

CONTENTS

Preface to the Second Edition	v
Preface to the First Edition	vii
List of Abbreviations	xi
Monotremata	1
Marsupialia	4
Insectivora	33
Dermoptera	63
Chiroptera	64
Primates	123
Edentata	177
Pholidota	182
Lagomorpha	183
Rodentia	213
Cetacea	410
Carnivora and Pinnipedia	425
Tubulidentata	509
Proboscidea	510
Hyracoidea	513
Sirenia	515
Perissodactyla	516
Artiodactyla	536
Index of Latin Names	653
Index of Vernacular Names	666

LIST OF ABBREVIATIONS

AA.	Anatomische Anzeiger
AB.	Archives de Biologie
AJA.	American Journal of Anatomy
AJP.	American Journal of Physiology
AM.	American Museum of Natural History, Bulletin
AMN.	American Midland Naturalist
AR.	Anatomical Record
AZ.	Arkiv for Zoologi
CE.	Contributions to Embryology, Carnegie Institute of Washington
CRAS.	Comptes Rendus de l'Académie des Sciences, Paris
CRSB.	Comptes Rendus de la Société de Biologie, Paris
CSIRO.	Commonwealth (Australia) Scientific and Industrial Research Organization
E.	Endocrinology
EPNA.	Exploracion du Parc National Albert
EPNG.	Exploracion du Parc National de la Garamba
EPNU.	Exploracion du Parc National de l'Upemba
HMCZ.	Harvard College Museum of Comparative Zoology, Bulletin
IAP.	International Académie Polonaise des Sciences et Lettres. Classe des Sciences, Mathématiques et Naturelles
JE.	Journal of Endocrinology
JM.	Journal of Mammalogy
JWM.	Journal of Wildlife Management
NAF.	North American Fauna
PRS.	Royal Society of London, Proceedings
PZS.	Zoological Society of London, Proceedings
QJMS.	Quarterly Journal of Microscopical Science
RSZ.	Revue Suisse de Zoologie
SB.	Senckenbergiana Biologica
TRS.	Royal Society of London, Transactions
TZS.	Zoological Society of London, Transactions
ZZ.	Zoologicheskii Zhurnal

PATTERNS OF MAMMALIAN REPRODUCTION

SECOND EDITION

Monotremata

TACHYGLOSSIDAE

Tachyglossus aculeata Shaw and Nodder

ECHIDNA

IN Tasmania the echidna breeds from near the end of June to the beginning of September (1). It breeds once a year and lays one egg, which is placed in the pouch by the mother. The pouch is visible in the embryo, but after hatching it is lost to sight until the beginning of the first heat (2). Normally, one egg is produced, although there were 4 exceptions in 140 cases. Both ovaries are functional, contrary to earlier statements. The passage through the oviduct is rapid. The first polar body is given off in the ovary and the second in the oviduct. With its coating of albumen and shell the egg is 4 mm. in diameter (1).

The oviduct consists of an infundibulum, a neck, and an albumen-secreting layer. The infundibulum is lined with tall, columnar, nonciliated secretory cells and with ciliated nonsecretory cells. The fluid from the secretory cells is clear and is secreted prior to ovulation. It is very plentiful and fills the space between the ovary and the infundibular membrane. The neck contains gobletlike mucoid cells, which secrete just before the egg leaves the follicle and early in fertilization. This secretion is not plentiful. The albumen-secreting cells produce two layers which cover the egg. The first layer is dense and is deposited during passage through the upper two thirds of the tube. It is produced mainly during the fertilization stages. The lower third produces the second layer of albumen, which is more fluid and is small in amount. This part of the oviduct is glandular, containing convoluted tubular glands as well as the secretory epithelium. The glands secrete before the ovum leaves the follicle, and the fluid is stored in the lumen (3).

At the tubo-uterine junction are more glands, which, with the glands of the tube, deposit the two-layered shell. The shell material is secreted by the basal part of the glands, while their middle and upper regions secrete a kind of uterine milk (3).

The intra-uterine period of incubation of the egg is said to be about 28 days (4), but the evidence only justifies the conclusion that it is between 12 and 28 days.

1. Flynn, T. T., and J. P. Hill. TZS., 24: 445–622, 1938–40.
2. Semon, R. W. In the Australian Bush. London, 1899.
3. Hill, C. J. TZS., 25: 1–31, 1941.
4. Broom, R. Proc. Linn. Soc., New South Wales, 10: 576–577, 1895.

ORNITHORHYNCHIDAE

Ornithorhynchus anatinus Shaw and Nodder

DUCK-BILLED PLATYPUS

The female platypus begins to breed when she is 16 inches long, but whether this is during the first or second season is uncertain. The adult length is about 18 inches. She breeds once a year, in July or August to mid-October. The breeding season is early in Queensland (July–August) and later in Victoria (September–October). Early in the season the testes, the crural glands, and the scent glands of the males enlarge. Copulation occurs in the water (1).

Only the left ovary is functional (2), and the eggs are always found in the left uterus (1). These are round and are 3 mm. in diameter at the time of fertilization; they immediately grow by imbibition of water and become oval. The number shed is 1 to 3, usually 2. As they pass down the oviduct, they receive a coat of albumen and a thin, leathery shell (3). In the uterus the eggs are apart, with the longer axis in the direction of the length of the uterus. At laying, the female squats in the nest on her rump with her tail between her legs and hands to receive the eggs (1). As it is laid the egg measures 16 to 18 mm. $\times$ 14 to 15 mm. (3). When two are laid they soon join, side by side (1). The sticky material which causes them to cement together is deposited in the oviduct (3). The average number of eggs or of young found in nests was 1.9 $\pm$ .05 (1).

A captive platypus in Australia laid eggs in two years. On one occasion the female mated for the first time on October 1 and laid an egg 15 days later. In the second year coitus occurred on October 11 and an egg was laid on October 25 or 26. Laying was thus about 14 days after first acceptance. On the earlier occasion incubation lasted from October 27 to November 7 or 8, an interval of 12 days. Lactation followed and lasted about 4 months (4).

Ovulation is followed by the folding of the follicular wall, which reduces the size of the cavity. This becomes partially filled with extravasated blood. The theca externa grows in to form lobule walls, and theca interna cells increase in numbers to form syncytial masses. The granulosa cells hypertrophy and produce the lutein tissue. The fully formed corpus luteum is 4.5 mm. × 3.5 mm. in diameter. Retrogression begins shortly before the eggs are laid and is accompanied by the invasion of numerous leucocytes. The formation of the corpus luteum in this, the most primitive mammal, is typical of that of most species (5).

1. Burrell, H. The Platypus. Sydney, 1927.
2. Flynn, T. T., and J. P. Hill. TZS., 24: 445–622, 1938–40.
3. Gatenby, J. B. QJMS., 66: 475–496, 1922.
4. Fleay, D. Victorian Nat., 61: 8–14, 29–37, 1944–5.
5. Hill, J. P., and J. B. Gatenby. PZS., 715–763, 1926.

Marsupialia

SOME progress has been made in the study of marsupial reproduction, chiefly in the family Macropodidae. Outstanding findings have been that the gestation period is shorter than the estrous cycle in the unmated female, and that development of blastocysts is delayed when young are in the pouch. Associated with short gestation is the apparent hormonic unimportance of the corpus luteum. The young are born in a very immature state and further development in the pouch tends to be prolonged. But, during lactation, blastocysts resulting from matings immediately after parturition remain dormant in the uterus for a long period. As there is no true implantation the term "delayed development" has been suggested as the appropriate one for this phenomenon.

Many species have a restricted breeding season. Litter size is related inversely to body size, with the smaller, rodentlike species having the larger litters. The number of young born frequently exceeds the number of teats so that considerable wastage exists. Pouch development is mainly under the influence of estrogens and, to a lesser extent, of progesterone. More work is needed on this problem.

DIDELPHIDAE

Philander opossum L.

WOOLLY OPOSSUM

The breeding season of this Central American opossum apparently lasts from March to August (1), but all the records mention pouch young and not uterine young. The number of pouch young varies from 2 to 7, with 5 by far the most frequent (2–6).

The writer has no means of determining whether all the references quoted deal with the same animal. The generic name seems to be far more variable than the data recorded.

1. Felten, H. SB., 39: 213–228, 1958.
2. Murie, A. Univ. Michigan Mus. Zool., Misc. Pub. 26, 1935.
3. Enders, R. K. Personal communication.
4. De Avilas Pires, F. D. Mus. Goeldi Hist. Nat., Belém, 19: 1–9, 1958.
5. Davis, W. B. JM., 25: 370–403, 1944.
6. Sanderson, I. T. PZS., 119: 755–789, 1949–50.

Dromiciops australis Philippi

MOUNTAIN OPOSSUM

Two to five young are produced at a birth but there are more fetuses, many of which perish before they are born (1, 2).

1. Mann, G. Invest. Zool. Chilenas, 2: 159–166, 1955.
2. Mann, G. Invest. Zool. Chilenas, 4: 209–213, 1958.

Marmosa

MOUSE OPOSSUMS

Marmosa cinerea Temminck. This opossum has no fixed breeding season and new-born young have been found in June, July, August, and October (1).

M. elegans Lund. During estrus and pregnancy the bursa ovarii is pulled downward and partially closed by the augmented weight of the genital tract. The oviduct and mesosalpinx fall over the opening, thus closing it altogether. There is no median vagina (2).

M. mexicana Merriam. This species probably breeds all year and has one litter each year (3). A female with 13 pouch young has been reported for June (4), and another with 10 for May (5).

M. murina L. A female with 10 pouch young was found (3).

1. Goodwin, G. G. AM., 87: 271–474, 1946.
2. Mann, G. Invest. Zool. Chilenas, 1(3): 11–16, 1951.

3. Enders, R. K. HMCZ., 78: 385–502, 1935.
4. Felten, H. SB., 39: 213–228, 1958.
5. Burt, W. H., and R. A. Stirton. Univ. Michigan Mus. Zool., Misc. Publ. 117, 1961.

Metachirus nudicaudatus Geoffroy

RAT-TAILED OPOSSUM

Pouch young (1 and 3) have been recorded in December (1).

1. Enders, R. K. HMCZ., 78: 385–502, 1935.

Didelphis marsupialis L.

COMMON OPOSSUM, VIRGINIA OPOSSUM

In Surinam this opossum has no set breeding season. Pouches of almost every female taken throughout the year are crammed with young, even those of individuals less than half grown (1). Two to three litters a year may be the rule, and the number of pouch young is up to 9, or more (2). Another record gives two seasons a year, June to July and the end of October (3). Counts of pouch young have varied from 2 to 9, the higher figures being the more frequent (4, 5). On the Mexican plateau breeding ceases in late fall and winter (6). The cycle lasts for about 7 days. The heat period, in the absence of the male, is not followed by a corpus luteum but by follicular atresia. (This is contrary to other reports.) The diestrous vaginal smear contains mainly leucocytes, which are not altogether absent at any time, but at the time of heat cornification of the wall sets in (7).

The remainder of this report covers the subspecies, *D. m. virginiana* Kerr, in the United States. In Texas the breeding season begins in January and lasts into October, with an anestrous period in November and December. The females are polyestrous and have at least two litters a year (8). In Florida heats begin in mid-January and most litters are born toward the end of that month and until February 10 (9). In southwest Georgia and northwest Florida one season extends from the second week of January to the first week of March and a second from the first week of April to the second week of June (10). In central California breeding extends from the third week of January to mid-October, and two litters a year is the rule (11). In Kansas breeding usually begins a little later, in February, with litters born mostly in the first half of March. A second season follows, with the young born mostly in the first half of June, but these young have difficulty in surviving the winter (12). In central Missouri breeding begins at about the same time as it does in Kansas and it is over by the beginning of September (13). In New York there are two well-defined seasons, the first from late January to mid-March and the second from mid-May to early July (14). The breeding season appears to be uniform throughout the range.

The entire population, including a majority of those less than a year old, take part in the first breeding season of the year (12). They are capable of breeding at the beginning of the following season regardless of their size at that time. This means an age at puberty of about 6 months for the females, but the males do not breed until they are about 8 months old (11).

The cycle recurs at about 28-day intervals and heat lasts from 1 to 2 days (15). Examination of vaginal smears gives an average cycle of 29.5 days, with a range of 22 to 38 days. Heat lasted a maximum of 36 hours but fertile matings took place only during the first 12 (11). Ovulation is spontaneous (16) and probably occurs early in heat (15), a fact that would account for the limited duration of fertility. Matings are mostly before 11 A.M. and usually before 7 A.M. (17). The average number of eggs shed is 22, and if one ovary is removed the remaining ovary sheds a larger number (16). The number of young is usually represented by counts of the pouch young so that the figures include survivors of various ages. Different reports give the following figures: 6.26 ($n = 50$) (9); 7.1 ± .18 (10); 7.2, range 4 to 11 (11); 8.9, mode 8, range 5 to 13 ($n = 42$) (13); 8.7, mode 9, which includes 22 per cent of all litters (14). In the last of these reports the proportion of males was 51.6 per cent.

The ova pass through the oviducts in 24 hours or a little less (16). The gestation period is about 12.5 days and the young are born very immature (16). Two observed periods lasted 12 days, 20.5 hours and 13 days, 4.5 hours respectively. A new heat period follows birth of the earlier litter of the year by an average of 97 days (11). No ovulations occur during lactation, which lasts for 70 to 80 days (17).

HISTOLOGY OF THE FEMALE TRACT

OVARIES. During anestrum the ovaries are very small and are packed with follicles, mostly in a state of atresia. No corpora lutea are present at this time. The corpus luteum reaches full size by 3 days after ovulation. It is maintained for 7 to 8 days, then it becomes a dirty yellow because of its lessened vascularity, and by 20 days it has almost disappeared (15). Six per cent of opossums have accessory ovarian tissue as stalked outgrowths from the ovaries (18). The ovaries are unique in the frequency of polynuclear ova, found in 93 per cent, and of polyovular follicles, found in 60 per cent of the ovaries studied (19). The ova are slightly elliptical and measure 0.165 × 0.135 mm. (17).

VAGINA. The vulva is within the rim of the cloacal aperture. There are a blind media vaginal canal and lateral ones shaped like interrogation marks. Spermatozoa travel up the lateral canals, and birth is by a temporary connection between the uterus and the median canal (16). The latter is lined during anestrum with an epithelium, 5 to 6 cells deep, of low columnar cells becoming flat next to the lumen. At proestrum this epithelium grows to 12 to 15 cells deep. Mitoses are abundant and no leucocytes are present. The vaginal smear consists mainly of large nucleated cells. At heat cornification sets in, with massive desquamation late in the period. There are no epithelial cells or leucocytes in the estrous smear, only cornified cells. During metestrum leucocytes invade the epithelium and appear in the smear. By 4 to 5 days epithelial cells have reappeared, and the smear also contains an occasional leucocyte as well as vacuolated epithelial cells. The lateral canals do not open until puberty. Sloughing in these organs is a more continuous process than it is in the median canal and is not limited to the end of heat. Cornification begins a little earlier; few leucocytes appear and most of the removal of debris is done by bacterial decomposition (15). During heat these canals are greatly distended by clear, thin, stringy mucus; just after heat they retrogress and are filled with a dry, cheesy mass of cell debris (16).

UTERUS. During proestrum the uteri are large, vascular, and turgid. After ovulation they swell and reach their maximum size at 5 to 6 days, while they are reduced by 13 to 15 days. In anestrum the glands are straight or slightly coiled, and the uterine epithelium is low-columnar or cuboidal. During heat the endometrium is more vascular and is edematous. Mitoses are present in the epithelium, which tends to become pseudostratified. The glands contain ciliated epithelium. In heat the glands are greatly coiled, their lumens are enlarged, and the epithelium is actively growing and is pseudostratified. During the lutein phase conditions are much the same, but there are fewer mitoses in the glands and the epithelial nuclei are basal. In later stages the capillaries are exceptionally large. At 15 days the endometrium is invaded by leucocytes, and by 20 days the uterus has returned to the resting stage (15).

The tubo-uterine junction has no folds of mucous membrane but is protected by a sphincter. The last 3 to 4 cm. of the tube contain numerous glands, and the lumen is much wider than it is in the Eutheria. This portion is extremely tortuous, and it enters the uterus at the side, opposite the mesentery (20).

PHYSIOLOGY OF THE FEMALE TRACT

Ovariectomy is always followed by the death of the embryos because of the collapse of the central layer of the endometrium (21).

The ovaries may be stimulated to secrete estrogens by the injection of P.M.S., if this is done from 100 days of age onward. Before this age the response given is the precocious conversion of embryonic egg nests to primordial follicles (22).

THE MALE

Spermatozoa are present in the testes of males at all times of the year (8). The testes increase in size from the age of 16 days to 60 days, then more slowly from 83 days to maturity. The seminiferous tubules are quiescent to day 100. Spermatozoa are present by 8 months, and androgens are secreted by 5 months. (23). A vaginal plug is formed at copulation and it lasts for less than 24 hours (17).

Castration at 22 days of age is followed by typical prostate differentiation (24), and the same is true for Cowper's glands and Bartholin's glands, at least for the first hundred days (25), but sex hormones do stimulate growth and histological differentiation (26). Young males treated with P.M.S. secrete androgens by the seventieth day of life or a little earlier (22).

As the young are born in a very immature condition, they afford a good opportunity to test the action of androgens and estrogens during what, in other species, is embryonic and fetal life. Androgens only mildly stimulate the Wolffian ducts in males, but in females the effect is greater and the ducts are preserved in an atypical manner. The effect on the Mullerian ducts is slight in the male, but extreme stimulation results in the female. Cessation of treatment is not readily followed by involution. The development of the prostate in males is markedly hastened and it is stimulated in the female, but there is little effect upon the testes and ovaries. Phallic stimulation is marked in both sexes, and the clitoris becomes a typical penis with a bifurcate glans (27).

Estrogens are decidedly toxic. The Wolffian ducts are stimulated in both sexes, more so in the male than in the female. There is localized stimulation of the Mullerian duct in the male. In the female it is precociously stimulated, with glandular development. Tremendous hyperplasia of the urogenital sinus occurs in both sexes with metaplasia and sloughing of cornified cells. There is no appreciable effect upon the testes or ovaries. Prostate development does not occur in males, but previously formed prostatic outgrowths grow to enormous proportions (27).

1. Sanderson, I. T. PZS., 119: 755–789, 1949–50.
2. Goodwin, G. G. AM., 87: 271–474, 1946.
3. Hill, J. P. QJMS., 63: 91–139, 1918.
4. Felten, H. SB., 39: 213–228, 1958.
5. Enders, R. K. HMCZ., 78: 385–502, 1935.
6. Dalquist, W. W. Mammals of the Mexican State of San Luis Potosi. Louisiana State University Press, 1953.
7. Martinez-Esteve, P. CRSB., 124: 502–504, 1937.
8. Hartman, C. G. J. Morph. and Physiol., 46: 143–215, 1928.
9. Burns, R. K., and L. M. Burns. RSZ., 64: 595–605, 1957.
10. McKeever, S. JWM., 22: 303, 1958.
11. Reynolds, H. C. Univ. California Publ. Zool., 52: 223–284, 1952.
12. Fitch, H. S., and L. L. Sandidge. Univ. Kansas Mus. Nat. Hist., Publ. 7: 309–338, 1952–5.
13. Reynolds, H. C. JM., 26: 361–379, 1945.
14. Hamilton, Jr., W. J. Cornell Univ. Agr. Exp. Sta., Mem. 354, 1958.
15. Hartman, C. G. AJA., 32: 353–421, 1923–4.
16. Hartman, C. G. Smithsonian Rep., 347–363, 1921.
17. McCrady, Jr., E. Am Anat. Mem., No. 16: 1938.
18. Hartman, C. G. CE., 380: 285–300, 1927.
19. Hartman, C. G. AJA., 37: 1–51, 1926.
20. Andersen, D. H. AJA., 42: 255–305, 1928.
21. Hartman, C. G. AJP., 71: 436–454, 1925.
22. Moore, C. R., and C. F. Morgan. E., 32: 17–26, 1943.
23. Moore, C. R., and C. F. Morgan. E., 30: 990–999, 1942.

24. Moore, C. R. AR., 80: 315–327, 1941.
25. Rubin, D. J. Exp. Zool., 94: 463–473, 1943.
26. Rubin, D. J. Morphol., 74: 213–282, 1944.
27. Moore, C. R. Physiol. Zool., 14: 1–45, 1941.

Didelphis paraguayensis Oken. In the Matto Grosso this opossum has been recorded with 10 and 7 well-developed pouch young in September (1).

1. Kühlhorn, F. Säugetierk Mitt., 1: 115–122, 1953.

Chironectes

WATER OPOSSUMS

Chironectes minimus Zimmermann. The young of this opossum are born in December and January (1).

C. panamensis Goldman. The male of this species has a pouch (2).

1. Cabrera, A., and J. Yepes. Historia Natural Ediar. Mamiferos Sud-Americanos. Buenos Aires, 1940.
2. Enders, R. K. HMCZ., 78: 385–502, 1935.

DASYURIDAE

PHASCOGALINAE

Phascogale (*Antechinus*) *flavipes* Waterhouse

MARSUPIAL MOUSE

This marsupial mouse reaches puberty at about 320 days, i.e., in its first season after birth. Reproduction is restricted to early August and the young are born early in September after a gestation lasting 31.5 days (range 30 to 33). Heat lasts for about 4.3 days (range 2 to 7) and mating is prolonged, lasting 5 hours (1). The females are apparently monestrous (2). The average litter size is 6.8 (range 3 to 8) (1), but actual births of 8, 10, 11, and 12 have been recorded (3).

The pouch develops about 16 days before parturition and the young are suckled for 90 days. The pouch retrogresses about 35 days after the young have been weaned.

1. Marlow, B. J. Australian J. Zool., 9: 203–218, 1961.
2. Marlow, B. J. In Sharman, G. B., Monogr. Biol., 8: 332–368, 1959.
3. Fleay, D. Victorian Nat., 65: 273–277, 1948–9.

Phascogale tapoatafa Meyer

MARSUPIAL RAT

The pouch area develops in May and the young are born early in July. The young are free at 9 weeks. In one instance 3 were born, in another, 4, but a wild female has been recorded as probably bearing 8 young, judging by the condition of the teats (1).

1. Fleay, D. Victorian Nat., 51: 89–100, 1934–5.

Phascogale (*Antechinus*) *apicalis* Gray. There is a record of a female with 7 young attached to her teats (1).

P. (*A.*) *maculatus* Gould. A record of the pigmy pouched mouse from the Cape York Peninsula gives a female with 5 pouch young in March (2).

P. (*Planigale*) *ingrami* Thomas. This species has a season limited to the wet period. Litters of 8 and 12 have been recorded (3).

1. Troughton, E. Furred Animals of Australia. New York, 1947.
2. Marlow, B. J. JM., 43: 433–434, 1962.
3. Davies, S. J. J. F. J. Roy. Soc., Western Australia, 43: 63–66, 1960.

Dasycercus cristicauda Krefft

This pouched mouse breeds from June to September and 7 young are usually born at a time. The female has practically no pouch; the young cling to her nipples for considerably over a month, and she staggers round with this burden (1). The gestation period lasts 29 to 31 days and the young remain attached to the teats for about 58 days (2).

1. Wood-Jones, F. The Mammals of South Australia. Adelaide, 1923–4.
2. Fleay, D. Victorian Nat., 78: 160–167, 1961.

Sminthopsis crassicaudata Gould

This pouched mouse breeds in June and July and gives birth to 6 young at a time (1).

1. Wood-Jones, F. The Mammals of South Australia. Adelaide, 1923–4.

Antechinomys laniger Gould

JERBOA POUCHED MOUSE

The jerboa pouched mouse has a litter of 6 to 8 young at a time (1).

1. Lucas, A. H. S., and W. H. D. Le Souef. The Animals of Australia. Melbourne, 1909.

DASYURINAE

Dasyurus viverrinus Shaw

MARSUPIAL CAT

The marsupial cat is monestrous and has one season a year, from late May or early June to early August, but the male does not appear to have a corresponding rutting season. Proestrum lasts 4 to 12 days and heat 1 to 2 days. Ovulation is spontaneous, and it occurs about 5 days after copulation or heat. The duration of pregnancy is therefore rather difficult to determine, but it probably lasts 8 to 14 days. In its absence pseudopregnancy ensues (1). The number of eggs shed varies from 20 to 35 with 20 to 25 as the usual number (2). The number of embryos is correspondingly high, and as the number of nipples in the pouch is usually six a considerable wastage of embryos results (1). The young remain attached in the pouch for 7 to 8 weeks (3). The diameter of the egg averages 0.24 mm. (2).

The ripe graafian follicles are borne on projecting bosses, and when they rupture they decrease in size. There is but little intrafollicular hemorrhage. The corpora lutea grow in 3 days, are yellowish white, and persist for the greater part of lactation, about 7 to 8 weeks, after which they decline (3). It is practically impossible to distinguish two thecal layers, the whole structure resembling theca externa. There are no mitoses in the developing corpus luteum, and the theca produces connective tissue only (4).

Proestrum is marked by an edematous swelling of the lips of the cloacal aperture, and the pouch enlarges somewhat and becomes tumid as the result of the enlargement of the sebaceous glands in this area. The sweat glands enlarge and coil, causing the interior of the pouch to become moist and somewhat sticky. The uteri enlarge, vascularity increases, and the glands are coiled in the basal portions of the endometrium. The uterine epithelium is low-columnar and is in a state of active mitosis. Leucocytes are abundant (1).

The changes during heat are a continuation of those observed in proestrum. The glandular epithelium is ciliated throughout. In the postestrous period, between heat and ovulation, the uterine condition is maintained, but the cloacal swelling gradually subsides. The first polar body is extruded while the egg is still in the ovary (1).

In pseudopregnancy the pouch continues to enlarge, and the development of the sweat and sebaceous glands reaches the level observed during pregnancy. Toward the end of the period the female often cleans out the pouch in the manner of the pregnant animal, as a preparation for the reception of the young. The uterine glands become very contorted and enlarged, so that the condition resembles the extreme lutein reaction observed in rabbits during pseudopregnancy. The epithelium is tall-columnar. At the end of the period much cellular desquamation into the gland ducts occurs, and the uterine epithelium undergoes the same change. Leucocytes are abundant but there is no extravasation of blood. The duration of pseudopregnancy is probably about two weeks (1).

Copulation is very prolonged, up to about two hours, and spermatozoa remain alive in the oviducts for about 2 weeks, an unusually long time which is probably related to the delayed ovulation in this species (1).

The long median vagina, which opens at the time of parturition, closes rapidly afterwards (5).

1. Hill, J. P., and C. H. O'Donoghue. QJMS., 59: 133–174, 1913.
2. Hill, J. P. QJMS., 56: 1–134, 1910–11.
3. Sanders, F. P. Proc. Linn. Soc., New South Wales, 28: 364–405, 1903.

4. O'Donoghue, C. H. Anat. Anz., 41: 353–368, 1912.
5. Hill, J. P. Anat. Anz., 18: 364–373, 1900.

Dasyurus quoll Zimmermann. The single breeding season a year lasts from the end of May to the first fortnight of August (1).

D. (*Dasyurinus*) *geoffroyi* Gould. This species is recorded as having 6 pouch young (2).

D. (*Dasyurops*) *maculatus* Kerr. Females of this species mate in mid-June and July, when the pouch develops. The gestation period lasts for 3 weeks (3), and the number of young is from 4 to 6 (4).

D. (*Satanellus*) *hallucatus* Gould. Findings include a September record of two females, one with 8 young attached and one with 4 (5), also two of 8 each for the same month. In the latter the pouch young were about one month old (6).

1. Hill, J. P., and C. H. O'Donoghue. QJMS., 59: 133–174, 1913.
2. Lucas, A. H. S., and W. H. D. la Souef. The Animals of Australia. Melbourne, 1909.
3. Fleay, D. Victorian Nat., 56: 159–163, 1939–40.
4. Troughton, E. Furred Animals of Australia. New York, 1947.
5. Tate, G. H. H. AM., 102: 199–203, 1953.
6. Fleay, D. Victorian Nat., 78: 288–293, 1962.

Sarcophilus harrisi Boitard

TASMANIAN DEVIL

One female has been recorded as mating in May. The pouch develops in April and the young are born in June and July. They are free from the pouch in slightly over 15 weeks. The first breeding is usually in their second season and the number of young born is from 3 to 4 (1). The gestation period lasts 31 days (2). One specimen is recorded with 21 newly discharged ova (3). In June spermatogonial mitoses were plentiful but further stages of spermatogenesis and mature spermatozoa were absent (4).

1. Fleay, D. Victorian Nat., 62: 100–105, 1935–6.
2. Kenneth, W. H. Gestation Periods. Edinburgh, 1943.
3. Flynn, T. T. Ann. Mag. Nat. Hist., 9(10): 225–231, 1922.
4. Sharman, G. B. Monogr. Biol., 8: 332–368, 1959.

THYLACININAE

Thylacinus cynocephalus Harris

TASMANIAN WOLF

Some females breed at any time of the year but most pups are born in May and from July to August. This suggests an extended season of about 4 months with two periods of activity separated by 2 months (1).

1. Guiler, E. R. JM., 42: 396–397, 1961.

MYRMECOBIINAE

Myrmecobius fasciatus Waterhouse

NUMBAT

This marsupial anteater mates from late midsummer to early autumn, i.e., from December to April. The young are usually born in summer, autumn, or early winter, from January to April or May. They are carried by the mother, who has no pouch, through the winter (1) and are detached from the teats in late August or September (2). The usual number of young is 4, but counts have varied from 2 to 4 (1).

In Western Australia dividing spermatogonia were present in the testes in September, but further stages of spermatogenesis and mature spermatozoa were absent (3).

1. Calabay, J. H. PZS., 135: 183–207, 1960.
2. Calabay, J. H. Australian Mus. Mag., 13: 145–146, 1960.
3. Sharman, G. B. Monogr. Biol., 8: 332–368, 1959.

PERAMELIDAE

Perameles nasuta Geoffroy

BANDICOOT

Two well-grown pouch young have been reported in September (1). The male has no vesiculae seminales and no ampulla to the vas deferens. Spermatozoa can be found in the urine (2). Electron microscopy of the spermatozoon has shown that the tail has two groups of fibrils. One consists of a complex axial filament and the other of peripheral fibrils (3).

1. Tate, G. H. H. AM., 102: 199–205, 1953.
2. Bolliger, A., and A. L. Carrodus. Med. J. Australia, 25: 1118–1119, 1938.
3. Cleland, K. W., and Lord Rothschild. PRS., 150B: 24–42, 1959.

Perameles eremiana Spencer. This species usually has 2 young at a time (1).

P. fasciata Gray. The females usually have 2 to 4 young (1).

P. myosura Wagner. The marl breeds in May or June and usually gives birth to 2 young, but occasionally to 3 (1).

1. Troughton, E. The Furred Animals of Australia. New York, 1947.

Thylacomys leucurus Thomas

BILBY

The bilby breeds from March to May but in the central area of South Australia rain and food supply regulate the breeding season. The number of young is 1 or 2 (1).

1. Wood-Jones, F. The Mammals of South Australia. Adelaide, 1924.

Chaeropus castanotis Ogilby

This bandicoot breeds in May and usually has 2 young (1).

1. Wood-Jones, F. The Mammals of South Australia. Adelaide, 1924.

Thylacis (Isoodon) macrourus Gould

This bandicoot is polyestrous and breeds all the year round. The females reach puberty when they are half grown, i.e., in their first year. The gestation period is less than 15 days, and up to 7 young may be born at a time. They remain attached to the teats for 7 weeks and continue to suckle for 10 days longer. One female produced 8 litters, at least 32 young, in 17 months (1).

1. Mackerras, M. J., and R. H. Smith. Australian J. Zool., 8: 371–382, 1960.

Thylacis (Isoodon) obesulus Shaw and Nodder. This bandicoot breeds in June and usually has 4 young at a birth (1). Another report gives 2 as the usual number, but occasionally up to 4 or 5 (2).

1. Wood-Jones, F. The Mammals of South Australia. Adelaide, 1924.
2. Rayment, F. Victorian Nat., 70: 194–196, 1953–4.

PHALANGERIDAE

PHALANGERINAE

Phalanger orientalis Pallas

GRAY CUSCUS

A female with 1 young has been found in June (1).

1. Tate, G. H. H. AM., 102: 199–203, 1953.

Trichosurus vulpecula Kerr

BRUSH-TAIL POSSUM

The possum has two main breeding seasons in the year, in March and in August (1), but most births occur in the autumn (April) season (2). Some females that have bred in the first season breed again in the later one (3). In the Adelaide area of South Australia births have been recorded for all months of the year except January and February, but most are in March and April (4). In New Zealand, where the species was introduced and now flourishes, the same pattern is evident but some reproduction is found all the year round (5). In the London zoo a total of 60 births were scattered throughout the year (6).

The time of first breeding in the females is at 250–300 days old and 1,700 g. weight (2). They are polyestrous, and ovulation is spontaneous. The estrous cycle varies in length from 22 to 58 days, but all cycles longer than 32 days began between April 20 and June 16. During the rest of the breeding periods the cycle lengths lay between 22 and 32 days, with a mean of 25.7 ± .3 days (3).

The gestation period lasts for 17.5 days and the number of young is usually 1 (3). In New Zealand 90.3 per cent of those examined had 1 young, 0.2 per cent had twins while 9.5 per cent were not pregnant (5). In the London zoo 6 sets of twins and 1 of triplets were found in the 60 births (6). Lactation inhibits estrous cycles and the females come in heat at an average of 8.0 ± .2 days after the suckling young are removed from the pouch (3). In the full-grown nullipara the pouch develops in January and February (7).

HISTOLOGY OF THE FEMALE TRACT

The ripe graafian follicle is 3 mm. in diameter. During formation of the corpus luteum theca mitoses and possibly granulosa mitoses may be found but no cells divide by mitosis after the fourth day. The corpus luteum reaches its maximum size of 4.1 mm. at 7 to 10 days after ovulation (3).

Epithelial cells are present in the vaginal smear at all times but they are fewest from 5 days after heat to 2 days before it. Occasional leucocytes are present during proestrum but they disappear entirely during heat. A few are present on the day after heat and they become abundant from the second to the fifth day. At no time is there much difference in the proportion of epithelial and cornified cells. During proestrum the vaginal epithelium consists of 2 or 3 layers of cells and some are cornified. At the time of heat the layers increase to 3–6, and the cornified cells are sloughing off (3).

At estrus the endometrium is edematous and the median vagina is distended with fluid (5). The luteal, glandular phase reaches its maximum at 7 to 8 days after heat and, in both the gravid and the nongravid uterus, it is over after the fifteenth day. During gestation the gravid uterus is much larger than the nongravid one. This phase lasts from 12 days after heat until after parturition. In the uterine epithelium mitoses are present until 5 days after heat. During the luteal phase no basophil granules are found in the cells at the periphery of the glands but they are present in those near the lumen (3).

The median vaginal passage opens for birth of the young through it; about 7 days later it closes. A hormonal influence is believed to control this (5).

The level of pregnanediol excretion in the urine is the same in pregnant and nonpregnant females (3).

One injection of 900 I.U. of estrone into the immature female caused the characteristic yellowish-brown discoloration of the pouch area. The latter is increased in size. In the nonbreeding adult 900 I.U. had no effect, but 12,000 I.U. caused pigment development (7). The injection of 5,000 to 50,000 I.U. of estrogen into mature females caused the pouch to thicken and pigment to be secreted. The median vagina swelled, and its epithelium was keratinized. In parous females 200 I.U. of estrogen caused the pouch pigment to develop (1).

In pouch young, 3 months old, the injection of 2 mg. of progesterone once a week caused the pouch rudimenta to increase in size and orange-brown pigment to be secreted, but retrogression set in after 3 months in spite of continued injections. Complete pouch formation did not result as it did if estrogens were injected. In nulliparous females the injection of 2 to 5 mg. of progesterone weekly caused the pouch to retrogress to such an extent that it practically disappeared. Similar injections of 5 mg. of progesterone weekly into a multiparous opossum with a large pouch caused it to retrogress in 3 weeks (8). The injection of progesterone into immature males caused the testes to ascend (9).

THE MALE

The male has no seminal vesicles or ampulla to the vas deferens, and spermatozoa may be found in the urine (10). They are fertile all year round (5) and the testis size does not undergo a seasonal variation (2). Estrogen given to a castrated, adolescent male transformed the scrotum into a permanent pouch. This did not happen in the full-grown male, nor did castration bring it about. The injection of gonadotrophic hormones hindered the change. Injected testosterone caused retrogression of the pouch that had been formed (11).

1. Bolliger, A., and A. L. Carrodus. J. Roy. Soc., New South Wales, 73: 218–227, 1940.
2. Dunnet, G. M. CSIRO. (Australia) Wildlife Res., 1: 1–18, 1956.
3. Pilton, P. E., and G. B. Sharman. JE., 25: 119–136, 1962.
4. Sharman, G. B. Monogr. Biol., 8: 332–368, 1959.
5. Tyndale-Briscoe, C. H. Australian J. Zool., 3: 162–184, 1955.
6. Zuckerman, S. PZS., 122: 827–950, 1952–3.
7. Bolliger, A., and A. L. Carrodus. J. Roy. Soc., New South Wales, 71: 615–622, 1938.
8. Bolliger, A., and A. L. Carrodus. J. Roy. Soc., New South Wales, 73: 228–232, 1940.
9. Bolliger, A., and A. L. Carrodus. Nature, 144: 671, 1939.
10. Bolliger, A., and A. L. Carrodus. Med. J., Australia, 25: 1118–1119, 1938.
11. Bolliger, A., and A. J. Tow. JE., 5: 32–41, 1947.

Trichosurus caninus Ogilby. In Victoria one young is born in May or June. This remains attached in the pouch for 6 weeks and finally leaves it at 4 months of age (1).

1. Fleay, D. Nat. Hist. (New York), 59: 352–355, 1950.

Acrobates pygmaeus Shaw

PIGMY GLIDER

Usually young numbering more than 4, which is the number of teats, are born at a time (1).

1. Troughton, E. Furred Animals of Australia. New York, 1947.

Cercaertus concinnus Gould

MUNDARDA

This marsupial is probably polyestrous, as pouch young have been recorded for January (1), March (2), July, and September. The number varies from 1 to 6, with 5 as the average (3).

1. Wood-Jones, F. The Mammals of South Australia. Adelaide, 1924.
2. Glauert, L. Western Australian Nat., 2: 54–57, 1949–51.
3. Troughton, E. Furred Animals of Australia. New York, 1947.

Petaurus breviceps Waterhouse

SUGAR GLIDER

The young, usually 2, are born in July (1), but later as one travels north from South Australia (2). Another account mentions two females each with 2 pouch young found in May (3), while in the London zoo 22 births have been spread throughout the year and almost every other birth was of twins (4).

1. Fleay, D. Victorian Nat., 49: 97–101, 1932–3.
2. Troughton, E. Furred Animals of Australia. New York, 1947.
3. Tate, G. H. H. AM., 102: 199–203, 1953.
4. Zuckerman, S. PZS., 122: 827–950, 1952–3.

Petaurus sciureus Shaw

GLIDER

In the London zoo 15 births were spread throughout the year and almost every other birth was of twins (1). The subspecies, *P. s. norfolcensis* Kerr, has no fixed season; 2 young are the rule and the gestation period lasts for 3 weeks (2).

1. Zuckerman, S. PZS., 122: 827–950, 1952–3.
2. Fleay, D. Victorian Nat., 70: 208–210, 1953–4.

Spilocuscus nudicaudatus Gould

CUSCUS

This species has 2 to 4 young at a birth (1).

1. Troughton, E. Furred Animals of Australia. New York, 1947.

TARSIPEDINAE

Tarsipes spenserae Gray

NOOLBENGER, HONEY POSSUM

This species has 1 to 4 young at a time (1).

1. Troughton, E. Furred Animals of Australia. New York, 1947.

PHASCOLARCTINAE

Phascolarctos cinereus Goldfuss

KOALA

The koala is monestrous (1) and it mates from September to late January (2, 3). It usually breeds every other year (3). The gestation period lasts for about 35 days and 1 young is the rule (2). In the San Diego zoo 1 young was born in mid-May after a gestation of between 25 and 30 days. It remained attached to the teats for 5½ months and was finally weaned at about 10 months (4). Another San Diego record gives a new pouch young in September (5). According to one report, however, corpora lutea of various ages are present in the ovary at one time, thus indicating that the female is polyestrous (6).

1. Lewis, F. Victorian Nat., 51: 73–76, 1934–5.
2. Troughton, E. Australian Mus. Mag., 11: 396–401, 1955.
3. Troughton, E. Furred Animals of Australia. New York, 1947.
4. Pournelle, G. H. JM., 42: 396, 1961.
5. Pournelle, G. H. Inter Zoo Year Book, 2: 83, 1960.
6. O'Donoghue, C. H. QJMS., 61: 433–473, 1916.

Pseudocheirus

RING-TAILED POSSUMS

Pseudocheirus convolutor Oken. In Tasmania the testes are shrunken and spermatogenesis is absent during January (1). Two females are recorded with 2 very small pouch embryos (2).

P. cooki Desmerest. This species breeds from May to August. Usually 6 young are born but there are only 3 functional teats. Identical twins are produced occasionally (3).

P. laniginosus Gould. The pouch develops in April and May and the young, 2 or 3, are born from May to July. They are free in 6 to 7 weeks and finally leave the pouch in about 12 weeks (4).

P. peregrinus Boddaert. Records have been given of two females with 2 young each and with 1 each in August, and also of two with 1 each in October (5). The species is polyestrous (1).

1. Sharman, G. B. Monogr. Biol., 8: 332–368, 1959.
2. Pearson, J., and J. M. De Bavay. Australian J. Zool., 7: 13–21, 1959.
3. Flynn, T. T. Ann. Mag. Nat. Hist., 9(10): 225–231, 1922.
4. Fleay, D. Victorian Nat., 44: 279–282, 1927–8.
5. Tate, G. H. H. AM., 102: 199–203, 1953.

Schoinobates volans Kerr

GLIDER-POSSUM

One young is born in July or August. It is free from the teat at 6 weeks of age and leaves the pouch in 4 months (1).

1. Fleay, D. Victorian Nat., 50: 135–142, 1933–4.

PHASCOLOMIDAE

Phascolomis (*Vombatus*)

WOMBATS

Phascolomis hirsutus Perry. This wombat breeds from April to June and the young are carried in the pouch until December (1).

P. mitchelli Owen. A July birth has been recorded in the London zoo (2).

P. ursinus Shaw. This species breeds all year round and has a single young (3).

1. Troughton, E. Furred Animals of Australia. New York, 1947.
2. Zuckerman, S. PZS., 122: 827–950, 1952–3.
3. Weindorfer, G., and G. Francis. Victorian Nat., 36: 165, 1919–20.

Lasiorhinus latifrons Owen

In South Australia this wombat appears to breed early in the summer (1).

1. Sharman, G. B. Monogr. Biol., 8: 332–368, 1959.

MACROPODIDAE

MACROPODINAE

Lagorchestes fasciatus Peron and Le Sueur

HARE WALLABY

This wallaby has 1 young at a time (1).

1. Troughton, E. Furred Animals of Australia. New York, 1947.

Petrogale

ROCK WALLABY

Petrogale inornata Gould. A female with 1 pouch young in September has been recorded (1).

P. pearsoni Thomas. The young leaves the pouch in November or December (2).

P. penicillata Griffith, Smith and Pidgeon. In the London zoo this wallaby has bred all the year, producing 1 young at a time (3).

P. xanthopus Gray. In the London zoo this wallaby has bred all the year and 2 sets of twins have been observed in 36 births (3).

1. Tate, G. H. H. AM., 102: 199–203, 1953.

2. Wood-Jones, F. The Mammals of South Australia. Adelaide, 1924.
3. Zuckerman, S. PZS., 122: 827–950, 1952–3.

Onychogalea frenata Gould

NAIL-TAILED WALLABY

A single record is available from the London zoo of 1 young born in September (1).

1. Zuckerman, S. PZS., 122: 827–950, 1952–3.

Thylogale

TAMMAR

Thylogale billardierii Desmarest. In the London zoo this wallaby has produced young in January, February (twice), and July; 1 young at a birth (1).

T. eugenii Desmarest. The anterior vaginal canals are greatly hypertrophied at the time of heat. One egg is fertilized and delayed development is the rule while the female is lactating. The species is polyestrous (2). In the London zoo births have occurred from April to October. One set of twins was observed in 14 births (1).

T. stigmatica Gould. Records of single pouch young have been obtained in April and May on four occasions, and 1 each in June, July, and September (3).

T. thetis Lesson. The pademelon has produced young once in the London zoo, in September (1). Embryo development is delayed during lactation (4).

1. Zuckerman, S. PZS., 122: 827–950, 1952–3.
2. Sharman, G. B. Australian J. Zool., 3: 156–161, 1955.
3. Tate, G. H. H. AM., 102: 199–203, 1953.
4. Pilton, P. E. Nature, 189: 984–985, 1961.

Protemnodon (Wallabia) rufogrisea Desmarest

BRUSH WALLABY

In Tasmania this wallaby breeds from October to November and 1, extremely rarely 2, pouch young may be found in March and April (1). The breeding season has also been recorded as occurring from August to September and from December to February (2). In the London zoo it has bred all the year round. In 51 births 6 sets of twins and 1 of triplets were recorded (3). The gestation period lasts for 40 days (3).

1. Weindorfer, G., and G. Francis. Victorian Nat., 37: 5–8, 1920–21.
2. Hill, J. P. QJMS., 63: 91–139, 1918.
3. Zuckerman, S., PZS., 122: 827–950, 1952–3.

Protemnodon agilis Gould. In each of two females 1 pouch young was found, 1 in July and 1 in August (1).

P. bicolor Desmarest. The black wallaby breeds all the year in the London zoo. One set of twins was recorded in 7 births (2).

P. *cangura* Muller. One female has been recorded with 1 young in September (1), and breeding has also been recorded in December (3).

P. dorsalis Gray. In the London zoo this wallaby breeds all the year. One young is born at a time (2). The presence of corpora lutea of different ages in the ovary at one time indicates that the species is polyestrous (4).

P. elegans Lambert. In the London zoo there has been one record of a single birth in September (2). In its native habitat there is no well-defined breeding season (5).

1. Tate, G. H. H. AM., 102: 199–203, 1953.
2. Zuckerman, S. PZS., 122: 827–950, 1952–3.
3. Caldwell, H. W. TRS., 178B: 463–486, 1887.
4. O'Donoghue, C. H. QJMS., 61: 433–473, 1916.
5. Troughton, E. Furred Animals of Australia. New York, 1947.

Macropus major Shaw

GRAY KANGAROO

In Australia the breeding season is from early September to early January (1). In the London zoo the species breeds all the year round. One young at a time is the rule; twins are rare (2). The females are polyestrous. The gestation period lasts for 38 or 39 days. The corpus luteum of pregnancy has become reduced in size by March and has disappeared by August. Delayed development is not a feature of this species (1) as it only breeds once a year (3).

1. Pilton, P. E. Nature, 189: 984–985, 1961.
2. Zuckerman, S. PZS., 122: 827–950, 1952–3.
3. Wood-Jones, F. The Mammals of South Australia. Adelaide, 1924.

Macropus rufus Desmarest

RED KANGAROO

In its native habitat this kangaroo has been recorded with newborn young in all months of the year (1). In the London zoo it also breeds all the year round and a single young is born (2). The gestation period lasts for 32 to 34 days (3). Delayed development of the embryos while young are in the pouch is probable (4).

1. Sharman, G. B. Monogr. Biol., 8: 332–368, 1959.
2. Zuckerman, S. PZS., 122: 827–950, 1952–3.
3. Sharman, G. B., and P. E. Pilton. PZS., 142: 29–48, 1964.
4. Sadlier, R. M., and J. W. Shield. Nature, 185: 335, 1960.

Macropus (Osphranter) robustus Gould

WALLAROO

In the London zoo the wallaroo breeds all the year and has 1 young (1). Delayed development during lactation is usual (2). In a single specimen the testes weighed 5.8 g. and the seminiferous tubules were 301 m. long. The length of a spermatozoon was 49.4 μ (3).

1. Zuckerman, S. PZS., 122: 827–950, 1952–3.
2. Sadlier, R. M., and J. W. Shield. PZS., 135: 642–643, 1960.
3. Knepp, T. H. Proc. Pennsylvania Acad. Sci., 12: 61–64, 1938.

Macropus fuliginosus Desmarest. The sooty kangaroo breeds once a year. The single young is born in January and it remains in the pouch until October (1).

1. Wood-Jones, F. The Mammals of South Australia. Adelaide, 1924.

Setonix brachyurus Quoy and Gaimard

QUOKKA

The female quokka breeds at 18 months, when she weighs about 2 kg. (1). The breeding season extends from late January to the end of October, with anestrum from November to January. Quokkas are polyestrous and heats are about 28 days apart. Ovulation is spontaneous, occurring about one day after copulation. One egg is shed at a heat period. The gestation period lasts from 24 to 28 days, but usually 26 to 27 days. Heat and ovulation follow immediately after parturition. This produces a corpus luteum of lactation which is maintained for 5 or 6 months and, during the first half of this time, the female is anestrous (2). If she is lactating, development of the uterine embryos may be delayed and the young may be born after an interval of 5 months (3).

HISTOLOGY OF THE FEMALE TRACT

The diameter of the mature follicle is 2.8 mm., and that of the egg, 120 to 130 μ. The corpus luteum measures 4 mm., or more (1). The vaginal smear contains partly cornified cells at the time of heat, and sloughing of the epithelium is heavy after heat is over. In the uterus epithelial and glandular cells are dividing at the time of proestrum and the epithelium is pseudostratified. Mitoses are present in some granulosa cells at the time of heat. The corpus luteum originates from the granulosa cells with some thecal elements (4). No ciliated cells are to be found in the oviducts (1), but the egg is able to reach the uterus in one day (4). The pseudovaginal canal remains open after the first parturition.

Behavioral estrus lasts for 12 hours and the uterine luteal phase, 18 days (4). The young leave the pouch at 25 weeks but they still take milk for 3 or 4 months. When they are too large to enter the pouch they feed from the exterior upon an elongated nipple (1).

The testes descend to the scrotum at 8 weeks (4).

1. Waring, H., G. B. Sharman, D. Lovat, and M. Kahan. Australian J. Zool., 3: 34–43, 1955.
2. Sharman, G. B. Nature, 173: 302, 1954.
3. Sharman, G. B. Australian J. Zool., 3: 56–70, 1955.
4. Sharman, G. B. Australian J. Zool., 3: 44–55, 1955.

Dendrolagus

TREE KANGAROOS

Dendrolagus matschiei Förster and Rothschild. This New Guinea kangaroo has borne young in the London zoo in January and December (1). Copulation may be necessary for ovulation to occur (2).

D. ursinus Schlegel and Müller. In the London zoo a single young has been born in April (1).

1. Zuckerman, S. PZS., 122: 827–950, 1952–3.
2. Matthews, L. H. PZS., 117: 313–333, 1947.

POTOROINAE

Bettongia cuniculus Ogilby

RAT KANGAROO

The rat kangaroo in Tasmania breeds at least from March to December, probably all the year, and is polyestrous. Ovulation is spontaneous, and one ovum is shed at a time (1). In the London zoo 26 births have been spread throughout the year. Of these, two were of twins (2). There is a post-parturition heat. Pregnancy is normally in alternate uteri, and while one is pregnant the other remains in a pseudopregnant state for the whole period. Lymph is expelled from this uterus just before parturition. The period of gestation is probably 6 weeks. Puberty is early, when the young are reasonably free from the pouch (1).

During heat mitoses are frequent in the endometrium, the epithelial cells show vacuolar degeneration, the gland cells are actively dividing, and the stroma is very edematous and is invaded by polymorphs. The epithelium and glands are strongly ciliated. In early pseudopregnancy (or pregnancy) there are no mitoses and fewer cilia and the glands are convoluted and actively secreting. Later, the epithelium is low-columnar, cilia are few, the number of capillaries increases greatly, and the glands straighten near the surface and remain convoluted near their bases. Then the epithelium disintegrates, a great invasion of leucocytes occurs, and connective tissue cells move up and replace the surface epithelium (1).

1. Flynn, T. T. Proc. Linn. Soc., New South Wales, 55: 506–531, 1930.
2. Zuckerman, S. PZS., 122: 827–950, 1952–3.

Bettongia lesueuri Quoy and Gaimard. The single young of this marsupial leaves the pouch in November, after a pouch life of nearly four months (1).

1. Wood-Jones, F. The Mammals of South Australia. Adelaide, 1924.

Aepyprymnus rufescens Gray

RAT KANGAROO

In the London zoo 15 births have been recorded. These were scattered throughout the year and included two sets of twins (1).

1. Zuckerman, S. PZS., 122: 827–950, 1952–3.

Caloprymnus campestris Gould

RAT KANGAROO

This marsupial has an irregular breeding season, with 1 young at a time being the rule (1). The male has no vesiculae seminales or ampulla to the vas deferens. Spermatozoa are found in the urine (2).

1. Troughton, E. Furred Animals of Australia. New York, 1947.
2. Bolliger, A., and A. L. Carrodus, Med. J. Australia, 25: 1118–1119, 1938.

Potorous tridactylus Kerr

POTOROO

In captivity the female potoroo breeds first at 1 year old, at a weight of 840 g., but in the wild at a lighter weight (1). The species breeds more or less all year but the favored times are from late August to early October and in January and February (2). The species is polyestrous and monovular with a cycle of 42 days (range 39 to 44). Heat lasts up to 12 days and ovulation is prabably coitus-induced. The gestation period is from 30 to 43 days from copulation to parturition and a new heat may occur at 42 days after the preceding one, i.e., soon after the end of pregnancy. Then development of the new embryo is delayed if a pouch young is present at the same time. The blastocyst may remain dormant for 4½ months. During this time further estrous cycles are suppressed. The pouch young is permanently attached to the teat for 64 days (3).

The percentage of cornified cells in the vaginal smear at the time of heat is more than 80 (3). Birth occurs through the lateral vaginal canal (4). Vascularity in the pouch increases a few hours before parturition takes place (3).

1. Guiler, E. R. Australian J. Sci., 23: 126–127, 1950.
2. Guiler, E. R. JM., 39: 44–58, 1958.
3. Hughes, R. L. Australian J. Zool., 10: 193–224, 1962.
4. Flynn, T. T. Proc. Linn. Soc., New South Wales, 47: xxviii, 1923.

Hypsiprymnodon moschatus Ramsay

MUSKRAT KANGAROO

This marsupial breeds from February to May, i.e., in the rainy season (1). Two is the usual litter size (2).

1. Troughton, E. Furred Animals of Australia. New York, 1947.
2. Lucas, A. H. S., and W. H. D. La Souef. The Animals of Australia. Melbourne, 1909.

Insectivora

THE insectivores need much more investigation, especially as the little that is known reveals some interesting patterns. This is particularly so in the families whose position is somewhat uncertain: those which are placed by some in this Order and by others in the Primates. In certain species of *Elephantulus,* for instance, much larger numbers of eggs are shed than embryos develop, the condition resembling that found in some marsupials, while the bleeding from the uterus, probably at the end of diestrum, may bear some relationship to the primate menstruation. Induced ovulation is probably rather frequent in some families, e.g., in the Soricidae. Another interesting point is that the corpus luteum of pregnancy does not seem to be of much importance in the maintenance of gestation. The interstitial gland in female European moles, with its growth parallel to that of the clitoris in the nonbreeding season, deserves further study. Its embryology and the possibility of its occurrence in other Talpidae are worth investigating. Altogether this Order appears to offer a more fertile field for study than any other.

TENRECIDAE

Tenrec ecaudatus Schreber

This Madagascar species is probably pregnant from November to January since young may be found from the end of January to April. Pregnant females have been taken in December and January (1). There is probably a long anestrum during the rainy season (2). Embryo counts have ranged from 10 to 32, with 16 to 18 as common figures (1, 3, 4).

1. Kaudern, W. AZ., 9(1): 1–22, 1914.

2. Feremutsch, K., and F. Strauss. RSZ., 56(suppl. 1): 1–110, 1949.
3. Rand, A. L. JM., 16: 89–104, 1935.
4. Goetz, R. H. Zeit. Ges. Anat., I, Zeit. Ent. Gesch., 106: 315–342, 1937.

Setifer (Ericulus) setosus Schreber

Pregnant specimens of this Madagascar insectivore have been taken in September and October, and also from December to early February (1). It is polyestrous, with cycles of 5 to 6 weeks and possibly a long anestrum in the rainy season (2). In 5 embryo counts the number has been either 2 or 3 (1).

The follicles are solid and fertilization takes place within the ovary. The fertilized ova move towards the surface; they are devoid of a corona radiata when they are shed. The corpus luteum is formed from the granulosa cells, which are extruded from the follicles and spread over the surface of the ovary (2). Neoformation of ova can occur but it is exceptional and probably not periodic. If it occurs it does so early in pregnancy. Some newly formed ova may be produced deep in the ovarian stroma and not from the germinal epithelium (3).

The bursa ovarica has a minute opening to the abdominal cavity (2). Its epithelium is ciliated; the cells divide during the ovulatory phase and secrete after it is over (4).

The uterine glands are poorly developed during anestrum, and the epithelium is cubical. In proestrum the glands become longer and more branched. The epithelium becomes columnar. During heat these growth phenomena continue and in metestrum there is much glandular development that is continued during the luteal phase. The epithelium becomes pseudostratified, with very tall cells (2). The gland cells contain mitoses in the ovulatory phase, not in the luteal phase, when they begin to secrete. In the ovulatory phase, too, mitoses are frequent in the stroma, which is edematous (4). The endometrium is edematous during proestrum (5).

1. Kaudern, W. AZ., 9(1): 1–22, 1914.
2. Feremutsch, K., and F. Strauss. RSZ., 65(suppl. 1): 1–110, 1949.
3. Kon, L. RSZ., 53: 597–623, 1946.
4. Feremutsch, K. RSZ., 55: 567–622, 1948.
5. Strauss, F. RSZ., 53: 511–517, 1946.

Hemicentetes semispinosus Cuvier

This Madagascar insectivore is polyestrous, with a cycle about 3 or 4 weeks long. Possibly there is a long anestrum during the rainy season. The follicle is solid; fertilization takes place within the ovary, and the fertilized ova move towards the surface. They are surrounded by a corona radiata when they are shed (1).

The bursa ovarii is ciliated; its cells divide during the ovulatory phase and secrete when it is over. Mitoses appear in uterine gland cells during the ovulatory phase and are absent in the luteal phase, when secretion begins. The stroma contains cells in mitosis and is edematous during the ovulatory phase (2).

1. Feremutsch, K., and F. Strauss. RSZ., 56(suppl. 1): 1–110, 1949.
2. Feremutsch, K. RSZ., 55: 567–622, 1948.

POTAMOGALIDAE

Potamogale velox Du Chaillu

One pregnant female is recorded for October (1).

1. Frechkop, S. EPNU., No. 14, 1944.

CHRYSOCHLORIDAE

Chrysochloris damarensis Ogilby

GOLDEN MOLE

The breeding season of this South West Africa mole is from April to July. New-born young are found in July and August. One record of a female gave 2 embryos (1).

1. Broom, R. Trans. South African Philos. Soc., 18: 283–311, 1907–9.

Chrysospalax villosus A. Smith

ROUGH-HAIRED GOLDEN MOLE

This mole becomes pregnant shortly after the rains have set in. Two young at a time is the usual number (1).

1. Roberts, A. Ann. Transvaal Mus., 4: 65–107, 1913–4.

ERINACEIDAE

ECHINOSORICINAE

Echinosorex gymnurus Raffles

MOONRAT

The moonrat of Borneo has been recorded with 2 small embryos in June (1).

1. Banks, E. J. Malay Br. Roy. Asiatic Soc., 9(2): 1–139, 1931.

ERINACEINAE

Aethechinus frontalis A. Smith

SOUTH AFRICAN HEDGEHOG

The time of birth is in summer (1), and the number of young is from 2 to 7.

1. FitzSimons, F. W. The Natural History of South African Mammals. London, 1919.

Erinaceus europaeus L.

HEDGEHOG

The hedgehog is said to be monestrous with two seasons a year (1) and litters are born in May–June and August–September (2). In captivity, however, there seems to be a continuous season from May to September (3), and this is probably what happens in the wild with peaks of births in the months cited. The gestation period seems to be variable as two reports give 34 to 49 days (4) and 35 to 40 days (5). In captivity the minimum has been 31–32 days, the usual, 34–35, and the maximum, 39 days (3). Ovulation is not spontaneous (6). The average litter size is 4.6 (3). The earlier-born young do not reach puberty in their first year (7).

HISTOLOGY OF THE FEMALE TRACT

The ovaries are roughly U-shaped, bent round a thick muscular hilus, and are exceptionally well vascularized. They are completely enclosed in a tough capsule, and the stroma is fibrous with some fat-containing cells. Pigment is very common in both the ovaries and uterus, especially during the breeding seasons. During the winter anestrum immature ova only are found and follicular degeneration is especially marked. Besides the usual types of atresia, another occurs in which several small follicles collapse producing glandular-looking cell masses. In April the ovaries become active (7).

There is a remarkable growth of the granulosa during the period of follicular expansion, and the preovulatory size of the follicle is 1.25 mm. In the absence of mating these cells fail to luteinize although normal vascularization occurs; blood is extravasated and fibrous tissue grows. The fully formed corpus luteum of ovulation is 0.7 to 1.0 mm. in diameter; it contracts to 0.5 mm. at the next ovulation and remains stationary at that size for some time, but the corpora remain separate, seldom becoming confluent as do those of pregnancy and of pseudopregnancy. After copulation pseudopregnancy is frequent, and the corpus luteum then formed is the usual type, with luteinization of the granulosa cells. This type grows to 1.1–1.4 mm. The corpus luteum of pregnancy is similar to that of pseudopregnancy. New follicles grow and reach the size usual at estrus, but they degenerate. The corpora of pregnancy and of pseudopregnancy often become confluent, and they are very slow to retrogress. After copulation the general vascular condition shows a marked increase. The ovum is 70 μ in diameter (7). There is no postparturient heat.

The vagina is large and muscular; it remains open at all times. In the upper part the lumen is large and the wall relatively thin; in the middle region it is more muscular, the lumen is much smaller, and a pair of large glands are present.

The lower vagina is still narrower, with large lymph nodes and numerous accessory glands embedded in the wall. The vagina is greatly dilated during heat; it does not subside during diestrum but does in pseudopregnant cycles. The inference may be drawn that the hedgehog is not truly monestrous, but only appears to be so in the wild state. The epithelial cycle is normal, with growth, cornification, and sloughing, which is most marked in the upper vagina. The vaginal glands regress during anestrum (7). There is considerable secretion of mucus by the glands during the breeding season (1).

The horns of the uterus are short, thick, and muscular and at their junction lie almost at right angles to the cervix and vagina. Well-defined cervical fornices are present. No striking changes occur in heat or in the noncopulatory cycle, but in proestrum new glands develop from epithelial ingrowths. In pseudopregnancy the epithelium is much folded, their is slight growth and secretion, and the stroma becomes edematous. In pregnancy one finds some early epithelial destruction and extravasation of blood and pigment into the lumen (7).

During heat the epithelial cells of the uterus measure 2.0 μ high, while in anestrum they are reduced to 5 to 7 μ. The gland cells are reduced from 16.5 to 10–12 μ. The stroma is edematous during estrus (8).

The oviduct is bent round so that it lies against the tip of the uterine cornu. There is a conspicuous parovarium. The ciliated epithelial cells grow as estrus approaches, and after ovulation some cells slough off and secretion is apparent, but none of the changes are marked. The tube becomes edematous during the breeding season (7).

During the winter anestrum follicles will ripen and the reproductive tract enters the estrous condition if females are kept in the laboratory at a temperature of 70 to 75° F. The temperature seems to be more important than increasing the daily amount of light. However, ovulation rarely occurs among these animals (9).

THE MALE

The male has a definite breeding season with spermatozoa from April to August. There is no true scrotum but the testes descend into a perineal pouch. The accessory organs are very large during the breeding season. In September or October the whole tract rapidly becomes smaller and, after a period of dormancy, again increases in March of the following year. The seminal vesicles increase sixtyfold in weight from anestrum to rut. In the sexually active animal the reproductive tract may represent 10 per cent of the body weight (2, 10, 11). The interstitial cells of the testis proliferate during the spring and begin to retrogress toward the end of July. In fall and winter this tissue is atrophic. The accessory organs parallel the interstitial cell development, and their involution is rapid after castration (12). Although spermatozoa are not produced during the nonbreeding season, some spermatocytes are always present, but the male does not respond to the stimulation of laboratory conditions as does the female (10).

1. Courrier, R. CRSB., 90: 808–809, 1924.
2. Marshall, F. H. A. J. Physiol., 43: 247–260, 1911.
3. Morris, B. PZS., 136: 201–206, 1961.
4. Herter, K. Zeit. f. Säugetierkunde, 8: 195–218, 1933.
5. Ranson, R. M. J. Hyg., 41: 131–138, 1941.
6. Zajaczek, S. Bull. Int. Acad. Polon. Sci. Lettr., Cl. Sci. Math. Nat., Ser. B, 939: 379–403, 1946.
7. Deanesly, R. TRS., 223B: 239–276, 1934.
8. Schütz, H. Zeit. f. Mikros.-Anat. Forsch., 59: 463–522, 1952–3.
9. Allanson, M., and R. Deanesly. PRS., 116B: 170–185, 1934.
10. Allanson, M. TRS., 223B: 277–303, 1934.
11. Schütz, H. AA., 101: 84–94, 1954.
12. Courrier, R. AB., 37: 173–334, 1927.

Atelerix albiventris Wagner

The gestation period is about 1 month long and 3 to 4 young are born at a time (1).

1. Jeannin, A. Les Mammifères Sauvages du Cameroun. Paris, 1936.

Hemiechinus auritus Gmelin

LONG-EARED HEDGEHOG

The litter size varies from 1 to 4 with records of 2 females giving birth to 2 and 4 young respectively in August. Other females had 2 young each, born from July to September, mostly in August and September (1). Another record gives a litter of 6, born in August (2).

1. Prakash, I. JM., 41: 386–389, 1960.
2. Gupta, B. B., and H. L. Sharma. JM., 42: 398–399, 1961.

Paraechinus micropus Blyth

Litters have varied from 1 to 5 and all records give births in August (1, 2).

1. Prakash, I. JM., 41: 386–389, 1960.
2. Gupta, B. B., and H. L. Sharma. JM., 42: 398–399, 1961.

MACROSCELIDIDAE

Macroscelides proboscideus Shaw

Pregnancies have been observed in August and September, and 8 females each bore 2 embryos. This is the usual number of young born (1). However, numerous eggs are shed from the ovaries at heat (2).

1. Shortridge, G. C. The Mammals of South West Africa. London, 1934.
2. van der Horst, C. J. JM., 25: 77–82, 1944.

Nasilio brachyrhynchus A. Smith

Pregnancies have been observed in January, June, and November (1); the January ones near term or just born (2), though another account gives the season of birth in Angola as about April (3). The number of fetuses or young has been 1 or 2 about equally (1, 2).

1. Loring, —. In G. C. Shortridge, The Mammals of South West Africa. London, 1934.
2. Lawrence, B., and A. Loveridge. HMCZ., 110: 1–80, 1953.
3. Hill, J. E. JM., 22: 81–85, 1941.

Elephantulus myurus Thomas and Schwann

ELEPHANT SHREW

The classification of the South African elephant shrews has presented some difficulties to the systematists. Allen (1) gives two subspecies of *Elephantulus rupestris* A. Smith, namely, *jamesoni* and *myuri*. Van der Horst gives the names of the subspecies with which he works as *E. myurus jamesoni*. He states that *E. rupestris* sheds only 2 eggs at a time (2), while *E. myurus jamesoni* sheds about 120, so the classification used by him is probably more nearly correct.

The age at puberty is about 5 weeks (3). The breeding season begins about the end of July and finishes in January, with a spread from December to March. In a state of nature, menstruation is seen only at the end of the breeding season because pregnancy occurs throughout the season. In captivity the species does not breed; the natural seasonal polyestrum is replaced by anestrum. Ovulation is spontaneous and the cycle is believed to last about 14 days. Ovulation may occur shortly after parturition but there is no lactation anestrum. Normally the ovulation process extends over 1 or more days and about 120 eggs are shed from the 2 ovaries. Animals that become pregnant in January tend to abort; the corpora lutea degenerate suddenly, a condition not found in the infertile cycle. As a rule there are only 3 pregnancies in the life of the female (3). The number of young at a birth is 1 to 2 (4) and the duration of pregnancy is about 8 weeks (5).

Although up to 60 eggs are shed only one is found in each uterine horn. They pass rapidly through the first part of the oviduct and collect within a swelling of the central part of this tube, remaining there for some time, after which they pass rapidly through the remainder of the tube. The central portion of the oviduct, termed the egg chamber, is nonciliated. Transport through the last part is by means of weak muscular contractions (6). The embryos remain at the 4-cell stage until they are implanted (7). Bleeding and desquamation occurs from a polyplike growth in the uterus apparently at the end of diestrum (8).

THE FEMALE TRACT

OVARY. The amount of fluid in the graafian follicles diminishes before ovulation until it remains as a narrow rim between the cumulus oöphoron and the granulosa. The latter layer enlarges and its cells begin to luteinize before ovulation (9, 10). The theca cells enlarge and multiply, but only in the deeper part of the follicle. As ovulation approaches, the granulosa cells at the cap of the follicle move toward the interior. At ovulation this attenuated cap tears off, and the contents of the follicle bulge out and spread over the surface of the ovary, so that the contents of the numerous everted follicles tend to fuse. The theca grows outwards as a core to each corpus luteum, and the whole structure becomes more convex, giving the peripheral layer a compact appearance. Germinal epithelium grows over the lutein masses. Gradually the thecal cores diminish in amount, leaving a layer of granulosa-lutein cells covered by one of germinal epithelium. These changes have not been accurately timed, but the last stage is reached before the uterine polyp is formed. The lutein tissue is fully encapsulated during the polyp stage, and the corpus luteum is degenerating when it begins to show necrotic changes (see below). The mode of vascularization of the lutein cells is believed to differ from that found in other mammals, as groups of 4 or more cells are arranged in close relation to a capillary loop in contrast to the usual investiture of each cell with a capillary network of its own (10).

UTERUS. In the immature elephant shrew the uterine epithelium consists of low-columnar to cuboidal cells with compressed nuclei arranged so that the epithelium has a pseudostratified appearance. The stroma has a narrow subepithelial cellular zone 15 to 20 μ thick, and the cells are basophilic. The glands are shallow. In anestrum the appearance is the same except that secretion is found in the lumen of the uterus, and the glands appear to be cystic (8).

The estrogenic phase at the beginning of the breeding season is marked by great edema of the endometrium, particularly in its basal portion. The glands increase in length, but remain straight; the cells divide and enlarge. As heat progresses, the edema spreads toward the lumen of the uterus, and the stromal cells become vacuolated; late in heat polymorphs invade the endometrium. At ovulation red blood cells are extravasated in the region of the thickened stroma, i.e., in the superficial layer (8).

During the lutein phase the edema slowly subsides, earliest from the superficial layers in which it has appeared last. The epithelium is high-columnar and pseudostratified; the glands become coiled. When the corpus luteum begins to degenerate, the uterine muscle contracts, and a polyplike body develops as part of the endometrium projecting into the lumen from its mesometrial pole. It is accompanied by considerable local edema, and the glands dilate. The epithelial cells become exceedingly tall, and the polymorphs which had previously invaded the endometrium disappear. Eventually the growth forms a pendulous mass hanging down into the uterus (8, 11). This structure does not develop if pregnancy and implantation occur so that it is essentially a part of the cyclical changes (12).

At a time which is evidently the end of diestrum, the capillaries in the polyp dilate and lacunae appear in the subepithelial layer. Bleeding occurs into the glands, cellular necrosis sets in, and the structure enlarges because of the edema and bleeding. Then the epithelium ruptures, and the whole necrotic mass pours out into the lumen of the uterus. Regeneration is rapid and it may overlap the estrogenic phase, so that late in the season the uterus of the shrew in heat may present a very complex series of changes. The writers cited believe that this very remarkable series of changes may be a forerunner of the menstrual changes found in the higher primates, and, in view of the disputed systematic position of this genus, the suggestion is interesting (8). Another interesting point is the partial resemblance between the formation and coalescence of the corpora lutea in this species and the way in which among New World monkeys they coalesce with the interstitial tissue by the disappearance of the thecal coat.

If coitus occurs and the female is fertilized, most of the ova begin their development, but, as the area of possible implantation is limited, only one is implanted and develops. The remainder are mostly imprisoned in the upper part of the uterus by a strong contraction of the circular muscle and degenerate at the 1-, 2-, or 4-cell stage (13).

THE MALE

The testes are abdominal and interstitial cells are few. The testes are active all the year and show no seasonal changes. The epididymis, too, undergoes no seasonal changes but the accessory glands are active from July to

January and then they retrogress. Interstitial cell size increases from June to August, is at a maximum in October, and then decreases to a minimum which is reached in April. Spermatogenesis is slow during the non-breeding season but it never ceases (14).

1. Allen, G. M. HMCZ., 83: 1–763, 1939.
2. van der Horst, C. J. JM., 25: 77–82, 1944.
3. van der Horst, C. J. TRS., 238B: 27–61, 1955.
4. Shortridge, G. C. The Mammals of South West Africa. London, 1934.
5. McKerrow, M. J. M. TRS., 238B: 62–98, 1955.
6. van der Horst, C. J., and J. Gillman. South African J. Med. Sci., 8: 41–49, 1943.
7. van der Horst, C. J., and J. Gillman. South African J. Med. Sci., 7: 47–71, 1942.
8. van der Horst, C. J., and J. Gillman. South African J. Med. Sci., 6: 27–47, 1941.
9. van der Horst, C. J., and J. Gillman. Nature, 145: 974, 1940.
10. van der Horst, C. J., and J. Gillman. South African J. Med. Sci., 7: 21–41, 1942.
11. van der Horst, C. J. Trans. Roy. Soc., South Africa, 31: 181–199, 1946.
12. van der Horst, C. J., and J. Gillman. South African J. Med. Sci., 7: 134–143, 1942.
13. van der Horst, C. J., and J. Gillman. AR., 90: 101–106, 1944.
14. Stock, Z. G. TRS., 238B: 99–126, 1955.

Elephantulus capensis Roberts. This elephant shrew sheds about 120 eggs at a time (1).

E. intufi A. Smith. This shrew probably breeds all the year (2). As a rule 2 eggs are shed at a time (1), and 2 embryos are found in the uterus (3).

E. renatus Kershaw. Lactation in December has been recorded (4).

E. rozeti Duvernoy. The pectoral glands develop at the same time as the testes (5). A female with 2 young fetuses has been recorded in May (6).

E. rupestris A. Smith. Embryos have been reported in September (3), April and May (7). Five counts of 2 embryos and six of 1 have been made. Only 2 eggs are shed at a time (1).

1. van der Horst, C. J. JM., 25: 77–82, 1944.
2. van der Horst, C. J. TRS., 238B: 27–61, 1955.
3. Shortridge, G. C. The Mammals of South West Africa. London, 1934.
4. Allen, G. M., and A. Loveridge. HMCZ., 75: 47–140, 1933–4.
5. Lang, H. JM., 4: 261–263, 1923.
6. Lataste, F. Act. Soc. Linn., Bordeaux, 39: 129–299, 1885.
7. Loring, —., and E. Heller. In G. C. Shortridge, The Mammals of South West Africa. London, 1934.

Petrodromus

Petrodromus robustus Thomas. There is a record of a pregnant female found in January (1).

P. rovumae Thomas. The single young is born in October (2).

P. tetradactylus Peters. One or 2 eggs at a time are released from the ovaries (3) and there is a record of 3 females each with a single embryo, collected in December (4).

1. Frechkop, S. EPNU., No. 14: 1944.
2. Loveridge, A. J. East Africa and Uganda Nat. Hist. Soc., 17: 39–69, 1922.
3. van der Horst, C. J. JM., 25: 77–82, 1944.
4. Heller, E. In G. C. Shortridge, the Mammals of South West Africa. London, 1934.

Rhynchocyon cirnei Peters

Specimens collected in Nyasaland in September had small embryos (1). One collected in January was lactating (2).

1. Lawrence, B., and A. Loveridge. HMCZ., 110: 1–80, 1953.
2. Allen, G. M., and A. Loveridge. HMCZ., 75: 47–140, 1933–4.

SORICIDAE

SORICINAE

Sorex araneus L.

COMMON SHREW

The common Old World shrew has a breeding season from May to September (1) or October (2). In northern Norway puberty occurs at about 8 g. weight or 70 mm. head and body length (3). The male may reach breeding condition in April following the spring of its birth but, in the laboratory, puberty may be reached in the first year (4). In Poland the number of females reaching puberty in their first year is low, about 0.3 to 2.0 per cent. No males attain puberty in the year of their birth (5). Heat lasts for a few hours (6).

The onset of the first heat of the season is gradual, and so is the season's cessation, with a greater number of failures to become pregnant at the post-parturition heat occurring as the year advances. Females that do not become pregnant pass into the lactation anestrum without the presence of corpora lutea in their ovaries. Breeding lasts longer in the southern parts of Great Britain than it does toward the north. No condition resembling pseudo-pregnancy has been observed. The average number of embryos is 6.45, and of corpora lutea 7.35. The litter size declines after July because of increased intra-uterine mortality, (2). In another account 7.7 was the average number of embryos observed in May, 7.7 in June, and 5.7 in August (5).

The duration of gestation is about 20 days and lactation lasts about 21, while there is usually a heat about 25 days after parturition (7). However, if the female is pregnant during lactation, delayed implantation may supervene (1). Under laboratory conditions 3 or 4 litters a year may be born (7). The mean litter size is largest, 8.2, at the beginning of the season but at the end it has dropped to five. Fetal regression increases in August and September but usually 1 or 2 young per litter are lost, not whole litters (1).

The ovary contains but few polyovular follicles. Ovulation is spontaneous; the number of eggs shed at a time is greatest in May, while the mode for the year is 8 (1).

The ripe graafian follicle is 350 μ in diameter and the ovum 72 μ. The corpus luteum has a central cavity for a while, but little blood is found in it. The mature corpus is about 500 μ in diameter, but retrogression sets in before parturition and is marked by the presence of mitoses in the luteal cells. At parturition the diameter is reduced to half. The ovary is surrounded by a closed capsule (2).

Only the lower part of the vagina is closed in immature females. The epithelium is intensely cornified during the first heat, and toward its end there is hyperemia and extravasation of leucocytes, which pass into the lumen. Cornification does not occur in the postparturition heats. The upper and lower parts of the vagina differ, with a circular fold at the point of change. The lower part is thinner and the wall more folded (2).

The uterus has a single cervix, and but little glandular tissue can be found in the endometrium. There is a characteristic fibrous zone between the mucosa and the muscle in the uterus of the nonparous female. The part of the oviduct proximal to the ovary is ciliated. Changes during the cycle are slight, and secretion occurs only while the ova are in transit (2).

Spermatogenesis begins in March and continues at least into November. The prostate is relatively large. The males do not breed in their year of birth; either do the females, though they tend to display precocious sexual activity in the fall. They apparently die at the end of their first breeding season (2).

1. Tarkowski, A. K. Ann. Univ. Mariae Curie Sklodowska, Sect. C, 10: 177–244, 1955–7.
2. Brambell, F. W. R. TRS., 225B: 1–62, 1935.
3. Hoyte, H. M. D. J. Anim. Ecol., 24: 412–425, 1955.
4. Wolska, J. Ann. Univ. Mariae Curie Sklodowska, Sect. C, 7: 497–539, 1952.
5. Crowcroft, P. The Life of the Shrew. London, 1957.
6. Pucek, Z. Acta Theriol., 3: 269–296, 1960.
7. Dehnel, A. Ann. Univ. Mariae Curie Sklodowska, Sect. C, 6: 359–377, 1952.

Sorex minutus L.

LESSER SHREW

The lesser European shrew begins to breed in mid-April in Great Britain. The season is at its height in June and is over in October. The young do not breed during their first year. The first heat is prolonged, with vaginal cornification and gradual hypertrophy of the sexual organs. The largest follicles average 330 μ in diameter, and the corpora lutea about 457 μ. The latter disappear quickly. There is a postparturient heat, and, if the female does not become pregnant at this time, lactation anestrum follows. Like *S. araneus,* breeding animals do not survive beyond their first season; unlike that species, no fibrous zone is found between the endometrium and the uterine muscle in young females. The average corpus luteum count is 6 to 8, and that of embryos is 6.2 (1).

In Poland from 4 to 10 per cent of females reach puberty in their first year. These are members of the first spring litter. No males become sexually mature in their first year (2).

1. Brambell, F. W. R., and K. Hall. PZS., 957–969, 1936.
2. Pucek, Z. Acta Theriol., 3: 269–296, 1960.

Sorex palustris Richardson

WATER SHREW

Pregnant and lactating females are found between February and August. Ovulation is possibly induced and there is a post-partum heat followed by pregnancy, though the embryos usually degenerate (1). Embryo counts have varied from 5 to 7; mean, 5.8; mode, 6 (1 to 4). The number of young born is given as 4 to 7 (5).

The ovaries develop and the uterine glands proliferate in January. The corpus luteum of pregnancy measures 0.5 to 0.6 mm. but it regresses before the end of gestation (1).

The male is not sexually active in the season of its birth but spermatogenesis begins in December and January, with active sperm present from February onward. In active males the testes weigh 100 to 200 mg. (1).

1. Conaway, C. H. AMN., 48: 219–248, 1952.
2. Borell, A. E., and R. Ellis. JM., 15: 12–44, 1934.
3. Long, W. S. JM., 27: 170–180, 1946.
4. Bailey, V. NAF., 55: 1936.
5. Rand, A. L. Canada Nat. Mus., Bul. 100, 1945.

Sorex vagrans Baird

In the salt water marshes of San Francisco this shrew gives birth to young from late February to early June, but mostly in April, and again to a limited extent in mid and late September; probably the later season represents mainly early young of the year (1). In Montana the season is from April to August (2). The gestation period is about 20 days, and the number of embryos averages 5.5. The number of young born averages 5.2; mode 5 and 6; range 2 to 9, while the average number of young found in nests was 4.7 (1). A count for Montana gave 2 to 9 embryos, with 6.4 average (2).

In Montana the testes enlarge in February and spermatozoa are present by the end of the month (2).

1. Johnston, R. F., and R. L. Rudd. JM., 38: 157–163, 1957.
2. Clothier, R. R. JM., 36: 214–221, 1955.

Sorex arcticus Merriam. This species is polyestrous and has several litters a year. The number of young is usually 6 to 9 (1) but there is one record of a female with 12 embryos (2). The testes diminish abruptly at the end of August (3).

S. cinereus Kerr. The common American shrew is polyestrous and breeds all the year except in winter. The litter size is from 4 to 10 (1). Nine embryo counts were scattered between 4 and 8, with 1.1 the mean (2, 4, 5).

S. dispar Batchelder. Two records each of 2 embryos found in August (1, 6), and one of 5 in May (4) have been recorded.

S. fumeus Miller. The smoky shrew breeds from late March to August. The gestation period lasts for 3 weeks or less and the average litter size is 5.5 (7). Embryo counts have varied from 2 to 7, with a mode of 6 (4, 8). Parturition is followed by mating (9).

S. longirostris Bachman. This species has 4 to 5 young at a time (10). One record is of a female with 4 fetuses in April (11).

S. merriami Dobson. This shrew probably mates from March to June. Records of 5, 6, and 7 embryos have been reported (12).

S. milleri Jackson. One record gives a lactating female taken early in July (13).

S. trowbridgei Baird. This shrew begins to breed during its first February and continues until late May and June, though occasional breeders may be found in August. The females reach puberty about two weeks later than the males. In eight instances the number of embryos was from 3 to 6, with an average of 5. The female has a post-partum heat and pregnancy (14).

1. Hamilton, Jr., W. J. The Mammals of Eastern United States. Ithaca, N.Y., 1943.
2. Quay, W. B. JM., 32: 88–99, 1951.
3. Bee, J. W., and E. R. Hall. Univ. Kansas Mus. Nat. Hist., Misc. Publ. 8, 1956.
4. Richmond, N. D., and H. R. Rosland. Mammal Survey of N. W. Pennsylvania, Harrisburg, Pa., 1949.
5. Harper, R. Univ. Kansas Mus. Nat. Hist., Misc. Publ. 12, 1956.
6. Tate, G. H. H. JM., 16: 213–215, 1935.
7. Hamilton, Jr., W. J. Zoologica, 25: 473–493, 1940.
8. Howell, J. C., and Conaway, C. H. J. Tennessee Acad. Sci., 27: 153–158, 1952.
9. Hamilton, Jr., W. J. JM., 30: 257–260, 1949.
10. Hall, E. R., and K. R. Kelson. The Mammals of North America. New York, 1959.
11. Dusi, J. L. JM., 40: 438–439, 1959.
12. Johnson, M. L., and C. W. Clanton. The Murrelet, 35: 1–4, 1954.
13. Baker, R. H. Univ. Kansas Mus. Nat. Hist., Misc. Publ. 9, 125–335, 1956.
14. Jameson, Jr., E. W. JM., 36: 339–345, 1955.

Microsorex hoyi Baird

PYGMY SHREW

This shrew has several litters in spring and summer, with 5 to 6 young as a rule (1). A record of 7 in July has been reported (2).

1. Hamilton, Jr., W. J. The Mammals of Eastern United States. Ithaca, N.Y., 1943.
2. Scott, T. G. JM., 20: 251, 1939.

Neomys fodiens Pennant

WATER SHREW

The young of this water shrew probably do not breed in the year of their birth, except perhaps for a few born early in the season (1) who breed possibly in July at a body weight of about 12–14 g. (2). The season extends from mid-April, or earlier, to September, and at least 2 litters a year are produced. Ovulation is probably not spontaneous. There is a post-partum estrus and probably pseudopregnancy occurs, but luteinization of interstitial tissue is found only after pregnancy. Fertilization takes place in the ovarian part of the oviduct while discus cells are still attached to the ovum. Pregnancy lasts probably for 24 days but it may be prolonged in the tubal stage if lactation supervenes. In some animals lactation anestrum is found, especially towards the end of the breeding season (1). Records of embryos include one with 13 unimplanted blastocysts (3) and two of 5 and 11 each (4).

The follicle at ovulation measures about 391 μ and the ovum 77 μ in diameter. The corpus luteum has a depressed rupture point, and its cavity may contain blood. At this early stage it measures about 395 μ. The ovarian interstitial tissue enlarges during pregnancy and merges with the corpus luteum, which becomes indistinguishable from it. At the 4-cell stage, the latest time it can be distinguished, the corpus luteum measures 1 mm. At estrus the uterine epithelium is tall-columnar; the nuclei are elongated and situated midway up the cells (1).

The testes reach their maximum weight, 100 to 220 mg., in April or May, and as one earlier specimen had testes weighing 9 mg., the increase must be rapid and very great. A copulation plug is formed and ejaculation of semen directly into the uterus is rare (1).

1. Price, M. PRS., 123B: 599–621, 1943–4.
2. Bazan, I. Ann. Univ. Mariae Curie Sklodowska, Ser. C., 9: 213–259, 1955.
3. Hoyte, H. M. D. J. Anim. Ecol., 24: 412–425, 1955.
4. Cantuel, P. Mammalia, 10: 140–144, 1946.

Blarina brevicauda Say

MOLE SHREW

The North American short-tailed shrew is sexually active from the first week in February, and the testes reach their greatest development in early April. The young are born from mid-April to the end of May, and probably a second litter may be born in August-September. The male has a side gland which develops with the testes (1). The litter size is 3 to 7, averaging 4.5 (2), but embryo counts from animals caught in the wild have given 5.2 ± .3 (1, 3, 4). The duration of pregnancy is 17 to 20 days from the end of sexual receptivity (2).

Ovulation is induced by coitus, and the female is in heat continuously for at least 33 days in the absence of the male. It requires several matings to induce ovulation, and, if sufficient are made, this occurs about 64 hours (55 to 71) from the time of the first mating. The follicles during heat are 0.35 to 0.45 mm. in diameter. The antrum is small and the granulosa cells are large. After copulation the follicles increase slightly in size, and the theca becomes more vascular. Ten hours before ovulation the basement membrane below the granulosa layer disappears. The first polar body is given off, and luteinization begins to push the ovum and a ball of granulosa cells surrounding it out of the follicle. Spermatozoa often penetrate the ovum before it is extruded from the follicle. The size of the ovum is 70 × 80 μ. The corpus luteum sometimes has a small cavity at the point of ovulation, and mitoses are frequent in the cells around the edge of the corpus. Its maximum size is 0.75 mm. in diameter, and it degenerates when the embryos are 7 mm. long. By the time of parturition it has almost or entirely vanished. The average number of corpora lutea found was 6.1, and of embryos 5.7 (2).

Pseudopregnancy lasts 9 to 10 days from the time of ovulation. There is no postparturition heat. Internal migration of blastocysts may occur; the ovarian capsule is usually closed (2).

During anestrum the vaginal lumen is usually closed and its lining consists of two layers of epithelial cells. At the time of heat it develops an abrupt flexure, sigmoid in the vertical plane, corresponding to that of the penis. The epithelium becomes stratified, and mitoses are abundant, both in proestrum and in heat. Cornification is well marked, and there is a leucocyte invasion at the end of heat. The vagina is cornified during lactation (2).

During heat the uterine epithelium is tall-columnar, with mitoses. Many leucocytes are to be found in the stroma, and they migrate to the lumen after copulation (2).

The oviduct is ciliated at the ovarian end, but few can be found in the mid-region and none at the uterine end. They are entirely absent during anestrum (2).

The penis is sigmoid and has horny ridges or rings on the glans. The testes produce no spermatozoa from November to January, and during this time the penis and accessory organs are small. Growth begins late in January, and by late February spermatozoa may be found. There is less regression in winter in males that are kept in the laboratory than in those obtained from the wild. Both males and females breed in their first year if they are born early enough; puberty in the male is reached in 50 to 80 days. The pectoral scent gland is developed in the spring, more so in the male than in the female. No copulation plug is formed at coitus (2).

1. Hamilton, Jr., W. J. JM., 10: 124–134, 1929.
2. Pearson, O. P. AJA., 75: 39–93, 1944.
3. Richmond, N. D., and H. R. Rosland. Mammal Survey of N. W. Pennsylvania. Harrisburg, Pa., 1949.
4. Hamilton, Jr., W. J. JM., 30: 257–260, 1949.

Blarina telmalestes Merriam. The Dismal Swamp shrew breeds from early spring to late September. It has 3 to 4 litters a year numbering from 3 to 7 (1).

1. Hamilton, Jr., W. J. The Mammals of Eastern United States. Ithaca, N.Y., 1943.

Cryptotis parva Say

SHORT-TAILED SHREW

The short-tailed shrew breeds from early March to November in the northern United States. In Florida it breeds the year-round (1). Heat lasts for 1 to 2 days. Five gestation periods lay between 21 and 23 days and, in 1 of 3, a post-partum heat was observed (2). A compilation of embryo and fetus counts gave the average litter size at 4.5 ± 3, with a range from 2 to 6 and a mode of 5 (1–4).

1. Hamilton, Jr., W. J. JM., 25: 1–7, 1944.
2. Conaway, C. H. JM., 39: 507–512, 1958.
3. Davis, W. B., and L. Joeris. JM., 26: 136–138, 1945.
4. Brimley, C. S. JM., 4: 263–264, 1923.

Notiosorex crawfordi Coues

Embryos have been found in April (1), August (2), and September (3), and a lactating female in May (1). Two of these had 5 embryos each and the other had 3.

1. Dixon, J. JM., 5: 1–6, 1924.
2. Hoffmeister, D. F., and W. W. Goodpaster. Illinois Biol. Monogr., 24: 1–152, 1954.
3. Clark, W. K. JM., 34: 117–118, 1953.

CROCIDURINAE

Crocidura

HOUSE SHREWS

Crocidura deserti Schwann. This shrew probably breeds all year round. Two November records gave 3 and 5 embryos each (1).

C. flavescens Geoffroy. In captivity young were born in January; the litter size was from 2 to 4 (2).

C. hildegardeae Thomas. A February record is of a female with nest young (3). Two females each with 4 fetuses have been found in September and November (4).

C. hirta Peters. A March record is of a female with 1 new-born young. Two lactating females were found in May; one of these had a single nestling and the other had 4 (3). There are also records of pregnancies in January and November (5).

C. jacksoni Thomas. One record gives 2 large embryos at the end of December (6).

C. leucodon Hermann. A female mated on April 10 and again on May 2. Thirty-one days after the second mating 3 newborn young were found dead (7).

C. luna Dollman. Pregnant specimens of this shrew have been caught in March and April. The first of these had 2 embryos (8).

C. occidentalis Pucheran. The giant swamp shrew has been found pregnant in January (9), March, and May (10). It has also been found nursing young in May and December (10). A long season is indicated. The number of young appears to be from 2 to 4.

C. poensis Fraser. One record of a female (*C. attila*) taken in January gave 3 embryos (11).

C. russula Hermann. In Sardinia pregnancies are found from March to September and the litter size varies from 2 to 6 (12). The Japanese subspecies, *dsi-nezumi* Temminck, probably breeds more or less continuously through the warm months. Specimens with 3 and 5 embryos and a lactating female have taken on Quelpart Island in September (13).

C. turba Dollman. One record gives a pregnant female taken in June (5).

1. Shortridge, G. C. The Mammals of South West Africa. London, 1934.
2. Marlow, B. J. G. PZS., 124: 803–808, 1954–5.
3. Allen, G. M., and A. Loveridge. HMCZ., 89: 147–214, 1941–2.
4. Lawrence, B., and A. Loveridge. HMCZ., 110: 1–80, 1953.
5. Frechkop, S. EPNU., 1944.
6. Hollister, N. AM., 35: 663–680, 1916.
7. Frank, F. Bonner Zool. Beitr., 4: 187–194, 1953.
8. Ansell, W. F. H. Ann. Mag. Nat. Hist., 12(10): 529–551, 1957.
9. Heller, E. In G. C. Shortridge, The Mammals of South West Africa. London, 1934.
10. Frechkop, S. EPNA., 1943.

11. Eisentraut, M. Zool. Jahrb., Syst., 85: 619–721, 1957.
12. Kahmann, H., and J. Einlechner. Zool. Anz., 162: 63–83, 1959.
13. Knox, Jr., J. J. Mammal. Soc., Japan, 1: 105–114, 1959.

Suncus murinus L.

MOUSE SHREW

In Malay 50 per cent of the males are fertile at a weight of 31.5 ± .1 g. and 95 per cent at 46.1 ± .1 g. Most pregnancies are found from October to December, but many occur in other months (1). On Guam they probably breed the year-round (2). The mean embryo number is 2.7, with a mode of 2 and 3, and range from 1 to 5. They are equally distributed in the two uterine horns (1).

1. Harrison, J. L. PZS., 125: 445–460, 1955.
2. Peterson, G. D. JM., 37: 278–279, 1956.

Suncus lixus Thomas. One record gives a female with 3 fetuses in January (1).

S. varilla Thomas. One record gives a female with 4 fetuses in October (2).

1. Ansell, W. F. H. Ann. Mag. Nat. Hist., 12(10): 529–551, 1957.
2. Lawrence, B., and A. Loveridge. HMCZ., 110: 1–80, 1953.

Sylvisorex

FOREST SHREWS

Sylvisorex granti Thomas. In Uganda a female with 2 suckling young was taken in January (1).

S. megalura Jentink. A female with 2 near-term fetuses was taken in October (2). Another with 2 was taken late in January (3).

S. sorella Thomas. There are two records each of 2 large or medium-sized fetuses in January (3).

1. Allen, G. M., and A. Loveridge. HMCZ., 89: 147–214, 1941–2.
2. Ansell, W. F. H. Nat. Mus. Southern Rhodesia, Occ. Papers, 3: 351–398, 1960.

3. Allen, G. M., and H. J. Coolidge, Jr. In R. P. Strong, The African Republic of Liberia and the Belgian Congo. Cambridge, Mass., 1930.
4. Hollister, N. AM., 35: 663–680, 1916.

Myosorex varius Smuts

MOUSE SHREW

A female with 3 embryos was found in July (1).

1. Cowles, R. B. JM., 17: 121–130, 1946.

Diplomesodon pulchellum Lichtenstein

SPOTTED SHREW

This shrew breeds once a year, so far as is known, and has 5 young born between April and August (1).

1. Heptner, W. G., L. G. Morosowa-Turowa, and W. I. Zalkin. Die Säugetiere in der Schutzwaldzone. Berlin, 1956.

TALPIDAE

DESMANINAE

Desmana moschata L.

This insectivore appears to breed all the year but the main season is in spring. Three to 5 young are born in a litter and the gestation period is 45 to 50 days (1).

1. Heptner, W. G., L. G. Morosowa-Turowa, and W. I. Zalkin. Die Säugetiere in der Schutzwaldzone. Berlin, 1956.

Galemys pyrenaicus Geoffroy

Spermatogenesis is in progress from November to May. At the beginning of January the genital tract is fully developed and continues thus until May. There is a resting period from June to October but spermatozoa persists until August. Heat periods continue from January to June, births occur from March to July, and lactation continues into August. The litter size varies from 1 to 5, with 4 the usual number (1).

The ovary does not bear an external spermatic segment as does the male, but the medullary zone resembles that of the male. Interstitial and thecal glands are present. The ovarian condition is regarded as that of a natural intersexuality (2).

1. Peyre, A. Mammalia, 20: 405–418, 1956.
2. Peyre, A. Soc. Zool. France, Bull. 77: 441–447, 1953.

TALPINAE

Talpa europaea L.

COMMON EUROPEAN MOLE

In Great Britain the common European mole breeds once a year, in spring or early summer (1), according to one report, but another gives the main season as lasting from the end of March, usually April, to early May. Occasionally later litters are born in August or September (2). The season begins a month later in Cheshire than it does in Suffolk (3). In France sexual activity extends from early March to June but 2 females captured in December were in full heat, with open vaginas but with follicles lacking antra (4). In Italy the season extends from February to April (5). In Russia it begins in mid-April and extends into August, though most young are born at the end of May (6). The females breed first in their second year, i.e., at a year after birth. The heat period probably lasts for 20–30 hours and, at this time, the utero-vaginal canal is filled with thick mucus. Normally the female has 1 litter a year but a few have a second heat. A few females caught in May and June have been found to be pregnant, but only 2 to 3 per cent of all the females are breeding at that time (7).

The average litter size is 3.8 ± .1, with a standard deviation of 0.84 and a mode of 4 (2). Another account gives the number of embryos as 3.8 ± .13, mode 4 and range 2 to 6 (3). The gestation period is from 30 to 40 days (8).

HISTOLOGY OF THE FEMALE

The ovaries are remarkable bodies. A type of tissue resembling interstitial tissue is massed separately from the true ovary. The capsule forms the outer layer of the interstitial part and pouches the ovigenous part, which is covered by germinal epithelium. There is also some interstitial tissue in the true ovary. Futhermore, in the uteri of a minority of moles, believed to be old ones, masses of tissue resembling interstitial cells are found. The ovarian interstitial gland has a definite cycle, which is opposite to that of the rest of the reproductive tract. It is least in size during heat and in pregnancy, which lasts about 4 weeks. The clitoris is peniform and grows or diminishes with the interstitial tissue. The corpus luteum persists throughout pregnancy but disappears quickly during lactation (1).

The vagina is closed except in the breeding season. Before it opens, the surrounding area becomes purplish blue and congested; when it is open this area swells. These changes commence in Great Britain in the first half of March and are complete by the end of the month. The vagina begins to close immediately after parturition. The vaginal canal is a solid epithelial strand, 5 cells thick, which increases to 6 to 15 cells as those at the center become stratified and desquamate to form the lumen. It closes by constriction of the surrounding muscle fibers. There is no os uteri, and the cornua open into a median uterovaginal canal which is open for about 15 mm. and closed for the caudal 5 mm. As this canal grows in proestrum, it doubles in length and assumes a sigmoid curve. It diminishes in length as soon as the animal becomes pregnant (1).

The oviducts lie mostly in the wall of the ovarian capsule. The uterine cornua increase in length during the breeding season, and, during heat, the blood vessels in the endometrium greatly develop, especially on the mesometrial side, and on this side, also, the circular muscle hypertrophies (1).

The placentas are not shed at parturition but are reabsorbed (3).

THE MALE

The testes weigh least in June. There is a slow increase to January, then rapid growth. Spermatogenesis is at its maximum in late February and early March when the testes weight 2 g. A rapid decline then sets in and they weigh only 600–700 mg. by late March (7). The interstitial tissue of the testis is large and is bathed in plasma from March to mid-April. At the end of this time these cells begin to be laden with yellowish-brown pigment. In May the plasma has disappeared, and there is little further change until the following January, except that the pigment does not appear to persist throughout the nonbreeding season. The prostate and the corpus spongiosum enlarge from the end of January and are at their maximum at the end of March and the beginning of April. They have diminished to their minimal size by the end of May (9). The level of lipids is high in the Leydig cells during the breeding season. At their maximum the seminiferous tubules measure 250 μ in diameter (10).

1. Matthews, L. H. PZS., 347–383, 1935.
2. Barrett-Hamilton, G. E. H. A History of British Mammals. London, 1910.
3. Godfrey, G. K. JM., 37: 438–440, 1956.
4. Godet, R. CRAS., 224: 498–499, 1947.
5. Balli, A. Riv. Biol., 29: 35–55, 1940.
6. Baskirov, I. S., and J. V. Zarkov. Ucen. Zap. Kazan Univ., 94: 1–66, 1934.
7. Godfrey, G. and P. Crowcroft. The Life of the Mole (*Talpa europaea* Linnaeus). London, 1960.
8. Heptner, W. G., L. G. Morosowa-Turowa, and W. I. Zalkin. Die Säugetiere in der Schutzwaldzone. Berlin, 1956.
9. Courrier, R. AB., 37: 173–334, 1927.
10. Lofts, B. QJMS., 101: 199–206, 1960.

Mogera latouchei Thomas

This mole mates in March. It has 4 to 6 young at a time and the gestation period is 6 weeks (1).

1. Ho, H. J. Contr. Biol. Lab., Sci. Soc. China, Zool. Ser. 11: 123–164, 1935.

SCALOPINAE

Neurotrichus gibbsii Baird

This mole breeds from February to September and the number of young is from 1 to 4 (1).

1. Dalquist, W. W., and D. R. Orcutt. AMN., 27: 387–401, 1942.

Parascalops breweri Bachman

HAIRY-TAILED MOLE

The hairy-tailed mole of eastern North America mates from the end of March to the first week of April and has one litter a year (1). The litter size is usually 4, but one female with 8 embryos has been reported (2). The gestation period lies between 4 and 6 weeks (1).

The vaginal orifice is closed except during the reproductive season; it is open before mating and is closed by September. The testes are enlarged early in March, decrease rapidly to June, and then gradually subside to their minimum size in September and October (1). These moles are sexually mature in the spring following their birth (3).

1. Eadie, W. R. JM., 20: 150–173, 1939.
2. Richmond, N. D., and H. R. Rosland. Mammal Survey of N. W. Pennsylvania. Harrisburg, Pa., 1949.
3. Hamilton, Jr., W. J. The Mammals of Eastern United States. Ithaca, N. Y., 1943.

Scapanus

WESTERN MOLE

Scapanus orarius True. In British Columbia this mole mates from January to early March and the young are born in March or April. Yearling females usually have 2 embryos, two-year-olds have 3, and older ones, 4 (1).

S. townsendii Bachman. The breeding season of Townsend's mole is in February and the one litter a year is born in late March. The number of young is usually from 2 to 4 (2). Embryo counts gave a mean of 3.0 ± .25, with a mode of 3, range 2 to 6 (3).

1. Glendenning, R. Canadian J. Anim. Sci., 39: 34–44, 1959.
2. Bailey, V. NAF., 55: 1936.
3. Moore, A. W. JM., 20: 499–501, 1939.

Scalopus aquaticus L.

COMMON AMERICAN MOLE

The breeding season of the mole is a limited one, with most females being bred in a period of 3 to 4 weeks. In Wisconsin the peak of breeding is in the last week of March and the first week of April. In Missouri it is about a month earlier, i.e., in February and the first week of March. In southern Indiana and Arkansas the season is also earlier. Young females do not breed in their year of birth. Usually there is 1 litter a year but possibly, though not certainly, some females may bear a second. Ovulation is probably coitus-induced. The gestation period probably lasts for 4 weeks or less. The number of postimplantation embryos averaged 3.91, with a marked mode of 4, while the average number of corpora lutea was 4.3, with a spread of from 3 to 7. Sometimes internal migration of blastocysts occurs and some embryonic mortality has been observed (1).

The vagina is closed until the ovary contains follicles with antra. At the time of heat the vaginal orifice is very white and avascular, in distinction to its appearance after ovulation, when the lips become reddish (1).

The testes of the juvenile males increase in size to June, after which they remain constant, about 100–200 mg. in weight, until early November. By the beginning of January they have increased tenfold in weight. They begin to decrease immediately until mid-July, when they weigh less than 400 mg. These changes occur a little later in Wisconsin. In juvenile males the interstitial cells increase greatly in size in the summer and they are smallest during the breeding season. These cells are foamy and vacuolated, with scattered pigment granules.

In Missouri the first spermatozoa appear in late December and regression of the seminiferous tubules is observed in June or July. The prostate is smallest in fall. The regressed gland may be distinguished from the infantile one by its greater content of connective tissue. Hypertrophy begins in late fall, and the gland is at its maximum by the end of December or January. Spermatozoa may be found in the epididymis for 2 weeks after spermatogenesis has ceased (1).

1. Conaway, C. H. JM., 40: 180–194, 1959.

CONDYLURINAE

Condylura cristata L.

STAR-NOSED MOLE

The star-nosed mole breeds in the first year of its life. The testes begin to grow in mid-January and enlarge to a length of 24 mm. They begin to reduce in May and by August have shrunk to 8 to 12 mm. long. Mature sperm are found by late February and the accessory organs are active by the middle of that month. Embryos may be found by mid-March and young are born from late March to early August. The number of embryos has ranged from 2 to 7 with an average of 5.4 ($n = 25$). Males and females alike have swollen tails just before the breeding season and during its height (1).

1. Eadie, W. R., and W. J. Hamilton, Jr. JM., 37: 223–231, 1956.

Dermoptera

CYNOCEPHALIDAE

Cynocephalus variegatus Audebert

FLYING LEMUR

DURING proestrum (?) the uterus of the Malayan flying lemur is hyperemic, the capillaries are congested, and blood is extravasated. There is a little local denudation of the epithelium, which may occur in any part of the uterus (1). One young is born at a time (2). The gestation period has been given as 60 days (3).

1. Herwerden, M. van. Tijdschr. Ned. dierk. Vereen., 10: 1–140, 1906.
2. Blanford, W. T. The Fauna of British India. Mammals. London, 1889–91.
3. Kenneth, J. H. Gestation Periods. Edinburgh, 1943.

Chiroptera

OUR knowledge of reproduction in bats has increased greatly in the past 15 years, and it is now possible to compare families in this respect.

Among those species that have been most thoroughly investigated, the Pteropidae have prolonged breeding seasons. Short seasons have been suggested for many species without much investigation. The gestation period is fairly long, 4 or 5 months, and the single young is general. There is no sign that one ovary functions more frequently than the other.

In Rhinopomatidae the single young is born after a short gestation. In Emballuronidae most species breed all the year. A single young is the rule and both ovaries are alive. In Nycteridae a long season, single young, and a short pregnancy prevail. In Megadermatidae the left ovary functions; possibly a prolonged season and a long gestation may be the pattern. In the Rhinolophidae and Hipposideridae the right ovary functions; the number of young is variable, 1 to 2 as a rule, mating is in the autumn, and ovulation is in the spring with storage of spermatozoa in the uterus during the winter. There may be a tendency for the corpus luteum to disappear during the short pregnancy.

In Phyllostomatidae two well-defined groups are found. In one group single young are usual, with a prolonged breeding season, possibly lasting all year. In the other group a restricted season with a long gestation and delayed development of the embryo may be the rule.

In Desmodontidae a single young, in either horn of the uterus, a long gestation, and year-round breeding are usual. Vespertilionidae have a restricted season in the temperate zone but usually breed all the year round in the tropics. Both ovaries are functional but implantation is frequently only in the right horn. *Miniopterus* is an exception as ovulation is from the right ovary. The number of young is variable, much more so than in any other family. From 1 to 4, but usually 1 or 2, young are born at a time. In temperate-zone species mating is in the fall and ovulation in the spring, with sperm storage in the uterus, and with a short gestation. Spring matings are also made but these are not essential for fertility.

The Molossidae have single young, usually implanted in the right horn. The breeding season may be either restricted or prolonged, depending on the species.

The fact that many bats are night flyers which during the day rest in places with very subdued light suggests an extreme sensitivity to light. The breeding patterns in the Rhinolophidae, Hipposideridae, and Vespertilionidae suggest that F.S.H. secretion decreases with the lower light values of fall months while L.H. secretion continues. Spermatogenesis ceases but the male accessory organs continue to function. Only in the spring, with increasing light, does ovulation occur—this at present seems somewhat paradoxical. Careful investigation of the bat pituitary gland and of its relations with the hypothalamus and with outside stimuli would be well worth while.

MACROCHIROPTERA

PTEROPIDAE

PTEROPINAE

Cynopterus sphinx Vahl

WHITE-EARED FRUIT BAT

There is some discrepancy in the descriptions of the breeding season of this Indian bat. One account places the mating period in October or early November, but with some spread, as some November and December specimens carry embryos in the same stage of development. Parturition follows in late February or early March after a gestation period of 115–125 days. One young at a time is the rule (1). The species breeds only once a year (2).

According to another account the testes are permanently abdominal and they, together with the accessory organs, enlarge in December in preparation for the breeding season (3). Birth takes place in March, with a post-partum heat and a new pregnancy in April. Two or more pregnancies follow in quick succession. Follicles ripen alternately in the right and left ovaries and the pregnancy is in the corresponding uterine horn. The corpus luteum persists until parturition and the gestation period is about 5 months (4). A record in the London zoo gives a single young born in March (5).

1. Moghe, M. A. Proc. Nat. Inst. Sci., India, 22B: 48–55, 1936.
2. Gopalkrishna, A., and M. A. Moghe. Proc. Nat. Inst. Sci., India, 26B (suppl.): 11–19, 1960.
3. Vamburkar, S. A. PZS., 130: 57–77, 1958.
4. Ramakrishna, P. A. Current Science, 16: 186, 1947.
5. Zuckerman, S. PZS., 122: 827–950, 1947.

Cynopterus brachyotis Müller. In India nearly all females are pregnant in the months June to September (1). The young are born in September (2).

1. Lyon, M. W. Proc. Nat. Mus., United States, 33: 547–571, 1908.
2. Baker, J. R., and Z. Baker. J. Linn. Soc., Zool., 40: 123–141, 1936.

Rousettus

FRUIT BATS

Rousettus aegyptiacus Geoffroy. Captive bats of this species have bred the year-round in Egypt. In the wild they mate in September and October and the single young is born in February or March (1, 2). According to another account mating takes place at the end of April and the young are born from February to November (3).

R. angolensis Bocage. In the Cameroons this bat mates in September and the young are born in December and January (4).

R. leachi A. Smith. In the London zoo this bat has given birth to young all year-round. In 73 births singles were usual, with twins at 1 birth in 4 (5).

R. leschenaulti Desmarest. Young are numerous in March; there is a second birth period in July and August but only a few are born at this time. None are born from September to February. A single young is the usual number (6).

R. seminudus Gray. In Ceylon this bat has 1 young at a birth. Gestation is believed to last for 15 weeks (7).

R. (Myonycteris) wroughtoni Andersen. Half-grown young have been recorded in September (8).

1. Baker, J. R., and Z. Baker. J. Linn. Soc., Zool., 40: 123–141, 1936.
2. Anderson, J. The Mammals of Egypt. London, 1902.
3. Kulzer, E. Zeit. Morph. Okol. Tiere, 47: 374–402, 1958.
4. Eisentraut, M. Mitt. Zool. Mus., Berlin, 25: 245–274, 1941.
5. Zuckerman, S. PZS., 122: 827–950, 1952–3.
6. Brosset, A. Mammalia, 26: 166–213, 1962.
7. Phillips, W. W. A. Ceylon J. Sci., 13B: 1–63, 1924–6.
8. Verschuren, J. EPNG., 1957.

Pteropus geddiei MacGillivray

FLYING FOX

This flying fox in the New Hebrides conceives in February and March, or possibly in June and July. The young are born from late August to early September. There are no pregnancies from November to January. The male is suddenly sexually mature at 600 g. body weight. Spermatogenesis occurs all year-round, but the testes weigh least in July and most in January (1). The interstitial cells of the testes are most numerous and are largest during the month of copulation and for three months previously, i.e., from October to February. From March to June they are moderate in number and size, and in July and August they are at their minimum (2).

1. Baker, J. R., and Z. Baker. J. Linn. Soc., Zool., 40: 123–141, 1936.
2. Groome, J. R. PZS., 110A: 37–42, 1940.

Pteropus giganteus Bruenn

INDIAN FRUIT BAT

In the Central Provinces this bat apparently mates in late August and early September. During this time, and not earlier, one horn of the uterus, and one horn only, contains spermatozoa. Pregnancy starts immediately and birth takes place at the end of January or early in February after a gestation of 140 to 150 days (1). In Ceylon conceptions occur from early December to early January, and birth takes place in late May or early June after a gestation of about 6 months (2). Another account puts birth in India from the end of March to the end of April (3). Copulations have been observed on January 25 in India (4). In the London zoo births have occurred from December to June (5). The number of young is always 1 and pregnancy occurs in either horn of the uterus with equal frequency (2). This bat breeds once a year (6).

Both ovaries and uterine horns are functional. One egg is released after the female has copulated. The progestation reaction starts while the ovum is still in the oviduct and is apparent only in the corresponding horn. The reaction is confined to the region of the uterus nearest the utero-tubal junction (7).

The epididymis contains spermatozoa all the year, but the number is greatest from August to December and least in April and May. The testes weight at the height of the season is 8.5 g. It is least, 4.1 g., in April (2).

1. Moghe, M. A. PZS., 121: 703–721, 1951–2.
2. Marshall, A. J. Proc. Linn. Soc., 159: 103–111, 1947.
3. Baker, J. R., and Z. Baker. J. Linn. Soc., Zool., 40: 123–141, 1936.
4. McCann, C. J. Bombay Nat. Hist. Soc., 42: 587–592, 1941.
5. Zuckerman, S. PZS., 122: 829–950, 1952–3.
6. Gopalkrishna, A., and M. A. Moghe. Proc. Nat. Inst. Sci., India, 26B (suppl.): 11–19, 1960.
7. Marshall, A. J. JE., 9: 42–44, 1953.

Pteropus alecto Temminck. Two females, each with 1 embryo, were captured in July. This is an Indonesian species (1).

P. ariel G. M. Allen. In this Maldive species the young are born in April (2).

P. conspicillatus Gould. This New Guinea species has been recorded as pregnant in August. The three females examined each had 1 embryo (1). It breeds about May (3).

P. eotinus Andersen. The remarks made under *P. geddiei* apply to this species (4).

P. gouldi Peters. This North Australian species mates in March and April. The young are born in September and October (4).

P. melanotus Blyth. In the Nicobar Islands this species mates in October and the young are born in February and March (4). One-third-grown young, 1 to a female, have been reported in March (2). In Borneo single nursing young may be found from December to March (5).

P. natalis Thomas. Females of this Christmas Island species have been reported as carrying fetuses, some of them near term, towards the end of

December. Well-grown young, still with the females, were found at the end of July (2).

P. niger Kerr. On Reunion Island the females mate in May and the young are born at the end of October (4). One young is born after a gestation of 4½ to 5 months (2).

P. ornatus Gray. In New Caledonia the number of pregnancies is greatest from June to September; the young are born by October after a gestation of about 6 months. One young is born at a time. Puberty is reached at 2 years (6).

P. poliocephalus Temminck. The red-necked fruit bat of Australia breeds in March or April and has its young in October (3). In the London zoo the birth of 1 young in October has been recorded (7).

P. rufus Geoffroy. An October record has been made of three females, each with 2 embryos (8), an unusual number for this genus.

P. scapulatus Peters. In Queensland this species mates in September and October (4). Birth is probably in April, after an October mating (3). There is a record of a female with 1 embryo taken in August (1).

P. subniger Kerr. On Reunion Island this species mates in May and the young are born at the end of October (4).

1. Tate, G. H. H. AM., 102: 199–203, 1953.
2. Andersen, K. Catalogue of the Chiroptera in the British Museum. London, 1912.
3. Ratcliffe, F. N. CSIRO., Australia, Bull. 53, 1931.
4. Baker, J. R., and Z. Baker. J. Linn. Soc., Zool., 40: 123–141, 1936.
5. Banks, E. J. J. Malay. Br. Roy. Asiatic Soc., 9(2): 1–139, 1931.
6. Sanborn, C. C., and A. J. Nicholson. Fieldiana Zool., 31: 313–338, 1950.
7. Zuckerman, S. PZS., 122: 829–950, 1952–3.
8. Rand, A. L. JM., 16: 89–104, 1935.

Epomophorus anurus Heuglin

Pregnant and lactating females of this African bat are found during most of the year if not the year-round. Spermatogenesis and spermatozoa were found in males captured from August to January. None were taken during the rest of the year. The female has 1 young at a time and the corresponding ovary contains the corpus luteum (1).

The ovary contains practically no interstitial cells, and no neogenesis of ova was found. At the beginning of gestation the corpus luteum contains blood at its center; this is gradually replaced by connective tissue. During pregnancy the chromophobes of the anterior pituitary swell, and much ribonuclein is deposited in them so that they may be confused with basophils. Many of them contain fine rose-colored granules, as shown by the Cleveland-Wolfe stains. These cells are transformed into erythrosinophils (1).

1. Herlant, M. Ann. Soc. Roy. Zool. de Belge, 84: 87–116, 1953.

Epomophorus angolensis Gray. The newborn single young is found in September and October. Twins are rare (1).

E. crypturus Peters. In Mozambique the young are born in October (2). One at a time is the rule (3).

E. franqueti Tomes. This bat has no restricted breeding season (4, 5).

E. gambianus Ogilby. There is a June record of pregnancies with single 2–3 cm. embryos (4).

E. labiatus Temminck. One female with a large fetus was taken in February (6).

E. wahlbergi Sundevall. One young is born at a time (7).

E. (Epomops) dobsoni Bocage. This epaulette bat has 2 young at a time (8).

E. (Hypsignathus) monstrosus H. Allen. In the Cameroons this species mates in the second half of October (9). Embryos have been found in May and December (10).

E. (Micropteropus) pusillus Peters. A single embryo has been reported in February and lactating females in May (10).

1. Shortridge, G. C. The Mammals of South West Africa. London, 1934.
2. Baker, J. R., and Z. Baker. J. Linn. Soc., Zool., 40: 123–141, 1936.
3. Loveridge, A. J. East Africa and Uganda Nat. Hist. Soc., 17: 39–69, 1922.
4. Aellen, V. Mém. Soc. Neuchateloise des Sci. Nat., 8(1): 1–121, 1950.
5. Lang, H., and J. P. Chapin. AM., 37: 497–563, 1917.
6. Allen, G. M., B. Lawrence, and A. Loveridge. HMCZ., 79: 31–126, 1935–7.
7. FitzSimons, F. W. The Natural History of South Africa. Mammals. London, 1919.
8. Hill, J. E. JM., 22: 81–85, 1941.
9. Eisentraut, M. Mitt. Zool. Mus., Berlin, 25: 245–274, 1941.
10. Verschuren, J. EPNG., 1957.

Casinycteris orgynnis Thomas

There is a record of a single fetus in a female captured in April (1).

1. Lang, H., and J. P. Chapin. AM., 37: 497–563, 1917.

Eidolon helvum Kerr

In the Cameroons this species mates in October and November, at the end of the rains. Birth takes place from mid-February to mid-March after a pregnancy of 4 months (1).

1. Eisentraut, M. Mitt. Zool. Mus., Berlin, 25: 245–274, 1941.

MACROGLOSSINAE

Macroglossus lagochilus Matschie

In India this species gives birth in August (1).

1. Baker, J. R., and Z. Baker. J. Linn. Soc., Zool., 40: 123–241, 1936.

Notopteris macdonaldi Gray

This bat usually has one young at a time; in New Caledonia most births occur in August (1).

1. Sanborn, C. C., and A. J. Nicholson. Fieldiana Zool., 31: 313–338, 1950.

MICROCHIROPTERA

RHINOPOMATIDAE

Rhinopoma

MOUSE-TAILED BAT

Rhinopoma hardwickei Gray. In India early embryos were found in May, and 3 females gave birth to their young in July (1). Another account gives June as the prevailing month of birth (2).

R. kinneari Wroughton. In Gwalior this bat breeds once a year, from April to July. Pregnancy may occur in either horn of the uterus (3). Another account gives the mating season in March. A single young is born (4). Birth is in August (1).

R. microphyllum Brünnich. The young are born in June (2).

1. Prakash, I. JM., 41: 386–389, 1960.
2. Brosset, A. Mammalia, 26: 166–213, 1962.
3. Srivastava, S. C. Proc. Zool. Soc., Bengal, 5: 105–131, 1952.
4. Gopalkrishna, A., and M. A. Moghe. Proc. Nat. Inst. Sci., India. 26B (suppl.): 11–19, 1960.

EMBALLURONIDAE

EMBALLURONINAE

Coleura afra Peters

SHEATH-TAILED BATS

This bat has 1 young at a time. The ovarian capsule is completely closed and the female has a very well marked prostate gland (1). There is a record of a December pregnancy (2).

1. Matthews, L. H. PZS., 111B: 289–346, 1941.
2. Verschuren, J. EPNG., 1957.

Rhynchonycteris naso Wied

SHARP-NOSED BAT

This Central American bat probably breeds more than once a year and the distribution of embryos suggests an irregular, or no, season. Pregnancies have been found in January, February, April (1), and May (2). One young at a time is the rule (1).

1. Dalquest, W. W. Texas J. Sci., 9: 219–226, 1957.
2. Burt, W. H., and R. A. Stirton. Univ. Michigan Mus. Zool., Misc. Publ. 117, 1961.

Saccopteryx

Saccopteryx bilineata Temminck. The white-lined bat of Central America has 1 young at birth and pregnancies have been observed in March, April, and May (1–4). Pregnancies have also been found from March to July and lactating females in June and July (5).

S. leptura Schreber. Pregnant females have been found in May (5).

S. (*Peropteryx*) *canina* Wied. One young at a time is the rule and pregnancies have been found in March and April (1).

S. (*P.*) *macrotis* Wagner. Five females, each with a single embryo, were taken in April (6).

S. (*Balanopteryx*) *io* Thomas. In Mexico lactating females were found at the end of July (7).

S. (*B.*) *plicata* Peters. In Central America pregnant females are found in March and April, while young are present in May (3). In Mexico embryos have been found in June (8), while in July they are advanced in pregnancy or carrying young, of which there is usually 1 (9).

1. Murie, A. Univ. Michigan Mus. Zool., Misc. Publ. 26, 1935.
2. Hall, E. R., and W. B. Jackson. Univ. Kansas Mus. Nat. Hist., Publ. 5: 646, 1953.
3. Felten, H. SB., 36: 271–285, 1955.
4. Enders, R. K. HMCZ., 78: 385–502, 1935.
5. Goodwin, G. G., and A. M. Greenhall. AM., 122: 187–302, 1961.
6. Dalquest, W. W., and E. R. Hall. JM., 30: 424–427, 1949.
7. Baker, R. H., and J. K. Greer. JM., 41: 413–415, 1960.

8. Davis, W. B., and R. J. Russell. JM., 33: 234–239, 1952.
9. Davis, W. B. JM., 25: 370–402, 1944.

Taphozous longimanus Hardwicke

INDIAN SHEATH-TAILED BAT

This species breeds the year-round and has a quick succession of pregnancies. The horns alternate in their function and the single ovum is shed from the ovary corresponding to the pregnant horn. In the mature follicle only 1 layer of cells surrounds the ovum (1).

1. Gopalkrishna, A. Proc. Nat. Inst. Sci., India, 21B: 29–41, 1955.

Taphozous mauritanus Geoffroy

In Madagascar pregnancies have been recorded for December and January (1). In the Belgian Congo fetuses near term were found in April and May, which is the usual time of birth in that region. A single young is the rule. The male reproduction tract is well developed in December, is intermediate in March and April, and is much reduced in July (2).

1. Kaudern, W. AZ., 9(1): 1–22, 1914.
2. Verschuren, J. EPNU., 1957.

Taphozous kachhensis Dobson. The single young is born between June 15 and July 15 (1).

T. melanopogon Temminck. The male reaches puberty at 6 months. The single young is born between April 15 and May 15 (1).

T. perforatus Hollister. One record of a single embryo was made in May (2). Parturition is in late May or early June. Many lactating females have been found in June. The species may be polyestrous (3) (?).

T. (*Saccolaimus*) *peli* Temminck. This African bat has been taken in June and December with single embryos (4).

T. (*S.*) *saccolaimus* Temminck. In Ceylon this bat has been found with single young in the months of September, October, and November (5).

1. Brosset, A. Mammalia, 26: 166–213, 1962.

2. Allen, G. M., B. Lawrence, and A. Loveridge. HMCZ., 79: 31–126, 1935–7.
3. Harrison, D. L. Durban Mus. Nov., 5(11): 143–149, 1958.
4. Lang, H., and J. P. Chapin. AM., 37: 497–563, 1917.
5. Phillips, W. W. A. Ceylon J. Sci., 13B: 1–63, 1924–6.

NOCTILIONIDAE

Noctilio

Noctilio labialis Kerr. One young is usually born at a time to the bulldog bat (1).

N. leporinus L. Thirty instances of February pregnancies have been recorded. Each had 1 embryo (2). In the Virgin Islands two May pregnancies have been reported (3). In Trinidad half the October males are in breeding condition with enlarged testes. Lactating females are found late in February and early in March (4).

1. Hall, E. R., and K. R. Kelson. The Mammals of North America. New York, 1959.
2. Allen, G. M. JM., 18: 514, 1927.
3. Goodwin, G. G. JM., 9: 104–113, 1918.
4. Goodwin, G., and A. M. Greenhall. AM., 122: 187–302, 1961.

NYCTERIDAE

Nycteris hispida Schreber

This bat of central and West Africa has an incomplete capsule around the ovary and very few glands in the endometrium. The separate cervical canals open on a single cervix in the vagina. The lumens of the canals contain no glands. The young are born early in December, and there is probably another pregnancy shortly afterwards but not so soon as in *N. luteola*. There is very little erectile tissue in the penis (1). The females mate early in January and there are many records of pregnancies in January and February, with fetuses near term in March and early April. Newborn young and lactating females have been recorded through April to June. In July the genital system is described as resting (2), but there is a record of pregnancy in August (3) and of newborn young, a few of them, in September. A single young is the rule (2).

1. Matthews, L. H. PZS., 111B: 289–346, 1941.
2. Verschuren, J. EPNG., 1957.
3. Lang, H., and J. P. Chapin. AM., 37: 497–563, 1917.

Nycteris luteola Thomas

In this central and east African bat the ovaries are enclosed in capsules, each with a large lateral slit. The cortex is thin except at the poles, and the graafian follicles are crowded together at these points. When ripe, the follicles are 400 μ in diameter, with a large antrum. The subepithelial coat of the vagina is in long folds and the epithelium above them forms a uniform coat which, therefore, varies from 100 to 35 μ in thickness. This becomes immensely thickened and is cornified during heat. The oviducts are very coiled and tortuous. The endometrium contains few glands. The females are probably polyestrous and have more than one pregnancy a year. These occur in quick succession. There is no trace of a corpus luteum in the ovary during pregnancy after an early stage. One young is born at a time (1), and the female has a postparturient heat (2).

The penis is covered with long, stiff hairs directed caudally. The prostate has lateroposterior parts only (1).

1. Matthews, L. H. PZS., 111B: 289–346, 1941.
2. Matthews, L. H. Nature, 143: 643, 1939.

Nycteris aethiopica Dobson. Nursing young have been found in December (1).

N. arge Thomas. Nearly adult young have been found in March and July, suggesting an extended breeding season (2).

N. capensis A. Smith. There are records of pregnant females taken in August (3) and of females, each with a single fetus, found in October (4).

N. damarensis Peters. There is a record of a female in December with a large fetus (1).

N. grandis Peters. This bat mates early in January and the young are born at the end of March and the beginning of April. A few births occur, also, in September (2).

N. nana Andersen. This bat mates early in December and as a rule the young, always single, are born in mid-March. One pregnant female has been reported in April (2).

N. pallida J. A. Allen. This bat probably breeds in December and the young are born in March. Some are born in September (2).

N. thebaica Geoffroy. A female with a fetus near term has been recorded for the end of January, and one suckling 2 young early in February (5).

1. Allen, G. M., and A. Loveridge. HMCZ., 75: 47–140, 1933–4.
2. Verschuren, J. EPNG., 1957.
3. Frechkop, S. EPNU., 1944.
4. Shortridge, G. C. The Mammals of South West Africa. London, 1934.
5. Allen, G. M., B. Lawrence, and A. Loveridge. HMCZ., 79: 31–126, 1935–7.

MEGADERMATIDAE

Megaderma (Lyroderma) lyra Geoffroy

INDIAN VAMPIRE BAT

The male attains puberty at age 15 months; the weight of the testes is then about 22 mg. The female breeds at 19 months. In July the testes begin to increase. They reach their maximum (to 316 mg.) in October, and after the mating season in November they decrease sharply in weight. Spermatozoa are found in the epididymis only during the breeding season. Conception takes place late in November and ovulation is always from the left ovary. One egg is released and implantation is always in the left horn. Birth takes place in April (1).

Follicles with antra are found in the ovaries in September. Late in October the ovaries are large and firm. They are enclosed in capsules which are hyperemic. At this time 5–8 follicles with much liquor folliculi are present. The endometrium and uterine glands have hypertrophied. Fertilization takes place in the anterior part of the oviduct. Embryonic development begins immediately and there is no storage of spermatozoa in the female tract (1).

The largest corpora lutea measure 900 μ in diameter. A small portion protrudes from the ovarian surface. Their decline is rapid, beginning soon after implantation. The luteal cells become elongated, vacuolated, and irregular in shape (1).

Breeding is once a year. Many multiovular follicles develop in the ovary, with up to 6 ova, probably due to the division of the oocyte (2).

1. Ramakrishna, P. A. J. Mysore Univ., 11B: 107–118, 1951.
2. Gopalkrishna, A., and M. A. Moghe. Proc. Nat. Inst. Sci., India, 26B (suppl.): 11–19, 1960.

Megaderma spasma L. The breeding record resembles that of *M. lyra.* Mating is late in November and births about April (1). In Ceylon the season may be later than in India as females with single fetuses are found in May (2).

M. (*Cardioderma*) *cor* Peters. This Ethiopian bat has a penis covered with hairs. The female usually has 1 young and the corpus luteum is exceptionally large. Each of two specimens examined had 1 fetus and 2 corpora lutea, one on each side (3). There is one report of a pregnant female taken in December (4).

1. Ramakrishna, P. A. J. Mysore Univ., 11B: 107–118, 1951.
2. Phillips, W. W. A. Ceylon J. Sci., 13B: 1–63, 1924–6.
3. Matthews, L. H. PZS., 111B: 289–346, 1941.
4. Verschuren, J. EPNG., 1957.

Lavia frons Geoffroy

This bat mates at the end of December and the young are born early in April. A pregnancy has also been reported for July and young from October onwards, so that the breeding season is probably indefinite (1, 2). Females with single fetuses have also been reported in July and August (3).

1. Verschuren, J. EPNG., 1957.
2. Lang, H., and J. P. Chapin. AM., 37: 497–563, 1917.
3. Frechkop, S. EPNU., 1944.

RHINOLOPHIDAE

Rhinolophus ferrum-equinum Schreber

GREATER HORSESHOE BAT

The European horseshoe bat is in heat toward the end of October, and it copulates at this time. Spermatozoa are stored in the uterus and vagina throughout the winter. Ovulation occurs in mid-April (1). There is also said to be a spring copulation, so the question whether or not conception can result from a fall copulation has not yet been settled (2).

The ovary is enclosed in a capsule, and the right ovary only is functional; consequently, pregnancy is always on the same side. The mature follicle measures 350 μ and the ovum 70 μ. The corpus luteum is buried in the ovary. When the bat is in heat, toward the end of October, the vaginal epithelium becomes cornified, but the follicles do not reach their maximum size until the time of ovulation. A copulation plug is formed by means of which most of the spermatozoa are stored until April, when the plug breaks up and is expelled. At the same time the vaginal epithelium, which has remained cornified, is desquamated. The corpus luteum is large and is not completely resorbed until near the end of lactation. Only one egg is released at a time, and one young is born in late June and July. Females have only one heat period in a year, and are 15 months old at the beginning of their first heat. The males are sexually mature at about the same time. The young are usually born in June (1).

Graafian follicles are present in the nonfunctional left ovary, but they degenerate and do not rupture (1).

The testes are active throughout the winter, and the spermatozoa are said to be stored in the female throughout this period in a pocket on the ventral surface of the vagina (3).

Ovulation and fertilization may be produced in January by the injection of a gonadotrophic extract of the pituitary gland (4).

1. Matthews, L. H. TZS., 23: 213–255, 1937.
2. Baker, J. R., and Z. Baker. J. Linn. Soc., Zool., 40: 123–141, 1936.
3. Rollinart, R., and E. L. Trouessart. CRSB., 52: 604–607, 1900.
4. Herlant, M. Bull. Acad. Roy. de Belgique, 20: 359–366, 1934.

Rhinolophus hipposideros Bechstein

LESSER HORSESHOE BAT

Adults of this European species mate in the fall and not in the spring, so spermatozoa are stored in the female tract throughout the winter (1). The reproductive tract of the females is in the estrous phase in January, and there is probably some spring mating of the first-year young, as some appear to give birth at 1 year. The 1½-year-old bats possess large follicles, and spermatozoa are found in the vagina at this heat. The (usually) single young is carried in the right horn of the uterus and is born in June or July. About 35 per cent of the females give birth to twins and the rest to single young (2).

The mature follicle measures 300 μ, and the ovum 70 μ. The corpus luteum is attached to the ovary by a peduncle (3).

In the male spermatogenesis begins in July, and the interstitial cells and accessory organs are in an active state. In October spermatogenesis has ceased, but the interstitial cells are still active, while the accessory organs are more so than they were in July. In winter there is no spermatogenesis, the interstitial cells are voluminous but not as vacuolated as in October, and the accessory organs continue to be well developed (4).

1. Baker, J. R., and Z. Baker. J. Linn. Soc., Zool., 40: 123–141, 1936.
2. Sluiter, J. W. Proc. Kon. Ned. Akad. Wetensch., 63C: 383–393, 1960.
3. Matthews, L. H. TZS., 23: 213–255, 1937.
4. Courrier, R. AB., 37: 173–334, 1927.

Rhinolophus geoffroyi A. Smith. This bat usually has 1 young, rarely 2 (1).

R. lepidus Blyth. In India the young are born between April 15 and May 15 (2).

R. luctus Temminck. In India the young are born in April (2). In Ceylon a female with 1 embryo was found in January (3).

R. megaphyllus Gray. In Australia the young are born in December (4).

R. rouxi Temminck. In India the young are born in April. The females have one parturition a year (2). In Ceylon the single young is born in April or May (3).

R. tragatus Hodgson. This Indian bat has 2 young, born towards the end of summer (5).

1. Shortridge, G. C. The Mammals of South West Africa. London, 1934.
2. Brosset, A. Mammalia, 26: 166–213, 1962.
3. Phillips, W. W. A. Ceylon J. Sci., 13B: 1–63, 1924–6.
4. Purchase, D., and P. M. Hiscox. CSIRO (Australia) Wildlife Res., 5: 44–51, 1960.
5. Blanford, W. T. The Fauna of British India. Mammals. London, 1888–91.

HIPPOSIDERIDAE

Hipposideros

HORSESHOE BATS

Hipposideros abae J. A. Allen. In this African bat lactation is general in April (1).

H. armiger Hodgson. This Himalayan leaf-nosed bat breeds once a year. The pairs of young are born towards the end of summer (2). They probably breed in August. The male has a very strong odor, absent in the female (3).

H. beatus Andersen. In this African bat birth is from March to April, and the females caught during late April were lactating (1).

H. bicolor Temminck. Three pregnant females were taken in March. One young is the usual number (4). According to another account they breed once a year and mate in March (5). The female has a brilliant coloration in the breeding season (6).

H. caffer Sundeval. In this African bat the penis is curved and is covered with hairs. The vas deferens has a conspicuous ampulla which is a blind diverticulum. The urethral glands are very well developed (7). There is always 1 young. Mating is probably in December and birth is in the third week of March (1).

H. cyclops Temminck. This bat mates at the end of December and the single young is born in mid-March (1). Embryos have been recorded in January and young in April (8).

H. galeritus Cantor. In Ceylon the single young is born early in April (4).

H. lankadiva Kelaart. In India the young are born in May or June (9).

H. nanus J. A. Allen. Births in this African species are in the period February to April (1).

H. speoris Schneider. The single young is born in May (9). A probable birth period in July has also been suggested, as several females with young were caught during the first days of August (4).

1. Verschuren, J. EPNG., 1957.
2. Blanford, W. T. The Fauna of British India. Mammals. London, 1888–91.
3. Hinton, M. A. C., and H. M. Lindsay. J. Bombay Nat. Hist. Soc., 31: 383–403, 1926.
4. Phillips, W. W. A. Ceylon J. Sci., 13B: 1–63, 1924–6.
5. Gopalkrishna, A., and M. A. Moghe. Proc. Nat. Inst. Sci., India, 26B (suppl.): 11–19, 1960.
6. Barrett-Hamilton, G. H. E. A History of British Mammals. London, 1910.
7. Matthews, L. H. PZS., 111B: 289–346, 1941.
8. Lang, H., and J. P. Chapin. AM., 37: 497–563, 1917.
9. Brosset, A. Mammalia, 26: 166–213, 1962.

Triaenops afer Peters

This bat, which is indigenous to Kenya, has 1 young at a time. No corpus luteum could be found in pregnant specimens, but the amount of ovarian stroma was slightly increased. The ovarian capsule is not closed, and the oviduct has a double sigmoid curve. The glans penis has flattened lobes. There is no scrotum (1). December embryos have been reported (2).

1. Matthews, L. H. PZS., 111B: 289–346, 1941.
2. Verschuren, J. EPNG., 1957.

PHYLLOSTOMATIDAE

CHILONYCTERINAE

Chilonycteris

LEAF-LIPPED BATS

Chilonycteris personata Wagner. In this Central American bat fertile males have been found in December and January (1). Females with 1 embryo each were taken in May (2).

C. rubiginosa Wagner. In Central America most females are pregnant with a single embryo in January (1). Pregnant females have also been taken in February (3), March (4), and May (2). In Trinidad young are present in May (5). A restricted season for these bats is probable.

1. Felten, H. SB., 37: 69–86, 1956.
2. Cockrum, E. L. Trans. Kansas Acad. Sci. 58: 487–511, 1955.
3. Bloedel, P. JM., 36: 232–235, 1955.
4. Burt, W. H., and R. A. Stirton. Univ. Michigan Mus. Zool., Misc. Publ. 117, 1961.
5. Goodwin, G. G., and A. M. Greenhall. AM., 122: 187–302, 1961.

Pteronotus

Pteronotus davyi Gray. Fertile males of this Central American bat have been recorded for December and January. March and September males were not fertile (1). Embryos have been reported in March and May (1, 2). Spermatozoa were found in a female taken in January (1). A single young is the rule (2).

P. suapurensis J. A. Allen. Fertile males have been found in December, and a female with 1 embryo in May. Spermatozoa were found in a female taken in January (1). In March single embryos have been reported for El Salvador (3).

1. Felten, H. SB., 37: 69–86, 1956.
2. Cockrum, E. L. Trans. Kansas Acad. Sci., 58: 487–511, 1955.
3. Burt, W. H., and R. A. Stirton. Univ. Michigan Mus. Zool., Misc. Publ. 117, 1961.

Mormoops megalophylla Peters

CINNAMON BAT

This Central American bat has been found with single embryos from March to June (1–5). In Trinidad lactating females have been found in March (6).

1. Cockrum, E. L. Trans Kansas Acad. Sci., 58: 487–511, 1955.
2. Burt, W. H., and R. A. Stirton. Univ. Michigan Mus. Zool., Misc. Publ. 117, 1961.
3. Hall, E. R., and K. R. Kelson. The Mammals of North America. New York, 1959.
4. Beatty, L. D. JM., 36: 290, 1955.
5. Baker, R. H. Univ. Kansas Mus. Nat. Hist., Publ. 9, 125–335, 1956.
6. Goodwin, G. G., and A. M. Greenhall. AM., 122: 187–302, 1961.

PHYLLOSTOMATINAE

Micronycteris

Micronycteris brachyotis Dobson. In Trinidad breeding males have been found in May and June (1).

M. hirsuta Peters. In Trinidad pregnant females have been taken in March and May. In May one had just given birth to a single young (1).

M. megalotis Gray. In Trinidad pregnant females have been taken in February and March. A lactating female was found in June (1). In El Salvador a pregnant female was found in March (2).

M. minuta Gervais. Pregnant and lactating females have been reported for Trinidad in May (1).

M. nicefori Sanborn. In Trinidad breeding males were found in October (1).

1. Goodwin, G. G., and A. M. Greenhall. AM., 122: 187–302, 1961.
2. Burt, E. R., and R. A. Stirton. Univ. Michigan Mus. Zool., Misc. Publ. 117, 1961.

Macrotus californicus Baird

LEAF-NOSED BAT

The males of this bat do not reach puberty in their first year but the females are impregnated in their first fall. The male sex organs are active in summer and fall but spermatogenesis has ceased by the end of November. By the end of September or October most females are pregnant with a single embryo (1). There is a slow growth of the embryo and the placenta through the winter; rapid growth sets in in March, with birth taking place in June. This is regarded as an instance of delayed development (2). One twin was recorded in 68 pregnancies (1).

1. Bradshaw, G. V. R. Mammalia, 25: 117–119, 1961.
2. Bradshaw, G. V. R. Science, 136: 645–646, 1962.

Macrotus mexicanus Saussure. Single embryos have been reported in February, March, and May (1).

1. Cockrum, E. L. Trans. Kansas Acad. Sci., 58: 487–511, 1955.

Lonchorhina aurita Tomes

LONG-EARED BAT

This bat has been found pregnant in February and March (1) and also in April (2).

1. Bloedel, P. JM., 36: 232–235, 1955.
2. Goodwin, G. G., and A. M. Greenhall. AM., 122: 187–302, 1961.

Macrophyllum macrophyllum Wied

The males of this Central American bat are fertile in December (1).

1. Felten, H. SB., 37: 179–212, 1956.

Tonatia bidens Spix

Pregnant females have been found in Trinidad in the month of May (1).

1. Goodwin, G. G., and A. M. Greenhall. AM., 122: 187–302, 1961.

Mimon crenulatum Geoffroy

Two pregnant females have been found in March (1).

1. Goodwin, G. G., and A. M. Greenhall. AM., 122: 187–302, 1961.

Phyllostomus discolor Wagner

JAVELIN BAT

Males in breeding condition have been found in Trinidad in January, August, and October (1). In Central America they have been reported for several months of the year (2). In Trinidad pregnant females have been taken in February, March, June, and August. Lactating females were found in August, September, and October, and females with young in August and October (1). In Central America the species probably breeds all year-round (2). In El Salvador young of various ages, and also a pregnant female, were obtained in December (3). In Colombia pregnant females were found in October (4).

1. Goodwin, G. G., and A. M. Greenhall. AM., 122: 187–302, 1961.
2. Felten, H. SB., 37: 179–212, 1956.
3. Burt, W. H., and R. A. Stirton. Univ. Michigan Mus. Zool., Misc. Publ. 117, 1961.
4. Valdivieso, D., and J. R. Tamsitt. JM., 43: 422–431, 1962.

Phyllostomus hastatus Pallas. Females advanced in pregnancy were found in Trinidad late in March and April. They were lactating in April and June, while a female with a single young was taken in September (1).

1. Goodwin, G. G., and A. M. Greenhall. AM., 122: 187–302, 1961.

Trachops cirrhosus Spix

A fertile male has been reported in December (1). Females with single embryos were taken in February (2).

1. Felten, H. SB., 37: 179–212, 1956.
2. Burt, W. H., and R. A. Stirton. Univ. Michigan Mus. Zool., Misc. Publ. 117, 1961.

Vampyrum spectrum L.

In Trinidad a lactating female was found in May (1).

1. Goodwin, G. G., and A. M. Greenhall. AM., 122: 187–302, 1961.

GLOSSOPHAGINAE

Glossophaga soricina Pallas

LONG-TONGUED VAMPIRE

This Central and South American bat breeds in the Matto Grosso once a year late in the spring, shortly before the rains (1). In El Salvador fertile males have been found from June to November, but not during the rest of the year. Pregnancies were found at most times, but at the height of the season, January to March, 43 per cent of the females are pregnant (2). In Mexico the single embryos have been found at all periods of the year (3). In Trinidad males in breeding condition were taken in February, March, and August. Pregnant females were found in January, April, May, June, and December, while lactating females were present in January, February, March, and June (4). Evidently in much of its range this species breeds at any time.

In one account the breeding season is reported to last for 2 to 3 weeks and ovulation is at this time. A nonpregnant female with a corpus luteum bleeds from a degenerating, hypertrophied endometrium. This bleeding is believed to occur at the end of pseudopregnancy (1).

1. Hamlett, G. W. D. AR., 60: 9–17, 1934.
2. Felten, H. SB., 37: 179–212, 1956.
3. Cockrum, E. L. Trans. Kansas Acad. Sci., 58: 487–511, 1955.
4. Goodwin, G. G., and A. M. Greenhall. AM., 122: 187–302, 1961.

Glossophaga longirostris Miller. In Trinidad pregnant females were found in February, March, April, and August. The single fetuses recorded for March varied from 14 to 23 mm. long so that they must have been of different ages. Females with young have been found in September and breeding males in March (1). A prolonged season is indicated.

1. Goodwin, G. G., and A. M. Greenhall. AM., 122: 187–302, 1961.

Lonchoglossa

Lonchoglossa ecaudata Wied. In Matto Grosso a new-born young was reported in February (1).

L. geoffroyi Gray. In Trinidad females were advanced in pregnancy in November; probably the young would have been born about a week later. Apparently all the females were pregnant at this time (2).

1. Kühlhorn, F. Säugetierk. Mitt., 1: 115–122, 1953.
2. Goodwin, G. G., and A. M. Greenhall. AM., 122: 187–302, 1961.

Choeronycteris mexicana Tschudi

LONG-TONGUED BAT

Single embryos have been found in March (1) and June (2). Lactating females have been reported for June (1, 3).

1. Baker, R. H. Univ. Kansas Mus. Nat. Hist., Publ. 9, 125–335, 1956.
2. Mumford, R. E., and D. A. Zimmermann. JM., 43: 101–102, 1962.
3. Campbell, B. JM., 15: 241–242, 1936.

Choeroniscus

Choeroniscus inca Thomas. In British Guiana pregnant females have been taken in February (1).

C. intermedius J. A. Allen and Chapman. In Trinidad a pregnant female has been taken in August (1).

1. Goodwin, G. G., and A. M. Greenhall. AM., 122: 187–302, 1961.

CAROLLIINAE

Carollia perspicillata L.

In the Amazon valley pregnant females have been found in August and January (1). In the Canal Zone (2) and in El Salvador (3) births of single young have been reported for March. Fertile males have been taken in November and January (4). In Trinidad breeding males and pregnant females have been taken in all months except November and December (5), so that a prolonged season is indicated.

1. Hamlett, G. W. D. AA., 79: 113–176, 1934.
2. Enders, R. K. HMCZ., 78: 385–502, 1935.
3. Burt, W. H., and R. A. Stirton. Univ. Michigan Mus. Zool., Misc. Publ. 117, 1961.
4. Felten, H. SB., 37: 179–212, 1956.
5. Goodwin, G. G., and A. M. Greenhall. AM., 122: 187–302, 1961.

Carollia castanea H. Allen. In El Salvador a few males are fertile in January, more in December and March. Pregnancies with single embryos have been reported from January to March and none from September to December (1). A December pregnancy has been reported from Mexico (2).

1. Felten, H. SB., 37: 179–212, 1956.
2. Lukens, P. W., and W. B. Davis. JM., 38: 1–14, 1957.

STURNIRINAE

Sturnira

Sturnira lilium Geoffroy. A suckling female has been recorded for June (1).

S. tildae de la Torre. In Trinidad a female with a single large fetus was taken in March (2).

1. Felten, H. SB., 37: 341–367, 1956.
2. Goodwin, G. G., and A. M. Greenhall. AM., 122: 187–302, 1961.

STENODERMINAE

Brachyphylla cavernarum Gray

A nursing female was found in July (1).

1. Anthony, H. E. Scientific Survey of Porto Rico and the Virgin Islands. New York Acad. Sci., 9: 1–96, 1927.

Uroderma bilobatum Peters

In Trinidad lactating females have been taken in February, and both pregnant and lactating ones in May. Nonbreeding males were found in May (1). Single embryos have been found in January (2), March, April (3), and May (4).

1. Goodwin, G. G., and A. M. Greenhall. AM., 122: 187–302, 1961.
2. Felten, H. SB., 37: 341–367, 1956.
3. Allen, J. A., and F. M. Chapman. AM., 9: 13–30, 1897.
4. Burt, W. H., and R. A. Stirton. Univ. Michigan Mus. Zool., Misc. Publ. 117, 1961.

Chiroderma

Chiroderma trinitatum Goodwin. A female with 1 large fetus has been recorded for March (1).

C. villosum Peters. In Trinidad a female with a half-grown fetus is recorded for August, and birth of a single young in September (1).

1. Goodwin, G. G., and A. M. Greenhall. AM., 122: 187–302, 1961.

Artibeus jamaicensis Leach

AMERICAN FRUIT BAT

In Trinidad males are in breeding condition at all times of the year, except perhaps in January and February. Pregnant females have been found from February to July and lactating ones into September (1). Single embryos have been recorded for February (2), March (3), and June (4). Lactating females have been found in March (5) and small young in June (6). A prolonged breeding season is indicated. In captivity the single young have been born from April to September (7). In El Salvador many embryos of various sizes were found in December (8).

The uterus is simplex and the oviducts enter near the midline (9).

1. Goodwin, G. G., and A. M. Greenhall. AM., 122: 187–302, 1961.
2. Jones, T. S. JM., 32: 223–224, 1951.
3. de la Torre, L. JM., 35: 113–116, 1954.
4. Cockrum, E. L. Trans. Kansas Acad. Sci., 58: 487–511, 1955.
5. Felten, H. SB., 37: 341–367, 1956.
6. Anthony, H. E. Scientific Survey of Porto Rico and the Virgin Islands. New York Acad. Sci., 9: 1–96, 1927.
7. Novick, A. JM., 41: 508–509, 1960.
8. Burt, W. H., and R. A. Stirton. Univ. Michigan Mus. Zool., Misc. Publ. 117, 1961.
9. Wislocki, G. B., and D. W. Fawcett. AR., 81: 307–314, 1941.

Artibeus lituratus Olfers

In Jamaica breeding males have been reported for practically the entire year, and pregnant females from February to July. Females with single young or lactating have been recorded from April to October (1). On the mainland single fetuses near birth and newborn young have been found in March (2, 3), but, in June, Mexican specimens are all pregnant with fetuses of various sizes. Thus an extended breeding season may be the rule (4).

1. Goodwin, G. G., and A. M. Greenhall. AM., 122: 187–302, 1961.
2. de la Torre, L. JM., 35: 113–116, 1954.
3. Bloedel, P. JM., 36: 232–235, 1955.
4. Lukens, P. W., and W. B. Davis. JM., 38: 1–14, 1957.

Artibeus cinereus Gervais. Breeding males have been found in March (1), and there is a July record of a female with a single embryo (2). For El Salvador single embryos of various, but fairly small, sizes have been reported in January (3).

A. hirsutus Andersen. Single embryos have been reported for February and May (4).

A. planirostris Spix. Females in advanced pregnancy or with single young have been recorded in June for Trinidad (5).

1. Goodwin, G. G., and A. M. Greenhall. AM., 122: 187–302, 1961.
2. de la Torre, L. JM., 35: 113–116, 1954.
3. Burt, W. H., and R. A. Stirton. Univ. Michigan Mus. Zool., Misc. Publ. 117, 1961.
4. Cockrum, E. L. Trans. Kansas Acad. Sci., 58: 487–511, 1955.
5. Jones, T. S. JM., 27: 327–330, 1946.

Centurio senex Gray

In Trinidad a breeding male was found in November, and a female gave birth to a single young in January (1).

1. Goodwin, G. G., and A. M. Greenhall. AM., 122: 187–302, 1961.

PHYLLONYCTERINAE

Phyllonycteris poeyi Gundlach

In the West Indies this bat has a single young born in June or July (1). In Cuba all the females are pregnant at the beginning of June and the single young is usually born about July 7 (2).

1. Allen, G. M. Bats. Cambridge, Mass., 1939.
2. Miller, G. S. Proc. Nat. Mus., United States, 27: 337–348, 1904.

Erophylla sezekorni Gundlach

This bat has 1 young at a time (1).

1. Hamlett, G. W. D. AA., 79: 113–176, 1934.

DESMODONTIDAE

Desmodus rotundus Geoffroy

VAMPIRE BAT

This South American bat has no well-defined season. The females are polyestrous and sometimes they experience post-partum heats (1). Pregnant females may be found at all times of the year (2). The gestation period lasts for at least 5 months and the single fetus is borne in either horn of the uterus. The ovum is released from either ovary, but a second one is occasionally released though it does not implant. Neogenesis of ova from the germinal epithelium is believed to occur. Spermatozoa are not stored in the uterus. Fertile males are found at all times of the year (2). In Trinidad the highest incidence of births is during April–May and October–November. The gestation period is longer than 84 days (3).

1. Wimsatt, W. A., and T. Trapido. AJA., 91: 415–446, 1952.
2. Felten, H. SB., 37: 341–367, 1956.
3. Goodwin, G. G., and A. M. Greenhall. AM., 122: 187–302, 1961.

Desmodus youngi Jentink. In Trindad breeding males were found in October. Lactating females with 1 young apiece have been recorded from August to October (1).

1. Goodwin, G. G., and A. M. Greenhall. AM., 122: 187–302, 1961.

Diphylla ecaudata Spix

VAMPIRE BAT

The single young is born at any time of the year (1). Fertile males have been reported for January, August, and September (2).

1. Dalquist, W. W. AMN., 53: 79–87, 1955.
2. Felten, H. SB., 37: 341–367, 1956.

NATALIDAE

Natalus

Natalus mexicanus Miller. Embryos have been recorded in January, April, May, and June. The number has always been a singlet (1–4).

N. tumidirostris Miller. In Trinidad this species is not breeding in November (5).

1. Cockrum, E. L. Trans. Kansas Acad. Sci., 58: 487–511, 1955.
2. Bloedel, P. JM., 36: 232–235, 1955.
3. Felten, H. SB., 38: 1–22, 1957.
4. Burt, W. H., and R. A. Stirton. Univ. Michigan. Mus. Zool., Misc. Publ. 117, 1961.
5. Goodwin, G. G., and A. M. Greenhall. AM., 122: 187–302, 1961.

VESPERTILIONIDAE

VESPERTILIONINAE

Myotis austroriparius Rhoads

RHOADS' BAT

This bat, indigenous to the eastern United States, breeds in the spring following the year of its birth. Males are in breeding condition, with enlarged epididymides, from October to April, few of them before December and practically all from February to April. On the west coast of Florida most of the mating is in the fall, but in peninsular Florida spring matings are the rule. The young are born in May, with the peak of births in the second week. About 90 per cent have 2 young at a time, the rest, 1 (1). At birth the young are extruded into a pocket formed by the interfemoral and alar membranes. They remain attached to the mother by the umbilical cord for several hours (2).

The head of the spermatozoon measures 4.6 μ (range, 4.4–4.9) $\times$ 1.9 μ (1.8–2.1); the midpiece, 12.8 μ (11.5–13.1) $\times$ 1.2 μ (1.0–1.4); the tail, 49.0 μ (47.5–54.0) (3).

1. Rice, D. W. JM., 38: 15–32, 1957.
2. Hamilton, Jr., W. J. The Mammals of Eastern United States. Ithaca, N.Y., 1943.
3. Hirth, H. F. J. Morphol., 106: 77–83, 1960.

Myotis capaccinii Bonaparte

In this bat of southern Europe and northwestern Asia spermatogenesis has ceased by November, but the epididymis is full of spermatozoa. The interstitial cells of the testis are variable, and the accessory organs are hypertrophied, though involution is beginning. In April some spermatozoa are still present, but the epididymis and accessory organs are in the resting state, while the interstitial cells are vacuolated (1).

1. Courrier, R. AB., 37: 173–334, 1927.

Myotis emarginatus Geoffroy

Females do not become sexually mature before their second summer. The adult females have spermatozoa in their uteri during the winter and so do many of the immature ones. The vaginal epithelium is cornified in both immature and mature specimens (1). A single follicle is present throughout the period of dormancy and the ovum is released spontaneously in the spring (2). The diameter of the follicle is about 475 μ, and of the ovum, 83.3 μ (3).

1. Sluiter, J. W., and M. Bowman. Kon. Ned. Akad. Wetensch., Proc., 54C: 594–602, 1951.
2. Sluiter, J. W., and L. Bels. Kon. Ned. Akad. Wetensch., Proc., 54C: 585–593, 1951.
3. Sluiter, J. W. Kon Ned. Akad. Wetensch., Proc., 57C: 696–700, 1954.

Myotis grisescens Howell

GRAY BAT

The gray bat of eastern North America mates in the fall, and 1 young is born in early June to early July (1), with some as early as May (2). The females have spermatozoa in their uteri in September, October, November, and December, the only months in which they were examined (3). One ovum is matured at a time and the diameter of the follicles with antra in September is about 200 μ. The number of such follicles decreases towards late September, when only one remains; this one persists through hibernation. The right horn of the uterus is much larger than the left one and implantation is always in the larger horn, but the large follicle is as often in the left ovary as it is in the right one. There is no evidence of unilateral ovarian development (4). The ovary is somewhat oval in shape and the round ligament is pigmented. The diameter of the tubal ovum is 110 μ (3).

Spermatogenesis occurs in the summer when the rest of the reproductive tract is quiescent. The epididymis is full of spermatozoa in fall, winter, and spring, and the accessory organs are full of secretion. The interstitial cells are largest in summer, retrogress in the fall, and remain small until spring. The male is sexually mature in its second summer (5).

1. Hamilton, Jr., W. J. The Mammals of Eastern United States. Ithaca, N.Y., 1943.
2. Rice, D. W. JM., 36: 89–90, 1955.
3. Guthrie, M. J. JM., 14: 199–216, 1933.
4. Guthrie, M. J., K. R. Jeffery, and E. W. Smith. J. Morphol., 88: 127–144, 1951.
5. Miller, R. E. J. Morphol., 64: 267–295, 1939.

Myotis keeni Merriam

The single young is born in July (1). In Illinois males taken in December have spermatozoa in the testes and the accessory organs are enlarged. In February spermatozoa are present in the uterus and oviduct (2).

The sperm head measures 4.4 μ (range, 4.0–4.6) × 1.9 μ (1.7–2.0); the midpiece, 15.4 μ (14.9–16.1) × 1.1 μ (0.9–1.3); and the tail, 62.0 μ (58.1–66.6) (3).

1. Hamilton, Jr., W. J. The Mammals of Eastern United States. Ithaca, N.Y., 1943.
2. Layne, J. N. AMN., 60: 219–254, 1958.
3. Hirth, H. F. J. Morphol., 106: 77–83, 1960.

Myotis lucifugus Le Conte

LITTLE BROWN BAT

The little brown bat of North America east of the Rocky Mountains has been much investigated, and the survival of spermatozoa in the uterus throughout the winter and fertilization by these spermatozoa in the spring have been proved beyond doubt.

Copulation occurs in the wild both in the fall and in spring. One young is born at a time, in mid-June to mid-July (1). Twins are rare and do not occur more frequently than in 1 per cent of births. The gestation period lies probably between 50 and 60 days. At parturition the female hangs with her head up and cups the tail and interfemoral membranes to catch the young, which are delivered by breech presentation (2). The sex ratio at birth is 50.6 per cent males (3). Spermatozoa survived in the uteri of females isolated in a refrigerator throughout the winter for at least 159 days, and cleaving eggs and blastocysts were found in the uteri in spring (4).

The theca of the graafian follicle does not divide into interna and externa. The large follicle during hibernation measures about 350 μ in diameter and the cells of the discus proligerus are hypertrophied. Just before ovulation, in spring, the antrum almost disappears, the volume of liquor folliculi decreases, and there is a preovulatory luteinization of the granulosa cells accompanied by a progestational reaction in the uterus. The wall of the rupturing follicle folds only to a small extent, and a cavity in the corpus luteum is rare. Ovulation occurs in late April or early May, during the passage from hibernation to summer quarters. When the females are kept in a warm laboratory, it is advanced to the middle of February. The diameter of the ovum is about 80 μ (5). Ovulation is spontaneous (6). The females reach sexual maturity at the end of their first summer. The embryo always develops in the right horn of the uterus (7), but the ripe follicle or corpus luteum is in the corresponding ovary in only 48 per cent of cases (5), or 49 per cent (8). In first-season females there is a preponderance of ripe follicles in the left ovary. The ovary is rounded and has a very inconspicuous neck (7).

In October the vaginal epithelium consists of many layers, but desquamation is continuous throughout the winter with the result that its thickness is reduced by April. In August intensive cornification sets in. The condition throughout hibernation may be described as one of submaximal estrus (9).

The round ligament of the uterus is not pigmented, in contrast to the condition in *M. grisescens* (7). The height of the epithelium is medium in October, and it is lowest in February. The uterine and glandular epithelia hypertrophy after ovulation so that the glandular epithelium appears to be irregular. The epithelium throughout is nonciliated and of a secretory type (9).

The tubal epithelium is ciliated and secretory. It is highest in October, is vacuolated during winter, and recovers in the spring. After ovulation, it is low and nonsecreting (9).

Spermatogenesis is in summer when the rest of the reproductive tract is quiescent. The epididymis is full of spermatozoa from fall to spring; the accessory organs are full of secretion and remain so until April. The interstitial cells are at their maximum size in summer, retrogress during the fall, and remain small until spring. The male is sexually mature in its second summer (10).

The sperm head measures 4.3 μ (range, 4.0–4.6) × 1.8 μ (1.5–2.0); the midpiece, 12.7 μ (12.0–13.2) × 1.1 μ (0.9–1.2); and the tail, 51.9 μ (44.0–56.5) (11).

1. Hamilton, Jr., W. J. The Mammals of Eastern United States. Ithaca, N.Y., 1943.
2. Wimsatt, W. A. JM., 26: 23–33, 1945.
3. Griffin, D. R. JM., 21: 181–187, 1940.
4. Wimsatt, W. A. AR., 88: 193–204, 1944.
5. Wimsatt, W. A. AJA., 74: 129–173, 1944.
6. Wimsatt, W. A., and F. C. Kallen. AR., 129: 115–131, 1957.
7. Guthrie, M. J. JM., 14: 199–216, 1933.
8. Guthrie, M. J., and K. R. Jeffers. AR., 71: 477–496, 1938.
9. Reeder, E. M. J. Morphol., 64: 431–453, 1939.
10. Miller, R. E. J. Morphol., 64: 267–295, 1939.
11. Hirth, H. F. J. Morphol., 106: 77–83, 1960.

Myotis myotis Borkhausen

COMMON EUROPEAN BAT

In this bat heat and insemination occur in the first autumn of life. Follicles do not ripen, however, in the first year (1). The usual time for mating is mid-September (2) and again in spring (3). Ovulation is at the end of March; the length of gestation is 50 days (2). One young is born at a time and the embryo is nearly always in the right horn of the uterus (3), but the corpus luteum is in the right ovary in only 55.7 per cent of cases (4). In one ovary a single follicle that is destined to rupture in the spring is present throughout the hibernation period. Ovulation is spontaneous but anything that raises the metabolic rate during the dormant period may induce it (5). In January the diameter of the ovum is 95.5 μ and of the follicle, 533.6 μ (6). Throughout winter spermatozoa may be found clinging to the uterine wall (7).

Spermatogenesis begins in May, when there are few spermatozoa in the epididymis; the interstitial cells appear to be active, but the accessory organs are large. The interstitial cells are pigmented. From December to February spermatogenesis has ceased, spermatozoa are present in the epididymis, the interstitial tissue is variable, and the accessory organs are well developed though there is some retrogression, less in the seminal vesicles than in the epididymis and prostate (2).

In the anterior pituitary gland orangophilic and carminophilic basophil cells hypertrophy at the time of ovulation. They are highly developed throughout gestation and they regress at parturition. It is suggested that these "basophil I" cells are active while the corpus luteum is active. Other basophilic cells ("II") hypertrophy during the autumnal mating season and persist during the prolonged winter estrus while the ovary shows only follicular growth. They regress during gestation. The granules of these cells are soluble in trichloracetic acid, and it is suggested that these cells are the origin of F.S.H. (8). The erythrosinophil cells increase during pregnancy at the expense of the chromophobes (9).

1. Sluiter, J. W., and M. Bouman. Kon. Ned. Akad. Wetensch., Proc., 54C: 594–601, 1951.
2. Courrier, R. AB., 37: 173–334, 1927.
3. Baker, J. R., and T. F. Bird. J. Linn. Soc., Zool., 40: 143–161, 1936.
4. Duval, M. J. de l'Anat., 31: 93–160, 1895.
5. Sluiter, J. W., and L. Bels. Kon. Ned. Akad. Wetensch., Proc., 54C: 585–592, 1951.
6. Sluiter, J. W. Kon Ned. Akad. Wetensch., Proc., 57C: 696–700, 1954.
7. Redenz, E. Zeit. Wiss. Biol., Abt. B, Zeit. Zellforsch. u. Mikros. Anat., 9: 734–739, 1929.
8. Herlant, M. AB., 67: 89–180, 1956.
9. Herlant, M. Ann. Soc. Roy. Zool. de Belge, 84: 87–116, 1953.

Myotis mystacinus Leisler

The females of this European bat are sexually mature in their second summer. In January the persistent follicle measures 398 μ, and the ovum, 77.5 μ in diameter (1). The number of young may be 1 or 2 (2), but another account gives a single young born in January or July (3).

1. Sluiter, J. W. Kon. Ned. Akad. Wetensch., Proc., 57C: 696–700, 1954.
2. Heptner, W. G., L. G. Morosowa-Turowa, and W. I. Zalkin. Die Säugetiere in der Schutzwaldzone. Berlin, 1956.
3. Mathias, P., and J. Séguéla. Mammalia, 4: 15–19, 1940.

Myotis sodalis Miller and Allen

This North American bat mates in September or October and spermatozoa can always be found in the uterus from October to February (1). Spermatozoa, however, are present in the epididymides of the males in December (2). They have survived in the uteri of isolated females up to 135 days, at least (3), and in December the uterus is turgid (2). The ovary is lobed and matures one ovum at a time. Ovulation in the laboratory under warm conditions has occurred in the middle of February. The round ligament is pigmented (1).

The sperm head measures 4.9 μ (range, 4.8–5.1) × 1.8 μ (1.7–2.0); the midpiece, 13.7 μ (13.0–14.0) × 1.0 μ (0.9–1.2); and the tail, 65.5 μ (62.0–69.9) (4).

1. Guthrie, M. J. JM., 14: 199–216, 1933.
2. Layne, J. N. AMN., 60: 219–254, 1958.
3. Wimsatt, W. A. AR., 88: 193–204, 1944.
4. Hirth, H. F. J. Morphol., 106: 177–183, 1960.

Myotis velifer J. A. Allen

The males of this bat do not produce spermatozoa until their second fall. Spermatogenesis takes place from late summer to early fall. After late September the testes are involuted, remaining quiescent until the next summer. The interstitial cells are largest in late summer. They reduce in size in early fall and remain small until spring. They and the accessory organs parallel the activity of the seminiferous tubules. However, spermatozoa remain in the epididymis through the winter (1). The single young is born during the last week in June (2, 3).

1. Krutzsch, P. H. AR., 139: 309, 1961.
2. Cockrum, E. L. Trans. Kansas Acad. Sci., 58: 487–511, 1955.
3. Glass, B. P., and C. M. Ward. JM., 40: 194–201, 1959.

Myotis adversus Horsefield. Young are born to this Australian species from mid-October to early November (1).

M. bechsteini Kühl. In Germany this bat gives birth to a single young early in May (2).

M. bocagii Peters. For this African bat a large embryo has been reported in early January and a very young male at the end of June, thus suggesting an extended season (3).

M. californicus Audubon and Bachman. In Mexico lactating females are found in June (4). Birth further north may occur from May to July (5). Males with enlarged testes are taken in September and October, but no pregnant females are found from mid-June until April. A single young is produced at a time (6).

M. chiloensis Waterhouse. This South American species has its single young in November (7).

M. daubentoni Kühl. This European bat has its single young in June or July (8).

M. evotis H. Allen. Two July records of this western American bat gave 1 fetus each (9). Another record gives single embryos in June (10), while early embryos have also been recorded late in May (11).

M. fortidens Miller and G. M. Allen. One pregnant female has been observed in May (12).

M. longicaudatus Ognev. On July 8 almost all females were with young (13).

M. nattereri Kühl. In this European bat the single young is born late in June or early in July (8).

M. nigricans Schinz. This bat has been reported as not breeding in Trinidad during September (14).

M. occultus Hollister. In New Mexico a specimen of this bat with 1 embryo has been taken in May (15).

M. peytoni Wroughton and Ryley. The young are born in March (16).

M. thysanodes Miller. In the southwestern part of North America the single young is born at the end of June (5).

M. tricolor Temminck. The young of this African bat are born late in December. The single embryo is always in the right horn of the uterus, but by November the whole uterus is occupied (17).

M. volans H. Allen. In this North American bat pregnant females, each with a single fetus, have been reported late in May (11) and June (5). Early June is probably the usual time of birth (18), but in Nevada it is delayed to mid-July (10).

M. yumanensis H. Allen. The single young of this western North American species are mostly born before mid-June but the season is longish; pregnancies have been recorded from mid-April to the end of June (19). Pregnant females have also been reported in mid-July (10). In New Mexico birth takes place at about June 25 (20). Birth may be later in the more northerly part of its range.

M. (Pizonyx) vivesi Menegaux. The fish-eating bat of Mexico has been reported as pregnant with a single embryo early in March (21) and early in June (22).

1. Purchase, D., and P. M. Hiscox. CSIRO (Australia) Wildlife Res., 5: 44–51, 1960.
2. Allen, G. M. Bats. Cambridge, Mass., 1939.
3. Lang, H., and J. P. Chapin. AM., 37: 497–563, 1917.
4. Baker, R. H. Univ. Kansas Mus. Nat. Hist., Publ. 9: 125–335, 1956.
5. Cockrum, E. L. Trans. Kansas Acad. Sci., 58: 487–511, 1955.
6. Krutzsch, P. H. JM., 35: 539–545, 1954.
7. Lataste, F. Act. Soc. Sci. de Chile, 1: 70–91, 1891.
8. Barrett-Hamilton, G. E. H. A History of British Mammals. London, 1910.
9. Cowan, I. McT., and C. J. Guiguet. British Columbia Provincial Mus., Handbook 11, 1956.
10. Hall, E. R. Mammals of Nevada. Berkeley, Calif., 1946.
11. Vaughn, T. A. Univ. Kansas Mus. Nat. Hist., Publ. 7: 513–582, 1954.
12. Dalquest, W. W., and E. R. Hall. JM., 29: 180, 1948.
13. Bogdanov, O. P. ZZ., 39: 1895–1896, 1960.
14. Goodwin, G. G., and A. M. Greenhall. AM., 122: 187–302, 1961.
15. Mumford, R. E. JM., 38: 260, 1957.
16. Brosset, A. Mammalia, 26: 166–213, 1962.
17. Harrison, L., and P. A. Clancey. Ann. Natal Mus., 12: 177–182, 1952.
18. Dalquest, W. W., and M. C. Ramage. JM., 27: 60–63, 1946.
19. Dalquest, W. W. AMN., 38: 224–247, 1947.
20. Commisaris, L. R. JM., 40: 441–442, 1959.
21. Reeder, W. G., and K. S. Norris. JM., 35: 81–87, 1954.
22. Burt, W. H. JM., 13: 363–365, 1932.

Lasionycteris noctivagans Le Conte

SILVERY BAT

This North American bat has 1 or 2 young, usually 2, born late in June or early in July (1, 2). In Utah the mating season is in August and September (2). Spermatozoa have been observed in the uterus and oviducts during February. At this time only a few are present in the seminiferous tubules of the male (3).

The sperm heads measure 5.5 μ (range, 5.1–5.8) × 2.0 μ (1.9–2.1); the midpiece, 8.9 μ (8.4–9.5) × 0.8 μ (0.7–1.0); and the tail, 73.4 μ (65.1–79.2) (4).

1. Hamilton, Jr., W. J. The Mammals of Eastern United States. Ithaca, N.Y., 1943.
2. Barnes, C. T. Utah Mammals. Salt Lake City, Utah, 1927.
3. Layne, J. N. AMN., 60: 219–254, 1958.
4. Hirth, H. F. J. Morphol., 106: 77–83, 1960.

Pipistrellus abramus Temminck

ASIATIC PIPISTRELLE BAT

In this bat spermatogenesis begins at the end of spring and ceases by October. From October to June the tubules are in repose, but spermatozoa are stored in the tail of the epididymis from October to May (1). No spring copulations have been observed, however. After the fall copulation the spermatozoa are stored in the uterus in such quantity that the organ swells (2). Ovulation takes place late in April and parturition early in July after a gestation of about 70 days. The litter size varies from 1 to 3, but 2 is the usual number (3). It has been shown that spermatozoa introduced into the uterus in the fall matings are capable of fertilizing the ova released in the spring (4).

1. Nakano, O. Folia Anat. Japonica, 6: 777–828, 1928.
2. Hiraiwa, Y. K., and T. Uchida. Sci. Bull. Fac. Agric., Kyushu Univ., 15: 255–266, 1955–56.
3. Uchida, T. Sci. Bull. Fac. Agric., Kyushu Univ., 12: 11–14, 1950.

4. Hiraiwa, Y. K., and T. Uchida. Sci. Bull. Fac. Agric., Kyushu Univ., 15: 565–574, 1955–6.

Pipistrellus pipistrellus Schreber

COMMON PIPISTRELLE

This common bat of Europe and Asia is said to copulate both in the fall and in the spring (1). The female mates in September, and spermatozoa may be found in the uterus until March. They cling to the uterine wall, and, since the glands are actively secreting during this time, they may derive nourishment from this source. Ovulation in France occurs in March and April (2), but in England it is said not to occur until May. No vaginal plug is formed, and the spermatozoa are stored in the uterus until after ovulation, when many are expelled. The female usually bears 1 young, and although ovulation may occur from either ovary, one writer has found that the pregnancy was in the right horn in 70 per cent of cases, 25 per cent were in the left, and the remaining 5 per cent were pregnant in both horns. When the embryo is in the horn opposite to the corpus luteum, transfer has been by migration across the body of the uterus (3). In England the females are pregnant or having their young in July (4).

The period of gestation is said to be about 44 days (3), but, as birth in France occurs in July to August while ovulation is said to be in March to April (2), its duration is still an open question. The female reaches puberty in her second year (3).

In the fall the vaginal epithelium becomes hyperplastic and keratinized, and it remains so all winter. The cornified tissue becomes so thick that the lumen is practically blocked during this period, which resembles a prolonged proestrum. Large graafian follicles can be found throughout, but they do not mature until near ovulation time (2).

The male does not reproduce until it is 2 years old. In mid-June spermatogenesis begins, but by the end of August it has ceased. The testes become smaller and by October they reach their minimum weight. The epididymis is small in August and enlarges in October, when its tail is larger than the testis and is full of spermatozoa. It regresses from May onward, and any remaining spermatozoa are phagocytized. The seminal vesicles and prostate are large from the end of August to December. In spring their involution is rapid, and at this time the epididymis is also regressing. The cycle of the interstitial cells is similar. Development begins in June and is at a maximum in August. They are still well developed in winter and appear to be glandular. In April involution sets in and the cells become pigmented (5).

A case has been described in which full spermatogenesis occurred at the usual time but in which the accessory organs remained atrophic. The interstitial cells of the testis were replaced by lymphoid cells (6).

1. Baker, J. R., and T. F. Bird. J. Linn. Soc., Zool., 40: 143–161, 1936.
2. Courrier, R. CRSB., 87: 1365–1366, 1922.
3. Deanesly, R., and T. Warwick. PZS., 109A: 57–60, 1939.
4. Whitaker, A. Naturalist, 74–83, 1907.
5. Courrier, R. AB., 37: 173–334, 1927.
6. Courrier, R. Compt. Rend. Assn. Anat., 21: 176–182, 1926.

Pipistrellus subflavus Cuvier

This bat of the eastern and southern United States is said probably to mate in August (1), but copulation has been observed in Indiana at the end of November (2), and it is said also to copulate in spring (3). Captured females all had spermatozoa in their uteri between November 11 and April 30. In the warm laboratory ovulation first occurred about March 7 (4). In the southern part of the range, birth is in late May, but further north it extends to late June and mid-July (1).

The ovary is rounded and has no neck. The number of mature follicles or tubal eggs was found to be $4.3 \pm .15$, mode 4 (4). Embryo counts have been given from 1 to 4. In one count ten bats had 2 embryos each. The other fourteen had 49 corpora lutea and 45 embryos (5). In another count there were 2 young in each of ten females (6). Three to 4 ova are usually shed and implant 2 to a horn. The pair nearest the ovaries lag in development and are reabsorbed (5).

The testes are quiescent in July and early August. Later in August spermatozoa are present and they may be found in the epididymis all winter. They have disappeared by May (6).

The sperm head measures 5.4 μ (range, 5.1–5.8) $\times$ 2.0 μ (1.9–2.1); mid-piece, width 0.8 μ (0.7–1.0); tail length, 72.1 μ (65.1–75.0) (7).

1. Hamilton, Jr., W. J. The Mammals of Eastern United States. Ithaca, New York, 1943.
2. Hahn, W. L. Biol. Bull., 15: 135–164, 1908.

3. Baker, J. R., and T. F. Bird. J. Linn. Soc., Zool., 40: 143–161, 1936.
4. Guthrie, M. J. JM., 14: 199–216, 1933.
5. Wimsatt, W. A. JM., 26: 23–33, 1945.
6. Layne, J. N. AMN., 60: 219–254, 1958.
7. Hirth, H. F. J. Morphol., 106: 77–83, 1960.

Pipistrellus (Nyctalus) noctula Schreber

This Old World bat copulates in September. Afterwards the cornified epithelium of the upper vagina sloughs off and forms a plug, which remains in place throughout hibernation (1). In Germany the young are born in early May (2). In England birth may be later than on the Continent; late June is given as the usual time (3). One is said to be the usual number in England, but on the Continent 1 and 2 are born at approximately equal numbers of births (2, 4). The period of gestation is probably about 49 days (5).

1. Grosser, O. Verh. d. Anat. Gesellsch., 17: 129–132, 1903.
2. Vogt, C. Compt. Rend. Assn. Adv. Sci., 655–662, 1881.
3. Whitaker, A. Naturalist, 74–83, 1907.
4. Allen, G. M. Bats. Cambridge, Mass., 1939.
5. Barrett-Hamilton, G. E. H. A History of British Mammals. London, 1910.

Pipistrellus ceylonicus Kelaart. Pregnant females have been found in the first week of September. Usually 2, but sometimes 1, young are born at a time (1). Birth is in the autumn in India (2).

P. coromandra Gray. In India births have been observed in April, May, and September. Two fetuses were found in a pregnant female in August (2).

P. culex Thomas. The young of this African bat are born in March (3).

P. fouriei Thomas. There is a record of this African bat with 2 fetuses in October (4).

P. hesperus H. Allen. The little canyon bat of North America has 2 young per litter as a rule, but occassionally only 1 is born. They are born in June and birth is little, if any, earlier in the southern part of its range than in the northern part (5). July has also been given as the month of birth (6).

P. kuhli Kuhl. This Old World bat usually has 2 young at a time (7). In North Africa the young are born by the end of May; in the Sahara they are born later (8). In February spermatozoa are in the uterus (9).

P. mimus Wroughton. In Ceylon this bat breeds all the year. It usually bears 2 young, but occasionally 1 is produced (1). In India no fetuses were observed in females taken in December and February (2).

P. nanus Peters. This African species is said rarely to bear 2 young, but a specimen with 2 was found in August (10) and there is a similar September record (11). Young bats have been seen in December, January (12), February (11), and April (3). The species probably reproduces all the year round (3).

P. nathusii Keyserling and Blasius. This European bat has 1 to 2 young at a time (13).

P. savii Bonaparte. This Eurasian bat has 2 young at a time (13).

P. (Nyctalus) leisleri Kuhl. The single young is born late in June (14). In India 2 embryos have been recorded. The females breed in December, at least, and they may be polyestrous (15).

1. Phillips, W. W. A. Ceylon J. Sci., 13B: 1–63, 1924–6.
2. Brosset, A. Mammalia, 26: 166–213, 1962.
3. Verschuren, J. EPNG., 1957.
4. Shortridge, G. C. The Mammals of South West Africa. London, 1934.
5. Cockrum, E. L. Trans. Kansas Acad. Sci., 58: 487–511, 1955.
6. Bailey, V. NAF., 55: 1936.
7. Blanford, W. T. The Fauna of British India. Mammals. London, 1888–91.
8. Lataste, F. Act. Soc. Linn., Bordeaux, 39: 129–299, 1885.
9. Lewis, R. E., and Harrison, D. L. PZS., 138: 473–486, 1962.
10. Lang, H., and J. P. Chapin. AM., 37: 497–563, 1917.
11. Lawrence, B., and A. Loveridge. HMCZ., 110: 1–80, 1953.
12. Allen, G. M., B. Lawrence, and A. Loveridge. HMCZ., 79: 31–126, 1935–7.
13. Heptner, W. G., L. G. Morosowa-Turowa, and W. I. Zalkin. Die Säugetiere in der Schutzwaldzone. Berlin, 1956.
14. Barrett-Hamilton, G. E. H. A History of British Mammals. London, 1910.
15. Ramaswamy, L. S. Half-yearly J. Mysore Univ., 7(2), 1933.

Eptesicus fuscus Peale and Beauvais

BIG BROWN BAT

This North American bat mates in the fall and its young are born in mid-June towards the north of its range, but earlier in the south (1). In Maryland births occur from May 15 to June 22 after an ovulation taking place about the first week of April (2). There is no doubt that spermatozoa survive in the uterus throughout the winter, since females collected in

December, kept in isolation at 40° F., and autopsied in May after a month at 75° F. were found to contain embryos. The spermatozoa had survived for at least 140 to 150 days (3). The number of young varies from 1 to 4; in the western part of its range the usual number of young is 1, while in the eastern part the number is 2 (4). Single embryos are recorded for the West Indies (5). There is considerable embryonic wastage; in 30 cases observed the number of blastocysts varied from 1 to 6, with an average of 3.9 and a mode of 4 (6).

The ovaries are small, rounded, and without a neck (7). Newborn bats develop antra to the follicles late in July and polyovular follicles are found only in the young of the year. Parous females are later than the young in developing follicles (2). The average number that enlarge during hibernation or that rupture is 4.2, and they are present in each ovary in almost equal numbers, 53 per cent in the left. It is impossible to differentiate between theca interna and theca externa in the follicle wall. The granulosa cells begin to luteinize before ovulation, and there is a decrease in the amount of follicular fluid, which penetrates the theca. Rupture is sudden, but there is little hemorrhage. It is usual to find a cavity in the corpus luteum (8). The interstitial cells of the ovary are large and vacuolated at parturition. They then decrease in size until October (2).

Spermatogenesis is in full progress by mid-August. It ceases by mid-October but the epididymis continues to be full of spermatozoa. All males reproduce in their first summer (2). By December there are still a few spermatozoa in the testes but no spermatocytes are spermatids. At this time the uterus contains a mass of spermatozoa, but there are none in the oviducts (9). According to another account spermatozoa may be found in the oviducts between December and March (10).

The head of the spermatozoon measures 5.0 μ (4.9–5.2) $\times$ 2.0 μ (1.9–2.1); midpiece, 9.0 μ (8.5–9.2) $\times$ 0.8 μ (0.7–1.0); tail length, 72.0 μ (68.1–75.1) (11).

1. Hamilton, Jr., W. J. The Mammals of Eastern United States. Ithaca, N.Y., 1943.
2. Christian, J. J. AMN., 55: 66–95, 1956.
3. Wimsatt, W. A. AR., 83: 299–306, 1942.
4. Cockrum, E. L. Trans. Kansas Acad. Sci., 58: 487–511, 1955.
5. Anthony, H. E. Scientific Survey of Porto Rico and the Virgin Islands. New York Acad. Sci., 9: 1–96, 1927.
6. Wimsatt, W. A. AR., 88: 193–204, 1944.
7. Guthrie, M. J. JM., 14: 199–216, 1933.
8. Wimsatt, W. A. AJA., 74: 129–173, 1944.
9. Evans, C. A. Am. Nat., 72: 480–484, 1938.

10. Layne, J. N. AMN., 60: 219–254, 1958.
11. Hirth, H. F. J. Morphol., 106: 77–83, 1960.

Eptesicus serotinus Schreber

SEROTINE BAT

This Old World bat copulates in September in France (1), and not again in the spring (2). But the testes begin to hypertrophy in May and spermatogenesis commences. In June spermatozoa are present and spermatogenesis continues until September. The epididymis is full of spermatozoa throughout hibernation, and the accessory organs are well developed as well as the interstitial tissue of the testis. In May the epididymis diminishes in size, the accessory organs atrophy, and the interstitial cells are pigment laden. In June the condition is the same except that pigment can no longer be found in the interstitial cells, which appear to be fully active by August and September (1). It would seem, therefore, that insemination is possible for a great part of the year in spite of the belief that it occurs only in the fall. The number of young, born late in May, is usually 1.

Castration of males in December and continuation of hibernation had no effect upon the epididymis and sperm storage during that season. The accessory organs remained active whether or not the epididymides were removed during the operation. But exposure of the castrated bats to higher temperatures, while not affecting the epididymis and spermatozoa, caused the accessory glands to atrophy. A castrate liberated and caught during the next hibernation season showed the usual castrate atrophy of the whole reproductive tract (1).

Females injected with a pituitary gonadotrophe in December ovulated 4 days afterwards. A similar experiment done in January was followed by pregnancy (3).

1. Courrier, R. AB., 37: 173–334, 1927.
2. Baker, J. R., and T. F. Bird. J. Linn. Soc., Zool., 40: 143–161, 1936.
3. Herlant, M. Bull. Acad. Roy. de Belgique, 20: 359–366, 1934.

Eptesicus capensis A. Smith. This African bat has been found in November with 2 fetuses (1).

E. garambae J. A. Allen. Birth in this African species is from the end of February to early March (2).

E. rendelli Thomas. Birth in this African species is from the end of March to early April (2).

E. tenuipinnis Peters. The birth of 1 young in March has been recorded in the Cameroons (3), and nursing young were found in Liberia in July (4).

E. (*Hesperotenus*) *tickelli* Blyth. In Ceylon this bat has an extended season; large and small embryos are present in most females taken in May. A December pregnancy and a single young born early in June have also been reported (5).

1. Shortridge, G. C. The Mammals of South West Africa. London, 1934.
2. Verschuren, J. EPNG., 1957.
3. Eisentraut, M. Mitt. Berlin Mus. Zool., 25: 245–274, 1941.
4. Allen, G. M., and H. J. Coolidge, Jr. In R. P. Strong, The African Republic of Liberia and the Belgian Congo. Cambridge, Mass., 1930.
5. Phillips, W. W. A. Ceylon J. Sci., 13B: 1–63, 1924–6.

Vespertilio murinus L.

PARTI-COLORED BAT

The nomenclature of this bat and that of *Myotis myotis* are so confused that much that has been recorded under the latter species may possibly be referred properly to *Vespertilio murinus*. So far as the writer can judge there is little difference in their reproduction.

Nycticeius humeralis Rafinesque

EVENING BAT

This North American bat may be found pregnant in April, May, and June (1). The females mate in August and the young are usually born late in May (2). The embryo counts range from 1 to 4, with 2 the usual number (1). There are no spermatozoa in the male tract by the end of April when the females are pregnant (3).

The head of the spermatozoon measures 5.2 μ (range, 4.9–5.5) $\times$ 2.1 μ (2.0–2.4); midpiece, 10.8 μ (10.0–11.1) $\times$ 2.1 μ (2.0–2.4); tail length, 77.3 μ (72.0–84.1) (4).

1. Cockrum, E. L. Trans. Kansas Acad. Sci., 58: 487–511, 1955.
2. Hamilton, Jr., W. J. The Mammals of Eastern United States. Ithaca, N.Y., 1943.
3. Layne, J. N. AMN., 60: 219–254, 1958.
4. Hirth, H. F. J. Morphol., 106: 77–83, 1960.

Rhogeessa

YELLOW BATS

Rhogeessa parvula H. Allen. This North American bat has been recorded as pregnant with 2 embryos in May (1).

R. tumida H. Allen. Twins are not unusual in this species (2).

1. Cockrum, E. L. Trans. Kansas Acad. Sci., 58: 487–511, 1955.
2. Goodwin, G. G., and A. M. Greenhall. AM., 122: 187–302, 1961.

Scotophilus temmincki Horsefield

LESSER YELLOW BAT

The breeding season of this Indian bat is very sharply defined, with mating about the third week of March. Fertilization is immediate and the young are born in the last week of June and the first week of July after a gestation of 105 to 115 days. The young bats breed during their first year of life (1). The number of young is usually 2, with corpora lutea in either ovary and pregnancy in both horns of the uterus. During late pregnancy the corpora lutea occupy most of the medullary region of the ovary but by September this tissue has degenerated (2).

The ripe follicle is about 277 μ in diameter, the ovum about 93 μ. The mature corpus luteum is about 390 μ. It never projects beyond the surface of the ovary and it contains extravasated blood in its center. New ova arise from the germinal epithelium (2).

The oviduct changes little during the cycle. Its ovarian portion is largely lined with glandular cells. These increase in size in February but they revert to the anestrous condition later in the month. The columnar cells of the uterine epithelium measure from 8 μ to 12 μ in height when they are resting. At the time of heat they increase to from 18 μ to 20 μ. At this time vascularity increases and the glands enlarge and are more plentiful. The endometrial epithelium ruptures due to congestion of blood in the subepithelial region, and often red blood cells may be found in the lumen. The vaginal wall undergoes cornification and desquamation during heat (2).

The males, like the females, breed before they are a year old. Weight of the testes increases from December and reaches its maximum in February when they weigh 4 or 5 times as much as they do in the nonbreeding season (3). This represents a rise for the two from 30 to 150 mg. The interstitial cell number is greatest when the testes are active. At the same time the prostate tubules and the bulbo-urethral glands enlarge and secrete (2).

1. Gopalkrishna, A. Proc. Indian Acad. Sci., 26B: 219–231, 1947.
2. Gopalkrishna, A. Proc. Indian Acad. Sci., 30B: 17–46, 1949.
3. Gopalkrishna, A. Proc. Indian Acad. Sci., 27B: 137–150, 1948.

Scotophilus heathi Horsefield

GREATER YELLOW BAT

In Ceylon the greater yellow bat is pregnant in October (1). In India the single young is born during the first half of August (2), but another account gives June as the month of birth (3). The undersides of pregnant females change from pale straw color to rich saffron (4).

1. Phillips, W. W. A. Ceylon J. Sci., 13B: 1–63, 1924–6.
2. Prakash, I. JM., 41: 386–389, 1960.
3. Brosset, A. Mammalia, 26: 166–213, 1962.
4. Blanford, W. T. The Fauna of British India. Mammals. London, 1888–91.

Scotophilus murino-flavus Heuglin. This African bat has been found in March carrying 2 large fetuses (1).

S. nigrita Schreber. This African bat gives birth to 1 or 2 young (2) in March (3).

1. Lang, H., and J. P. Chapin. AM., 37: 497–563, 1917.
2. Shortridge, G. C. The Mammals of South West Africa. London, 1934.
3. Verschuren, J. EPNG., 1957.

Chalinolobus

Chalinolobus argentatus Dobson. Births in this African species are usually in March and April. The number of young varies from 1 to 2 (1, 2). Embryos have been recorded in January (3).

C. gouldi Gray. In January 22 of 25 females were either pregnant or lactating (4).

C. humeralis J. A. Allen. Embryos have been recorded in February (3).

1. Lang, H., and J. P. Chapin. AM., 37: 497–563, 1917.
2. Allen, J. M., B. Lawrence, and A. Loveridge. HMCZ., 79: 31–126, 1935–7.
3. Verschuren, J. EPNG., 1957.
4. Simpson, K. G. Victorian Nat., 78: 325–327, 1962.

Lasiurus borealis Müller

RED BAT

This North American bat mates late in August or early in September (1); the young are born late in May and in June. The litter size varies from 1 to 4, with 3 as the usual number (2, 3). About 63 per cent of the females with embryos had 3, and of those with young 56 per cent had 3 (3). The males have spermatozoa in the epididymis in August and also in March and May. None are present in June (4).

The head of the spermatozoon measures 5.3 μ (5.0–5.7) $\times$ 2.0 μ (1.9–2.3); midpiece, 11.2 μ (10.5–12.2) $\times$ 0.8 μ (0.7–1.0); tail length, 67.1 μ (64.0–74.0) (5).

1. Stuewer, F. W. JM., 29: 180–181, 1948.
2. Jones, C. J. JM., 42: 538–539, 1961.
3. Cockrum, E. L. Trans. Kansas Acad. Sci., 58: 487–511, 1955.
4. Layne, J. N. AMN., 60: 219–254, 1958.
5. Hirth, H. F. J. Morphol., 106: 77–83, 1960.

Lasiurus cinereus Peale and Beauvois. The hoary bat of North America mates in August (1) and the young are usually born towards the end of June or early in July. The number of young is usually 2 (1, 2).

L. seminolus Rhoads. This bat has its young, usually 2, late in May or in June (2).

L. semotus H. Allen. This Hawaiian bat has been found in May carrying 2 fetuses (3).

L. (Dasypterus) ega Peters. A record of this Mexican bat gives 2 as the number of young (4).

L. (D.) floridanus Miller. This bat has 2 to 3 young late in May and early June. The testes contain no spermatozoa from February to September. They are plentiful from September to December while they may be found in the epididymides from September to February (5).

1. Bailey, V. NAF., 55: 1936.
2. Cockrum, E. L. Trans. Kansas Acad. Sci., 58: 487–511, 1955.
3. Baldwin, P. H. JM., 31: 455–456, 1950.
4. Allen, J. A. AM., 22: 191–262, 1906.
5. Sherman, H. B. Quart. J. Florida Acad. Sci., 7: 193–197, 1944.

Barbastella barbastellus Schreber

BARBASTELLE BAT

The number of young born to this Eurasian bat is usually 2 (1).

1. Heptner, W. G., L. G. Morosowa-Turowa, and W. I. Zalkin. Die Säugetiere in der Schutzwaldzone. Berlin, 1956.

Plecotus auritus L.

LONG-EARED BAT

The reproductive cycle in this Old World bat is essentially similar to that of other hibernating species. Spermatogenesis is at its height in August, and the interstitial cells of the testis are large at that time (1). It does not occur in spring, and spermatozoa survive throughout the winter in the uterus (2). Copulation is said to take place only in the fall (3). The young, one usually, are born in June and July (4).

1. Courrier, R. AB., 37: 173–334, 1927.

2. Redenz, E. Zeit. Wiss. Biol., Zeit. Zellforsch. u. Mikros. Anat., 9: 734–749, 1929.
3. Baker, J. R., and T. F. Bird. J. Linn Soc., Zool., 40: 143–161, 1929.
4. Barrett-Hamilton, G. E. H. A History of British Mammals. London, 1910.

Plecotus (Corynorhinus) rafinesquii Lesson

LUMP-NOSED BAT

The female of this western North American species mates at 4 months. Mating begins in October and continues all winter. No vaginal plug is formed. Ovulation takes place from February to April, probably a few days after the bats have left their winter tunnels. In 49 of 50 observed pregnancies the single embryo was in the right horn, although both ovaries were equally active. The spermatozoa survive throughout the winter, for at least 76 days and probably to 108 days, but they soon disappear after ovulation. Pregnancy lasts about 73 days but its length varies with changes in temperature (1). Birth is usually towards the end of June (2).

The ovaries are well encapsulated and the round ligament is pigmented. A medium-sized follicle in October measures 250–300 μ in diameter. There is some evidence that copulation provides a stimulus needed for its growth. The largest follicle, with a maturation spindle in the ovum, is 400 μ in diameter. The antrum is small and the ovum is surrounded by an enormous ball of cumulus cells, all of which are ejected at ovulation. This ball of cells, together with the ovum, enters the oviduct but the cells are soon shed. At ovulation the first polar body and the second spindle have been formed. Fertilization takes place in the upper part of the oviduct. The ovum is 73 μ in diameter. Transfer to a warm room does not readily cause the bats to ovulate earlier than mid-February but an injection of pregnancy urine is effective during the winter. The mature corpus luteum is 725 μ in diameter. There is almost always 1 corpus luteum; a second has been observed once but the two were of different ages. In the early stages of implantation the cytoplasm of the lutein cells is foamy; later the cells are vacuolated. The interstitial cells of the ovary are large in the nonbreeding season; they enlarge still further in pregnancy and are largest during lactation. Occasional reabsorption of embryos has been observed (1).

In winter the vaginal smear contains many nucleated epithelial cells but very few cornified ones and no leucocytes. At the time of ovulation the smear contains mostly nucleated epithelial cells and few cornified ones. At no time do sections of the vagina show cornification, but the epithelium is deeply stratified (1).

THE MALE

The testes are active in the first fall of the bats' life but they begin to retrogress in November. The accessory glands are small. In the mature bat the testes begin to enlarge in April and they continue to grow through the summer. Spermatozoa are present in September and October. The testes then shrink abruptly. The epididymis gradually decreases to a minimum at the end of April, but spermatozoa have been found in this organ as late as April 27. The accessory glands begin to enlarge in August, are largest in late October and early November, and remain enlarged throughout the winter. Copulation occurs in the fall and winter when the interstitial cells are small; the relation between the size of these cells and the accessory gands is an inverse one. The total volume of interstitial cells in the testis shows the same relationships. Castration in November does not disturb the epididymis but it does cause the accessory glands to shrink. Abundant vigorous spermatozoa remain in the epididymides of the castrates (1).

The head of the spermatozoon measures 3.8 μ (range 3.7–4.0) $\times$ 1.9 μ (1.7–2.0). This is the smallest for all bats that have been measured. The mid-piece measures 8.9 μ (8.5–9.5) $\times$ 1.1 μ (0.9–1.3) and the tail, 50.0 μ (46.1–57.3) (3).

1. Pearson, O. P., M. R. Kofoid, and A. K. Pearson. JM., 33: 273–320, 1952.
2. Mohr, C. E. JM., 15: 49–53, 1934.
3. Hirth, H. F. J. Morphol., 106: 77–83, 1960.

Idionycteris phyllotis G. M. Allen

Lactating specimens of this bat have been found in June. At this time the males are not producing spermatozoa (1).

1. Jones, C. J. JM., 42: 538–539, 1961.

Euderma maculatum J. A. Allen

SPOTTED BAT

In New Mexico lactating specimens have been taken in June. At this time none of the males contained spermatozoa (1).

1. Jones, C. J. JM., 42: 538–539, 1961.

MINIOPTERINAE

Miniopterus australis Tomes

In the New Hebrides this bat copulates at the end of August or the beginning of September, i.e., in the spring, and the development of the embryo begins at once. One embryo develops, nearly always in the right horn. The testes are small in September, but spermatozoa are still in the epididymis, from which they disappear by December (1). The young are born at the end of December or early in January (2). In April the testes begin to grow, and by August they attain their maximum weight (1).

1. Baker, J. R., and T. F. Bird. J. Linn. Soc., Zool., 40: 143–161, 1936.
2. Sanborn, C. C., and A. J. Nicholson. Fieldiana Zool., 31: 313–338, 1950.

Miniopterus minor Peters

This bat from Zanzibar has its young earlier than December, the usual time of pregnancy for other species from the same caves. The ovarian capsules are open. The cornua are almost at right angles to the body of the uterus. The vaginal epithelium mucifies during pregnancy (1).

1. Matthews, L. H. PZS., 111B: 289–346, 1941.

Miniopterus natalensis A. Smith

This bat, a native of South Africa, has the ovary enclosed in a capsule with a small opening. The ovarian stroma contains much interstitial tissue.

The corpus luteum is large and, when young, has a blood clot in the center. One young is born at a time, and in 8 pregnant females all the embryos were in the right horn of the uterus and all the corpora lutea in the left ovary (1). However, late in pregnancy the whole uterus is occupied. Birth occurs before the third week of December (2). Embryos have been reported in July and November (3).

The glans penis is minute; the vas deferens has an enlarged ampulla consisting of a mass of connective tissue with numerous diverticula. The prostate is large (1).

1. Matthews, L. H. PZS., 111B: 289–346, 1941.
2. Harrison, L., and P. A. Clancey. Ann. Natal Mus., 12: 177–182, 1952.
3. Verschuren, J. EPNG., 1957.

Miniopterus schreibersii Kuhl

This bat, a native of southern Europe and Algeria, ovulates in October or November and mates and is fertilized at that time. Ovulation is almost invariably from the right ovary (1). A single young is usual (2).

In August the testes are in full spermatogenesis; in September this condition has ceased, but the interstitial tissue is well developed and the accessory organs are very active. The epididymis is very voluminous and is full of spermatozoa. In November some testes have retrogressed while others are still well developed, but, although spermatozoa are present, their formation has ceased. The interstitial tissue is reduced. This state continues into January and April, but the number of spermatozoa diminishes. In April, however, the interstitial tissue is showing signs of recovery. At the time of fertilization the testes are inactive and the spermatozoa must, therefore, come from the epididymis (1).

1. Courrier, R. AB., 37: 173–334, 1927.
2. Heptner, W. G., L. G. Morosowa-Turowa, and W. I. Zalkin. Die Säugetiere in der Schutzwaldzone. Berlin, 1956.

Miniopterus macrocneme Revilliod. In New Caledonia no pregnancies are found until October. They are at their peak in December, and young are born at the end of December and early in January (1).

M. rufus Sanborn. A pregnant female has been reported in September (2).

1. Sanborn, C. C., and A. L. Nicholson. Fieldiana Zool., 31: 313–338, 1950.
2. Verschuren, J. EPNG., 1957.

KERIVOULINAE

Kerivoula

Kerivoula cuprosa Thomas. This African bat has been recorded as with young in October (1).

K. lanosa A. Smith. This African bat has been reported to have 2 young at a time (2).

K. picta Pallas. This Asiatic bat has been reported as mating in June. A pregnant female was found in October. One young is usual (3).

1. Verschuren, J. EPNG., 1957.
2. Allen, G. M. Bats. Cambridge, Mass., 1939.
3. Phillips, W. W. A. Ceylon J. Sci., 13B: 1–63, 1924–6.

NYCTEROPHILINAE

Antrozous pallidus Le Conte

PALE BAT

This North American bat mates from October to February; and the young are born from the middle of May to the end of June after a gestation period of about 9 weeks (1). The number of embryos is from 1 to 3, with 2 in at least 67 per cent of cases (2, 3).

1. Orr, R. T. Proc. California Acad. Sci., Ser. 4, 28: 165–246, 1954.
2. Cockrum, E. L. Trans. Kansas Acad. Sci., 58: 487–511, 1955.
3. Baker, R. H. Univ. Kansas Mus. Nat. Hist., Publ. 9, 125–335, 1956.

MOLOSSIDAE

Molossops greenhalli Goodwin

Single fetuses have been found in females taken in June. The young are born in that month (1).

1. Goodwin, G. G., and A. M. Greenhall. AM., 122: 187–302, 1961.

Tadarida braziliensis Geoffroy

FREE-TAILED BAT

In Florida this free-tailed bat mates in mid-February to late March, at ovulation time. One young is born in late May to late June (1). In Texas (2) and Oklahoma (3) the season of birth continues until mid-July. The gestation period lasts about 11 to 12 weeks. The females are sexually mature at 9 months. The males have a short rutting season, from February to mid-April (1). The cells of the discus proligerus do not hypertrophy as they do in *Myotis lucifugus* (4).

Fairly large graafian follicles are present in the ovaries at all times of the year except when small embryos are present, i.e., for about 2 weeks following ovulation. These are almost always in the right ovary only. Spermatozoa are not stored in the uterus throughout winter and are found only from mid-February to March. The single embryo is almost invariably in the right horn. In the male spermatozoa are present in the testes and epididymis in February and March. Atrophy occurs in April and lasts until September, when proliferation begins in the testes tubules. No spermatozoa, however, can be found until the end of January, but they have not yet reached the epididymis (5).

In summer the testes are abdominal and in spring they are partly interfemoral (6). The males are sexually mature at from 18 to 22 months (7).

1. Hamilton, Jr., W. J. The Mammals of Eastern United States. Ithaca, N.Y., 1943.
2. Cagle, F. R. JM., 31: 400–402, 1950.
3. Twente, Jr., J. W. JM., 37: 42–47, 1956.
4. Wimsatt, W. A. AJA., 74: 129–173, 1944.
5. Sherman, H. B. JM., 18: 176–187, 1937.
6. Krutzsch, P. H. JM., 36: 236–242, 1955.
7. Short, H. L. JM., 42: 533–536, 1961.

Tadarida aegyptiaca Geoffroy. Births occurred in September but not in March (1).

T. femorosacca Merriam. This North American bat has been found with single embryos at the end of April (2) and the end of June (3). Births probably occur late in June and early in July (4).

T. midas Sundevall. No lactating or pregnant females were found in February but very small embryos were present in March. Females near term were found in April, and newborn young in October, so two seasons a year are probable (5).

T. molossa Pallas. This bat has been found with single embryos in May (3, 6).

T. teniotis Rafinesque. At the end of May single embryos near term were found. At this time the testes were not enlarged (7).

T. (Chaerophon) limbata Peters. This African bat has been found with large fetuses in January (8). Lactating females have been found in May and the young are born in early or mid-April (5).

T. (C.) pumila Cretzschmar. Females with single large fetuses have been found in November (9) and lactating females pregnant with single fetuses in May (10).

T. (C.) russatus J. A. Allen. Of 20 females taken in September 6 had one fetus each (11).

T. (Mops) congica J. A. Allen. This African bat has been found with single large fetuses in September (11).

T. (M.) faradjius J. A. Allen. This African bat probably breeds all the year; females near term have been found in April and June, lactating females in June, and some with small embryos in September (5).

T. (M.) occipitalis J. A. Allen. The only record for this species is of half-grown young in September (12).

T. (M.) thersites Thomas. Embryos have been found in April (5).

T. (M.) trevori J. A. Allen. A female with a very large fetus has been found in September (11).

T. (Nyctinomis) ansorgei Thomas. The only record for this African bat is the absence of pregnant females from January to July (5).

T. (N.) condylura A. Smith. This African bat mates in February and gives birth to its young in mid-May. Very small embryos were reported in April and large ones in May (5).

T. (N.) ochracea J. A. Allen. Eleven of 14 females taken in March carried an embryo in the right uterine horn (11).

T. (Otomops) wroughtoni Thomas. This species of southern India has been recorded with late fetuses and new-born young in December. The females are not pregnant in May (1).

1. Brosset, A. Mammalia, 26: 166–213, 1962.
2. Benson, S. B. JM., 21: 26–29, 1940.
3. Cockrum, E. L. Trans. Kansas Acad. Sci., 58: 487–511, 1955.

4. Gould, P. J. JM., 42: 406–407, 1961.
5. Verschuren, J. EPNG., 1957.
6. Borell, A. E. JM., 20: 65–68, 1939.
7. Lewis, R. E., and D. L. Morrison. PZS., 138: 473–486, 1962.
8. Lawrence, B., and A. Loveridge. HMCZ., 110: 1–80, 1953.
9. Allen, G. M., and A. Loveridge. HMCZ., 89: 147–214, 1941–2.
10. Harrison, D. L. Mammalia, 22: 592–594, 1958.
11. Lang, H., and J. P. Chapin. AM., 37: 497–563, 1917.
12. Braestrup, F. W. Ann. Mag. Nat. Hist., 10(11): 269–274, 1933.

Molossus (*Eumops*) *perotis* Schinz

MASTIFF BAT

The young of this American bat are born usually in June (1) though in some districts the period of birth is prolonged from June to August (2). Usually there is 1 embryo but occasionally 2 have been found (3). The testes are enlarged during the spring and small during the summer (3). The males have glands on the lower part of the throat which enlarge in the breeding season. They are active from December to March and are small in summer (4).

The oviduct joins the uterus at an angle of 120°. It enters by a papilla which is muscular and nonglandular (5).

1. Cockrum, E. L. Trans. Kansas Acad. Sci., 58: 487–511, 1955.
2. Cockrum, E. L. J. Arizona Acad. Sci., 1: 79–84, 1960.
3. Krutzsch, P. H. JM., 36: 407–414, 1955.
4. Howell, A. B. JM., 1: 111–117, 1919.
5. Andersen, D. H. AJA., 42: 255–305, 1928.

Molossus ater St. Hilaire. Pregnant females were found in March and lactating ones from July to September. Males were not breeding in October but were in breeding condition in December and also from March to August (1).

M. major Kerr. Lactating females with single young were found in September. Males in breeding condition were found from March to August (1).

M. nigricans Miller. This Central American bat has been recorded as pregnant in May (2) and November (3). On each occasion a single embryo was present.

M. (*Eumops*) *underwoodi* Goodwin. Pregnant females have been taken in May and July. Other females have been giving birth to young or lactating in July, which seems to be the usual month for births (4, 5).

M. (Promops) centralis Thomas. Lactating females have been recorded in April (1).

1. Goodwin, G. G., and A. M. Greenhall. AM., 122: 187–302, 1961.
2. Cockrum, E. L. Trans. Kansas Acad. Sci., 58: 487–511, 1955.
3. Felten, H. SB., 38: 1–22, 1957.
4. Cockrum, E. L., and A. L. Gardner. JM., 41: 510–511, 1960.
5. Constantine, D. G. JM., 42: 404–405, 1961.

Primates

THE main inquiry in regard to primate reproduction has centered around menstruation and its significance. Our knowledge of the lower primates, especially of the Lemuroidea, is very scanty, and, as a means of providing links in reproductive behavior between the higher primates and other mammals, work on these forms is highly desirable. There is some suggestion of a restricted breeding season in many of the lower forms and of varying intensity of reproduction with the season in some of the higher ones. Also, bleeding into the uterine cavity, often of small amount, at the end of the life of the corpus luteum appears to be widespread, but it is not sufficient in quantity to produce overt menstruation except in the Old World primates. The small amount of bleeding implies that there is less tissue destruction in the endometrium.

Small litters, one or two offspring at a birth, are the rule throughout the Order. The gestation period is unusually prolonged, considering the size of the mothers.

TUPAIIDAE

TUPAIINAE

Tupaia glis Diard

GREATER TREE SHREW

In Malaya 50 per cent of the males are fertile at a weight of 114.5 ± .5 g., and 95 per cent at 190.6 ± 1.0 g. (1). Mating is from January to March and it declines in April and May. Gestation lasts for 46 to 50 days and there is a post-partum heat. In 7 births 6 were twins and the other a single (2). Another account gives the number of embryos as 2 to 4, with 3 the usual number (3).

1. Harrison, J. L. PZS., 125: 445–460, 1955.
2. Hendrickson, J. R. Nature, 174: 794–795, 1954.
3. Schultz, A. H. Am. J. Physical Anthropol., N.S., 6: 1–23, 1948.

Tupaia javanica Horsfield

TREE SHREW

The Javan tree shrew is said to experience proestrous bleeding (1), but it is possible that this bleeding and uterine desquamation is menstruation-like in nature (2, 3.)

1. Stratz, K. H. Der geschlechtsreife Säugethiereierstock. Hague, 1898.
2. Herwerden, M. van. Bijdrage tot de kennis van menstrueelen cyclus en puerperium. Leiden, 1905.
3. van der Horst, C. J., and J. Gillman. South African J. Med. Sci., 6: 27–47, 1941.

Tupaia castanea Miller. A female with 2 embryos was found in August (1).

T. minor Gunther. In Malaya 50 per cent of the males of the lesser tree shrew are fertile at a weight of 163.3 g., and 95 per cent at 449.7 ± 173 g. Four females each carried 2 embryos (2). Young are found in September (3).

T. tana Raffles. A pregnant female has been found in November (3).

1. Miller, Jr., G. S. Proc. United States Nat. Mus., 31: 247–286, 1907.
2. Harrison, J. L. PZS., 125: 445–460, 1955.
3. Banks, E. J. Malay Br. Royal Asiatic Soc., 9(2): 1–139, 1931.

Urogale everetti Thomas

PHILIPPINE TREE SHREW

Eleven births in zoos have given a gestation period of either less than 50 days (1) or approximately 56 days (2). Of these births 4 have been of twins and the rest singles.

1. Snedigar, R. JM., 30: 194–195, 1949.
2. Wharton, C. H. JM., 31: 352–354, 1950.

LEMURIDAE

LEMURINAE

Hapalemur griseus Link

The season of birth in Madagascar is December and January (1). The single young is born after a gestation period of 160 days (2).

1. Schlegel, H., and F. P. L. Pollen. Fauna Madagascarensis. Leyden, 1867–74.
2. Petter-Rousseaux, A. Mammalia, 26(suppl. 1): 1–88, 1962.

Lemur catta L.

RING-TAILED LEMUR

Puberty in this and other species of the genus is at 2½ years (1). Mating is from the end of March to early July and pregnancy continues until October or later, when the single young is born. Young are found from November until March (2). According to another report most births are in the second half of March and rarely after mid-April (3). In the London zoo births have occurred in March, April, June, and September, suggesting a much more extended season (4). The gestation period lasts about 4½ months (1).

1. Petter-Rousseaux, A. Mammalia, 26(suppl. 1): 1–88, 1962.
2. Kaudern, W. AZ., 9(1): 1–22, 1914.
3. Deschambre, E. Bull. Mus. Hist. Nat., Paris, Ser. 2, 7: 315–319, 1935.
4. Zuckerman, S. PZS., 122: 827–950, 1952–3.

Lemur macaco L.

BLACK LEMUR

In Madagascar this lemur mates from February to May; the single young is born from October to January (1). In the London zoo births have taken place mostly from March to early May, with a few in June and September. Births in the Giza zoo have been from March to June, with a few in July (2). In Buffalo, New York, births have occurred in April and June (3). The gestation period lasts about 5 months. Twins are born in about 10 per cent of births, and there has been one record of a triplet birth (2).

In 1,000 counted follicles there were 20 per cent of binovular, 6 per cent of triovular, and 4 per cent of multiovular follicles (4).

1. Kaudern, W. AZ., 9(1): 1–22, 1914.
2. Zuckerman, S. PZS., 122: 827–950, 1952–3.
3. Perkins, R. M. JM., 20: 503–504, 1939.
4. Harrison, R. J. Nature, 164: 409, 1949.

Lemur fulvus Geoffroy. The brown lemur has a heat period lasting for 5 to 6 days (1).

L. mongoz L. In Madagascar the mongoose lemur is found pregnant from July to October; young have been observed in November and December (2). A few records from the London zoo give births in April and May (3). Usually a single young is born (4).

L. rubriventer Geoffroy. The number of young this lemur bears at a time is usually 2 (4).

1. Cowgill, U. M., A. Bishop, R. J. Andrew, and G. E. Hutchinson. Proc. Nat. Acad. Sci., 48: 238–241, 1962.
2. Kaudern, W. AZ., 9(1): 1–22, 1914.
3. Zuckerman, S. PZS., 122: 827–950, 1952–3.
4. Rand, A. L. JM., 16: 89–104, 1935.

Lepilemur mustilenus I. Geoffroy

GENTLE LEMUR

In Madagascar the mating season is from February to May; it includes the latter half of the rainy season and possibly the beginning of the dry season. New-born young are found in October (1). One young is born at a time (2). Another account gives the mating season as at the end of June with births in October and November after a 4 to 5-month gestation (3).

1. Kaudern, W. AZ., 9(1): 1–22, 1914.
2. Hill, W. C. O. Primates, vol. I. Edinburgh, 1953.
3. Petter-Rousseaux, A. Mammalia, 26(suppl. 1): 1–88, 1962.

CHEIROGALEINAE

Cheirogaleus major Geoffroy

MOUSE LEMUR

A female was found with 2 full-term fetuses in January. The usual litter consists of triplets (1). From June to September the clitoris enlarges, the vaginal opening elongates, and the external genitalia become pink-colored. During the rest of the year they are quiescent and pale. The vaginal opening increases and unites with the urethra in a 25-day cycle. Leucocytes are most numerous in the smear during the period of lengthened clitoridal opening (2).

Heat lasts for 4 to 5 days and the closing period for 3 days. In the latter phase no cornified cells are present in the vaginal smear. Epithelial cells predominate; leucocytes come in at the end of the period (3).

1. Kaudern, W. AZ., 9(1): 1–22, 1914.
2. Petter-Rousseaux, A. CRAS., 239: 1083–1085, 1954.
3. Petter-Rousseaux, A. Mammalia, 26(suppl. 1): 1–88, 1962.

Cheirogaleus medius Geoffroy. This lemur becomes very fat in the non-breeding season. In Paris spermatozoa were present from July on (1).

1. Petter-Rousseaux, A. Mammalia, 26(suppl. 1): 1–88, 1962.

Microcebus murinus Miller

MOUSE LEMUR

In Madagascar this lemur probably has two annual seasons; fetuses have been found in November, January, and February and young in December (1). Another report gives the females as all pregnant in October and November, each containing 2 or 3 embryos (2). Triplets are usually produced, but there is a record of one female with 4 embryos (1).

Spermatogenesis begins in July and is at its maximum in October (3). In Paris it occurs at the corresponding season of the year (4).

In the female the region of the clitoris swells and reddens but the vagina closes. There is no bleeding. This period lasts for 4 to 6 days. Then the vagina opens and the heat period lasts for 2 to 5 days. Mating is usual at the third day. The vaginal smear consists of large epithelial cells, more or less keratinized. Leucocytes appear towards the end of this period. Then the vagina closes during a period of from 5 to 10 days. During this time keratinized cells are absent. The smear consists of epithelial cells, with more leucocytes at the end of the period. The interval between heats is variable, from 36 to 78 days, but 18 of 23 were between 45 and 55 days. The gestation period lasts from 59 to 62 days (4).

1. Kaudern, W. AZ., 9(1): 1–22, 1914.
2. Rand, A. L. JM., 16: 89–104, 1935.
3. Sphuler, O. Zeit. Zellforsch., 23: 442–463, 1935.
4. Petter-Rousseaux, A. Mammalia, 26(suppl. 1): 1–88, 1962.

Microcebus coquereli Grandidier. In the London zoo birth has occurred on one occasion in August (1).

1. Zuckerman, S. PZS., 122: 827–950, 1952–3.

Phaner furcifer Blainville

In one Madagascar specimen spermatogenesis was in progress during July (1).

1. Petter-Rousseaux, A. Mammalia, 26(suppl. 1): 1–88, 1962.

INDRIIDAE

Lichanotus (Avahi) laniger Gmelin

WOOLLY LEMUR

In Madagascar this lemur is pregnant in June or July (1). The single young is found from September to November (2). Spermatogenesis begins at the end of August and the young are born in the same month (3).

1. Rand, A. L. JM., 16: 89–104, 1935.
2. Kaudern, W. AZ., 9(1): 1–22, 1914.
3. Petter-Rousseaux, A. Mammalia, 26(suppl. 1): 1–88, 1962.

Propithecus verreauxi Grandidier

SIFACA

In Madagascar the sifaca has heats from November to March and pregnant females are found from March to July (1). One young is usual (2). They mate in February and the young are born from June to August after a gestation of 5 months (3).

1. Kaudern, W. AZ., 9(1): 1–22, 1914.
2. Rand, A. L. JM., 16: 89–104, 1935.
3. Petter-Rousseaux, A. Mammalia, 26(suppl. 1): 1–88, 1962.

Indri indri Gmelin

INDRI

The gestation period of the indri lasts for 60 days (1).

1. Kenneth, J. H. Gestation Periods. Edinburgh, 1943.

DAUBENTONIIDAE

Daubentonia madagascariensis Gmelin

AYE-AYE

According to the natives of Madagascar the aye-aye gives birth to its young in February and March. One young is born at a time (1).

1. Kaudern, W. AZ., 9(1): 1–22, 1914.

LORISIDAE

LORISINAE

Loris tardigradus L.

SLENDER LORIS

This loris breeds twice a year. The single young, occasionally twins, are born in late April or May and in November or December (1). Judging by cornified smears there are two breeding periods, in June–July and September–November, but only 1 litter a year is usual since the gestation period lasts for over 5 months (2). One report gives 171 days (3). Proestrum lasts for 7 to 10 days and is marked by enlargement of the clitoris. Heat lasts for 7 days (3).

1. Hill, W. C. O. Primates, vol. I. Edinburgh, 1953.
2. Ramaswami, L. S., and T. C. A. Kumar. Naturwissenschaften, 5: 115–116, 1962.
3. Nicholls, L. Nature, 143: 246, 1939.
4. Rao, C. R. N. J. Mysore Univ., 1: 57, 1927.

Nycticebus coucang Boddaert

SLOW LORIS

This loris may breed all the year round in its native habitat but fertility may increase towards the end of the year (1). One young is usual (2), after a gestation period of 174 days (3) or 90 days (4).

1. Zuckerman, S. PZS., 1059–1075, 1933.
2. Hill, W. C. O. Primates, vol. I. Edinburgh, 1953.
3. Nicholls, L. Nature, 143: 246, 1939.
4. Kenneth, J. H. Gestation Periods. Edinburgh, 1943.

Nycticebus intermedius Dao van Tien. In Vietnam a litter of 2 was born in January (1).

1. Dao van Tien. ZZ., 40: 139–141, 1961.

Perodicticus potto Müller

POTTO

A birth has been recorded in February (1) and suckling young in the same month (2). The vagina is always open and does not swell at estrus. Judging by the disappearance of leucocytes from the vaginal smear, the interval between heats is about 50 days (3).

1. Hill, W. C. O. Primates, vol. I. Edinburgh, 1953.
2. Hollister, N. Nat. Hist. Mus., United States, Bul. 99: 1–4, 1924.
3. Petter-Rousseaux, A. Mammalia, 26(suppl. 1): 1–88, 1962.

GALAGINAE

Galago crassicaudatus Geoffroy

BUSH-BABY

In South West Africa all females were found to be pregnant with a single fetus in early September (1). In the London zoo births have been recorded in May, June, and September (2), and in the Washington, D.C., zoo in July (3). In Paris spermatogenesis probably begins in January. It is in full swing in August (4).

1. Pitman, C. R. S. A Game Warden Among His Charges. London, 1931.
2. Zuckerman, S. PZS., 122: 827–950, 1952–3.
3. Davis, M. JM., 41: 401–402, 1960.
4. Petter-Rousseaux, A. Mammalia, 26(suppl. 1): 1–88, 1962.

Galago senegalensis Geoffroy

BUSH-BABY

The central African bush-baby appears to have a restricted breeding season lasting from December to July, with heat periods about 6 weeks apart. The distribution of births in the London zoo also suggests a limited breeding season, with December and January as its beginning (1) and with most births from April to May (2). Some have been recorded scattered through other months of the year (2, 3). Heat lasts for 5 to 6 days and is accompanied by a colorless discharge from the vulva. The usual number of young is 2, range 1 to 2. The period of gestation is 4 months. The female has a long, pendulous clitoris (4). Few multiovular follicles are present in the ovaries (5).

A more recent report gives cycles as recurring all the year round at 4- to 6-week intervals, and there is a post-partum heat. However, in the Nuba Mountains sexual activity is most intense in mid-December (6). In anestrum the vaginal wall consists of a stratum granulosum with but little cornification. At estrus there is massive cornification and desquamation. The mature follicle measures 600–700 μ in diameter, and the ovum, 70 μ. The mature corpus luteum measures 1.5–2.0 mm. in diameter and it bulges above the surface of the ovary. But when the embryo is 4.0 mm. in crown-rump length the corpus luteum is reduced to 1.0 mm., and by the time the embryo measures 12.0 mm. it has become a corpus albicans devoid of luteal cells. This is at about one-third of the way through pregnancy, and there are no accessory corpora lutea (6).

Measured intervals between heats have been between 36 and 42 days, while heats have lasted for 5 to 6 days. For 5 days no leucocytes are present in the vaginal smear. At the sixth day they appear quickly and the cornified cells disappear (7).

The male reaches puberty at 20 months (4). In Paris spermatogenesis begins in May (7).

1. Zuckerman, S. PZS., 1059–1075, 1933.
2. Zuckerman, S. PZS., 122: 827–950, 1952–3.
3. Butler, H. PZS., 129: 147–149, 1957.
4. Lowther, F. de L. Zoologica, 25: 433–462, 1940.
5. Harrison, R. J. Nature, 164: 409, 1949.
6. Butler, H. PZS., 135: 423–430, 1960.
7. Petter-Rousseaux, A. Mammalia, 26(suppl. 1): 1–88, 1962.

Galago alleni Waterhouse. In Nigeria and the Cameroons this bush-baby breeds all the year round (1).

G. demidovii Fischer. The breeding season is probably early in October and the gestation period 3 months or less (2). But the single young have been born in September and April (2, 3). There are also records of pregnancies in December and July (1).

1. Sanderson, I. T. TZS., 24: 623–725, 1938–40.
2. Cansdale, G. S. J. Soc. Preservation Fauna of the Empire, 50: 7–12, 1944.
3. Lilford, Lord. PZS., 542, 1892.

Euoticus elegantulus Le Conte

This lemuroid has no fixed season, but in the north Cameroons more appeared to be breeding from June to October (1).

1. Sanderson, I. T. TZS., 24: 623–725, 1938–40.

TARSIIDAE

Tarsius spectrum Pallas

SPECTRAL TARSIER

The Malayan tarsier is polyestrous and breeds all the year round with apparently little seasonal variation (1). The cycles average 23.5 ± .7 days. There is no bleeding from the vagina, and, at the time of heat, the vulva swells and the vaginal smear becomes entirely cornified. The stage of full cornification lasts for 24 hours and is followed by an invasion of leucocytes with recession of the vulva (2). Changes in the uterus have been described, but it is not clear to what stage of the cycle they belong. They consist of swelling of the glands, mitoses of the epithelial cells followed by localized congestion, and extravasation of blood, which is phagocytized and thus carried to the uterine cavity. There is no extensive desquamation of uterine epithelium (3). The ovary contains few multiovular follicles (4).

1. Zuckerman, S. PZS., 1059–1975, 1933.
2. Catchpole, H. R., and J. F. Fulton. JM., 24: 90–93, 1943.
3. Herwerden, M. van. Monatsschr. f. Geb. u. Gyn., 24: 730–748, 1906.
4. Harrison, R. J. Nature, 164: 409, 1949.

Tarsius syrichta L.

Observations in the London zoo give a cycle of about 1 month. Occasional bleeding takes place. At the active phase the labia minora enlarge and open so that the glans clitoridis can be seen. During sexual activity in both sexes the yellow color of the lower, nonflexible part of the ear concavity becomes more intense. A bright orange powdery secretion may be found there and also on the eyelids and chin. When the female is in heat the feces have a strong odor. The seminal vesicles are enormous and a waxy copulation plug is produced (1).

1. Hill, W. C. O., A. Porter, and M. D. Southwick. PZS., 122: 79–119, 1952–3.

ANTHROPOIDEA

CEBIDAE

AOTINAE

Aotes trivirgatus Humboldt

DOUROUCOULI

In Panama this monkey possibly breeds in December and the young are born in June (1). According to accounts from zoos in the temperate zone births have occurred in February, April, May, July, August, and November. Reproduction may take place at any time of the year, with a trend towards parturition in the summer (2). Birth of twins is frequent (3).

1. Enders, R. K. HMCZ., 78: 385–502, 1935.
2. Hill, W. C. O. Primates, vol. IV. Edinburgh, 1960.
3. Sanderson, I. T. The Monkey Kingdom. London, 1957.

ALOUATTINAE

Alouatta palliata Gray

HOWLER MONKEY

Not much is known of the reproduction of the howler monkeys of South America, but there appears to be no fixed breeding season (1). A single young is usually produced at a birth (2). The ovarian wall thins during follicular enlargement, and a well-defined stigma is formed on the follicle. The theca interna is strongly marked. There is little luteinization of granulosa cells in the ruptured follicle before the ingrowth of capillaries, but some luteinization of the interstitial tissue is already occurring. Fat is deposited in the cells of the theca interna and of the granulosa. The latter luteinize more slowly than the former, but eventually they are larger. The corpus luteum contains a cavity which disappears late in the lutein phase. The wall of the corpus luteum disappears so that this body becomes indistinguishable from the interstitial tissue (3).

The epithelium of the vagina keratinizes to a variable degree, and this change does not bear much relation to the stage of the cycle (3).

The endometrium during the follicular phase has low-columnar epithelium with basal nuclei. The glands are straight, with many mitoses. In the luteal phase the epithelium is tall, the glands are coiled, and mitoses are frequent. Late in this phase there is marked degeneration and some desquamation of the endometrium. Red blood cells are extravasated, but these changes are not so intense as those which have been described for *Macaca* (3). The oviduct has a very narrow lumen as it passes through the uterine muscle. It has no sphincter, and no folds or villi (4).

The testes descend near the time of puberty (5). The average testis weight is 25 g. (6).

1. Wislocki, G. B. CE., 414: 173–192, 1930.
2. Leopold, A. S. Wildlife of Mexico. Game Birds and Mammals. Berkeley, Calif., 1959.
3. Dempsey, E. W. AJA., 64: 381–405, 1939.
4. Andersen, D. H. AJA., 42: 255–305, 1928.
5. Wislocki, G. B. Human Biol., 8: 309–347, 1936.
6. Schultz, A. H. AR., 72: 387–394, 1938.

Alouatta caraya Humboldt. The black howler monkey has been seen with very young babies in August and September (1); births are more numerous in the spring months (2).

A. seniculus L. In Surinam a late pregnancy has been recorded in October (3). Twins are rare but one pair has been recorded (4). The gestation period is about 139 days (5).

1. Miller, F. W. JM., 11: 10–22, 1930.
2. Hill, W. C. O. Primates, vol. V. Edinburgh, 1962.
3. Zuckerman, S. PZS., 1059–1075, 1932.
4. Schultz, A. H. Zoologica, 12: 243–262, 1921.
5. Kenneth, J. H. Gestation Periods. Edinburgh, 1943.

CEBINAE

Cebus apella L.

WEEPING CAPUCHIN

The cycle of this South American monkey is from 16 to 20 days in length. Ovulation occurs about 9 days before the onset of menstruation. Copulation takes place at, or just before, the time of ovulation. Menstruation, which can be detected by the examination of vaginal lavages, is at the time of minimal vaginal desquamation, while at ovulation desquamation is maximal. The species breeds all year, but most births occur in May–June and October–November (1). In the London zoo births have been recorded in June and December after a gestation period of about 6 months (2). A set of twins has been recorded (3).

The injection of 3,500 R.U. of estrogen over a period of 13 days was not accompanied by, or followed by, menstruation. A few red cells were found in the lavage, but they are believed to have been due to trauma (4).

1. Hamlett, G. W. D. AR., 73: 171–181, 1939.
2. Zuckerman, S. PZS., 122: 827–950, 1952–3.
3. Stott, Jr., K. JM., 34: 385, 1953.
4. Zuckerman, S. J. Physiol., 84: 191–195, 1935.

Cebus albifrons Humboldt. The injection of 9,000 R.U. of estrogen in 11 days was not followed by menstruation upon cessation of the injections.

While they were in progress epithelial cells in the vaginal smear increased in proportion to leucocytes, and after injections ceased the latter increased in numbers. Menstruation has not been observed in the normal female (1).

C. capucinus L. In the Rotterdam zoo two females mated in January. One young was born in July and the other at the end of August (2). A July birth has been recorded in the London zoo (3).

1. Zuckerman, S. J. Physiol., 84: 191–195, 1935.
2. Bemmell, A. C. V. Zool. Garten, 24: 246–247, 1959.
3. Zuckerman, S. PZS., 122: 827–950, 1952–3.

Saimiri sciurea L.

SQUIRREL MONKEY, TITI

This species breeds all the year and has 1 young (1). The gestation period lasts 6 months (2). Ovaries in the late follicular stage contained 7 per cent of binuclear ova and 2 per cent of trinuclear ones. Twelve per cent of the follices were binovular and 3 per cent triovular. In the luteal phase there were fewer multiple follicles and more atretic ones (3).

1. Husson, A. M. Studies on the Fauna of Suriname and Other Guyanas, 1: 13–40, 1957.
2 Hill, W. C. O. PZS., 138: 671–672, 1962.
3. Harrison, R. J. Nature, 164: 409, 1949.

Saimiri oerstedi Reinhardt. This monkey has a limited breeding season. There is no sexual skin coloration (1).

1. Wislocki, G. B. CE., 414: 173–192, 1930.

ATELINAE

Ateles geoffroyi Kuhl

SPIDER MONKEY

The South American spider monkey has no restricted breeding season (1), and the female menstruates periodically for 3 to 4 days at intervals of 24 to

27 days (2). The ovarian cycle is similar to that described for *Alouatta palliata.* In both species the corpus luteum becomes indistinguishable from the interstitial tissue in the lutein phase of the cycle (1, 3).

The vaginal wall cornifies in the follicular stage, and a series of ridges and denticles develop. These, together with much of the cornified layer, are sloughed off late in this stage and for a short time after ovulation, but, though the denticles disappear entirely, some cornified tissue is always present (3). The uterine changes resemble those described for *Alouatta palliata* (3).

The testes descend early in life, and the penis is covered with black, cornified barbs oriented backward from the head to the base of the penis, which may engage with the denticles which are present in the vagina of the female during the follicular phase. The clitoris is described as gigantic for primates, in which group it is usually small (4).

1. Wislocki, G. B. CE., 414: 173–192, 1930.
2. Goodman, L., and G. B. Wislocki. AR., 61: 379–387, 1935.
3. Dempsey, E. W. AJA., 64: 381–405, 1939.
4. Wislocki, G. B. Human Biol., 8: 309–347, 1936.

Ateles paniscus L. The gestation period of the black spider monkey has been recorded as 139 days (1). The ovaries contain but few multiovular follicles (2).

1. Kenneth, J. H. Gestation Periods. Edinburgh, 1943.
2. Harrison, R. J. Nature, 164: 409, 1949.

Lagothrix

WOOLLY MONKEY

Lagothrix cana Geoffroy. The gestation period is 8 months and 10 days; puberty is attained at 4 years (1).

L. lagotricha Humboldt. The gestation period has been given as 139 days (2). In a young male the testes weighed 0.7 g. The length of the seminiferous tubules was 226 m. (3).

1. Hill, W. C. O. Primates, vol. V. Edinburgh, 1962.
2. Kenneth, J. H. Gestation Periods. Edinburgh, 1943.
3. Knepp, T. H. Proc. Pennsylvania Acad. Sci., 13: 58–62, 1939.

CALLITHRICIDAE

Callithrix (Hapale) jacchus L.

MARMOSET

Little is known of the reproduction of the South American marmoset although it breeds readily in captivity all the year round (1). Puberty in the female occurs at 14 months of age, and the gestation period lasts 140 to 150 days (2). Combined data give the litter size as 1 to 3, with a mean value of 2.0 ± .06. The chorionic blood vessels anastomose, but no modification of the sex of twins has been observed (3).

1. Zuckerman, S. PZS., 122: 827–950, 1952–3.
2. Lucas, N. S., E. M. Hume, and H. H. Smith. PZS., 447–451, 1927.
3. Wislocki, G. B. AJA., 64: 445–483, 1939.

Callithrix (Mico) argentata L. In the Edinburgh zoo this marmoset has borne a single young and twins (1).

1. Hill, W. C. O. Primates, vol. III. Edinburgh, 1957.

Leontocebus rosalia L.

SILKY MARMOSET

Records of births in November and April have been found (1, 2). Twins are usual (2, 3), and there is a record of a pair that produced a total of 17 young in their lifetime (4). Another pair had 4 sets of twins in 2½ years (5). Two litters a year with a post-partum heat and a gestation period of 132 to 134 days are also reported (6).

1. Zuckerman, S. PZS., 122: 827–950, 1952–3.
2. Ditmars, R. L. New York Zool. Soc., Bul. 36: 175–176, 1933.
3. Zukowski, L. Zool. Garten, 9: 61–62, 1937.
4. Mann, L. Q. From Jungle to Zoo. London, 1935.
5. Rabb, G. B., and J. E. Rowell. JM., 41: 401, 1960.
6. Ulmer, F. A. JM., 42: 253–254, 1961.

Leontocebus (*Oedipomidas*) *spixi* Reichenbach. In 35 of 40 pregnant uteri twins were found; there were 2 sets of triplets and 3 singles. There were usually 2 corpora lutea (1). In Panama one member of this species bred in February and 2 young were born in June (2).

L. (*Tamarin*) *midas* L. In the London zoo a mating was first observed in September and a single young was born in June (3).

L. (*T.*) *nigricollis* Spix. This tamarin has been reported with 2 young (4).

1. Wislocki, G. B. AJA., 64: 445–483, 1939.
2. Enders, R. K. HMCZ., 78: 385–502, 1935.
3. Hill, W. C. O. Primates, vol. III. Edinburgh, 1957.
4. Rabb, G. B., and J. E. Rowell. JM., 41: 401, 1960.

CERCOPITHECIDAE

CERCOPITHECINAE

Macaca cyclopis Swinhoe

FORMOSAN MONKEY

Sexual skin swelling is marked in this species, especially in the root of the tail, circumcallosities, and vulval region. The modal cycle length is 28 days, mean 29.9 days, and range 16 to 49 days. Cycles tend to be suppressed in summer. Menstruation lasts 3.34 days. The changes in vaginal secretions resembled those observed in *M. fuscata*. In one pregnancy the interval from menstruation to parturition was 163 days (1).

1. Asakura, S. Japanese Assn. Zoological Gardens and Aquaria, 2: 85–94, 1960.

Macaca fuscata Blyth

JAPANESE MONKEY

The mean cycle length is 24.4 days, mode 22 days, and range of 17 observations, 16 to 37 days. Menstruation lasted 3.5 days. The sexual skin of the young adult female swells but it does not in the older ones. In the latter there is, however, marked flushing of the sexual skin, nipples, and face. During the follicular stage of the cycle the vaginal secretions were transparent, viscid mucus. This type was found up to the time of maximum sexual skin activity. In the luteal phase the secretions were opalescent and flaky. The mean gestation period was 5.43 months, range of 13 was from 5.0 to 6.0, with a mode of 5.3 months. Births were concentrated in the months of June and July, with conceptions in January and February. The rectal temperature falls just before ovulation and then rises. The follicular stage lasts for 12.0 days, and the lutein stage 10.7 days (1). In the London zoo there is a single record of a January birth (2).

1. Asakura, S. Japanese Assn. Zoological Gardens and Aquaria, 2: 85–94, 1960.
2. Zuckerman, S. PZS., 122: 827–950, 1952–3.

Macaca irus Cuvier

(=*M. cynomolgus* Boddaert)

CRAB-EATING MACAQUE

The notes given here upon *Macaca irus* refer also to the similar species *M. fascicularis mordax* (1). According to Elliott, three separate species, *M. irus, M. fascicularis,* and *M. mordax* exist, but all are Malayan species and they are often included under the one specific name. There is no reason to believe that these forms differ in any details of their reproduction.

In a state of nature the crab-eating macaque appears to breed at any time of the year, but the peak of conceptions is in October and November (1, 2). In the London zoo births have taken place at all times of the year (3). The cycle length is 25 to 29 days (1); 32.4 ± .5 days, mode 30, range 23 to 42 days (4); or 34.4 ± 1.6 days, range 24 to 52 days (5). The period of gestation is from 160 to 170 days (4). In the laboratory there tends to be an amenorrheic season in summer (5). The period of menstruation lasts from 2 to 13 days, usually 2 to 6 days. The color of the sexual skin intensifies (becomes redder) following menstruation, reaches its height at the mid-interval, and then diminishes (5). The time of greatest frequency of coitus occurs at 7 to 11 days after the beginning of menstruation (1).

The reproductive tract has been minutely compared with that of man. The labia minora are represented by fine membranous folds in the anterior portion of the vulva, and the labia majora are often incomplete, especially the ventral portion. The vaginal epithelium is papillated and many lymph nodes are present. In the early interval the cells are stratified and cornified; later, denudation sets in. Most leucocytes are found just after ovulation. The time of greatest denudation is from 9 to 19 days, and of greatest leucocyte flow at 2 days and again 1 day before menstruation, at the time of the decline in the color of the sexual skin. The vaginal secretions are acid (1).

The cervical canal contains a pronounced fold, absent in man, which renders the passage tortuous. The uterine muscle contains a strong network of crossing fibers, absent in man, between the circular and longitudinal layers. At the myoendometrial boundary is a layer of cell islands. These are undifferentiated in the follicular phase, but at 12 to 14 days glycogen appears and the cells become flattened, so that the layer takes on the appearance of a muscularis mucosae. Dedifferentiation occurs at the end of the cycle. There are two main areas in which the blood vessels branch, the first at the level of these myoblasts, and the second near the surface of the mucosa. This second layer is mostly concerned in menstruation. Of the part of the endometrium which is shed at this time, two thirds is denuded early, and the last third after a slight interval. From days 5 to 14 regeneration occurs. The glands and blood vessels are straight at this time. From 8 to 14 days mitoses are common and the glands are sprouting. From 12 to 14 days the glands become tortuous, the capillary net is well developed, and plasma begins to flow into the stroma. From 15 to 18 days the glands are more tortuous, their epithelium is higher, and the nuclei are pressed to the bases of the cells while secretion begins. At 18 to 24 days mitoses have ceased, and the secretion gives glycogen and mucin reactions. Just before menstruation one finds hyperemia of the blood vessels and an outpouring of leucocytes. The vessels suddenly break down with loss of their epithelium, and blood pours into the stroma; denudation follows owing to the loosening of the tissues by the previous edema. All these changes closely correspond with those found in man (1).

In the ovariectomized female the injection of 250 to 1,000 R.U. of estrogen daily in increasing doses over a 12-day period (6,500 R.U. were injected in all) caused swelling of the sexual skin, and menstruation followed cessation of the injections (6).

Testis weight averages 30.8 g. (7).

1. Joachimowitz, R. Biol. Gen., 4: 447–540, 1928.
2. Herwerden, M. van. Tijdschr. Ned. Dierk. Vereen., 10: 1–140, 1906.
3. Zuckerman, S. PZS., 122: 827–950, 1952–3.

4. Spiegel, A. Zent. f. Gyn., 22: 1762, 1931.
5. Corner, G. W. AR., 52: 401–410, 1932.
6. Zuckerman, S. J. Physiol., 84, 191–195, 1935.
7. Schultz, A. H. AR., 72: 387–394, 1938.

Macaca mulatta Zimmermann

BENGAL RHESUS MONKEY

In its native habitat in the Himalayas the rhesus monkey is said to mate mainly in September and to give birth in March (1). At lower altitudes young tend to be born from March to May and from September to October (2). In the Yale Obstetric monkey colony, with far better feeding and management than is usual, breeding occurs at all times of the year and menstruation is regular throughout the year. In this colony the menarche has occurred at about 2 years and effective breeding about a year later (3). In other colonies the age at puberty is about a year later, but reproduction is not regular until about 4½ years. Before this, menstruation may be irregular and it frequently occurs without ovulation (4). Puberty is reached gradually, with reddening of the buttocks as the first sign. This occurs, on the average, at 3,200 g. of weight, and the first menstruation at 3,600 g. (5), or 3,350 g. (range 3,040 to 3,850 g.) (4, 6), but the average weight at first conception is 5,000 g. (4). The cycle length for a series from several laboratories gave a mode of 28 days; a mean of 27.36 ± .17 days, and a standard deviation of 5 to 7 days, with 75 per cent of cases falling between 23 and 33 days (4, 5, 7). The usual duration of menstruation is 4 to 6 days, with a spread from 2 to 11 days (7).

The time of ovulation, which is spontaneous (8), has been determined by rectal palpation. It varies from day 9 to day 20 of the cycle, with a pronounced mode at 13 days. The mean is also 13 days, and 77 per cent of ovulations fall from days 11 to 15 inclusive. The postovulatory, or lutein, phase of the cycle is by far the most invariable in its length (3). Ovulation occurs from the right ovary in 50.8 per cent of the cases, which suggests that each sheds one egg with equal frequency. Lactation amenorrhea lasts, on the average, for 7 months (4).

The female permits coitus at any time in the cycle, but desire is greatest about 2 days before ovulation. No premenstrual rise has been observed (9), in contrast to the condition in man. The majority of conceptions occur between days 11 and 14, with a spread from day 9 to day 18 (4). Another record, which includes 160 pregnancies, gives the greatest fertility between noon of the eleventh and noon of the twelfth day of the cycle (10). A single mating on day 6 gave 14 per cent success; on day 11, 30 per cent; and on day 17, 1.3 per cent (11). Twinning occurs at about the same rate as in man, about once in 90 to 100 births (3).

The period of gestation in 29 cases ranged from 146 to 180 days, calculated from the probable date of ovulation. The mean was 163.7 ± 1.0 days, and the standard deviation, 8.0 days (4). Another record of 34 pregnancies gave a range from 158 to 175 days (12). The sex ratio of 36 births was exactly even (4).

HISTOLOGY OF THE FEMALE TRACT

OVARY. The corpus luteum of the cycle grows mainly by hypertrophy of the granulosa cells, with some growth by theca interna cells in the folds of the collapsed follicular wall. There is very little extravasation of blood at ovulation, and the fully formed corpus luteum is a solid structure with a prominence at the rupture point. Lipid degeneration of the cells occurs at the time of menstruation, and traces of the corpora persist for several cycles. The maximum diameter of the corpus luteum of the cycle is about 5.5 mm. (4, 13). Three types have been described: the first is the normal one; the second is a corpus luteum aberrens, which does not seem to produce progesterone and which contains large numbers of theca-lutein cells. This is a normal corpus luteum degenerating in an atypical manner. The third is similar but is derived from luteinized unruptured follicles (14). The nature and fate of the lipids in the cells have been carefully followed (15). The size of the ovum without the zona pellucida is 80 μ in diameter; with the zona it is 109 μ (4). The number of ova in one ovary ranges from $50{,}120 \pm 10{,}600$ to $60{,}810 \pm 16{,}470$. The right and left ovaries contain about equal numbers (16).

Before the follicle ruptures, the granulosa cells become loosened and assume a radial arrangement. Their nuclei stain very densely. Blood vessels invade the granulosa layer about 2 days after ovulation. The organization of the corpus luteum is completed by days 7 to 9, and the structure begins to degenerate about day 13 from ovulation. Before menstruation sets in, extreme lipid degeneration of the lutein cells begins but the theca cells retain their identity (17).

In the graafian follicle alkaline phosphatase may be found in both the theca interna and the granulosa (18).

VAGINA. The vaginal epithelium has waves of growth and cornification which reach their height at the time of ovulation, and of desquamation, which is greatest just after ovulation (7). The vaginal lavage follows a similar curve with the greatest number of desquamated cells at the expected time, i.e., just after ovulation. Leucocyte counts fall near the mid-cycle and rise just before menstruation. On the whole the smear is not a good indication of the reproductive state, but the number of live leucocytes, i.e., those that resist the entrance of dye in a fresh lavage, appears to be significant (4). The cells of the vaginal wall are filled with glycogen (7).

UTERUS. The cervix uteri has several diverticuli in the canal and is more contorted than that of man. The line of transition between the cervical and vaginal epithelium is clear, but it varies in position in different animals (19).

During the follicular phase of the cycle the surface of the endometrium is intact. Stromal mitoses and an increase in epithelial mitoses are observed. At ovulation the stroma is dense. Stromal mitoses remain as before, but epithelial ones decline in frequency. A feature of this period is the large size of the superficial stromal cells. In the pregravid endometrium there are more epithelial mitoses and this layer contains tall, pseudostratified cells with much glycogen. The stroma is edematous and the sinuous glands become progressively more dilated with a dense secretion. The basement membrane of the surface epithelium has no capillaries in contact with it. In the menstrual phase the capillary bed collapses, edema fluid is lost, and there are minute extravasations of blood and necrobiosis. Stromal cells are large and there is extensive infiltration of white blood cells. About one third of the total endometrium is sloughed off. During the repair stage the surface epithelium is low and squamous in type. The stroma is looser and no mitoses are to be found near the surface (20). The uterine mucosa after the postmenstrual repair has a low surface epithelium; the gland tubules are straight and slightly dilated. Glycogen is most abundant in the cells at this time. In the lutein phase the glands become spiral and greatly dilated, and their epithelium is high, with roughened surfaces. The surface epithelium is high, the stroma becomes spongy, and mitoses are absent throughout the period. During menstruation blood is extravasated in the superficial layer of the endometrium, and the resulting hematoma lift off the epithelium and much of the mucous layer. Regeneration starts from the glandular epithelium (7).

Menstruation without ovulation has been described, and it appears to be especially frequent in adolescent monkeys. It occurs through the partial destruction of a preovulatory type of endometrium, and the usual premenstrual changes do not occur at all (21).

A few red blood cells may be found in the vaginal lavage at about ovulation time in many monkeys. Their appearance is much more frequent in winter, at the height of the breeding season, than in summer. Their origin does not seem to have been ascertained, and, in view of the occurrence of postestrous and proestrous bleedings in lower forms, further work might lead to results of interest (4).

Two types of arteries have been described in the endometrium. One type consists of longer arteries coiled for the entire height of the endometrium, with many elastic fibrils in their coats. The other type consists of smaller arteries which supply the pars basalis. They are devoid of elastic fibrils. The first type degenerates after ovariectomy but the second does not (22). The first type is involved in the breakdown that occurs in menstruation (23).

The cilia in the oviducts are active at all times during the cycle (7). Mitoses in the epithelium are most abundant when the largest follicles are present in the ovaries, and they are most frequent at the ovarian end. Estrogen injections cause mitoses to appear in the tubal epithelium of ovariectomized females (24).

PHYSIOLOGY OF THE FEMALE TRACT

There is an evanescent swelling of the sexual skin in new-born monkeys of both sexes (25). It recalls the so-called genital crisis, or activity of the reproductive tract, which has been found in guinea pigs, and is believed to be due to the placental transmission of maternal hormones.

Bilateral ovariectomy at the beginning of menstruation is not followed by another period. If the operation is performed 72 hours or more after its beginning, or in the preovulatory phase, menstruation follows 5 to 6 days later even though the regular menstruation had not ceased at the time of the operation (26). These results are significant in relation to the occurrence of menstruation without ovulation. Ovariectomy in the postovulatory phase is also followed by bleeding (27). Hysterectomy has no effect upon the ovaries or on the cycle, as judged by vaginal desquamation or by sexual skin swelling (28, 29).

In contrast to the results of ovariectomy, hypophysectomy early in the cycle is not followed by bleeding. If it is done in mid-cycle, when the sexual skin is well developed, bleeding follows after an interval of 2 to 4 days (30).

Abrupt changes in the color of the sexual skin do not occur during the cycle, but the color reaches its maximum intensity during the third week and diminishes before the end of the cycle. The swelling follows a similar series of changes (31). The skin is brilliantly colored during pregnancy. As the brilliance is known to be produced by the action of estrogens, this is to be expected (4).

The pH of the vaginal secretions of immature monkeys is 6.8 to 7.2. Under the influence of estrogen injections it fell to 5.5 to 6.0 (32). In mature ovariectomized females a fall from a level of 6.0 to 5.0 followed the injection of estrogens (33).

The erythrocyte count in the blood is lowest at the beginning and end of the cycle and highest at the middle. In the ovariectomized female it rises if estrogens are injected (34).

In the nonpregnant female, androgen excretion is at the rate of 1.6 I.U. daily; the effect of ovariectomy is irregular (35). The excretion of androgens during pregnancy is at a maximum level of 7 to 10 I.U. daily. In pregnancy, estrogens rise to a level in the urine of 75 to 450 I.U. daily by 60 to 110 days. They remain at a high level until a few days before parturition, then the level drops slightly. The excretion of estrogens drops rapidly after parturition, but androgens continue to be excreted for a month (36). There is no estriol in the urine of the pregnant monkey (37).

Ovariectomy may be performed as early as the twenty-fifth day of pregnancy without causing abortion. The bright red color of the sexual skin is maintained throughout gestation and lactation, thus giving further proof that during pregnancy estrogens are secreted by tissues other than the ovaries (38). When hypophysectomy was performed between days 27 and 156 of pregnancy, half went to term with normal duration. Six of these ten were delivered alive (39).

The female is sexually very interested in the male at from 16 to 25 days of pregnancy (40). The placental sign, extravasation of blood from the uterus, which may be detected in the vaginal lavage, is first given at from 15 to 20 days, and it lasts 20 to 22 days (41). The pelvic ligament relaxes at 5½ months of pregnancy and is firm again within 3 to 4 weeks after parturition. These changes also occur in the ovariectomized pregnant monkey (42). Gonadotrophic hormones may be detected in the urine for a few days, i.e., from days 19 to 25, during pregnancy (43).

The activity of the oviducts during the cycle has been investigated. Spontaneous activity is lowest during the follicular phase. At ovulation the contraction rate is 8 to 13 per minute; it is not very regular and is rather undulatory in character. Contractions are slower during the lutein phase but are more regular, with a rate of 3 to 8 per minute (44). The degree of spontaneous activity is said to be greatest at this time and during menstruation than it is in the follicular phase (45).

Free progesterone in the plasma is low during and following menstruation. It is high near mid-cycle; then it decreases, but rises in the latter half. It is again low just before menstruation. Bound progesterone is low at all times (46, 47). The placental level of progesterone is 30 to 40 times less than it is in man (37).

By calculating the minimal amount of estrogen needed to maintain the sexual skin in the ovariectomized female and to prevent the onset of menstruation, it has been deduced that the minimal daily output of estrogen by the female macaque is 150 to 200 I.U. (48). The immature female requires 8 to 10 R.U. of estrogen daily to produce some reddening of the sexual skin, but 20 R.U. daily rising to 80 R.U. caused swelling as well as reddening. It took a week for the reddening to become appreciable, and by 22 days it was maximal. The epithelial cells in the vaginal smear increased and leucocytes decreased. The uterus grew, also the cilia in the oviducts, but the ovaries weighed less than normal (49, 50).

In the mature ovariectomized macaque the injection of estrogens causes growth of the endometrium to the preovulatory stage. If they are given in threshold doses, i.e., from 75 to 150 I.U. daily, varying with the animal, and injections are continued at this level, bleeding occurs at intervals of from 5 to 8 weeks. Higher doses prevent the bleeding from occurring. It is believed, therefore, that the threshold of estrogen activity rises and falls in a rhythmic manner (51). These facts have given rise to the "estrogen withdrawal" theory of menstruation, which states that growth of the endometrium is due to the action of estrogen and that when the growth stimulus is removed degeneration sets in with consequent bleeding. The average dose that will build up the uterus to the stage where bleeding sets in after its cessation is about 125 I.U. daily for 10 days. The endometrium can be maintained after the building up by the daily injection of 50 to 100 I.U. of estrogen or by 0.5 mg. (= 0.5 I.U.) of progesterone (48, 52). While 0.5 Rab.U. of progesterone will inhibit bleeding for a time, it will not do so indefinitely; for this to occur at least 1 Rab.U. is required (53).

The threshold single dose of estradiol benzoate that is followed by bleeding is 0.75 to 1.00 mg. However, in continued tests upon one animal the threshold appears to be lower if the initial dose is high and later injections are decreased, and higher if the initial injection is low and the threshold is reached by increasing the amount (54).

By use of the colchicine technique, which causes the accumulation of mitotic figures, the degree of sensitivity of various parts of the tract to estrogens has been worked out. Epithelial tissues are most sensitive, then stroma cells, and, lastly, muscle cells (55). The response of sexual skin to estrogens does not depend upon the nervous system, as denervated skin gives the usual reaction (56).

The bleeding that follows estrogen deprivation may also be prevented by the injection of testosterone propionate at the rate of 5 mg. daily. This amount will not prevent the bleeding that follows progesterone withdrawal, but 25 mg. daily has this effect (57). This amount injected into normal females inhibits ovulation and menstruation and causes the clitoris to enlarge (58). The level of hormones required to produce progestational proliferation of the endometrium is 150 I.U. of estrogen daily together with 0.5 mg. of progesterone (53, 59).

The ovaries of hypophysectomized monkeys will grow follicles if 50 I.U. of pregnant mares' serum is injected daily. If this is followed for 3 days by 100 I.U., they may be brought to the ovulation level (60). P.M.S. also appears to be the best substance now available for causing ovulation in the summer anestrous period, but it is easy to produce overstimulation and luteinization. A dose of 200 I.U. daily has given the best results (61), but the time at which injections are made is important. The best time is from the sixth to eighth day of the cycle (62).

In the anterior pituitary F.S.H. is highest on days 9 to 11 of the cycle. L.H. is also highest at that time but the ratio of L.H. to F.S.H. is highest—at 10:1—from days 11 to 15. Immature animals had lower levels than adults (63).

THE MALE

The sexual skin of the male is much less apparent than that of the female, but males vary a good deal in the intensity of coloration (13). The male excretes 1.0 to 4.7 I.U. of androgens and 1.1 to 2.5 I.U. of estrogens each 24 hours (64). In the adult hypophysis the amounts and ratio of L.H. and F.S.H. are as in the female at the beginning and end of the cycle (63).

The testes regress after birth and, especially, dedifferentiation of the interstitial cells may be noted. Spermatogonia increase by the end of the first year and a slow development continues until the end of the third. Then follows rapid growth with differentiation of the Leydig cells. The earliest spermatozoa were found at 2 years, 11 months; the latest age for this attainment was 3 years, 5 months (65).

In the presence of the testes or during the injection of testosterone the contractility of the vas deferens and seminal vesicles is inhibited, but the injection of estrogens stimulates them (66). The castrated adult requires the daily injection of 5 mg. of testosterone propionate to restore the sexual skin to its normal color (however, see above) and to produce normal accessory organs. When they are produced, they may be maintained with weekly injections of 17.5 mg. (67). In the hypophysectomized male 200 I.U. of P.M.S. gave some repair to spermatocytes. Given immediately after hypophysectomy, the same dose reduced the rate of degeneration of the testes, maintaining spermatogenesis for 20 days (60).

1. Hingston, R. W. G. A Naturalist in the Himalayas. London, 1920.
2. Prakash, I. JM., 41: 386–389, 1960.
3. van Wagenen, G. Personal communication. 1963.
4. Hartman, C. G. CE., 433: 1–161, 1932.
5. Allen, E. CE., 380: 1–44, 1927.
6. Smith, R. M., and B. B. Rubenstein. E., 26: 667–679, 1940.
7. Corner, G. W. CE., 332: 75–101, 1923.
8. Corner, G. W. PSEBM., 29: 598–599, 1932.
9. Ball, J., and C. G. Hartman. Am. J. Obst. Gyn., 28: 117–119, 1935.
10. van Wagenen, G. AR., 91(4): 42, 1935.
11. van Wagenen, G. E., 40: 37–43, 1947.
12. Krohn, P. L. PZS., 134: 595–599, 1960.
13. Corner, G. W., G. W. Bartelmez, and C. G. Hartman. AJA., 59: 433–457, 1936.
14. Corner, G. W. CE., 541: 85–96, 1942.
15. Rossman, I. CE., 541: 97–109, 1942.
16. van Eck, G. J. V. AR., 125: 207–224, 1956.
17. Corner, G. W. CE., 557: 117–146, 1945.
18. Corner, G. W. CE., 592: 1–8, 1948.
19. Sandys, O. C., and S. Zuckerman. J. Anat., 72: 352–357, 1938.
20. Bartelmez, G. W., G. W. Corner, and C. G. Hartman. CE., 592: 99–144, 1951.
21. Corner, G. W. J. Am. Med. Assn., 89: 1838–1840, 1927.
22. Okkels, H., and E. T. Engle. Acta Path. Microbiol. Scand., 15: 150–168, 1938.
23. Daron, G. H. AJA., 58: 349–419, 1936.
24. Allen, E. Am. J. Obst. Gyn., 35: 873–875, 1938.
25. Zuckerman, S., and G. van Wagenen. J. Anat., 69: 497–500, 1935.
26. van Wagenen, G., and S. B. D. Aberle. AJP., 99: 271–278, 1931.
27. Allen, E. AJP., 85: 471–475, 1928.

28. Burford, T. H., and A. W. Diddle. Surgery, Gyn. Obst., 62: 701–707, 1936.
29. van Wagenen, G., and H. R. Catchpole. PSEBM., 46: 580–582, 1941.
30. Smith, P. E., H. H. Tyndale, and E. T. Engle. PSEBM., 34: 245–247, 1936.
31. Zuckerman, S., G. van Wagenen, and R. H. Gardiner. PZS., 108A: 385–401, 1938.
32. Weinstein, L., N. W. Wawro, R. V. Worthington, and E. Allen. Yale. J. Biol. Med., 11: 141–148, 1938.
33. Dow, D., and S. Zuckerman. E., 25: 525–528, 1939.
34. Guthkelch, A. N., and S. Zuckerman. J. Physiol., 91: 269–278, 1937.
35. Dorfman, R. I., B. N. Horwitt, R. A. Shipley, W. R. Fish, and W. E. Abbott. E., 41: 470–488, 1947.
36. Dorfman, R. I., and G. van Wagenen. Surgery, Gyn. Obst., 73: 545–548, 1941.
37. Short, R. V., and P. Eckstein. JE., 22: 15–22, 1961.
38. Hartman, C. G. PSEBM., 48: 221–223, 1941.
39. Smith, P. E. E., 55: 655–664, 1954.
40. Ball, J. AR., 67: 507–512, 1937.
41. Hartman, C. G. Bull. Johns Hopkins Hospital, 44: 155–164, 1929.
42. Hartman, C. G., and W. L. Straus. Am. J. Obst. Gyn., 37: 498–500, 1939.
43. Hamlett, G. W. D. AJP., 118: 664–666, 1937.
44. Seckinger, D. L., and G. W. Corner. AR., 26: 299–301, 1923.
45. Li, R. C. Chinese J. Physiol., 9: 315–328, 1935.
46. Forbes, T. R., C. W. Hooker, and C. A. Pfeiffer. PSEBM., 73: 177–179, 1950.
47. Bryans, F. E. E., 48: 733–740, 1951.
48. Corner, G. W. Bull. Johns Hopkins Hospital, 67: 407–413, 1940.
49. Allen, E. J. Morphol. and Physiol., 46: 479–519, 1928.
50. Allen, E., and A. W. Diddle. Am. J. Obst. Gyn., 29: 83–87, 1935.
51. Zuckerman, S. PRS., 123B: 441–456, 1937.
52. Corner, G. W. AJP., 124: 1–12, 1938.
53. Hisaw, F. L., and R. O. Greep. E., 25: 1–14, 1938.
54. Zuckerman, S. JE., 2: 438–443, 1940–41.
55. Worthington, R. V., and E. Allen. Yale J. Biol. Med., 12: 137–153, 1939.
56. Zuckerman, S. PRS., 118B: 22–33, 1935.
57. Engle, E. T., and P. E. Smith. E., 25: 1–6, 1939.
58. Zuckerman, S. Lancet, 233: 676–680, 1937.
59. Engle, E. T. Cold Spring Harbor Symposia, 5: 111–114, 1937.
60. Smith, P. E. E., 31: 1–12, 1942.
61. Hartman, C. G. CE., 541: 111–126, 1942.
62. Pfeiffer, C. A. AR., 91(4): 33, 1945.
63. Simpson, M. E., G. van Wagenen, and F. Carter. PSEBM., 91: 6–11, 1956.
64. Dorfman, R. I., and G. van Wagenen. PSEBM., 39: 35–36, 1938.
65. van Wagenen, G., and M. E. Simpson. AR., 118: 231–251, 1954.
66. Martins, T., J. R. Valle, and A. Porto. CRSB., 129: 1126–1129, 1938.
67. Zuckerman, S., and A. S. Parkes. J. Anat., 72: 277–279, 1938.

Macaca nemestrina L.
PIG-TAILED MACAQUE

Puberty in the female is reached at about 50 months of age (1). In the London zoo breeding is year-round and the gestation period lasts about 170 days (2). The mean cycle length is 32.5 ± 1.5 days, the follicular phase lasting 14 days and the luteal phase 15 to 16 days (1). The sexual skin is well marked; it begins to swell during menstruation and regresses shortly before the middle of the cycle. The body weight increases in the first half of the cycle and begins to decline with the sexual skin. This seems to be due to the edema which is associated with the swelling. Less fluid is taken in during the swelling phase, but the urine output is much higher during the involutionary or luteal phase (3). The sexual skin responds to estrogens when it has been denervated (4). The red cell count in the blood is lowest at the beginning and end of the cycle, and highest at the middle (5). The period of gestation in one case was 171 days, and there was no menstruation or sexual skin swelling during lactation (6).

1. Zuckerman, S. PZS., 315–329, 1937.
2. Zuckerman, S. PZS., 122: 827–950, 1952–3.
3. Krohn, P. L., and S. Zuckerman. J. Physiol., 88: 369–387, 1937.
4. Zuckerman, S. PRS., 118B: 22–33, 1935.
5. Guthkelch, A. N., and S. Zuckerman. J. Physiol., 91: 269–278, 1937.
6. Zuckerman, S. PZS., 593–602, 1931.

Macaca radiata Geoffroy

BONNET MACAQUE

The bonnet monkey of southern India has a cycle length typical for the genus; four cycles ranged from 25 to 36 days and averaged at 29.5 days. The duration of gestation from the probable date of conception, in 3 cases, was 153, 166, and 169 days (1). The sexual skin is confined to the labia majora and the circumanal region, and is mostly an inconspicuous dark purple with a little red at the margins of the labia and lateral streaks of red from the sitting pads and the tail base. There is a slight swelling of the area in the follicular stage (1).

The ovaries are more elongated than they are in the rhesus monkey. Interstitial tissue consists of a small amount of theca interna and abundant hyaline remains of atretic follicles. The cervix uteri is large and complex compared with that of the rhesus monkey, and it secretes much more mucus. Cyclical epithelial desquamation in the vagina is scanty (1). These facts greatly resemble those recorded for the related *M. pileatus,* which has not been treated separately.

Relaxation of the pelvic ligament as a preparation for parturition begins at about 5½ months (2). After ovariectomy the injection of 500 M.U. daily of estrogen induces a mucus flow, but there is little difference in the color of the sexual skin, which also changes but little during the cycle (3).

1. Hartman, C. G. JM., 19: 468–474, 1938.
2. Hartman, C. G., and W. L. Straus. Am. J. Obst. Gyn., 37: 498–500, 1939.
3. Parkes, A. S., and S. Zuckerman. J. Anat., 65: 272–276, 1931.

Macaca silenus L.

LION-TAILED MONKEY

In this species the sexual skin swells. Cycle lengths have varied from 29 to 69 days, with a mean of 39.6 days. The follicular phase lasted 20.5 days, sexual swelling 14.5 days, and the lutein phase 17.0 days. Menstruation lasted 2.5 days (1). In southern India the young are born in September (2).

1. Asakura, S. Japanese Assn. Zoological Gardens and Aquaria, 2: 85–94, 1960.
2. Webb-Peploe, C. G. J. Bombay Nat. Hist. Soc., 46: 629–644, 1946–7.

Macaca sinica L.

TOQUE MACAQUE

The toque macaque of Ceylon has been treated here as a separate species from the bonnet macaque of Bengal, following modern usage.

The vagina is imperforate until puberty, which is reached at an age of from

2½ to 3 years. At first, mucus is discharged, later bleeding sets in regularly in gradually increasing quantities. Puberty is accompanied by the reddening of some areas on the face. The sexual skin does not become edematous and it fades from bright purple to brownish after menstruation. There is a very large colliculus in the cervix uteri. The latter secretes large quantities of mucus during the cycle, which is usually 29 days in length. Menstruation lasts 1 to 4 days. The single young may be born at any time of the year (2).

1. Hill, W. C. O. Ceylon J. Sci., 5D: 21–36, 1939.
2. Phillips, W. W. A. Ceylon J. Sci., 13B: 261–283, 1924–6.

Macaca sylvanus L.

BARBARY APE

The cycle of the Barbary ape is usually from 27 to 33 days, with menstruation lasting 3 to 4 days. The sexual skin is at a maximum at from 9 to 11 days and remains so for 4 days, after which regression is gradual. The vaginal smear is not very informative, but just before menstruation it is caseous and richer in epithelial cells. The injection of 310 R.U. of estrogen daily into an immature female causes the sexual skin and uterus to develop. If this is followed by 24 Rab.U. daily of progesterone, the uterine glands develop to the premenstrual level (1). The gestation period is about 210 days, and 1 young is usually born at a time (2).

1. Courrier, R. Bruxelles Med., March 1, 1936.
2. Panouse, J. B. Trav. de l'Inst. Scient. Chérifien, Ser. Zool., 5: 1957.

Macaca maurus Cuvier. This monkey from Celebes has a 30- to 40-day cycle. The color changes in the sexual skin are striking. At the height of swelling, at mid-cycle, the under surface of the short tail is expanded so that it forms a bright red spherical swelling continuous with an elongated vulval swelling (1).

1. Zuckerman, S. PZS., Ser. A: 315–329, 1937.

Cercocebus

MANGABEY

Cercocebus albigena Gray. This mangabey has a cycle length of about 29 days. It menstruates, and the sexual skin periodically swells (1).

C. torquatus Kerr. The sooty mangabey of West Africa has a menstrual cycle averaging 33.4 ± .1 days; range, 28 to 46; mode, 30 days (1). Swelling of the sexual skin subsides between days 14 and 20. Menstruation was not suppressed by the continued injection of estrogens unless they were started earlier than day 18 of the cycle (3). In the London zoo births have been in May, June, July, and November (3).

1. Zuckerman, S. PZS., 315–329, 1937.
2. Zuckerman, S. PRS., 118B: 13–21, 1935.
3. Zuckerman, S. PZS., 122: 827–950, 1952–3.

Papio comatus Geoffroy

CHACMA BABOON

The chacma baboon of South Africa breeds at any time of the year (1). The menarche was reached at an average of 3.2 years in 15 females. The range was 2.5 to 3.6 years. Regular cycles were established between the fourth and the seventh cycle and pregnancy followed 2 to 4 years later (2).

The average length of 404 cycles was 35.6 ± .16 days with a spread from 29 to 42 days. The period of turgescence averaged 19.4 ± .76 days, with a spread from 6 to 25 days (3). The modal point of the cycle length appears to be about 34 days but the length is rather irregular (4).

The area of external swelling extends from the root of the tail to a curved line drawn through the junction of the abdominal wall with the front of the thighs. The changes in this area may be divided into three periods: perineal rest, perineal turgescence, and perineal deturgescence. The period of perineal rest begins 1 to 5 days before menstruation and continues through menstruation and afterwards for 7 to 25 days. There is a sudden diminution in the degree of swelling 24 hours before the maximum is reached, then a sudden great increase which lasts 1 day. It is believed that the sudden fall coincides with ovulation. Perineal deturgescence is usually rapid, but it shows great variability. In winter turgescence lasts a shorter time, and the rest period is longer, than at other times of the year. The degree of swelling is greatest in April and May, i.e., during the fall. At the beginning of turgescence a thick white discharge oozes from the vagina. This becomes thicker until maximum turgescence is reached, and at this time it forms a false vaginal plug. About 8 days before the onset of menstruation the secretion becomes scanty, thin, and clear (4). The period of deturgescence coincides with corpus luteum activity (5). Serum mucoprotein and plasma fibrin levels rise with turgescence and fall when it subsides. Fibrinolytic activity increased with deturgescence and remained elevated until the onset of menstruation when it declined precipitously (6).

Menstruation lasts overtly for 4 to 9 days, though red blood cells may be detected microscopically in the vaginal lavage for 4 to 19 days, but usually for 9 days (4). One, rarely 2, young are born at a time and the period of gestation has varied from 173 to 193 days, though all but two were between 184 and 193 days, with a mean of 187 days for 14 births (5). There is no ovarian activity during lactation (7).

The changes in the internal organs have been well worked out and they are described as resembling those found in *Comopithecus hamadryas.* The diameter of the resting follicle is about 1 mm., and of the mature follicle, 6.3 mm. There is a wave of growth at the beginning of the follicular phase, but it is not necessarily the largest follicle that ovulates. No conelike extrusion forms on the follicle that is destined to rupture. At ovulation some blood is extravasated, and it remains mostly at the periphery of the developing corpus luteum. After rupture the follicle shrinks to 4 mm., and all its layers are much folded over so that the developing corpus luteum is lobulated. Full development is reached very slowly, and theca cells do not share in the luteinization. During menstruation the cells of the corpus luteum become vacuolated, and connective tissue increases enormously. The lutein cells measure 11.8 μ at their maximal development, which is reached after 7 days in the corpus luteum of the cycle. The maximal size of the corpus luteum during pregnancy is 6.7 mm., with the cells averaging 21.2 μ. There is no evidence for waves of follicular growth and atresia during pregnancy. The corpus luteum persists throughout this period but disappears rapidly after parturition. Hyaline atresia of follicles is found but it is not so frequent as in *Macaca irus* (8). There are few multiovular follicles (9).

Growth and keratinization of the stratified vaginal epithelium begins during menstruation. It continues throughout the follicular phase and produces a characteristic corrugation of the epithelium. Cornification is intensified at ovulation, and there is considerable sloughing after it. At this time leucocytes are scarce, but they increase throughout the luteal phase (8).

During the luteal phase the endometrium is 5.4 mm. thick. The epithelium is high-columnar, with oval vesicular nuclei. Few leucocytes are present. The glands are very coiled, with numerous basal buds, and they secrete during the mid-luteal phase. The stroma is edematous in its middle zone. At menstruation about two thirds of the endometrium is sloughed off (8).

The cervix is coated with stratified epithelium for 3 to 4 mm. from the os uteri, then there is an abrupt change to columnar epithelium, with glands which enlarge enormously during pregnancy. There is little morphological change during the cycle, but much secretion occurs in the mid-luteal period (8).

The medial third of the oviduct is straight; the rest is coiled. The fimbriated end is large and deep red. The oviduct as a whole increases in vascularity immediately after ovulation (8).

If progesterone is injected in the first half of the cycle, deturgescence results and menstrual bleeding may occur. The minimal dose for deturgescence is 3 mg., and for bleeding, 20 mg. in a single dose, or 15 mg. if it is given in 5-mg. doses at 3 to 4 days apart (10). Deturgescence is also caused by the injection of estradiol benzoate early in the cycle; 0.1 mg. produces slight deturgescence, but with 1.0 mg. it is complete. Since the cycles were lengthened, it is believed that the effect is produced by an action upon the ovaries (11). In the spayed baboon the minimal daily dose for 15 days sufficient to produce slight turgescence was 0.01 mg. of estradiol benzoate. Daily injections of 0.02 to 0.04 mg., usually 0.04 mg., produced an effect similar to that found in the normal baboon. The daily dose needed to build the endometrium to a state in which bleeding followed the cessation of injections was .01 to .06 mg., usually 0.01 to 0.02 mg. (12).

1. Zuckerman, S. PZS., 325–343, 1931.
2. Gilbert, C., and J. Gillman. South African J. Med. Sci., 25: 99–103, 1960.
3. Gillman, J., and C. Gilbert. South African J. Med. Sci., 11: 1–54, 1946.
4. Gillman, J. South African J. Sci., 32: 342–355, 1935.
5. Gilbert, C., and J. Gillman. South African J. Med. Sci., 16: 115–124, 1951.
6. Gillman, J., R. A. Pillay, and S. S. Naidoo. JE., 19: 303–309, 1959.
7. Zuckerman, S. PZS., 593–602, 1931.
8. Zuckerman, S., and A. S. Parkes, PZS., 139–191, 1932.
9. Harrison, R. J. Nature, 164: 409, 1949.

10. Gillman, J. E., 26: 80–87, 1940.
11. Gillman, J. E., 29: 633–638, 1941.
12. Gillman, J. E., 31: 172–178, 1942.

Papio cynocephalus L. The mean cycle length of the yellow baboon was 33.3 ± .7 days, the range was 25 to 41 days, and the mode was 31 and 32 days (1). In the London zoo births have occurred in July, October, and December (2).

P. doguera Pucheran. This baboon is probably the one described as *P. anubis*. If so, the following record is relevant. The mean length of 20 cycles was 34.7 ± .5 days, with a range of 28 to 38 days and a mode of 35 days. Menstruation is observed in this species (1). The average testis weight is given as 117 g. (3).

1. Zuckerman, S. PZS., 315–329, 1937.
2. Zuckerman, S. PZS., 122: 827–950, 1952–3.
3. Kinsky, M. AA., 108: 65–82, 1960.

Comopithecus hamadryas L.

SACRED BABOON

The sacred baboon breeds all the year round, but the height of the season is from May to July and the lowest point is in January and February (1). The cycle is 31.4 ± .4 days (2) or 36.2 ± 1.5 days. Earlier cycles tend to be longer than those observed after the animals have settled down to laboratory life. The modal length of the cycle is 33 days in both series, and the follicular phase lasts 15 to 19 days. The duration of the lutein phase is 14 to 15 days (3). The number of young is usually 1, rarely 2 (4), and the period of gestation is about 170 days, range 154 to 183 days (5).

The perivulvar swelling grows from the tenth to the fourteenth day of the cycle. It remains large for 2 to 3 days and then disappears in 6 or 7 days. Altogether it lasts for 3 weeks. Menstruation begins about 3 to 5 days after the tumefaction disappears; it lasts for 5 days. Tumefaction begins about 2 to 4 days after the disappearence of the menses. Vaginal desquamation is seen from the end of maximal tumefaction and it lasts for 2 or 3 days. From the end of tumefaction to the beginning of the menses an abundant milky fluid may be found in the vagina. The amount of mucus is greatest in the cervix at the height of tumefaction; it is more fluid and transparent on days 19 and 20 of the cycle, just before the vaginal desquamation. It disappears at the twenty-eighth day. A new cycle begins about 10 to 20 days after the young has been born (6).

The reproductive tract is similar to that of *P. comatus,* but the organs are smaller. The endometrium in the early lutein phase is 3.1 mm. deep, with high-columnar epithelial cells which have granular cytoplasm and large, oval, vesicular nuclei, usually at the base of the cells. There is a good basement membrane; the glands are tortuous and are without the basal buds that have been described in man. The glands are barely secreting; there is no edema and few polymorphs. In the region of the isthmus the glands are shallower, wider, and mostly empty. At the end of menstruation the extravasated blood is localized, the glands straight, long, and thin, with little secretion. There are numerous leucocytes and also a good basal membrane. In the follicular phase the endometrium is 3.8 mm. thick. The free edge of the columnar epithelium is covered by a thin layer of secretion. The nuclei are large with numerous mitoses. The glands are simple, straight tubules, few in number and with little secretion. Numerous mitoses and many wandering cells are present. The stroma is open, with numerous mitoses and many leucocytes (2).

Ovariectomy causes an immediate reduction of the color of the sexual skin below the resting level. From 60,000 to 80,000 M.U. of estrogen are needed to cause complete swelling of this area (7).

After castration the male loses its cape of gray fur and the sexual skin regresses, both on the snout and on the buttocks. The injection of testosterone at the rate of 100 mg. per week restored all these features. The snout began to color in 2 weeks, the buttocks were enlarged and began to color in 1 month, and the cape was restored in 5 months (8). The average testis weight is 34.15 g. (9).

1. Hartman, C. G. JM., 12: 129–142, 1931.
2. Zuckerman, S., and A. S. Parkes, PZS., 139–191, 1932.
3. Zuckerman, S. PZS., 315–329, 1937.
4. Shortridge, G. C. The Mammals of South West Africa. London, 1932.
5. Zuckerman, S. PZS., 122: 829–950, 1952–3.
6. Seguy, J., and P. Bullier, Arch. Mus. Hist. Nat., Paris, Ser. 6, 12: 309–311, 1935.
7. Parkes, A. S., and S. Zuckerman. J. Anat., 65: 272–276, 1931.
8. Zuckerman, S., and A. S. Parkes. JE., 1: 430–439, 1939.
9. Kinsky, M. AA., 108: 65–82, 1960.

Mandrillus leucophaeus Cuvier

DRILL

The drill has a menstrual cycle of 32.6 ± .9 days (1). The injection into an immature male of 50 mg. of testosterone twice weekly, working up to 300 mg. weekly, caused the premature development of secondary sexual characters. Within a week the circumanal region of red extended and became more intense; this condition was followed by swelling. At the end of the first month the face began to change color, and, by the third month, it resembled that of an adult male. In three months the scrotum began to color. Removal of one testis proved the immaturity, and when the injections were stopped the colors faded and the swelling subsided (2).

1. Zuckerman, S. PZS., 315–329, 1937.
2. Zuckerman, S., and A. S. Parkes, JE., 1: 430–439, 1939.

Mandrillus sphinx L. The mandrill has a gestation period variously reported as 220 or 270 days (1).

1. Kenneth, J. H. Gestation Periods. Edinburgh, 1943.

Theropithecus gelada Rüppell

GELADA BABOON

The female of this Ethiopian baboon reaches puberty at 5 years. The cycle ranges from 32 to 36 days. There is sexual skin around the neck as well as in the perineal region. Like the latter it is subject to periodic swelling and reddening. The color is lost as menstruation approaches but the swelling comes up as a series of beads and vesicles around the neck. These regress and menstruation follows (1).

Testis weight has been reported as 21.5 g. (2).

1. Matthews, L. H. TZS., 28: 543–552, 1953–6.
2. Kinsky, M. AA., 108: 65–82, 1960.

Cercopithecus aethiops L.

GRIVET MONKEY

The African grivet monkey probably breeds all the year-round (1), but in the Congo the seasons of the rains in November and May seem to be the most favorable (2). In zoos and in the wild all birth records have been between January and July (2–4), while a large fetus has been reported in January (5). The gestation period is about 7 months (1). Usually 1 young is born and twins are rare (1), but one such birth has been recorded in the San Diego zoo (4).

There is no sexual skin or external swelling. Menstruation recurs at intervals of about 31 days (6). The ovary contains few multiovular follicles (7). Injection of large amounts of estrogen into an ovariectomized female did not cause menstruation or withdrawal bleeding, and no external changes were observed (6).

1. Shortridge, G. C. The Mammals of South West Africa. London, 1934,
2. Verheyen, R. Inst. Parcs Nat. Congo Belge, 1–161, 1951.
3. Zuckerman, S. PZS., 122: 827–950, 1952–3.
4. Stott, Jr., K. JM., 27: 394, 1946.
5. Allen, G. M., and A. Loveridge. HCMZ., 89: 147–214, 1941–2.
6. Zuckerman, S. J. Physiol., 84: 191–195, 1935.
7. Harrison, R. J. Nature, 164: 409, 1949.

Cercopithecus diana L. This monkey has been reported with a fairly large fetus in July (1).

C. l'hoesti Sclater. A female with a single embryo has been reported in February (2).

C. mitis Wolf. The Pluto monkey menstruates and has a probable cycle length of about 30 days (3). Fetuses have been recorded in October (4) and February (5), while a nursing baby has been found in January (5).

C. mona Schreber. A female with a single embryo has been found in April (2).

C. nictitans L. A female with an embryo and two females carrying babies have been reported for March (5). The species has been reported as breeding all the year with a preference for December to April (6).

1. Allen, G. M., and H. J. Coolidge, Jr. In R. P. Strong, ed., The African Republic of Liberia and the Belgian Congo. Cambridge, Mass., 1930.

2. Eisentraut, M. Zool. Jahrb., Syst., 85: 619–672, 1957.
3. Zuckerman, S. PZS., A: 315–329, 1937.
4. Lawrence, B., and A. Loveridge. HCMZ., 110: 1–80, 1953.
5. Allen, G. M., B. Lawrence, and A. Loveridge. HCMZ., 79: 31–126, 1935–7.
6. Haddon, A. J. PZS., 122: 297–394, 1952–3.

Allenopithecus nigroviridis Pocock

In the San Diego zoo a pair of these monkeys have had a single young on three occasions, once in June and twice in July (1).

1. Pournelle, G. H. JM., 43: 265–266, 1962.

Erythrocebus patas Schreber

RED MONKEY

In one specimen the testes weighed 7.2 g., and the length of the seminiferous tubules was 439 meters. Spermatozoa were 38.7 μ long (1).

1. Knepp, T. H. Proc. Pennsylvania Acad. Sci., 13: 58–62, 1939.

COLOBINAE

Presbytis (Semnopithecus) entellus Dufresne

LANGUR

The Indian langur probably reaches puberty at 6 to 7 years of age. There is a fairly well-defined season of birth which lasts from November to March, though some young are born in April. In the Aln hills and the Gujarat most young appear to be born after January (1). In Ceylon there is no definite breeding season (2).

Menstruation occurs each month and lasts about 4 days. Extravasation of blood begins on the dorsal side of the uterus. The epithelium of the cervix is not cast off during menstruation, and the glands in this region actively secrete during this time. In general, the uterine changes recall those found in the macaque. Ovulation probably occurs in the interval but menstruation may happen without it (3).

In one instance the gestation period lasted 196 days (4). Twins are not infrequent (5). An immature male was injected with testosterone twice weekly. By 4 weeks the glans protruded from the penis and the sexual skin reddened and became enlarged (6).

1. McCann, C. J. Bombay Nat. Hist. Soc., 36: 618–628, 1933.
2. Phillips, W. W. A. Ceylon J. Sci., 13B: 261–283, 1924–6.
3. Heape, W. TRS., 185B: 411–471, 1894.
4. Hill, W. C. O. Ceylon J. Sci., 20: 369–389, 1936–7.
5. Blanford, W. T. The Fauna of British India. Mammals. London, 1888–91.
6. Zuckerman, S., and A. S. Parkes. JE., 1: 430–439, 1939.

Presbytis aygula L. In the London zoo this langur has given birth to young all the year-round (1).

P. cristatus Raffles. The average testis weight is 5.5 g. (2).

P. johni Fischer. This langur has no definite season (3). In South India young have been observed in June and September (4).

P. senex Erxleben. In Ceylon this langur has no special breeding season, but the young are generally born from February to March. Twins are very occasionally produced (5).

1. Zuckerman, S. PZS., 122: 827–950, 1952–3.
2. Schultz, A. H. AR., 72: 387–394, 1938.
3. Pocock, R. I. Fauna of British India. Mammals. London, 1939–41.
4. Webb-Peploe, C. G. J. Bombay Nat. Hist. Soc., 46: 629–644, 1946–7.
5. Phillips, W. W. A. Ceylon J. Sci., 13B: 261–283, 1924–6.

Nasalis larvatus Wurmb

PROBOSCIS MONKEY

This monkey has no fixed breeding season. The gestation period is believed to last about 166 days (1).

1. Sanderson, I. T. The Monkey Kingdom. London, 1957.

Colobus

Colobus badius Kerr. Lactating females have been found in December (1), and two females, one with a small fetus and the other with a large one, in September (2).

C. polykomos Zimmermann. This monkey has no fixed season for breeding and no preferred months (3), 4).

1. Allen, G. M., and A. Loveridge. HCMZ., 75: 47–140, 1933–4.
2. Allen, G. M., and H. J. Coolidge, Jr. In R. P. Strong, ed., The African Republic of Liberia and the Belgian Congo. Cambridge, Mass., 1930.
3. Haddon, A. J. PZS., 122: 297–394, 1952–3.
4. Pournelle, G. H. Inter Zoo Year Book, 2: 83–84, 1960.

PONGIDAE

HYLOBATINAE

Hylobates hoolock Harlan

WHITE-BROWED GIBBON

In one specimen the first menstruation occurred at 7 months. Ten periods averaged 27.5 ±3.4 days, and the duration of menstruation, 2 to 4 days, with 2.6 ± .7 days as the average. The turgescence of the genitalia was permanent and the vaginal smear showed no rhythmic fluctuation with the stage of the cycle (1). The number of young is usually one (2).

1. Matthews, L. H. PZS., 116: 339–364, 1946–7.
2. Blanford, W. T. The Fauna of British India. Mammals. London, 1888–91.

Hylobates lar L.

GIBBON

The average length of 17 cycles in two adult females of the gibbon was 29.8 ± .6 days, with a range from 21 to 43 days. Bleeding lasted 2 to 5 days, averaging 2.4 days. There is no sexual skin swelling, but the labia vary in their degree of extrusion and coloring (1). Puberty is probably reached at the age of 8 to 10 years, and there is no regular breeding season (2). One young is born at a time.

1. Carpenter, C. R. AR., 79: 291–296, 1941.
2. Carpenter, C. R., and A. H. Schultz. Comp. Psychol. Monog., 16: 1–212, 1940.

Hylobates concolor Harlan. The black gibbon reaches puberty at about 7 years, according to one observation. Menstruation recurs at intervals of a little over a month and lasts about 2 to 3 days (1).

1. Pocock, R. I. PZS., 169–180, 1905.

Symphalangus syndactylus Raffles

SIMANG

The single newborn young have been seen at the end of January (1).

1. Miller, G. S. Proc. United States Mus., 26: 437–483, 1903.

PONGINAE

Pongo pygmaeus Hoppius

ORANG-UTAN

The cycles of the orang-utan last about 29 days (1). Sexual skin swelling is restricted to pregnancy (2). The gestation period lasts about 8 to 9 months (3). In one zoo specimen it was 263 days (4). In two adults the testes averaged 35.3 g. (5).

1. Zuckerman, S. PZS., A: 315–329, 1937.
2. Schultz, A. H. JM., 19: 363–366, 1938.
3. Brown, C. E. JM., 17: 10–13, 1936.
4. Graham-Jones, O., and W. C. O. Hill. PZS., 139: 403–410, 1962.
5. Schultz, A. H. AR., 72: 387–394, 1938.

Pan satyrus L.

CHIMPANZEE

The chimpanzee breeds all the year round. Puberty is gradual with swelling of the sexual skin as the first sign of its approach. This swelling may last for some months, and then it subsides; menstruation follows in a few days (1). The mean age at first menstruation (menarche) is 8 year 11 months. The young females are not immediately fertile, and they may not become pregnant for from 4 months to 2 years, with a usual interval of from 1 to 1½ years. The mean cycle length during the first few months is 50 ± 4.4 days; later, when the females are fully adult, the mean length becomes 34 to 35 days, which may be taken as the true mean cycle length. In adolescents, during winter, there is a tendency to a prolonged preswelling stage with amenorrhea. Isolation, copulation, and the sex of the cage mate have no effect upon the cycle length. The most constant parts of the cycle are the postswelling stage and menstruation. The latter lasts about 4 days in adolescents and 2 to 3 days in adults (2). The spread is from 1 to 7 days (3).

The sexual skin does not show such marked color changes as those observed in macaques, but the swelling is very great, amounting in some cases to an increase of 1,400 cc. in volume. It is caused by the accumulation of intercellular fluid. For several days after menstruation the sexual skin area is quiescent, then swelling begins, the time depending upon the length of the cycle. At first the labial and circumanal regions are involved, then it spreads anteriorly through the prepudendal region and laterally to the callosities. Usually the maximal swelling is reached at about the fifteenth day of the cycle, and it subsides about 11 days before menstruation begins, but there is often a partial renewal a day before menstruation. Detumescence is rapid, occupying about 48 hours (3). The two periods of maximal swelling coincide with, or just precede, the periods of greatest estrogen excretion in the urine.

Ovulation occurs usually from the sixteenth day of the cycle onward, as judged by the dates of fertile matings, but it is believed that a more accurate dating can be made by referring it to the probable date of the next menstruation. By this method it is believed to occur about 14 days before the next missed period, a time which is in close agreement with that in macaques and man (4). A definite rhythm of sexual receptivity has been observed. This is zero during menstruation and in the immediate postmenstrual period. Maximal receptivity occurs between the middle and the end of the time of maximal swelling, then it falls to zero and remains so for the rest of the cycle, i.e., it falls abruptly at days 18 to 22 (5).

The vaginal smear changes somewhat during the cycle. The number of epithelial cells falls just before menstruation and increases early in the lutein phase. Leucocytes are very variable (6). A vaginal plug is formed after copulation (7).

The number of young born is usually 1; six pairs of twins were born in 120 parturitions. The mean duration of gestation is 226.8 days from the probable date of conception. The range is from 196 to 260 days, and the standard deviation is 13.3 days (8).

The excretion of estrogens in the urine amounts, at its maximum, to 200 to 400 I.U. per 24-hour sample. The peak is reached at the time of maximal genital swelling and just before menstruation. The excretion of androgens is about 4 to 8 I.U. daily, and it is subject to irregular variation. Males excrete about twice as much androgen and somewhat less than half as much estrogen as females (9, 10).

The injection of 3,000 R.U. of estrogen over 3 days into an immature ovariectomized female caused sexual skin swelling to begin. It reached its maximum by 15 days, when 16,000 R.U. had been given (11).

The Aschheim-Zondek urine reaction for the diagnosis of pregnancy is given by the chimpanzee, but it appears to be usable only for a limited period toward the beginning of pregnancy (12). This reaction depends upon the excretion of gonadotrophic hormones. In three specimens the average testis weight was 118.8 g. (13).

1. Schultz, A. H., and F. F. Snyder. Bull. Johns Hopkins Hospital, 57: 193–205, 1935.
2. Young, W. C., and R. M. Yerkes. E., 33: 121–154, 1943.
3. Elder, J. H., and R. M. Yerkes. AR., 67: 119–143, 1936.
4. Elder, J. H. Yale J. Biol. Med., 10: 347–364, 1938.
5. Yerkes, R. M. Human Biol., 11: 78–111, 1939.
6. Tinkelpaugh, O. L., and E. van Campenhout. AR., 48: 309–322, 1931.
7. Tinkelpaugh, O. L. AR., 46: 329–332, 1930.
8. Peacock, L. M., and C. M. Rogers. Science, 129: 959, 1959.
9. Allen, E., A. W. Diddle, T. H. Burford, and J. H. Elder. E., 20: 546–549, 1936.
10. Fish, W. R., W. C. Young, and R. I. Dorfman. E., 28: 585–592, 1941.
11. Zuckerman, S., and J. F. Fulton. J. Anat., 69: 38–46, 1934.
12. Zuckerman, S. AJP., 110: 597–601, 1935.
13. Schultz, A. H. AR., 72: 387–394, 1938.

Gorilla gorilla Savage and Wyman

GORILLA

In the Columbus, Ohio, zoo a female reached puberty at about 5 years (1). The cycles last about 45 days (2), and the gestation period was about 257 to 259 days in the case of a baby born in captivity in December (1). A single testis weighed 20.6 g. with the epididymis; without it the weight was 11.6 g. (3).

1. Thomas, W. D. Zoologica, 43: 95–104, 1958.
2. Noback, C. R. AR., 73: 209–225, 1939.
3. Hall-Craggs, E. C. B. PZS., 139: 411–414, 1962.

HOMINIDAE

Homo sapiens L.

MAN

The mean age at the menarche for about 115,000 cases, mostly of whites in many countries, is 15.45 ± .006 years; standard deviation, 2.15; and mode, 14.5 years. It is generally held that it is earlier in races living in warm climates than in those inhabiting colder ones. The evidence hardly supports this view, and doubtless other factors, especially that of nutrition, modify the data, which have been summarized in graphic form (1, 2). More recent data indicate that differences are being obliterated and that a mean age of about 14.6 years is now practically universal (3). The menarche occurs much less frequently in summer than in winter in New York (4). The human subject probably frequently experiences menstrual periods without ovulation in the earlier years of reproductive life, and a few such cases have been recorded (5).

Accurately recorded cycle lengths are few and far between, and the variability is greater than is usually realized. Material gathered from several sources gave, after abnormally long cycles had been eliminated, following our practice with other mammals, a mean length of 28.32 ± 0.6 days, a standard deviation of 5.41 days, and a modal length of 27 and 28 days. Eighty per cent of the cycles lay between 21 and 34 days. These figures are very close to similar ones calculated for the rhesus monkey; the mean length is one day more in man. The duration of menstruation normally varies from 2 to 8 days, with a mode at 5 days (6).

The time of ovulation in relation to the cycle has occasioned much controversy. Several lines of evidence agree in dating it near the middle, counting from the first day of menstruation (7). According to some workers (8) it occurs invariably 15 days before menstruation, but most are unwilling to agree that it is so definitely fixed. It would constitute a most unusual biological phenomenon if there were not some variation. Attempts have been made to solve the problem by plotting the times at which a single coitus within the cycle must have been the cause of a pregnancy (9). These have shown that conception may result at any time but that it is most likely in mid-cycle. Work of this nature is open to the criticism that it depends upon statements that may be unreliable. Biologically it is only an approximation because the data are plotted on a standard 28-day cycle, which is merely the average duration. The data cannot be plotted exactly since the fact that pregnancy has ensued interrupts the cycle. One statement is to the effect that ovulation usually occurs 14 ± 2 days before the appearance of the next menstrual period but that there is a spread from the eighth to the twentieth day counting from the beginning of the menses (10).

A phenomenon which, so far, has been found definitely only in man is the menopause, or cessation of ovarian function, at a fairly definite time of life. This is a very gradual process, as, toward the end of reproductive life, the cycles become more irregular. The modal age at the menopause has been given at 49 years (1). The ovaries become less responsive to gonadotrophic hormones at this time, and the amount of follicular atrophy increases together with the amount of connective tissue. The amount of F.S.H. in the anterior pituitary increases and a gonadotrophe consisting almost entirely of F.S.H., or its equivalent, appears in the urine. These changes in the activity of the pituitary and excretion of F.S.H. are also found in ovariectomized women soon after the operation (11). Probably, if we had more information upon the latter part of life in other primates, similar findings would be recorded for them. The cessation of reproduction in rats is a much more gradual process, some ceasing early and others continuing to be in good reproductive condition to the end of a very long life. They do not show a cessation of reproduction at a fairly definite age as does man.

Fecundity has been thoroughly studied and the results may be found in books on population statistics. Twinning usually occurs once in about 83 births; triplets are said to occur once in 83^2 and higher numbers in increasing powers, an approximation which gives a good idea of their relative frequency. The frequency of twin births varies with the race. Negroes have a high rate, 1 : 70.7 for births in the United States. For the same period of years (1922–36) the rate among whites was 1 : 88.6 births (12). Yellow races have a very low rate of twin births; 1 : 145 is a figure quoted for Japan (13). About 25 to 29 per cent of twins born to white parents are monozygotic twins, i.e., they are derived from a single fertilized ovum. Most of the difference between the Mongolian and the white frequencies is due to a relative deficiency of dizygotic twins among the former. In Japan, the frequency of dizygotic twins is only one-fourth to one-third that in American whites or Negroes, but that of monzygotic twins is much the same in each race. Similarly in India and Ceylon the proportion of monozygotic twins to dizygotic twins is 2 to 1 (14). The genetic tendency to produce monozygotic twins is a human characteristic and it varies little with race. On the other hand the tendency to produce dizygotic twins does vary and it may, therefore, depend upon genes that are not uniformly distributed throughout the human species. As an illustration of this tendency to produce twins the character has been traced for several generations in some families, in one for 8. In this pedigree the tendency was inherited as a recessive (15). It is difficult, for obvious reasons, to obtain evidence on the biological frequency of births or the effects of various factors on fecundity in man. There is evidence, however, that only children are no less fertile than average (16), and that the number of children to the family is less for fathers at the highest and lowest intelligence levels than in the middle levels. The intelligence level of the mothers seems to have little influence in this respect (17). Probably sexual ability is involved in these relationships.

The sex proportion for a very large body of statistics is 51.4 per cent males. Racial and national differences have been observed, but it is not clear how far these are due to imperfections of registration (18). A compilation of 46 million births in Western Europe gave a proportion of 51.34 per cent males, very close to the figures quoted above for the United States (19).

The duration of gestation is usually calculated from the first day of the last menses. One set of data calculated in this way gave 280.2 ± .4 days, standard deviation 9.2 days. Another set calculated from the last day of the last menses gave 274.0 ± .4 days, standard deviation 10.4 days. When allowance is made for the duration of menstruation, these averages are in close agreement. If the average data of ovulation is 13 days from the last menses, the average actual duration of gestation is 267 days. The duration of gestation in man is more variable than it is in any other species for which records are available (20).

THE FEMALE

The histology of the tract of women is similar to that of the rhesus and the crab-eating monkey and the changes during the cycle are essentially the same. Since accurately dated nonpathological material is scanty, the literature is not reviewed in detail here.

In the ovary neogenesis of ova has been suggested as taking place in adult life, principally from the germinal epithelium (21). The theca interna of the mature ovum contains alkaline phosphatase which disappears after ovulation. There is none in the granulosa (22). In the developing corpus luteum capillaries invade the granulosa on day 2 and reach the cavity by day 4. The theca interna and the granulosa are indistinguishable from each other on day 3. Fibroblasts invade the central cavity on day 5. On day 6 it becomes again possible to distinguish the theca interna and granulosa elements. On day 7 thin-walled venules line the border of the central cavity and communicates with engorged collecting veins. By day 8 a connective-tissue layer lines the central cavity. From days 9 to 12 a progressive exhaustion of the secretory cells sets in; the granulosa elements shrink and a widespread vacuolization may be observed. The central cavity enlarges. On days 13 and 14 "mulberry" cells may be observed with the oil immersion lens (23).

In the oviduct the number of secretory cells increases towards the uterus. During the follicular phase of the cycle the epithelial cells are at their highest and the secretory cells are lowest, as this is the time of their maximum secretory activity. There seems to be no cyclical variation in the number or height of the cilia. At ovulation the amount of alkaline phosphatase and supranuclear glycogen decreases while nonspecific esteras increases (24, 25). Contractions during the mid and late interval are rapid and variable in height, while in the premenstrual and menstrual phases they are slower and more uniform in amplitude (26).

During the proliferative phase uterine contractions increase in frequency as ovulation approaches. They last for about 10 to 15 seconds and their amplitude varies. At this time the motility of the isthmus and cervix increases. The peak of motility is reached at ovulation. In the secretory phase the frequency decreases. In the isthmus, and especially in the corpus, contractions of long duration and frequently of very high amplitude but with long refractory periods are found. At menstruation the activity is low and there is focuses on the corpus. The amplitude is high, duration short, with long refractory periods during which the muscle is relaxed (27, 28). By the balloon method rapid feeble contractions have been found throughout the cycle with a marked increase in frequency at ovulation. Slower and more powerful contractions were found only during the postovulatory period and at ovulatory menstruation. The evidence suggests that the first type is under the influence of estrogens and the second, of progesterone (29). The utero-tubal junction is not guarded by a fold of mucosa (30). Air can be forced from the uterus into the tube at a pressure of about 100 mm. mercury, but when the flow is established the pressure falls to 40 mm. (11). The vaginal smear can be used with care for distinguishing the phases of the cycle, but the changes are not clear-cut. The percentage of cornified cells rises sharply at about the time of ovulation, and there is some indication that it does so also just before menstruation in ovulatory cycles. Leucocytes are more abundant just after ovulation. The amount of desquamation is less in man than it is in the rhesus monkey (31, 32).

The pH of the vaginal secretions varies from 4.0 to 5.0 depending upon the stage of the cycle. The highest degree of acidity is found in mid-cycle, at about the time of ovulation (33). The reaction in man is considerably more acid than that found in any other species, including the macaque, in which the acidity is also somewhat unusual.

The levels of sex hormones in the blood and urine have been intensively investigated, and charts have been made illustrating the changes with the cycle. Recently the findings have been summarized (10). A peak in estrogen excretion in the urine, about 1,000 I.U. in 24 hours, occurs at mid-interval, near ovulation time. A second peak is reached a few days before the onset of menstruation. Most of it is present as estriol (34, 35). In the blood, estrogens are low until near ovulation, about 30 I.U./l., then there is an abrupt rise to a level of near 60 I.U./l. which is maintained until day 22, when it abruptly falls (36).

Progesterone is excreted in the urine as pregnanediol. The level of the former in the corpus luteum is about 20 μg./g. of tissue; it is highest in the 6-day-old corpus luteum and finally disappears at 18 days (37). In the urine pregnanediol appears about 24 hours after ovulation, reaches a peak in a few days, and disappears rather abruptly 1 to 3 days before menstruation (38).

Small quantities of gonadotrophic hormones are found in the urine during the cycle. There is a peak of F.S.H. excretion at mid-cycle and L.H. is also high at this time. The level falls during the luteal phase and rises towards its end, a rise that continues into menstruation (39). There is also a gradual rise in the level throughout the reproductive life. This rise accelerates with the approach of the menopause and continues for several years afterwards (40).

In pregnancy estrogens appear in considerable amounts in the urine. The level is low for the first 8 weeks, 200 to 1,000 I.U. daily; and it rises steadily to term, when 15,000 to 40,000 I.U. are excreted each day (35). Corresponding changes have been observed in the level of estrogens in the blood. Pregnanediol is found in the urine during pregnancy, but it is not excreted in greater amounts than in the luteal phase of the cycle until the sixty-ninth day. It then rises steadily until delivery, when the amount excreted falls abruptly (41). Progesterone in the peripheral venous blood increases from 11 to 35 weeks of pregnancy from 4 μg./100 ml. to 26 μg./100 ml. Then it increases more rapidly. It may be noted that the half-life of progesterone in circulating blood is 5 minutes or less. The corpus luteum and placenta must be secreting very actively (42).

Gonadotrophic hormones have been found in the blood and urine of the newborn, and they are believed to have originated in the mother and to have crossed the placenta. Their appearance is brief and inconstant and may be related to the "genital crisis," or brief uterine hemorrhage, found in some newborn babies (11). They reappear in appreciable quantities at the menarche and have been observed to fluctuate with the cycle, the amount reaching a peak in the mid-interval, but the individual variation is great. It has been stated the F.S.H. activity is more apparent in preovulatory blood than it is later. During pregnancy the amount of gonadotrophic hormone in the urine, principally L.H. in type, rises rapidly from about the time of the first missed period to the sixtieth day. Then it gradually declines, but the amount is still appreciable at parturition (11). The human female differs from other species in this respect. Such gonadotrophins are found for a limited time in the urine of the pregnant chimpanzee, and, for a still more limited period, in that of the pregnant macaque. Other species that have been tested, except the giraffe, have given negative results. The hormone is of placental origin. The anterior pituitary is exceptionally high in F.S.H. and low in L.H. (43).

At the time of ovulation there is a rise of approximately 1° F. in the basal body temperature. The elevated temperature is maintained until the beginning of the next menses and seems to be associated with the action of progesterone (44). By applying the benzidine test to vaginal secretions it has been found that 94 per cent of ovulating women showed a little blood at the time of, or just after, the rise in body temperature (45).

The amount of estrogen needed to build the endometrium of the ovariectomized woman to the point at which bleeding follows cessation of the injections is 1.5 to 2.0 mg. of estradiol benzoate weekly (46), or 4,200 I.U. daily. The amount of progesterone secreted by the corpus luteum may be estimated by the amount of pregnandiol excreted. If this is a reliable indication, the methods of assay are not yet very accurate, the corpus luteum of the cycle secretes between 5 and 20 mg., daily (5). At the third month of pregnancy the amount is about to 10 to 20 mg. daily (46). It is not certain, however, that all of this substance is produced by the corpus luteum; the fact that pregnancy may be continued after ovariectomy as early as the second month is an indication of secretion by another organ, probably the placenta.

THE MALE

The normal ejaculate of the male is about 3 to 5 cc., containing 100,000,000 to 150,000,000 spermatozoa per cc. (47). The pH varies from 6.9 to 7.36, with an average of 7.19. Prostate fluid ranges from 6.3 to 6.6, average 6.45, and that of the seminal vesicles averages 7.29 (48). Human semen is exceptionally well buffered (49), which may explain why the spermatozoa are able to survive the exceptionally acid reaction of the vagina. It is liquid when it is ejaculated and rapidly coagulates, but liquefies in a few minutes. The prostatic fluid is rich in an enzyme, fibrinolysin, which liquefies fibrin (50, 51). Prostatic fluid is high in Na, K, and Ca, and that of the seminal vesicles is high in acid-soluble phosphate (48). Other constituents of the semen include 100 μg./l. of estrogens, mainly estrone (52), inositol, 58 mg./100 ml. (53), and sialic acid to the extent of 124 mg./100 ml. Prostatic fluid contained 61 mg./100 ml., and seminal-vesicle fluid, 231 mg./100 ml. (54).

Androgen excretion in the urine averages 13.8 mg. daily, but the range is considerable, being from 8.1 to 22.6 mg. Excretion of these substances does not follow a cyclical pattern. The castrate excretes about half the amount of androgen excreted by the normal male, and, in normals, the amount decreases with advancing age (11). Normal females excrete about the same quantities of androgens as do normal males; and the male excretes estrogens, but without cyclical variation (11).

Spermatozoa take 19 to 23 days for their passage through the epididymis (56). Their length of life in the female tract is for a limited time only, but reports of maximum survival vary from 2 to 5 days. The vagina is the least healthy environment for them and the cervix the best (11).

In isolated tissues spermatozoa travel through the uterus in about 27 minutes and through the oviduct in 42 minutes, giving a rate from cervix to infundibulum of 65 to 75 minutes (57).

Successful artificial inseminations have been carried out from day 10 to day 20 of the cycle, indicating the amount of spread in ovulation time. Most successes were from days 10 to 15 (58). The previous cycles tended to be short for the early successes and long for the late ones.

In one testis the length of seminiferous tubule was 250 meters (59).

1. Dickinson, R. L., and H. H. Pierson. J. Am. Med. Assn., 85: 1113–1117, 1925.
2. Mills, C. A. Human Biol., 9: 43–56, 1937.
3. Backman, G. Acta Anat., 4: 421–480, 1948.
4. Engle, E. T., and M. C. Shelesnyak. Human Biol., 6: 431–453, 1934.
5. Corner, G. W. The Hormones in Human Reproduction. Princeton, 1942.
6. King, J. L. CE., 363: 79–94, 1926.
7. Hartman, C. G. Time of Ovulation in Women. Baltimore, 1936.
8. Knaus, H. Periodic Fertility and Sterility in Woman. Vienna, 1934.
9. Asdell, S. A. J. Am. Med. Assn., 89: 509–511, 1927.
10. Velardo, J. T. In Velardo, J. T., ed., The Endocrinology of Reproduction. New York, 1958.
11. Hoffman, J. Female Endocrinology, Including Sections on the Male. Philadelphia, 1944.
12. Strandskov, H. H. Am. J. Physical Anthropol., N.S., 3: 49–55, 1945.
13. Komai, T., and G. Fukuoka. Am. J. Physical Anthropol., 21: 433–447, 1936.
14. Sarkar, S. S. Trans. Bose. Res. Inst., 16: 1–9, 1944–6.
15. Fetscher, R. Arch. f. Rass. Ges. Biol., 20: 432–433, 1928.
16. Asdell, S. A. Fertility and Sterility, 2: 312–318, 1952.
17. Asdell, S. A. Ann. Ostetrica e Gin., 3(special number): 105–110, 1953.
18. Ciocco, A. Human Biol., 10: 36–64, 1938.
19. Lampe, P. H. J. Kon. Ned. Akad. Wetensch., Proc., 57C: 643–651, 1954.
20. Asdell, S. A. J. Agric. Sci., 19: 382–396, 1929.
21. Schwartz, O. H., C. C. Young, and J. C. Crouse. Am. J. Obst. Gyn., 58: 54–64, 1949.
22. Corner, G. W. CE., 575: 1–8, 1948.
23. Corner, Jr., G. W. AJA., 98: 377–401, 1956.
24. Snyder, F. F. Bull. Johns Hopkins Hospital, 35: 141–146, 1924.

25. Fredricsson, B. Acta·Anat., 38 (suppl. 37): 1–23 ,1960.
26. Seckinger, D. L., and F. F. Snyder. Bull. Johns Hopkins Hospital, 39: 371–378, 1926.
27. Karlson, S. Acta Obstet. Gyn. Scand., 33: 253–263, 1954.
28. Posse, N. Acta Obstet. Gyn. Scand., 37 (suppl. 2): 1–24, 1958.
29. Garrett, W. J. J. Physiol., 132: 553–558, 1956.
30. Andersen, D. H. AJA., 42: 255–305, 1928.
31. Papanicolaou, G. N. AJA., 52: 519–637, 1933.
32. de Allende, I. L. C., E. Schorr, and C. G. Hartman. CE., 557: 1–26, 1943.
33. Rakoff, A. E., L. G. Feo, and L. Goldsteen. Am. J. Obst. Gyn., 47: 467–494, 1944.
34. D'Amour, F. E. J. Clin. Endocrinol., 3: 41–48, 1943.
35. Browne, J. S. L., and E. H. Venning. AJP., 116: 18–19, 1936.
36. Markee, J. E., and B. Berg. Stanford Med. Bull., 2: 55–60, 1944.
37. De Wit, D. Klin. Woch., 21: 459–460, 1942.
38. Browne, J. S. L., and E. H. Venning. E., 21: 711–721, 1937.
39. McArthur, J. W., J. Worcester, and F. M. Ingersoll. J. Clin. Endocrinol. Metab., 18: 1186–1201, 1958.
40. Albert, A., R. V. Randall, R. A. Smith, and C. E. Johnson. In E. T. Engle and G. Pincus, eds., Hormones and the Aging Process. New York, 1956.
41. Browne, J. S. L., J. S. Henry, and E. H. Venning. Am. J. Obst. Gyn., 38: 927–955, 1939.
42. Short, R. V., and B. Eton. JE., 18: 418–425, 1959.
43. Witschi, E. E., 27: 437–446, 1940.
44. Siegler, S. L., and A. M. Siegler. Fertility and Sterility, 2: 287–301, 1951.
45. Bromberg, Y. M., and B. Bercovici. Internat. J. Fertility, 3: 60–66, 1958.
46. Allen, W. M. Bull. New York Acad. Med., 17: 508–518, 1941.
47. Pratt, J. P. In E. Allen, ed., Sex and Internal Secretions. Baltimore, 1939.
48. Huggins, C. Proc. 3rd Annual Conference on Biology of the Spermatozoa, New York, 60–65, 1942.
49. Willett, E. L., and G. W. Salisbury. Cornell Univ. Agric. Exp. Sta., Mem. 249, 1942.
50. Huggins, C., and W. Neal. J. Exp. Med., 76: 527–541, 1942.
51. Huggins, C., and V. C. Vail. AJP., 139: 129–134, 1943.
52. Diefalusy, E. Acta Endocrinol., 15: 317–324, 1954.
53. Hartree, E. F. Bioch. J., 66: 131–137, 1957.
54. Warren, L. J. Clin. Invest., 38: 755–761, 1959.
55. Fraser, R. W., A. P. Forbes, F. Albright, H. Sulkowitch, and E. C. Reifenstein. J. Clin. Endocrinol., 1: 234–256, 1941.
56. Brown, R. L. J. Urol., 50: 786–788, 1943.
57. Brown, R. L. Am. J. Obst. Gyn., 47: 407–411, 1944.
58. Farris, E. J. Acta Endocrinol., Suppl. 28: 114–130, 1956.
59. Bascom, K. F., and H. L. Osterud. AR., 31: 159–169, 1925.

Edentata

(XENARTHRA)

MYRMECOPHAGIDAE

Myrmecophaga tridactyla L.

GIANT ANTEATER

A RECORD of a female with a single embryo was obtained in Brazil (1) The gestation period is 190 days (2).

1. Miller, F. W. JM., 11: 10–22, 1930.
2. Kenneth, J. H. Gestation Periods. Edinburgh, 1943.

Tamandua tetradactyla L.

There is a record of a female that gave birth to a single young (1).

1. Webb, C. S. The Odyssey of an Animal Collector. New York, 1954.

Cyclopes didactylis Gray

TWO-TOED ANTEATER

This anteater produces 1 young at a birth. In Panama it breeds in December and January (1).

1. Enders, R. K. HCMZ., 78: 383–502, 1935.

BRADYPODIDAE

Bradypus

SLOTH

Bradypus cuculliger Wagler. The three-toed sloth mates from March to April and the single young is born in July or August, mostly about August 1 (1).

B. griseus Gray. The single young are usually born from July to September (2). Breeding is probably year-round but mostly early in the dry season. The gestation period lasts from 4 to 6 months (3).

1. Beebee, W. Zoologica, 7: 1–67, 1926.
2. Goodwin, G. G. AM., 87: 271–474, 1946.
3. Britton, S. W. Quart. Rev. Biol., 16: 13–34, 1941.

Choloepus

TWO-TOED SLOTH

Choloepus didactylus L. A single young was born in the Detroit zoo in February after a gestation of at least 263 days (1).

C. hoffmanni Peters. In December both lactating and pregnant females were found in Panama (2). In the London zoo a birth of a single young has occurred in May (3).

1. Stone, W. D. JM., 38: 419, 1957.
2. Enders, R. K. HCMZ., 78: 385–502, 1935.
3. Zuckerman, S. PZS., 122: 827–950, 1952–3.

(CINGULATA)

DASYPODIDAE

DASYPODINAE

Euphractus sexcinctus L.

SIX-BANDED ARMADILLO

In the London zoo births of 1 or 2 young at a time have taken place in widely scattered months of the year (1). A female with 3 embryos has been recorded for July (2), and another in the same month with 2 fetuses near birth (3).

1. Zuckerman, S. PZS., 122: 827–950, 1952–3.
2. Sanborn, C. C. JM., 11: 61–68, 1930.
3. Kühlhorn, F. Säugetierk Mitt., 2: 66–72, 1954.

Euphractus villosus Giebel

HAIRY ARMADILLO

In the London zoo births of 1 or 2 young have taken place from March to September (1). The gestation period has been given as 65 days (2). In the Argentine this species mates in September and the 2 newborn young are found at the end of November (3).

1. Zuckerman, S. PZS., 122: 827–950, 1952–3.
2. Kenneth, J. H. Gestation Periods. Edinburgh, 1943.
3. Krieg, H. Zeit. f. Morphol. u. Okol. Tiere, 14: 166–190, 1929.

Priodontes giganteus Geoffroy

A single young, occasionally 2, is usual (1).

1. Krieg, H. Zeit. f. Morphol. u. Okol. Tiere, 14: 166–190, 1929.

Tolypeutes

Tolypeutes conurus Geoffroy. A female pregnant with a single fetus near parturition was found at the end of November (1).

T. tricinctus L. A well-developed embryo has been reported for July and a less-developed one in August (2).

1. Krieg, H. Zeit. f. Morphol. u. Okol. Tiere, 14: 166–190, 1929.
2. Sanborn, C. C. JM., 11: 61–68, 1930.

Dasypus novemcinctus L.

NINE-BANDED ARMADILLO

This armadillo of the southern parts of North America and Central America first breeds at about 1 year. Ovulation is from June to August, not at a very precisely defined time (1). Birth is usually in March, a few in April (2). One follicle ovulates spontaneously and a single corpus luteum is formed. In ten days this grows until it forms 75 to 90 per cent of the ovary. The passage of the ovum through the oviduct takes 7 days. Polyembryony is usual; the critical division occurs at the time of implantation, which is delayed for about 3 or 4 months. True gestation lasts about 4 months and birth is from February to May (1). The ovary is partially surrounded by a bursa. The vaginal epithelium consists of columnar mucus-secreting cells. At no time in the cycle does it become cornified. (3).

The corpus luteum seems to be functional and secretory immediately on its formation, according to one account (1). According to another (4), the cells appear to be inactive until implantation, when secretory droplets appear. However, this account limits the implantation delay to at least 3 weeks. Serum progesterone increases in amount just before ovulation; it continues high afterwards and there is another marked increase after implantation (1).

Bilateral ovariectomy at the middle of the 3- to 4-month delay in implantation is followed by earlier implantation than usual, about 30 to 34 days after the operation. If it is done later, implantation follows at the normal time but resorption or abortion follows. In early true gestation removal of the corpus luteum has a similar effect, but if it is removed after the first third of the period gestation continues (5).

1. Talmage, R. V., and G. D. Buchanan. Rice Inst., Pamphlet 41(2), 1954.
2. Newman, H. H. Am. Nat., 47: 513–539, 1913.
3. Enders, A. C., and G. D. Buchanan. Texas Rpts. Biol. Med., 17: 323–401, 1959.
4. Hamlett, G. W. D. Quart. Rev. Biol., 10: 432–447, 1935.
5. Buchanan, G. D., A. C. Enders, and R. V. Talmage. JE., 14: 121–128, 1956.

Dasypus hybridus L. The mulita armadillo is believed, like *D. novemcinctus,* to have a gestation period that is prolonged by delayed implantation. In Argentina implantation occurs about June 1, and birth is in October. The embryos are free for at least two months. Polyembryony is the rule, as many as 12 identical twins being produced from one ovum (1, 2).

1. Fernandez, M. Morphol. Jahrb., 39: 302–333, 1909.
2. Hamlett, G. W. D. Quart. Rev. Biol., 10: 432–447, 1935.

Pholidota

MANIDAE

Manis

MANIS crassicaudata Gray. The Indian pangolin usually has 1 young, occasionally 2. A medium-sized fetus has been found in Ceylon in July (1). In India one was born in November (2).

M. javanica Desmarest. The single young is born at any season of the year (3).

M. pentadactyla L. This Indian pangolin has its single young, rarely 2, from January to March in the Deccan, and in July in the Shevroy Hills (4).

M. (Phataginus) tricuspis Rafinesque. This African pangolin has 1 young at a time. An 80-mm. fetus has been reported in January and one of 280-mm., near birth, which occurs at 290-mm., in November (5).

M. (Smutsia) gigantea Illiger. This African pangolin has been recorded as pregnant with a single fetus in November and December (5).

M. (S.) temmincki Smuts. This African pangolin bears 1 young at a time (6).

1. Phillips, W. W. A. Ceylon J. Sci., 13B: 285–289, 1924–6.
2. Prakash, I. JM., 41: 386–389, 1960.
3. Banks, E. J. Malay Br. Roy. Asiatic Soc., 9(2): 1–139, 1931.
4. Blanford, W. T. The Fauna of British India. Mammals. London 1888–91.
5. Hatt, R. T. AM., 66: 643–672, 1933–4.
6. FitzSimons, F. W. The Natural History of South African Mammals. London, 1919.

Lagomorpha

COITUS-INDUCED ovulation seems to be general in the females of this Order. An interesting suggestion is that it leads to superfetation in the hares.

OCHOTONIDAE

Ochotona princeps Richardson

PIKA

The breeding season of this pika is from May to September (1), and two litters a year are possible (2). The gestation period lasts for 31 days and is followed by a post-partum heat (3). Embryo counts have varied from 2 to 4, with 3 the most frequent number.

Two sets of corpora lutea have been found in females taken in June and July. The lutein cells measure about 30 μ and they have a granular cytoplasm until after parturition. The diameter of the mature follicular ovum is 100 to 120 μ. In the follicle the theca interna is poorly developed and interstitial cells are absent from the ovary (4). The outside diameter of the tubal ovum is 151 μ, and the inside, 140.4 μ. As many as 50 spermatozoa have been found in the perivitelline space (5).

1. Orr, R. T. Mammals of Lake Tahoe. San Francisco, 1949.
2. Dice, L. R. JM., 8: 228–231, 1919.
3. Severaid, J. H. JM., 31: 356–357, 1950.
4. Duke, K. L. AR., 112: 737–760, 1952.
5. Anderson, S. Univ. Kansas Mus. Nat. Hist., Publ. 9: 405–414, 1959.

Ochotona alpina Pallas. Two or 3 litters are born in the summer but reproduction is over by the end of July. The litter varies from 1 to 5, with 3 the usual number. The young of either sex do not reproduce in the summer of their birth (1).

O. collaris Nelson. In the Yukon this pika has about 4 young, born early in the summer (2). There is possibly a second litter, the first in May or June and the second in July or August (3).

O. daurica Pallas. This Manchurian pika has 2 litters a year (4).

O. hypoborea Pallas. In Siberia reproduction is finished by the first 19 days of August. One litter of 2 to 6 young is the rule. Placental scar counts give an average of 4.8 young (5). In Japan the species breeds from May to June. Three counts give 2 embryos each and one gave 3 (6).

O. pusilla Pallas. This pika apparently has 2 litters a year and may have up to 12 young per litter (7).

O. roylei Ogilby. A single record gives 4 embryos in a female of this Central Asiatic pika (8).

1. Khmelevskaya, N. V. ZZ., 40: 1583–1584, 1961.
2. Rand, A. L. Canada Nat. Mus., Bull. 100, 1945.
3. Cowan, I. McT., and C. J. Guiguet. British Columbia Provincial Mus., Handbook 11, 1956.
4. Loukashkin, A. S. JM., 21: 402–405, 1940.
5. Kapitonov, V. I. ZZ., 40: 922–933, 1961.
6. Haga, R. JM., 41: 200–212, 1960.
7. Heptner, W. G., L. G. Morosowa-Turowa, and W. I. Zalkin. Die Säugetiere in der Schutzwaldzone. Berlin, 1956.
8. Blanford, W. T. The Fauna of British India. Mammals. London, 1888–91.

LEPORIDAE

PALEOLAGINAE

Pronolagus

ROCK HARE

Pronolagus crassicaudatus Geoffroy. A female has been found with 2 fetuses in January (1), and 1 to 2 is the usual litter size (2).

P. randensis Jameson. A female taken late in June had a single large leveret in the uterus (3).

1. Roberts, A. In G. C. Shortridge, The Mammals of South West Africa. London, 1934.
2. Haagner, A. K. South African Mammals. London, 1920.
3. Jameson, H. L. Ann. Mag. Nat. Hist., 7(20): 404–406, 1907.

Romerolagus diazi Diaz

VOLCANO RABBIT

This Mexican species breeds in early spring, from March to April. Probably more than 1 litter a year is born (1). The number of young averages 3, with a range from 1 to 4, born after a gestation period of about a month (2).

1. Davis, W. B. JM., 25: 370–403, 1944.
2. Leopold, A. S. Wildlife of Mexico. The Game Birds and Mammals. Berkeley, Calif., 1959.

LEPORINAE

Lepus arcticus Ross

ARCTIC HARE

On Southampton Island the Arctic hare breeds probably in March and April, perhaps earlier, and the young are born in late June and July. This is based on the finding of males with enlarged testes early in April (1), but the interval between breeding and birth is long compared with the 42-day gestation common in the genus. Another account gives the time of breeding in east Greenland as early May (2). On Baffin Island the young are born late in June or early July (3), as on Southampton Island (1). In the Adelaide Peninsula, North West Territory, two females with 4 fetuses each were found at the end of May (4). The number of young varies from 4 to 8 (1, 2), and these numbers seem to be common throughout the range.

1. Sutton, G. M., and W. J. Hamilton, Jr. Mem. Carnegie Mus., 12(2): 1–111, 1932.
2. Howell, A. H. JM., 17: 315–337, 1936.
3. Soper, J. D. Canada Nat. Mus., Bull. 53, 1928.
4. Macpherson, A. H., and T. H. Manning. Canada Nat. Mus., Bull. 161, 1959.

Lepus americanus Erxleben

SNOWSHOE HARE

The snowshoe rabbit or hare of North America mates first in March. In Manitoba it continues breeding until early June (1), but further south it may continue to breed until August (2). It is polyestrous and reaches puberty during the second year of life. The litter varies from 1 to 7, with a mean size of 2.8, and a mode at 3 (3), but it may have 8 to 10 young in years of plague (1). In northern Alberta the litter size increases as the season progresses. Fecundity is highest in the higher latitudes. The average annual number of litters is 2.75, with an average embryo number of 3.8 (4). In captivity 4 litters a year are possible and have been produced (5). There is a postparturient heat (6).

The vulva is swollen and red when the female is in heat; the first litter is born in April, the maximum number of litters are born in May, and the season declines in June and July. The gestation period is 38 days. The mating season is advanced by at least 30 days if the hares are exposed to 18 hours of light a day in winter. Masking the eyes prevents this effect, and transfer from 18 to 9 hours of light causes the developing gonads to regress (7).

When the hare is brown in color there is a large amount of gonadotrophic hormones in the blood; when it is white the level is low. The injection of gonadotrophic hormones into physiologically white animals caused the hair to become much darker and provoked a copious shedding of the winter hair (7).

In January the testes are small and abdominal. They increase in size during February and descend near the end of the month. They reach their maximum weight in March and rapidly decrease from July onwards. In September their weight is lowest and they return to the abdomen. The penis is pale, almost white, in the nonbreeding season. Just before the breeding season it becomes red at the tip, and the color soon spreads to the whole penis. This organ is pale throughout the first year of life, and the males do not mate until their second year (3). Reduction of the amount of light delays testis growth in the spring (7).

1. Seton, E. T. Life Histories of North American Animals. New York, 1909.
2. Hamilton, Jr., W. J. The Mammals of Eastern United States. Ithaca, N.Y., 1943.

3. Aldous, C. M. JM., 18: 46–57, 1937.
4. Rowan, W., and L. B. Keith. Canadian J. Zool., 34: 273–281, 1956.
5. Severaid, J. H. JWM., 9: 290–295, 1945.
6. Severaid, J. H. The Snowshoe Hare, Its Life History and Artificial Propagation. Augusta, Maine, 1942.
7. Lyman, C. P. HCMZ., 93: 393–461, 1943.

Lepus californicus Gray

JACK RABBIT

The California jack rabbit has several litters a year. It breeds the year-round but tends to cease in December and January. The females breed at about a year old and may have as many as 5 or 6 litters a year, though the usual number is 4. In captivity the litter size averaged 1.8 young (1). In the wild an average of 2.6 fetuses was found. Eight per cent of all embryos were reabsorbing and 8 per cent of litters were being entirely reabsorbed. There was some reabsorption in 15 per cent of all litters. A comparison of corpus luteum counts and healthy embryos gave a 16 per cent loss (2). The age of the doe has no perceptible effect upon litter size but the number does tend to increase in the latter part of the season (1). In another study the litter size in the wild was 2.2 ± .1 (3). Another account gives a loss of 13 per cent of the ova shed. Of this, 6.7 per cent was preimplantation loss and 6.2 per cent, postimplantation. In this sample there were few entirely lost litters. The average number of young a doe produced in a year was 9.8. The duration of gestation was 41 to 47 days, with a mean of 43 days. There is a post-partum heat (4).

The ripe follicle measures 1.5 to 2.0 mm., and the corpus luteum, 1.2 mm. It is pink in color when it is functional. Blood follicles are rare but 7 per cent of follicles do not ovulate; nevertheless, they form corpora lutea. Ovulation is probably coitus-induced. Transabdominal migration of ova was observed (4).

The vaginal epithelium and smears do not exhibit cyclical changes, and leucocytes are always present. The vulva is very variable but the changes are not related to the cycle. After parturition the uterus returns to the resting state very rapidly. Those that have been pregnant show longitudinal striations. Pseudopregnancy exists but it is not always followed by lactation (4).

Male puberty is reached at 1,900 g. weight, or at 5 to 7 months, but spermatozoa are not abundant until the bucks weigh 2,100 g. Males that reach these weights in the nonbreeding season may be delayed in attaining puberty. There is a definite, but slight, testis cycle which begins late in November. A peak is reached in January and some retrogression sets in during June; spermatozoa may be retained in the epididymis although many rabbits have regressed seminal vesicles and bulbo-urethral glands (4). The mean weight of both testes in adults is from 10 to 15 g. The number of spermatozoa in the testes lessens from July or August and few are present in December, but this varies from year to year. Sperm counts in the epididymis follow the same trend (2).

1. Haskell, H. S., and H. G. Reynolds. JM., 28: 129–136, 1947.
2. Bronson, F. H., and O. W. Tiemeier. JWM., 22: 409–414, 1958.
3. Vorhies, C. T., and W. P. Taylor. Arizona Agr. Exp. Sta., Tech. Bull. 49, 1933.
4. Lechleiter, R. R. JM., 40: 63–81, 1959.

Lepus europaeus Pallas

COMMON HARE

On the Ciscaucasian steppes the breeding season of this hare lasts from midwinter until the middle of summer (1), but, although they are capable of breeding at 8 months, few of those born in March have young in the same year (2). Drought causes the cessation of breeding in this region (1). In England breeding begins in November or December, is at a maximum in March, and virtually ceases in July, but litters have been found at all times of the year (3). In southern Ontario, where the hare was introduced in 1912, breeding begins in January, reaches a peak from the second week of February to the third week of June, and declines in August. At the peak season all the females are pregnant (4). Introduced into New Zealand, the species breeds from July to December, or from early spring to midsummer (5). The average embryo number and litter size on the steppes in relation to the season are given as follows (1):

	Embryos	Litter size
Winter	1.7	1.5
Spring	3.5	3.3
Summer	3.7	3.1
Autumn	3.0	2.0

In Ontario early litters averaged 1.6 (January to March), and late ones, 3.8

(April to June) (4). In New Zealand the litter size varies from 1 to 5, with 2 as the mode (5), and that appears to be the modal number in Central Europe (6).

Ovulation is coitus-provoked; this may happen during pregnancy so that superfetation is a common phenomenon (7). Gestation lasts for 42 days (6), though it may be as short as 35 days (8), and the embryos all develop in one horn of the uterus (7).

1. Kolosov, A. M. ZZ., 20: 154–172, 1941.
2. Rieck, W. Zeit. Jagdwiss., 2: 49–90, 1956.
3. Barrett-Hamilton, G. E. H. A History of British Mammals. London, 1910.
4. Reynolds, J. K., and R. H. Stinson. Canadian J. Zool., 37: 627–631, 1959.
5. Wodzicki, K. A. New Zealand DSIR Bull. 98, 1950.
6. Hediger, H. Physiol. Comp. Oecol., 1: 46–62, 1948.
7. Stieve, H. Zool. Anz., 148: 101–114, 1952.
8. Strauss, F. RSZ., 65: 434–441, 1958.

Lepus saxatilis Cuvier

MOUNTAIN HARE

This South African hare breeds from October to April (1). It is polyestrous and has 2 to 5 young to a litter (2), but 2 seems to be the usual number. However, there are records of fetuses in May (3) and June (4). The gestation period is about a month (5).

1. Shortridge, G. C. The Mammals of South West Africa. London, 1934.
2. FitzSimons, F. W. The Natural History of South African Mammals. London, 1919.
3. Bradfield, R. D. Field Notes on S. W. African Mammals. nd.
4. Lancaster, D. G. In letter to G. C. Shortridge (1).
5. Wilhelm, J. H. S. W. Africa Sci. Soc., 6: 51–74, 1933.

Lepus timidus L.

VARYING HARE

This hare mates first in early February and has about 2 litters a year, in spring and summer (1). The spring litters consist of from 1 to 5 young and the average is 2.15. Summer litters are larger, with up to 8 young and averaging 3.24 young (2). The gestation period is about 42 days, but mating and ovulation can occur from the thirty-ninth day on (3). In the male spermatogenesis is found from February into May (4).

1. Heptner, W. G., L. G. Morosowa-Turowa, and W. I. Zalkin. Die Säugetiere in der Schutzwaldzone. Berlin, 1956.
2. Höglund, —. Villrevy (Stockholm), 1: 267–282, 1957.
3. Hediger, H. Wild Animals in Captivity. London, 1950.
4. Naumov, S. P. Trans. Cent. Lab. Biol. Game. Anim., 6: 4–44, 1944.

Lepus alleni Mearns. In the southwestern part of North America this hare breeds all the year, but to a lesser extent in summer and winter. Probably 4 litters a year are produced. The litter size averages 1.94 ± .09. One or 2 is the usual litter size, but it ranges to 5 (1).

L. capensis L. This hare is polyestrous and breeds in late winter or early spring (2). It has 3 or 4 litters a year. The first averages 1.8 young; the second, 4.2; and the third, 4.3 (3).

L. mexicanus Lichtenstein. No adult females were pregnant in June or December, but one was lactating in June (4).

L. nigricollis Cuvier. The black-naped hare of southern Asia breeds all year and has usually 2 young to a litter. The gestation period is about 1 month and the females first breed at 6 months (5).

L. townsendi Bachman. This jack rabbit of North America mates in April; the young, 3 to 6, but usually 4, are born from May (7) to early July (6).

L. (*Poelagus*) *marjorita* St. Leger. There is a record of this African grass hare with a single large embryo in August (8).

1. Vorhies C. T., and W. P. Taylor. Arizona Agr. Exp. Sta., Tech. Bull. 49, 1933.
2. Haagner, A. K. South African Mammals. London, 1920.
3. Perevalov, A. A. ZZ., 35: 141–154, 1956.
4. Davis, W. B., and P. W. Lukens, Jr. JM., 39: 347–367, 1958.
5. Phillips, W. W. A. Ceylon J. Sci., 14B: 209–293, 1927–8.
6. Seton, E. T. Life Histories of Northern Animals. New York, 1909.
7. Bailey, V. NAF., 53: 1931.
8. Hatt, R. T. AM., 76: 457–604, 1939–40.

Sylvilagus aquaticus Bachman

SWAMP RABBIT

In Texas this species breeds from January to August, and perhaps a little later. There is a sudden increase in breeding intensity in mid-January which reaches a peak from February to March (1). In Georgia breeding is from March to September (2). In Missouri the females undergo an anestrous period from November to the end of January. Conceptions probably begin about February 4, as lactating females were first found on March 16. However, some males with spermatozoa may be found in any month (3).

The Texas report gives 1 to 5 as the range of embryos, with an average of 2.8 and a mode of 2 and 3 (1). Another report gives 1 to 4, with 2.6 average (2). The Missouri series gives a range of 2 to 6 for corpus-luteum counts, the average being 3.7. For a more limited series of embryo counts the average was 2.8 and for placental scars, 3.4 (3). A sample taken in Mississippi from March 9 to April 4 gave the corpus-luteum count as from 3 to 6, average, 3.7. There was no total resorption of litters in normal times but after severe floods 12 of 18 females were found thus and the tendency remained high for the later pregnancies that spring. Two to 4 per cent of ova are lost before implantation (4).

An observed gestation period lasted for 39–40 days (1).

1. Hunt, T. P. JM., 40: 82–91, 1959.
2. Lowe, C. E. JM., 39: 116–127, 1958.
3. Toll, J. E., T. S. Baskett, and C. H. Conaway. AMN., 63: 398–412, 1960.
4. Conaway, C. H., T. S. Baskett, and J. E. Toll. JWM., 24: 197–202, 1960.

Sylvilagus audubonii Baird

In Arizona this rabbit does not breed from September to December; the testis weight increases in December and decreases in September. The average litter size is 2.9 with a mode of 3 and range, 2–4. Embryonic resorption seems to be rare (1). The young are able to breed in the year of their birth (1). In Death Valley breeding probably ceases at the same time as in Arizona (2). In Mexico breeding begins earlier, as embryos have been found in December (3). The gestation period is 28 to 30 days (4).

1. Sowls, L. K. JM., 38: 234–243, 1957.
2. Grinnell, J. Proc. California Acad. Sci., Ser. 4, 23: 115–169, 1935–47.
3. Baker, R. H. Univ. Kansas Mus. Nat. Hist., Publ. 9: 125–335, 1956.
4. Dice, L. R. JM., 10: 225–229, 1929.

Sylvilagus bachmani Waterhouse

BRUSH RABBIT

This rabbit of western North America breeds from December to the end of June. There is probably a post-partum heat. The average number of embryos is 4, and the range 3 to 6. In counts 55 corpora lutea and 53 embryos were found, but 5 of the latter, possibly 8, were being resorbed. The gestation period is probably 27 ± 3 days. The ovary contains many interstitial cells (1).

The males are in breeding condition from November to June and possibly into July. Their greatest activity is from January to April and possibly into May. Spermatozoa are found in December, but by late June few may be found in the testes though they are still present in the epididymis. The testis weight rises from its low mean of 1.4 g. in October to its high of 5.6 g. in February (1).

1. Mossman, A. S. JWM., 19: 177–184, 1955.

Sylvilagus braziliensis L.

This rabbit probably breeds all the year. Young have been observed from September to April in Central America (1). In the Matto Grosso pregnant females and suckling young have been reported from May to July (2). Four counts of 4 embryos each are on record (3).

1. Goodwin, G. G. AM., 87: 271–474, 1946.
2. Kühlhorn, F. Säugetierk. Mitt., 2: 66–72, 1954.
3. Enders, R. K. Personal communication, 1945.

Sylvilagus floridanus Allen

EASTERN COTTONTAIL RABBIT

This rabbit, which lives in North America east of the Rocky Mountains, is polyestrous and breeds from mid-January into August, though the season may be curtailed for a month at either end in the northern part of its range. The young reach puberty at 40 weeks, and in the breeding season several litters are produced (1). The average litter size is 4.5 (2); or 5.0 ± .2, with a range from 2 to 7, and a mode of 4 to 5 (3). Embryo counts have yielded the following figures: 4.7 ± .2 (4), 4.8 ± .2 (5), and 5.6 ± .2 (6). The last of these included embryo counts, placental scars, and nest young in one combined figure. Lactation does not increase the rate of fetal resorption (6). In another count of 175 females the average litter size was 5.56 ± .9. There was 5 to 8 per cent resorption, but whole litters were not affected (7). If the does do not conceive at the beginning of the season they wait for 15 or 16 days before becoming pregnant. This suggests a pseudopregnancy interval. Ovulation rates in this sample were about 4.0 to 4.26, depending on the year. There was a preimplantation loss of 0.1 to 0.33 ova per litter (8). The gestation period has been observed as 26.5 days (9), but is usually given as 30 days (1).

The testes enlarge late in December and soon descend. They remain in the scrotum until September, at which time they still contain spermatozoa (1). Motile spermatozoa are present in the testes by mid-January but there is no heavy accumulation in the epididymides until late in February. The numbers are maintained until well into August. Young males of the previous year reach breeding condition later than older ones (6). It is exceptional for males to be capable of breeding in the year of their birth, though occasional instances have been found (10). The weight of a single mature testis was 9.6 g., and in this instance the seminiferous tubules measured 339 meters in length (11).

The female prostate is large and its content increases considerably in the breeding season (12). The level of gonadotrophic hormones in the pituitary of the male is at a minimum from October to January and at a maximum in March, when it is 600 per cent of the minimal level. There is a rapid decrease in early fall. The percentage of basophils rises from 4.4 to 13.8 during the transition from the nonbreeding to the breeding state. The weights of the testes, seminal vesicles, and prostate undergo corresponding changes. In the female there are no significant seasonal changes in the gonadotrophe level or in basophil numbers (13).

1. Trippensee, R. E. Proc. North Am. Wildlife. Conf., 1936.
2. Hamilton, Jr., W. J. JM., 21: 8–11, 1940.
3. Beule, J. D., and A. T. Studholme. Pennsylvania Game News, 13: 6–7, 28–29, 1942.
4. Llewellyn, L. M., and C. O. Handley. JM., 26: 379–390, 1945.
5. Sheffer, D. E. JWM., 21: 90, 1957.
6. Ecke, D. H. AMN., 53: 294–311, 1955.
7. Lord, R. D. JWM., 25: 28–33, 1961.
8. Conaway, C. H., and H. M. Wight. JWM., 26: 278–298, 1962.
9. Hendrickson, G. O. JM., 24: 273, 1943.
10. Cooley, M. E. JM., 27: 273–274, 1946.
11. Knepp, T. H. Zoologica, 24: 329–332, 1939.
12. Elchlepp, J. G. AR., 99: 656, 1947.
13. Elder, W. H., and J. C. Finerty. AR., 85: 1–16, 1943.

Sylvilagus cunicularis Waterhouse. A record for the highland rabbit gave 5 embryos in June. Young were observed in the same month and a lactating female in August (1). A male with enlarged testes was found in June (2).

S. nuttallii Bachman. This western cottontail has up to 6 young, born from April to July (3). The gestation period is 28 to 30 days (4).

S. palustris Bachman. The marsh cottontail probably breeds all the year and has litters of 3 to 5 young (5). Another account suggests that the breeding season is limited to the period from March to the end of August (6).

S. transitionalis Bangs. This New England cottontail has a limited breeding season beginning in mid-April. It has 3 to 4 litters a year. When some of these rabbits were treated nightly with artificial light, beginning in December, increased sexual activity was observed in January, 23 days later (7).

1. Davis, W. B. JM., 25: 370–403, 1944.
2. Davis, W. B., and P. W. Lukens, Jr. JM., 39: 347–367, 1958.
3. Bailey, V. NAF., 55, 1936.
4. Cowan, I. McT., and C. J. Guiguet. British Columbia Provincial Mus., Handbook 11, 1956.
5. Tompkins, I. R. JM., 16: 201–205, 1935.
6. Blair, W. F. JM., 17: 197–207, 1936.
7. Bissonnette, T. H., and A. G. Csech. Biol. Bull., 77: 364–367, 1939.

Oryctolagus cuniculus L.

DOMESTIC RABBIT

The rabbit will breed more or less at any time of the year but there are no cycles in the strict sense of the word. Suggestions of cycle lengths have been conflicting, in one 4 to 6 days (1), and in another 7 days or multiples thereof (2). These suggestions were based on behavior. Vaginal cycles, dealt with later in this article, usually suggest a 14-day cycle and are evidently based upon the pseudopregnant period, which may also affect conclusions drawn from behavior. The female ovulates only after coitus or under strong sexual excitement. The usual interval between stimulation and ovulation is 10 hours, and during the first hour of this period sufficient F.S.H. is released from the pituitary to cause the preovulatory ripening of the graafian follicles. Follicles reach the estrous stage in waves, and the female may remain in heat for a month or more at a time in the absence of the male. If coitus does not result in pregnancy, which lasts for 30 to 32 days, corpora lutea are formed which last for about 16 days. At the end of this time the doe makes a nest by plucking fur from her breast, and her mammary glands are in condition for lactation, just as if she were pregnant. This condition is called pseudopregnancy. During the lutein phase glandular development in the uterine mucosa is very great, so that a transverse section appears like open lacework. The vaginal smear is not a reliable indication of the reproductive state.

REPRODUCTION IN GENERAL

The age of puberty in rabbits is affected by breed differences, but there is very little information on this point. It is also affected by the time of birth. Does born in the fall reach puberty (fertility) in about 5½ months, but those born in the spring require about 8½ months. They tend to copulate 1 to 2 months before they are capable of ovulation. In England, does kept in unheated rooms often experience anestrum from October to March, but the tendency is not absolute. The optimum season for reproduction is from May to July (3). In the United States it has been the writer's experience that

July to September, when the temperature is high, are the poorest months. As does pass into the anestrous condition, copulation may occur but graafian follicles do not reach the state at which they are capable of rupturing (3). During the breeding season, in the absence of the male, the doe remains in heat almost indefinitely, and during this time waves of follicles mature, last for about 7 to 10 days, and then become atretic (4). If does are kept in continuous light or in total darkness, this rhythm of growth is not affected except that more large follicles are present in the does kept in the light (5). Atresia may be due to degeneration and phagocytosis of the granulosa, with fibrotic ingrowth of the theca interna, or to extravasation of blood and the formation of blood follicles (6). If coitus is permitted, the mature follicles begin to grow by copious secretion of liquor folliculi, and rupture occurs about 10 hours afterward (7).

The stimulus for this growth is apparently emotional since intense excitement, such as the act of mounting or being mounted by another doe, evokes it and since local anesthesia of the vagina and vulva just before coitus does not prevent its occurrence (8). It is due to the release of a gonadotrophic hormone from the anterior pituitary since removal of that organ within an hour post coitum prevents ovulation. If the operation is performed after that time, ovulation occurs as usual and corpora lutea are formed. Evidently the pituitary releases sufficient hormone in a little more than an hour after coitus (9, 10). The pituitary may also be caused to release the stimulating hormones by electrical stimulation of the head or of the lumbar region of the cord, but the action is somewhat slower (11); by the intravenous injection of 0.9 to 1.1 mg. of picrotoxin (a hypothalamic stimulant) per kg. of body weight (12); and by salts of copper and cadmium (13). This stimulus is necessary for the formation of corpora lutea since rupture of mature follicles by pricking is ineffective (14, 15). The suggestion has crept into the literature that ovulation may be provoked by mechanical stimulation of the cervix, as in the cat, but this is incorrect.

The developing follicle takes about 18 days to mature. In the nonbred rabbit this development is followed by a phase of atrophy lasting for 9½ days (16).

The ovaries produce corpora lutea with approximately equal frequency; a series of records gave 51.4 per cent of corpora lutea in the left ovary (3).

The fact that ovulation usually occurs only after coitus enabled the pioneer worker to differentiate clearly between the estrogen and lutein phases of the cycle. Pseudopregnancy, the condition when sterile coitus, the injection of gonadotrophes, or the use of one of the methods described above, has caused ovulation and corpus luteum development, lasts normally for 16 to 17 days (3, 17). But pseudopregnancy in the doe whose uteri have been removed lasts from 24 to 29 days (18). The corpora lutea of the cycle in the guinea pig similarly treated remain for a much longer period, but those of the rat are not prolonged. It is an open question whether removal of the uterus removes an impulse which normally brings an end to the life of the corpus luteum (in contrast to the action during pregnancy), or whether this organ uses a substance needed by the corpus luteum. More work is needed on the problem.

The doe rabbit matures follicles and is ready to ovulate immediately after the end of pregnancy or of pseudopregnancy (3).

In order to test the length of life of the ovum by artificial insemination at different times, it is necessary to arrange that eggs shall be shed. Thus, in such an experiment, sterile coitus with a vasectomized buck is allowed with the knowledge that ovulation will occur about 10 hours later. This is a great advantage in work of this nature, and also in obtaining dated material for embryological work. The time of ovulation in the rabbit can be obtained with greater accuracy than is possible in any other species. It has been found that, after allowing for the time taken by the spermatozoa to reach the top of the oviduct, the eggs have a fertile life of not more than 6 hours after ovulation. They are capable of being fertilized only while they are in, or as they leave, the plug of extruded liquor folliculi at the top of the oviduct. As they move down the tube and acquire a layer of albumin, they can no longer be fertilized. The percentage of successful inseminations and the litter size decrease at a constant rate with inseminations from 5 hours before to 2 hours after ovulation (19). The ova retain their capacity for fertilization for a maximum of 8 hours after they are shed from the ovary but the chance of fertilization is sharply reduced after 4 hours (20). Ova just liberated from the follicles are surrounded by cumulus cells of the stratum granulosa. If spermatozoa are present, these cells are dispersed by the first ones to penetrate into the mass. This happens normally within 1 to 1½ hours after ovulation. If spermatozoa are not present, it takes about 17 hours for the separation to occur. The passage through the oviduct takes about 62 to 82 hours (21).

The litter size of rabbits varies with the strain or breed. Thus, a fecund inbred strain has given a mean litter size of 8.1 ± .2, with a mode of 9, a range of 1 to 13, and a standard deviation of 2.7; but a strain of small (Polish) rabbits gave as mean 4.0 ± .1, mode 4, range 1 to 7, and standard deviation, 1.54. About 20 per cent of the eggs shed fail to develop to term. This is partly due to loss of ova and partly to fetal atrophy (19). The latter is caused in large measure by inherited factors acting through the mother (22). The mean litter size rises until the doe is at least 3 years old (23). Experimentally it is reduced if artificial insemination is performed 24 or more hours before (24), or fewer than 5 hours before, ovulation. In the first case too few spermatozoa survive, and in the second case too few eggs (19). The sex ratio is 53.1 per cent males (25).

The duration of gestation is usually 30 to 32 days. An extensive series gave 31.08 ± .02 days, standard deviation, 0.83 days; and mode, 31 days (26). It is affected by breed: Polish rabbits had a mean gestation of 30.4 days and an inbred albino strain, 32.9 days. It does not vary with the season or the age of the mother (27) but tends to be higher within the breed when small litters are carried (28). The same is true if the smaller litter is experimentally produced (29). In French lop-eared rabbits the gestation period has averaged 33.4 days and the litter size, 7.7. In the Vienna white the corresponding figures were 33.5 days and 6.6 young (30).

In a doe which is allowed to become pregnant immediately after parturition and which is suckling a large litter, above 3 to 4, implantation is not delayed but the embryos die in the blastocyst stage. If smaller litters are suckled, the pregnancy continues and lasts the usual time. A suckling doe goes out of heat in about a week because of follicular atresia and the loss of about half the normal weight of the ovaries. The uteri also atrophy and almost resemble the state found in an ovariectomized doe. New follicles mature rapidly after weaning (3).

THE WILD RABBIT

In west Wales the main breeding season lasts from January to June; there is less breeding in other months and very little in November. Embryo counts were lowest in January and highest in late May to early June, with an overall average of 4.36. A few more fetuses were found in the left uterus than in the right. For fetuses over 20 days old there was a very slight preponderance of females (31). It is believed that 0.23 per cent of the follicles that ovulate produce 2 ova and 2 embryos, and also that transperitoneal migration of ova occurs in 0.37 per cent. The diameter of the corpus luteum towards the end of pregnancy was 2.5 to 3.0 mm. By one day after parturition it had shrunk to 1.4 mm. (32).

The preimplantation loss of ova amounted to 9.5 per cent. It was least at the height of the season and when the number of corpora lutea was as high as 5 or 6 (33). A type of embryonic mortality frequently encountered was that in which all the embryos had died and were being resorbed. In these instances the young were usually 11 to 15 days old when they died. The amount of mortality declined steeply with increasing maternal weight. Of all litters, 41 per cent suffered some loss and at least 11 per cent of ova may be accounted for in this way. The proportion lost was highest in actively lactating does (34).

In Australia the rabbit is reproductively quiescent in midsummer; most of the breeding is in the spring and autumn. A few litters are produced in winter. The mean corpus-luteum count was 4.37, and the amount of fetal atrophy, 13 per cent, but the many litters entirely lost were not counted in this figure (35).

In the Hawkes Bay region of New Zealand the breeding season starts in June and the pregnancy rate is high until November. Young of the year may extend this season by a further 8 or 10 weeks. Puberty is reached in the male at 9 months, and in the female at 5 months. The testes are smallest in January and largest in September. Then they decrease rapidly until they weigh less than half of their maximum, which is about 4 g. The males weigh 1.1 kg. at puberty. The mean number of ova shed is 5.83, with a mode of 5 and a range from 3 to 11 (S.D. = 1.76). The number varies with the age of the doe, ranging from 5.0 in young ones to 6.2 in old ones. In surviving litters 15 per cent of ova were lost, and 18 per cent of litters were lost entirely in the second week of gestation. The mean number of young per litter was 5.0, and the number of young per adult in a season was 26 to 26 in their first year and 18 to 21 in their second (36).

HISTOLOGY OF THE FEMALE TRACT

OVARY. Soon after birth the nests of cells which include the germ cells tend to atrophy. They are largely replaced by the production of epithelial invaginations, which begin at about 35 days of age, are at their maximum from 51 to 60 days, and then fall off (37). It is said by some that there are further waves of ovogenesis at 8 and 14 months and that at these times the amount of interstitial tissue diminishes (38), but these waves are probably not so intense as the earlier ones, since others have found little neogenesis of ova in mature rabbits. According to the latter workers, atresia of oöcytes is small, only 10 per cent at any time, although in larger follicles it amounts to 60 per cent (39).

As ovulation can be so easily timed, it has been observed and photographed on numerous occasions. The diameter of the mature follicle at coitus is about 1.5 mm. It is relatively flat and has a bluish tinge. Most growth occurs between 3 and 6 hours after coitus, and the follicle protrudes slightly from the surface of the ovary. At 9 hours the vascularity of the theca increases, and there is some slight hypertrophy of the epithelial cells of the theca interna. The first sign of approaching ovulation is the gradual formation of the macula pellucida, a small clear area projecting from the follicle and surrounded by a network of capillaries. As the macula increases in height the blood vessels rupture, and a small blood clot forms below the ultimate point of opening. At this time the follicle is about 1.8 mm. in diameter. Rupture and extrusion of liquor folliculi and the ovum through a small opening at the apex take about 7 seconds. The liquor folliculi does not flow much but forms a small cone. The greater part of it remains in the cavity, which does not collapse to any great extent. Later, the floor of the follicle approaches the surface so that the corpus luteum, with the point, which becomes luteinized, projects from the surface of the ovary more than the follicle from which it was derived (15, 40).

In the ruptured follicle the granulosa cells are nearly all retained, but there is rarely any blood clot except near the point of rupture. Hypertrophy of these cells and ingrowth of the theca interna begin immediately. By 6 hours the ingrowth of connective tissue in the form of spindle-shaped theca cells is marked, and by the fourth day the central cavity, filled with liquor folliculi, has been almost entirely obliterated. By 8 days the lutein cells reach their full development, and are about eight times their former diameter. The corpora lutea of pregnancy persist throughout gestation and are reabsorbed gradually after its termination, but somewhat more rapidly if the doe is lactating. Those of pseudopregnancy begin to degenerate at 18 days. A visible sign of degeneration is an abrupt change from a pinkish vascular state to a chalky yellow avascular one (3, 41).

The size of the ovum is $123.0 \pm 1.9\ \mu$ without, or $188.6 \pm 2.0\ \mu$ with the zona pellucida (42), a comparatively large size for mammals. There is one polyovular follicle to 200 monovular ones (43). Alkaline phosphatase is found in the granulosa but not in the theca interna (44). This is unusual and may be related to the coitus-provoked ovulation found in this species.

VAGINA. The vulva of the rabbit in heat is usually purple to reddish pink and somewhat swollen. This is a good, but not infallible, sign of heat since at other times it is pale. The vaginal epithelium is more or less stratified, with many mucous cells, especially near the cervices, and it does not undergo characteristic changes with heat. Similarly, smears are not reliable as indications of the reproductive state (45, 46), a fact which is common in those species in which mucous cells are frequent. Cornified cells are most abundant at 4- to 6-day intervals. There are no changes in leucocyte numbers. Blood estrogens were higher when the number of cornified cells rose (47). Examination of vaginal smears gave a 14.8-day cycle (48), a length that suggests pseudopregnancy.

UTERUS. The cervices are not plugged at any time by a seal of mucus. The endometrium is very vascular during the follicular phase, but the glands are not much developed. In the lutein, or progestational, phase the glands undergo marked development. The lumens are much expanded, with the result that the appearance resembles that of open lace. During the stage of involution, when the corpora lutea are retrogressing, extensive extravasation of blood has been described (49). The uterus is ciliated in a zone extending from 1.5 cm. from the os uteri anteriorly. In the region of the cervix ciliation is abundant, but this type of epithelium is scanty elsewhere until the tubo-uterine junction is reached. Ciliated cells do not vary in number with the physiological state of the animal (50). The uterine epithelium is said to become a syncytium during the life of the corpus luteum (51).

The oviduct is ciliated, the tubo-uterine junction has four primary folds which project into the uterus, and the muscle forms a slight sphincter. These folds are devoid of glands. According to one account it is difficult to force fluid from the uterus into the oviduct (52), but another worker found it fairly easy (50). Possibly the stage of the "cycle" influences the ease of passage; further work is needed to decide the question.

PHYSIOLOGY OF THE FEMALE TRACT

The corpora lutea are essential for the maintenance of pregnancy. Early removal by ovariectomy causes reabsorption of the embryos; later, abortion is the rule (3, 53). It is easy to shell out the corpora lutea, and the question arises, How many are needed to maintain pregnancy, and what is their output of progesterone? The first question has not yet been answered directly. A summary of available information suggests that one corpus luteum is necessary to bring about progestational proliferation (the lacelike condition of the glands in the uterus), and, as 0.2 mg. daily of progesterone produces the same effect in the ovariectomized rabbit, this must be approximately the daily ouput of a corpus luteum (54). It has been shown that to prolong the proliferation for longer than 5 days, more than two corpora lutea are necessary, and to sensitize the uterus sufficiently to produce deciduomata, at least four are necessary (55). As 0.5 to 1.0 mg. of progesterone are needed to maintain pregnancy, these figures tally very well (54), though it must be remembered that smaller litters with fewer corpora lutea are successfully brought to term in the normal female. Other estimates of the amount of progesterone required for implantation have varied. One gave 1.25 mg. per day (56), another, 2.0 mg. (57), and another, 6 to 19 mg. (58). Yet another account gave 1.0 mg. daily for implantation and 3 to 5 mg. daily to maintain pregnancy, depending on the number of embryos (59).

Relaxin can be detected in the blood serum as early as 3 days after coitus. It increases rapidly from the twelfth to the twenty-fourth day, when it remains constant for the remainder of pregnancy. It declines rapidly at parturition and has practically disappeared within 3 days. At its maximum, 1 G.P. unit is present in 0.1 cc. of serum. It can also be detected in increasing amounts in the urine throughout gestation (60).

The uterine muscle gives strong, but irregular, contractions when the doe is in heat. At 22 hours after ovulation these contractions change rather abruptly to small irregular ones, and, instead of being sensitive to pituitrin, the muscle becomes insensitive (61). Removal of the corpora lutea quickly restores the sensitivity (62).

The pituitaries of 10-day-old rabbits contain no gonadotrophic hormone, but at 15 to 21 days follicle-stimulating hormone can be detected. At 28 days luteinizer also is present (63). In view of the fact that coitus causes the release of hormones from the pituitary and the ripening of follicles, the content of gonadotrophic hormone before and after coitus has attracted much attention. Before coitus the rabbit's pituitary contains about 30 ovulating units per gland. Half an hour after coitus this has dropped 80 per cent, and, in another trial, after 24 hours it has risen from 1.5 to 5 units (64, 65). The ovulating unit used here is the amount required to ovulate a rabbit per kg. body weight. Some authors have used this term for the amount required to cause ovulation in a rabbit weighing about 3 kg. After ovulation the content rises rapidly during pseudopregnancy, reaching a maximum of 60 units at 10 days and falling to 15 at the end. The curve during pregnancy is similar, but the peak is reached at 16 days, and the minimum at 25 days (64). The maximum during pseudopregnancy varies with the season; it is low in fall and winter and high in the spring (65). After mating there is a general reduction of the level of hormones in the pituitary, F.S.H., L.H., thyrotrophic, and adrenotrophic (66), but lactogenic hormone does not change (67). Occasionally a pseudopregnant rabbit will copulate. When this occurs ovulation does not follow and the pituitary does not lose gonadotrophic hormone (68). The number of granulated acidophils and basophils in the pituitary is greatest during heat. It falls rapidly after coitus, the basophilic degranulation being greatest, and then rises parallel with the gonadotrophic hormone content (69). Removal of the gonads increases the potency of the pituitary, and the effect is more marked in the female than in the male. The potency in the female increases 100 per cent in 3 months, and it is accompanied by a strikingly increased basophilia (70). However, another investigation did not record such a great increase in potency (71).

The estrous rabbit excretes about 5 R.U. of estrogen per liter of urine. Excretion rises to a maximum of 25 R.U. at 11 to 15 days during pseudopregnancy. In pregnancy the level at this time is 100 R.U. and a maximum of 180 R.U. is reached at 16 to 20 days, after which the amount declines rapidly to 10 R.U. at term (72).

Free progesterone appeared in the peripheral blood 100 minutes after mating; then its level fluctuated widely and with very little regularity for a day and a half, after which it remained unchanged for a week (73). Progesterone and its metabolites are present in the ovarian vein of the rabbit during heat and their levels rise after ovulation (74). In mid-pregnancy the ovarian-vein progesterone level was 2.36 μ/ml.; then there was a gradual decline to 0.41 μ/ml. by 2 days before parturition. The levels in the peripheral blood followed the same trend (75).

The prolactin content of the pituitary of immature females is 140 B.U. per gram of tissue; in the mature female it rises to 420 B.U. (76). In pseudopregnant rabbits it rises to a peak at 5 days, and if the does are suckling the level is 66 per cent higher than if they are not suckling (77).

Immature 4-week-old rabbits do not give an ovarian response to pituitary hormones (78). In adult rabbits the minimal ovulating dose of anterior pituitary extract is 0.1 R.U., and of prolan, 2.0 R.U., but in the rabbit hypophysectomized immediately before injection 25 per cent and 50 per cent more were needed, respectively. The latter figures should approximate the minimum amount released from the pituitary after coitus (79). Another report gives 5 M.U. of prolan as the optimal ovulating dose in the intact animal (80). During pregnancy 10 R.U. of prolan are needed to cause ovulation. If the injection is made at 11 days, there is no interruption of pregnancy, but at 17 days all gestations are interrupted (81). By properly spaced injections of prolan (82) or anterior pituitary extracts (83, 84) toward the end of pseudopregnancy, this condition can be greatly prolonged, since new waves of corpora lutea are developed.

In the mature rabbit superovulation is not produced by the injection of P.M.S., probably because the latter causes the immediate rupture of those follicles which are already mature and the luteinization of others (85). Superovulation may be caused in both juvenile and adult does with F.S.H. prepared by tryptic digestion of anterior pituitary extracts. The number of embryos may thus be greatly increased, but few of them survive, with the result that the number alive in late pregnancy is less than usual (86). The use of horse pituitary extract, which is very high in F.S.H., has produced similar results (87).

Ovulation after electrical stimulation through the head may be inhibited by the injection of 0.5 mg. of estradiol benzoate or 10 mg. of testosterone propionate (88). The injection of 150 R.U. of estrogen daily at 3 to 4 days after coitus prevents implantation, although if the injections are delayed to 5 to 6 days, it is not prevented. Smaller doses at 3 to 4 days reduced the number of implantations (89). The activity of the corpus luteum in producing glandular hypertrophy is inhibited by the injection of 600 I.U. of estrogen; 360 I.U. prevent nidation of the ova, but at 5 to 6 days 720 I.U. are needed to terminate the pregnancy. From 12 days on, 240 I.U. cause abortion (90). If the injection of estrogens is performed within the first 5 days of pseudopregnancy 1,000 R.U. are needed to terminate this condition, and, since 675 R.U. suppress the progestational activity of 3 Rab.U. of progesterone, it may be concluded that the rabbit produces more than 3 Rab.U. of progesterone in the first 5 days of pseudopregnancy (91).

Uterine motility in the doe in heat is rapidly inhibited by the injection of progesterone. It ceases in 4 hours with 0.3 Rab.U., in 2 hours with 0.6 Rab. U., and in 55 minutes with 1.2 Rab.U. (92). On the other hand the muscle in the ovariectomized doe is quiescent by 2 to 3 days after operation. The injection of 2 to 5 R.U. of estrogen per kg. of body weight restores the motility within 24 hours, but during pseudopregnancy 1,090 R.U. per kg. are ineffective (93). This response to estrogen can first be observed 10 hours after the injections are made, and the metabolic rate of the tissue rises with the restoration of motility. Hyperemia of the vascular bed was maximal within 30 minutes of injection (94). With the lack of response to pituitrin as a criterion of progesterone activity it was found that 5 Rab.U. gave complete inhibition in all cases, whereas less than 1 Rab.U. never gave complete inhibition. Above this minimum the number of rabbits which failed to respond to pituitrin gradually increased (95). At 1 to 3 days post coitum 5 μ of estrogen given for 3 days interrupts the pregnancy. From 2 to 5 days 10 μ are needed, and from 3 to 7 days the amount is increased to 15 μ daily (96).

Carbonic anhydrase is low in the endometrium in the estrous doe. The amount rises on the fourth day after mating and reaches its maximum at 8 days. This enzyme was not found in the endometria of several other species examined (97).

The pseudopregnant uterine proliferation is maintained by 0.2 to 0.4 Rab.U. of progesterone daily for the first 6 days; thereafter 1.5 Rab.U. are required daily to 11 days. After this time increased dosages did not maintain the proliferation, but if 20 R.U. of estrogen were also injected it was maintained. It was also found that in the normal animal the injection of estrogens will maintain the pseudopregnant corpora lutea for longer than the usual 16 days (91). This work has led to the view that the corpus luteum needs estrogen for its maintenance. It has been found that 1 μg. of estradiol will maintain luteal function in rabbits (98), and that, after removal of the gravid horn in unilaterally pregnant rabbits, 3 μg. of estradiol benzoate a day prevent the involution of the corpora lutea, even beyond the usual term (99). This reaction does not depend on the pituitary since almost similar doses are effective in the hypophysectomized rabbit (100). It is believed, also, that one of the functions of the uterus is to produce estrogens, which maintain the corpora lutea during the second half of pregnancy. After hysterectomy in the first half of this period, they last for the usual pseudopregnant time; but if the operation is performed in the second half they decline precipitously, though they are preserved if estrogens are injected (101).

The blood supply to the uterus is believed to depend to some extent upon the acetyl-choline content of the tissue. It rises in the uteri of spayed does treated with estrogens, but the rise is transient, lasting only 6 hours or less (102.)

The injection of large amounts of estrogen causes the tube-locking of ova, a condition which also occurs under similar circumstances in the rat. On the other hand, the injection of progesterone hastens their passage (103). Our knowledge of tubal motility leaves a good deal to be desired.

Certain data upon the effects of progesterone upon the uterus were given in the discussion of the output of the corpus luteum. However, it is known that estrogen is equally necessary, as the following results show: A dose of 0.5 mg of progesterone twice daily is necessary for implantation after ovariectomy; after 11 days 1.0 mg. twice daily will carry the pregnancy to term

(104). According to another account 2.0 I.U. daily are needed from 11 to 15 days, and after that 4.0 I.U. (104), or 5.0 mg. (106); but if a trace (1:1,600) of estrone is added, only 0.5 mg. is needed (107). During pregnancy the serum progesterone increases rapidly for 12 days. Then a slow increase follows until term when the level reaches 10 μg./ml. (108). Pregnanediol appears in the urine immediately after mating and it may be detected for 1 to 3 days. Then it appears again at 8 to 10 days and remains detectable until just before parturition (109).

Parturition is delayed if a new set of corpora lutea is caused to develop late in pregnancy by the injection of gonadotrophic hormones. The injection of 1.5 I.U. of progesterone daily at, and after, the normal time of parturition has the same effect. The fetuses continue to grow but do not live beyond the thirty-fifth day *in utero*. Parturition usually becomes abnormal (110). The injection of estrogens at the rate of 150 I.U. daily toward the end of pregnancy also maintains the corpora lutea and produces the same effect, but if the doe has been ovariectomized parturition follows despite the injections (111).

In the hypophysectomized rabbit it has been found that injected progesterone and estrogens have the usual effects upon the endometrium, but the effect is not so great as in the ovariectomized, injected rabbit whose pituitary is intact (112). The same results were obtained with estrogen alone (113) and with progesterone alone (114), although it has also been found that a greater amount of progesterone is needed under these conditions to produce a given effect (115). Pregnancy may also be prolonged in the hypophysectomized rabbit by the combined injection of 1.9 to 2.5 mg. of progesterone and 2 μg. of estradiol benzoate daily (116).

The pH of fluids secreted in isolated segments of the uteri of rabbits brought into heat by the injection of stilbestrol averaged 7.8, with a range from 7.73 to 7.90 (117). During heat the amount of tubal secretion is 0.79 ml./24 hours, per tube and at 46.0 cm. H_2O pressure. It is reduced late in pregnancy and after ovariectomy (118). The amount secreted decreases in 2 to 3 days and levels off at 50 per cent of the rate in estrus (119). This fluid flows towards the ovary during the 3 or 4 days in which the utero-tubal junction is closed after ovulation. This flow is instrumental in aiding the rapid transport of spermatozoa through the tube. The junction is closed as long as estrogens are active. When progesterone nullifies their action the junction opens (120, 121). The upper vagina has a pH of 8.1; the cervix, 7.9; the uterine horn, 7.6; the isthmus, 7.4; and the infundibulum, 7.8. At the same time the fluid in the ripe follicle has a pH of 7.3 (122).

Within the uterus the spermatozoa travel at the rate of 0.5 to 3.3 mm. per minute and they probably travel by their own power (123). They reach the oviduct at about 3 hours after coitus. By 6 hours 5,000 are present in the whole oviduct. This number remains fairly constant. The utero-tubal junction is an effective barrier to the entry of great numbers (124).

The dewlap is apparently a secondary sexual character in rabbits, as it is rarely present in the males but usually present in the females of certain breeds. Its development may be inhibited or its involution may be caused by ovariectomy. In males its development may be induced by the injection of estrogens (125).

The anal and inguinal scent glands are the same in both sexes and are capable of developing under the influence of both estrogens and androgens, except that the inguinal ones react differently to the two hormones (126).

THE MALE

The average ejaculate of the buck rabbit is 0.32 cc. with 200 million spermatozoa per cc. (127). The amount and concentration vary with the breed, probably because of differences in body size. Thus the small Polish buck produced a mean of 0.48 g. and a total of 44.2 million spermatozoa, while the Flemish giant and blue Vienna bucks produced 0.94 g. and 139.8 million spermatozoa (128). The gelatinous mass in the semen varies from 0.34 g. to 4.29 g., the average being 1.67 ±.36 g. This substance contains but little fructose but does contain an appreciable amount of estrogens (129). The inositol content of the semen is 28 mg./100 ml. (130). Obtained from the vas deferens the spermatic fluid has a pH of from 6.2 to 7.35. The average pH of semen is 7.3 to 7.4 (range 6.75 to 8.1) (131). Another average gives the semen pH as 7.2 (132).

The number of spermatozoa needed to fertilize some ova depends somewhat upon the amount of fluid introduced. If fewer than 1 million are inseminated the chance of fertilization is reduced; if the number used is below 10,000 fertilization does not occur at all (133). If 0.2 ml. of fluid is used, 90,000 spermatozoa are sufficient to give maximum fertility (134).

Spermatozoa must remain in the female tract for about 5 hours before they attain their full capacity for fertilization. This is known as "capacitation." The egg is mature at ovulation. The spermatozoa take about 4 hours to reach the egg (135).

There is no compensatory hypertrophy of the testis after unilateral castration, and the number of spermatozoa is related to the testis weight. Thus, in the animal with one testis only half the usual number of spermatozoa are produced (136).

The spermatozoa pass through the epididymis in 4 to 7 days (137, 138). No vaginal plug is formed at copulation. The spermatozoa find their way into the uterus by penetration, not by suction or peristalsis (137), but they are transported up the uterus to the tubo-uterine junction in a short while, less than 5 minutes (139), and this is said to involve peristaltic action (140).

The amount of testosterone needed to cause a buck castrated early in life to copulate is 10 mg. per day (141). The threshold dose of testosterone propionate to maintain the accessory organs in the castrate is between 0.25 and 2.5 mg. daily (142). The excretion of 17-ketosteroids per 48 hours in the male is $2.3 \pm .1$ μg. In the castrate it is $1.35 \pm .1$ μg., and in the cryptorchid, $1.40 \pm .97$ μg. (143).

The normal male pituitary contains more gonadotrophic hormones than that of the female; after castration the potency increases 36 per cent in 3 months, in contrast to the 100 per cent increase recorded for the female (70). The pituitary of the normal male is higher than that of the female in both F.S.H. and L.H. content. At 10 days of age there is no gonadotrophic hormone at all. At 15 to 21 days there is follicle stimulator only, although from 25 days onward both hormones are present (63). The prolactin content of the pituitary is increased if estrogen is injected into males, but 5,000 I.U. injected over a period of 10 days is less effective than 500 or 1,000 I.U. (144).

The length of seminiferous tubules in one testis was 70 meters (145). The prenatal proportion of males, from 20 days to term, was $50.74 \pm .76$ (146).

1. Myers, K., and W. E. Poole. Australian J. Zool., 10: 225–267, 1962.
2. Myers, K., and W. E. Poole. Nature, 195: 358–359, 1962.
3. Hammond, J., and F. H. A. Marshall. Reproduction in the Rabbit. Edinburgh, 1925.
4. Hill, M., and W. E. White. J. Physiol., 80: 174–178, 1934.
5. Smelser, G. K., A Walton, and E. O. Whetham. J. Exp. Biol., 11: 353–363, 1934.
6. Marshall, F. H. A. Essays in Biology, in Honor of Herbert M. Evans. Berkeley, Calif., 1943. Pp. 381–385.
7. Ancel, P., and P. Bouin. CRSB., 67: 497–498, 1909.
8. Fee, A. R., and A. S. Parkes. J. Physiol., 70: 385–388, 1930.
9. Fee, A. R., and A. S. Parkes. J. Physiol., 67: 383–388, 1929.
10. Smith, P. E., and W. E. White. J. Am. Med. Assn., 97: 1861–1863, 1931.
11. Marshall, F. H. A., and E. B. Verney. J. Physiol., 85: 12P, 1935.
12. Marshall, F. H. A., E. B. Verney, and M. Vogt. J. Physiol., 97: 128–132, 1939.
13. Emmens, C. W. JE., 2: 63–69, 1940.
14. Loeb, L. J. Am. Med. Assn., 53: 1471–1474, 1909.

15. Walton, A., and J. Hammond. Brit. J. Exp. Biol., 6: 190–204, 1928.
16. Desaive, P. AB., 59: 31–146, 1948.
17. Templeton, G. S. United States Dept. Interior, Wildlife Circ. 4, 1940.
18. Asdell, S. A., and J. Hammond. AJP., 103: 600–605, 1933.
19. Hammond, J. J. Exp. Biol., 11: 140–161, 1934.
20. Chang, M. C. J. Exp. Zool., 121: 351–382, 1952.
21. Pincus, G. PRS., 107B: 132–167, 1930.
22. Hammond, J. Zuchtungskunde, 3: 523–547, 1928.
23. Kopec, S. AR., 27: 95–118, 1924.
24. Hammond, J., and S. A. Asdell. Brit. J. Exp. Biol., 4: 155–185, 1927.
25. Asdell, S. A. Brit. J. Exp. Biol., 1: 473–486, 1924.
26. Nachtsheim, H. Zeit. Zucht., 33B: 343–408, 1935.
27. Rosahn, P. D., H. S. N. Greene, and C. K. Hu. Science, 79: 526–527, 1934.
28. Martin, E. A. Hojas Divulgadoras, Madrid, 34(23): 1942.
29. Wishart, J., and J. Hammond. J. Agric. Sci., 23: 463–472, 1933.
30. Ocetkiewicz, J., J. Kawinska, and T. Bednorowski. Rocz. Nauk Rolniczych, 77B: 889–903, 1961.
31. Stephens, M. N. PZS., 122: 417–434, 1952–3.
32. Allen, P., F. W. R. Brambell and I. H. Mills. J. Exp. Biol., 23: 312–331, 1947.
33. Brambell, F. W. R., and I. H. Mills. J. Exp. Biol., 24: 192–210, 1947.
34. Brambell, F. W. R., and I. H. Mills. J. Exp. Biol., 25: 241–269, 1948.
35. Poole, W. E. CSIRO (Australia), Wildlife Res., 5: 21–43, 1960.
36. Watson, J. S. New Zealand J. Sci. Tech., 38B: 451–482, 1956–7.
37. Duke, K. L. J. Morphol., 69: 51–81, 1941.
38. Pinto-Nunes, J. CRSB., 111: 598–599, 1932.
39. Pincus, G., and E. V. Enzmann. J. Morphol., 61: 351–383, 1937.
40. Hill, R. T., E. Allen, and T. C. Kramer. AR., 63: 239–245, 1935.
41. Togari, C. Folia Anat. Japonica, 4: 337–363, 1926.
42. Diomidova, N. A., and N. A. Kuznecova. Biol. Z., 4: 243–245, 1935.
43. Desaive, P. AB., 60: 357–407, 1949.
44. Corner, G. W. CE., 575: 1–8, 1948.
45. Snyder, F. F. AR., 32: 242, 1926.
46. Kunde, M. M., and T. Proud. AJP., 88: 446–452, 1929.
47. Hamilton, C. E. AR., 110: 557–571, 1951.
48. Imai, I. Hiroshima J. Med. Sci., 9: 61–81, 1960.
49. Hammond, J., and F. H. A. Marshall. PRS., 87B: 422–440, 1914.
50. Parker, J. H. TRS., 219B: 381–419, 1931.
51. Klein, M. Bull. d'Histol Appl., 10: 327–354, 1933.
52. Andersen, D. H. AJA., 42: 255–305, 1928.
53. Bluazzi, M. Arch. Chem. Biol., 18: 409, 1933.
54. Corner, G. W. Cold Spring Harbor Symposia, 5: 62–65, 1937.
55. Brouha, A. AB., 45: 571–609, 1934.
56. Chambon, Y. CRSB., 143: 1172–1175, 1949.
57. Hafez, E. S. E., and G. Pincus. PSEBM., 91: 531–534, 1956.
58. Chang, M. C. E., 48: 17–24, 1951.
59. Kehl, R., and Y. Chambon. CRSB., 142: 674–676, 1948.

60. Marder, S. N., and W. L. Money. E., 34: 115–121, 1943.
61. Knaus, H. Arch. f. Gyn., 138: 201–216, 1929.
62. Klein, M., and L. Klein. CRSB., 112: 821–824, 1933.
63. Saxton, J. A., and H. S. N. Greene. E., 24: 494–502, 1939.
64. Hill, R. T. J. Physiol., 83: 129–136, 1934.
65. Friedman, M. H., and G. S. Friedman. E., 24: 626–630, 1939.
66. Saxton, J. A., and H. S. N. Greene. E., 30: 395–398, 1942.
67. Friedman, M. H., and S. R. Hall. E., 29: 179–186, 1941.
68. Makepeace, A. W., G. L. Weinstein, and M. H. Friedman. E., 22: 667–668, 1938.
69. Wolfe, J. M., D. Phelps, and R. Cleveland. AJA., 55: 363–405, 1934.
70. Smith, P. E., A. E. Severinghaus, and S. L. Leonard. AR., 57: 177–195, 1933.
71. Wolfe, M. J. AJA., 50: 351–357, 1932.
72. Beerstecher, E. E., 31: 479–480, 1942.
73. Forbes, T. R. E., 53: 79–87, 1953.
74. Hilliard, J., E. Endröczi, and C. H. Sawyer. PSEBM., 108: 154–156, 1961.
75. Mikhail, G., M. W. Noall, and W. M. Allen. E., 69: 504–509, 1961.
76. Reece, R. P., and C. W. Turner. Missouri Agric. Exp. Sta., Res. Bull. 266, 1937.
77. Meites, J., and C. W. Turner. E., 31: 340–344, 1942.
78. Hertz, R., and F. L. Hisaw. AJP., 108: 1–13, 1934.
79. White, W. E., and S. L. Leonard. AJP., 104: 44–50, 1933.
80. Paduceva, A. L. Trud. Dinam. Razvit., 11: 187–197, 1939.
81. Snyder, F. F., and H. Koteen. PSEBM., 41: 432–434, 1939.
82. McPhail, M. K. J. Physiol., 79: 118–120, 1933.
83. Asdell, S. A. Unpublished work.
84. Parkes, A. S. PRS., 104B: 189–197, 1936.
85. Cole, H. H. AJA., 59: 299–331, 1936.
86. Warwick, E. J., R. L. Murphree, L. E. Casida, and R. K. Meyer. AR., 87: 279–296, 1943.
87. Parkes, A. S. JE., 3: 268–279, 1942–4.
88. Zondek, B., and J. Sklow. E., 28: 923–925, 1941.
89. Pincus, G., and R. E. Kirsch. AJP., 115: 219–228, 1936.
90. Reynaud, R. Thesis, Algiers, 1934.
91. Allen, W. M. Cold Spring Harbor Symposia, 5: 66–83, 1937.
92. Allen, W. M., and S. R. M. Reynolds. Am. J. Obst. Gyn., 30: 309–317, 1935.
93. Reynolds, S. R. M., and W. M. Allen. AJP., 102: 39–55, 1932.
94. McLeod, J., and S. R. M. Reynolds. PSEBM., 37: 666–668, 1938.
95. Makepeace, A. W., G. W. Corner, and W. M. Allen. AJP., 115: 376–385, 1936.
96. Greenwald, G. S. J. Exp. Zool., 135: 461–482, 1957.
97. Lutwak-Mann, C. JE., 13: 26–38, 1955–6.
98. Robson, J. M. Quart. J. Exp. Physiol., 29: 159–164, 1939.
99. Klein, M. CRSB., 130: 929–931, 1939.
100. Robson, J. M. J. Physiol., 95: 83–91, 1939.
101. Greep, R. O. AR., 80: 465–477, 1941.
102. Reynolds, S. R. M. J. Physiol., 95: 258–268, 1939.
103. Anderes, E. Schweiz Med. Woch., 71: 364–366, 1941.
104. Pincus, G., and N. T. Werthessen. AJP., 124: 484–490, 1938.

105. Allen, W. M., and G. P. Heckel. AJP., 125: 31–35, 1939.
106. Courrier, R., and A. Jost. CRSB., 130: 726–729, 1939.
107. Jost, A. Ann. Physiol. Physicochem. Biol., 15: 1065–1086, 1939.
108. Zarrow, M. X., and G. M. Neher. E., 56: 1–8, 1955.
109. Verly, W. G., I. F. Somerville, and G. F. Marrian. Bioch. J., 46: 186–190, 1950.
110. Heckel, G. P., and W. M. Allen. Am. J. Obst. Gyn., 35: 131–137, 1938.
111. Heckel, G. P., and W. M. Allen. E., 24: 137–148, 1939.
112. Asdell, S. A., and H. R. Seidenstein. PSEBM., 32: 931–933, 1935.
113. Robson, J. M. J. Physiol., 84: 148–161, 1935.
114. Robson, J. M. J. Physiol., 84: 296–301, 1935.
115. Reynolds, S. R. M., W. M. Firor, and W. M. Allen. E., 20: 681–682, 1936.
116. Robson, J. M. J. Physiol., 97: 517–524, 1940.
117. Shih, H. E., J. Kennedy, and C. Huggins. AJP., 130: 287–291, 1940.
118. Bishop, D. W. AJP., 187: 347–352, 1956.
119. Mastroianni, L., and R. C. Wallach. AJP., 200: 815–818, 1961.
120. Black, D. L., and S. A. Asdell. AJP., 192: 63–68, 1958.
121. Black, D. L., and S. A. Asdell. AJP., 197: 1275–1278, 1959.
122. Zimmermann, W. Zool. Anz., Suppl. 24: 143–149, 1960.
123. Adams, C. E. JE., 13: xxi–xxii, 1959.
124. Braden, A. W. H. Australian J. Biol. Sci., 6: 693–705, 1953.
125. Hu, C. K., and C. N. Frazier. PSEBM., 38: 116–119, 1938.
126. Coujard, R. Rev. Canadienne Biol., 6: 3–26, 1947.
127. Macirone, C., and A. Walton. J. Agric. Sci., 28: 122–134, 1938.
128. Frölich, A., and O. Venge. Acta Agric. Suecana, 3: 83–88, 1948.
129. Mukherjee, D. P., M. P. Johari, and P. Bhattacharya. Nature, 168: 422–423, 1951.
130. Hartree, E. F. Bioch. J., 66: 131–137, 1957.
131. Bishop, D. W., and H. P. Matthews. Science, 115: 209–211, 1952.
132. Sergin, N. P. Probl. Zivotn., No. 12: 100–122, 1935.
133. Walton, A. PRS., 101B: 303–315, 1927.
134. Cheng, L. P., and L. E. Casida. PSEBM., 69: 36–39, 1948.
135. Austin, C. R., and A. W. H. Braden. Australian J. Biol. Sci., 7: 179–194, 1954.
136. Edwards, J. PRS., 128B: 407–421, 1939.
137. Florey, H., and A. Walton. J. Physiol., 74: 5P–6P, 1931.
138. Nishikawa, Y., and Y. Waida. Bull. Nat. Inst. Agric. Sci., Japan, Ser. G: 69–81, 1952.
139. Krebhiel, R. H., and H. P. Carstairs. AJP., 125: 571–577, 1939.
140. Rossman, I. AR., 69: 133–149, 1937.
141. Fremery, P. de, and M. Tausk. Acta Brevia Neerl. Physiol., 7: 164–165, 1937.
142. Bern, H. A. AJA., 84: 231–277, 1949.
143. Kumeldorf, D. J. E., 43: 83–88, 1948.
144. Meites, J., and C. W. Turner. PSEBM., 49: 190–193, 1942.
145. Bascom, K. F., and H. L. Osterud. AR., 31: 159–169, 1925.
146. Mills, I. H. JE., 12: ix–xi, 1955.

Brachylagus idahoensis Merriam

PYGMY RABBIT

This rabbit has from 5 to 8 young and probably 2 litters a year (1) usually born from late May to early August (2).

1. Bailey, V. NAF., 55, 1936.
2. Hall, F. R., and K. R. Kelson. The Mammals of North America. New York, 1959.

Rodentia

THE rodents comprise a large number of species, and their habits are so diverse that it is not surprising to find a variety of reproductive patterns among them. In many cases it is very difficult to observe their reproduction owing to their nocturnal or subterranean habits.

In the Sciuridae probably most species are polyestrous during the spring and summer, with an anestrous period during hibernation. One species, the thirteen-lined ground squirrel, is known to ovulate only after coitus, and probably many others have the same habit. Apparently there is no heat or ovulation until lactation is finished, and this reduces the possible annual number of litters.

The Geomyidae seem to have a bewildering variety of patterns, even in one genus. Some seem to be polyestrous all the year, some are for a limited period, and some are said to have only one litter a year. Possibly the habitat may have much influence upon their behavior. Many of them live in deserts or on their fringes.

The Heteromyidae are nearly all polyestrous for most of the year, but reproduction appears to be in abeyance in late summer and fall.

The Cricetidae and Muridae are alike in being mainly polyestrous all the year, especially under laboratory conditions. In the wild, however, litters tend to be few in July and August and again in the winter. At the latter time the short duration of light appears to be a more injurious factor than the lowered temperature. In species such as these, with an intense rate of metabolism, food supply is probably a major factor in reproduction. A post-parturition heat is the general rule in both families but does not occur in all species, even within a genus. This is especially true in *Peromyscus*. When pregnancy occurs at this time and the females are lactating, delayed implantation and prolongation of gestation is the rule, especially among the smaller species. If the female does not become pregnant, lactation anestrum is the rule. Ovulation is spontaneous, and the corpus luteum of the cycle is nonfunctional, though it is not clear whether this is so in the Gerbillinae. Pseudopregnancy has been recorded in both families, but very few species have been investigated to ascertain whether it is general. Apparently confinement reduces the litter size as, in general, the litter size of the wild female is larger than that of the laboratory or zoo female.

The Caviidae and related families have a longer cycle than the Cricetidae and the Muridae. Probably in all of them the corpus luteum of the cycle is functional, as it is in the guinea pig. The gestation period, in all species in which it is known, is exceptionally long for their size.

APLODONTIDAE

Aplodontia rufa Rafinesque

MOUNTAIN BEAVER

The female breeds first in her second year, but she probably has a sterile cycle in her first year. There is one heat period, in February, and 1 litter a year, born after a gestation period of 28 to 30 days. Ovulation is spontaneous. The vaginal smear is cornified at the time of heat (1).

The ripe follicle measures about 3 mm., and the ovum, 90 to 100 μ outside the zona pellucida. Examination of 12 pregnant females gave averages of 2.8 corpora lutea and 2.4 embryos. The ovulations were evenly spread between the two ovaries, but 18 embryos were in the left horn of the uterus and 11 in the right. The whole reproductive tract is hypertrophied at estrus. There is no bursa ovarii, but at heat the fimbria enlarge; at this time, too, the vulva is greatly swollen. The corpus luteum increases in size nearly to term; then it regresses quickly (1).

The testes and accessory organs are largest in midwinter and they begin to retrogress in February and March. In the second year of life these organs remain larger than they were in the first. The testes are abdominal except in the breeding season, when they are semiscrotal. The spermatozoa measure 100 μ in length (2).

1. Pfeiffer, E. W. JM., 39: 223–235, 1958.
2. Pfeiffer, E. W. AR., 124: 629–637, 1956.

SCIURIDAE

SCIURINAE

Sciurus carolinensis Gmelin

GRAY SQUIRREL

In Illinois this squirrel has two breeding seasons in the year—about January 1 to 10 and June 15 to 25 in the southern part of the state, and January 20 to 30 and July 5 to 15 in the northern part (1). In Texas the winter season begins in December, the summer one in late May or early June (2). In Kentucky the times of breeding are similar, with larger litters often in the winter season (3). Introduced into Great Britain the species thrives. It experiences anestrum from September to January and has two breeding peaks, in March and June (4). Introduced into British Columbia the species breeds from March 3 to April 8 and again from June 15 to July 18, about the same as in Great Britain but later than in more southern regions (5). In South Africa, also, there are 2 litters a year, in spring and summer, with the main season from October to January (6).

The young do not breed during their first year (6), but there is a difference between the seasons of those born in the previous spring and summer. Males of the former group remain sexually active for 6 to 8 months while the summer-born males are active for only 3 months (7). In mature squirrels the ovary does not regress much during anestrum.

In a limited series the number of ripe follicles at the time of heat was found to average 4.7; the corpora lutea of pregnancy, 3.6; and the number of embryos, 3.6 (4). In Texas the litter size averaged 2.7, range 1 to 4, but usually 2 to 3, and the period of gestation was 44 days (2). In Great Britain the average spring litter size was 2.5 and the autumn, 3.2. Embryo counts in spring gave 2.2, mode 2, and in autumn the mode was 3. Implantations were evenly distributed between the uterine horns (6).

The size of the mature follicle is about 1.1 mm., and that of the corpus luteum of pregnancy 1.0 to 1.3 mm. The latter begins to regress at mid-pregnancy. The ovum size is 95 μ. There is very little interstitial tissue in the ovary, but a large amount of fibrous tissue is present (4).

The vaginal orifice is closed before puberty and during anestrum. The vulva swells during heat and forms a papilla 0.5 cm. wide and 1 cm. long, which probably remains for 2 weeks. During heat the vaginal epithelium hypertrophies, cornifies, and sloughs off, followed by an invasion of leucocytes. A vaginal plug is formed at copulation (4).

The uterus is bicornuate and the outer coats of the cornua anastamose for 0.5 cm. before the fusion of the circular muscle, which is unusually thick. During anestrum the stroma is dense and poorly vascularized, the glands are few and small, the epithelial cells are 7 to 12 μ high, and hyalinized blood vessels are found in the serosa. During heat all the tissues grow, the stroma becomes edematous, and the cell nuclei enlarge. Vascularity increases, the epithelial cells increase to 15 to 20 μ in height, and the lumen fills with fluid. In metestrum the endometrium becomes folded, the epithelial cells increase further to 25 to 30 μ, and the glands proliferate and become coiled though they are not secreting; they open into depressions between the epithelial folds. Changes in the cervix are similar to those in the uterine horns (4).

The oviduct is 50 to 60 mm. long and is much coiled. Some cilia are found during anestrum; during heat they increase in numbers, the epithelium is higher, and the interciliary cells are actively secreting. (4).

During the anestrous period the reduction in testis size is not great and these organs are readily retractable at any time. In January they weigh about 6.2 g. and in May, 5.7 g. (1). The seminal vesicles are relatively small, slightly branched, and much coiled; they open into the urethra. The prostate and bulbo-urethral glands are large (8). In January the latter measure 20 mm. in diameter but they decrease to 1 to 2 mm. in August (9). They begin to develop in October and reach their maximum size by the end of December (1).

In Great Britain there is no true anestrous period in the male. The seminal vesicles are bent back upon the prostate. The bulbo-urethral glands are spirally wound. The penis is sharply reflexed at the distal end. There is no fat accumulation in the interstitial cells of the testis (10).

1. Brown, L. G., and L. E. Yeager. Illinois Nat. Hist. Survey, 23: 448–536, 1945.
2. Goodrum, P. D. Texas Agric. Exp. Sta., Bull. 591, 1940.
3. Hibbard, C. W. JM., 16: 325–326, 1935.
4. Deanesly, R., and A. S. Parkes. TRS., 222B: 47–78, 1933.
5. Robinson, D. J., and I. McT. Cowan. Canadian J. Zool., 32: 261–282, 1954.
6. Shorten, M. PZS., 121: 427–459, 1951–2.
7. Kirkpatrick, C. M., and R. A. Hoffman. JWM., 24: 218–221, 1960.
8. Mossman, H. W., L. W. Lawlah, and J. A. Bradley. AJA., 51: 89–155, 1932.

9. Packard, R. L. Kansas Mus. Nat. Hist., Misc. Publ. 11, 1956.
10. Allanson, M. TRS., 222B: 79–96, 1933.

Sciurus niger L.

FOX SQUIRREL

In Illinois most old females and those that were born in the previous spring breed in December, January, and early February. Then there is little breeding until the females born in the previous summer and the older females that produce a second-season litter come in heat, usually in May and June. These seasons are about 3 weeks later in the northern part of the state than in the south. Thus, the peaks of the southern seasons are December 15 to 25 and May 25 to June 5, while in the north the corresponding dates are January 5 to 15 and June 15 to 25 (1). In Michigan the season begins in late December or early January. After a decline more breeding takes place from March to May. It increases in June but falls off in July (2). In Mexico there are also two seasons in the year. Lactating females have been taken in late March and early April, and pregnant females have been found in June (3).

In Illinois the average number of fetuses in both seasons is 2.4. Farmland squirrels have a high litter size, about 3.4, while those living on oak-hickory upland averaged 2.5 (1). In Michigan an average of 3.0 ± .1; mode, 3; and range 1 to 6 was found. The gestation period is about 45 days (2).

The male is able to breed in all months except August, but the frequency of males with degenerating testes is twice as many in summer and fall than in winter and spring. Males probably stay in breeding condition for two or more seasons (4). The degenerating bulbo-urethral gland is characterized by massive amounts of collagenous connective tissue which surrounds isolated lobules of very small inactive tubules. This is the best criterion for distinguishing between juveniles and adults in nonbreeding condition. The changes in the accessory glands are not absolutely correlated with the condition of the testes. Prostate and bulbo-urethral glands are most subject to change, and the seminal vesicles are less affected (5). The bulbo-urethral glands are large—diameter, 20 mm.—from November to early June. Then they become reduced to 1 to 2 mm. (6).

The length of the sperm head is 12.1 μ (range, 10.0 to 15.0), and its width, 11.0 μ (9.0–14.0). The midpiece width is 1.9 μ (1.3–2.5), and the tail length is 110.0 μ (98.9–115.0) (7).

1. Brown, L. G., and L. E. Yeager. Illinois Nat. Hist. Survey, 23: 448–536, 1945.
2. Allen, D. L. Michigan Dept. of Conservation, Game Division, Publ. 100, 1943.
3. Baker, R. H. Univ. Kansas Mus. Nat. Hist., Publ. 9: 125–335, 1956.
4. Kirkpatrick, C. M. AJA., 97: 229–255, 1955.
5. Mossman, H. W., R. A. Hoffman and C. M. Kirkpatrick. AJA., 97: 257–301, 1955.
6. Packard, R. L. Univ. Kansas Mus. Nat. Hist., Misc. Publ. 11, 1956.
7. Hirth, H. F. J. Morphol., 106: 77–83, 1960.

Sciurus vulgaris L.

EUROPEAN RED SQUIRREL

The common European red squirrel breeds at 8 to 10 months old (1). In Russia the breeding seasons are from January to March and from mid-May to June, but in the south four ovulations a year may occur, extending the breeding season through March and April and occasionally into August and September. The inference may be drawn that this species is polyestrous under favorable conditions (2). In England pregnant females have been found in every month from December to July, but in August most of the females taken indicated by the state of the reproductive tract that anestrum was approaching (3). The observed gestation period is 38 days; in this instance 6 young were born (4).

Sexually active males are found in England from November to July inclusive, possibly longer (3).

1. Heptner, W. G., L. G. Morosowa-Turowa, and W. I. Zalkin. Die Säugetiere in der Schutzwaldzone. Berlin, 1956.
2. Labacev, S. V. ZZ., 13: 280–291, 1934.
3. Rowlands, I. W. PZS., 108B: 441–443, 1938.
4. Eibl-Eibesfeldt, I. Zeit. Tierpsychol., 8: 370–400, 1951.

Sciurus aberti Woodhouse. This squirrel breeds from May to August, varying with the altitude. It probably has 2 litters a year, with 3 to 4 the usual litter size (1, 2).

S. alleni Nelson. Females with 2 embryos each have been found in March and lactating ones in April (3). Specimens with 4 embryos each have also been reported in July and August (4).

S. apache Allen. A record, dated July, of a female with 3 embryos has been found (1), also two in the same month with 2 and 3 fetuses each (2).

S. aureogaster Cuvier. In January males with enlarged testes have been found and, early in March, a female with 2 fetuses near term (5).

S. deppei Peters. This Central American squirrel has been recorded with 3 embryos (6).

S. gerrardi Gray. All females of this Central American squirrel taken in June were pregnant and the males were producing spermatozoa. The number of young varies from 1 to 3, but 2 is the usual number (7).

S. griseus Ord. The California gray squirrel has its young in January and February in Oregon (8), and from May to June on the Mexican border (1). In California heat may occur at any time from January to June. One litter a year of 2 to 4 young are born after a gestation period of 43 days or more (9).

S. hoffmanni Peters. This Central American squirrel has been recorded with 3 embryos (10).

S. negligens Nelson. The young of this Mexican squirrel are born in the first week of July (11).

S. poliopus Fitzinger. One record for this squirrel gives 2 embryos, one in each horn. Lactating and pregnant females have been found in August (12).

S. variegatoides Ogilby. This squirrel probably has from 4 to 6 young at a time (13).

1. Mearns, E. A. United States National Mus., Bull. 56, 1907.
2. Leopold, A. S. Wildlife of Mexico. The Game Birds and Mammals. Berkeley, Calif., 1959.
3. Baker, R. H. Univ. Kansas Mus. Nat. Hist., Publ. 9: 125–335, 1956.
4. Dice, L. R. Univ. Michigan Sci. Ser. 12: 245–268, 1937.
5. Hall, E. R., and K. R. Kelson. The Mammals of North America. New York, 1959.
6. Murie, A. Univ. Michigan Mus. Zool., Misc. Publ. 26, 1935.
7. Enders, R. K. HCMZ., 78: 385–502, 1935. Also personal communication, 1945.
8. Bailey, V. NAF., 55: 1936.
9. Ingles, L. G. California Fish and Game, 33: 139–157, 1947.
10. Enders, R. K. Personal communication, 1945.
11. Dalquest, W. W. Mammals of the Mexican State of San Luis Potosi. Baton Rouge, La., 1953.
12. Davis, W. B. JM., 25: 370–403, 1944.
13. Goodwin, G. G. AM., 87: 271–474, 1946.

Microsciurus alfori Allen

PYGMY SQUIRREL

This squirrel breeds from April to June, at least. Specimens have been found in Panama with active testes in June (1).

1. Enders, R. K. HCMZ., 78: 385–502, 1935.

Sciurillus pusillus Desmarest

Pregnant females with 2 embryos have been found in June (1).

1. Olalla, A. M. Rev. Mus. Paulista, 19: 425–430, 1935.

Tamiasciurus hudsonicus Erxleben

AMERICAN RED SQUIRREL

This squirrel ranges from the Rocky Mountains eastward. In Manitoba it mates from late March and early April and has 1 litter a year (1). In New York State two seasons are found. The first is from February to March and the second from June to July. Some adult females have 2 litters a year, others may not breed until midsummer. The yearling spring young breed at the February to March season, but most of those born in the summer wait until the second season to breed. Most males are fertile from February to September (2). The litter size is 3 to 6, and the gestation period lasts about 40 days (3). Embryo counts have given an average of 4.2 young (4).

The female has an unusually long, coiled vagina during heat. In the male Cowper's glands are minute and open to the urethra in the bulb without a penile duct. There is no bulbo-urethral gland. The seminal vesicles are excessively large. The penis is long and filiform without a baculum. The urethra has a diverticulum in the bulbar region (5). The testes descend in January, but there are no spermatozoa in the epididymis until January 20.

They decrease in size in June but may enlarge again in July. Spermatozoa are still present in September and October (4).

The dimensions of the sperm head are 14.0 μ (12.0–16.5) × 12.2 μ (9.5–14.0); the midpiece width is 2.1 μ (1.5–2.5); and the tail length is 117.0 μ (108.0–120.9) (6).

1. Seton, E. T. Life Histories of North American Animals. New York, 1909.
2. Layne, J. N. Ecol. Monogr., 24: 227–267, 1954.
3. Hamilton, Jr., W. J. The Mammals of Eastern United States. Ithaca, N.Y., 1943.
4. Hamilton, Jr., W. J. AMN., 22: 732–745, 1939.
5. Mossman, H. W., L. W. Lawlah, and J. A. Bradley. AJA., 51: 89–155, 1932.
6. Hirth, H. F. J. Morphol., 106: 77–83, 1960.

Tamiasciurus douglasii Bachman. The orange-bellied chickaree has its young in June and July. From 4 to 7 young are usual (1, 2), mostly 5 or 6. Mating is in early spring (3).

T. fremonti Audubon and Bachman. The young of this chickaree are born in spring or early summer. They number 1 to 4, and one litter a year is the rule (4).

1. Ingles, L. G. Mammals of California and Its Coastal Waters. Stanford, Calif., 1954.
2. Bailey, V. NAF., 55, 1936.
3. Leopold, A. S. Wildlife of Mexico. The Game Birds and Mammals. Berkeley, Calif., 1959.
4. Bailey, V. NAF., 53, 1931.

Funambulus palmarum L.

PALM SQUIRREL

In the vicinity of Bangalore the males are in a condition of sexual activity all the year and pregnant females may be found at any time, but most litters are born late in May and in September or January (1). In Ceylon most are born from October to May (2). From 2 to 4 is the usual litter size (3).

The weight of the testes of the adult is about 2.5 g. and there is no seasonal variation. They become functional when the pair weigh 1.12 g. Interstitial cells, also, show little seasonal variation. They grow until the testes weigh 1.8 g. At this point the seminal vesicles begin to grow and increase in proportion to the testis weight (1).

1. Prasad, M. R. N. J. Mysore Univ., 11B: 89–105, 1951.
2. Phillips, W. W. A. Ceylon J. Sci., 14B: 209–293, 1927–8.
3. Blanford, W. T. The Fauna of British India. Mammals. London, 1888–91.

Funambulus pennanti Wroughton

The females reach puberty at from 6 to 8 months. Heat lasts about a day. Three litters, averaging 3 young apiece, are produced each year. Litters of 2 and 4 are rare. The gestation period lasts 40 to 42 days. Lactation lasts for 2 months, and the females do not come into heat during this period (1).

During proestrum, which lasts for 2 days, the vaginal contents are fluid and contain mucus which tends to flow out of the vagina. The ovaries contain growing follicles. In estrus the smear has fewer squamous epithelial cells, which are the only ones present in proestrum. More cornified cells are present. Ovulation occurs at the end of this stage, when leucocytes make their appearance in moderate numbers. This period lasts for 1 day. In metestrum, which lasts for about 4 days, large numbers of round cells and oval epithelial cells appear late in the period. Postestrum is characterized by a great increase in the number of leucocytes. At the end of this period only cell debris may be found (2).

1. Banerji, A. J. Bombay Nat. Hist. Soc., 54: 335–343, 1957.
2. Arslan, M. Biologia, 5: 74–83, 1959.

Funambulus layardi Blyth. Females with 2 and 3 large fetuses have been taken in September and February (1).

1. Phillips, W. W. A. Ceylon J. Sci., 14B: 209–293, 1927–8.

Ratufa macroura Pennant

In Ceylon 3 or 4 young are born at a time (1). Another record is of a single young born in February after a gestation period that was probably 4 weeks long (2).

1. Phillips, W. W. A. Ceylon J. Sci., 14B: 209–293, 1927–8.
2. Hill, W. C. O. Spolia Zeylandica, 21: 189–191, 1938–9.

Protoxerus stangeri Waterhouse

GIANT SQUIRREL

A female with 2 minute embryos was found in December (1).

1. Allen, G. M., and A. Loveridge. HCMZ., 89: 147–214, 1941–2.

Funisciurus

Funisciurus auriculatus Matschie. In the Cameroons a female with a single embryo has been reported in February or March (1).

F. congicus Kuhl. A male with enlarged testes was found in June (2).

E. pyrrhopus Cuvier. The young, usually 2, are born at least from June to September (3).

1. Eisentraut, M. Zool. Jahrb., Syst., 85: 619–672, 1957.
2. Hill, J. E., and T. D. Carter. AM., 78: 1–211, 1941.
3. Allen, G. M., and H. J. Coolidge, Jr. In R. P. Strong, ed., The African Republic of Liberia and the Belgian Congo. Cambridge, Mass., 1930.

Paraxerus cepapi A. Smith

BUSH SQUIRREL

This African squirrel has been found with large and small embryos in January (1). It has also been found pregnant in August, October, and February (2). In one of these specimens 2 embryos were found. The litter size has been reported as from 2 to 4 (3).

1. Lawrence, B., and A. Loveridge. HCMZ., 110: 1–80, 1953.
2. Frechkop, S. EPNU., 1944.
3. Shortridge, G. C. The Mammals of South West Africa. London, 1934.

Paraxerus emini Stuhlmann. There is a December record of this squirrel with a single medium-sized fetus (1).

P. flavivittis Peters. The young are born in September (2).

P. ochraceus Huet. This squirrel has been found with 2 fetuses, in November and December (3).

1. Allen, G. M., and A. Loveridge. HCMZ., 89: 147–214, 1941–2.
2. Loveridge, A. J. East Africa and Uganda Nat. Hist. Soc., 17: 39–69, 1922.
3. Abbott, —. In G. C. Shortridge, The Mammals of South West Africa. London, 1934.

Heliosciurus

Heliosciurus gambianus Ogilby. This squirrel has from 2 to 4 young at a time (1).

H. rufobrachium Waterhouse. In December specimens with 1 and 2 large fetuses have been found. Towards the end of January very young animals were seen (2).

1. Verheyen, R. Inst. Parcs Nat. Congo Belge, Upemba, 1951.
2. Allen, G. M., B. Lawrence, and A. Loveridge. HCMZ., 79: 31–126, 1935–7.

Callosciurus

Callosciurus caniceps Gray. The females of this Malayan squirrel reach puberty at 240 g. The mean embryo count was 2.2, mode 2, range 1–5. They are equally distributed between the uterine horns. Of the males, 50 per cent are fertile at 210 ± .1 g., and 95 per cent at 259 ± 1.0 g. (1).

C. erythraeus Pallas. This squirrel has a litter of from 3 to 4 young (2).

C. hippurus Geoffroy. No pregnancies were found from April to August (3).

C. lokroides Hodgson. A female caught in March had 4 fetuses (4).

C. lowi Thomas. Three females with 2 embryos each and 1 with 3 have been found (1).

C. nigrovattus Horsefield. Females reach puberty at 190 g. or more. The mean embryo count was 2.2, mode 2, range 1 to 4. Most pregnancies were found from April to September. Of the males, 50 per cent were fertile at 149 ± .4 g., and 95 per cent at 267 ± .7 g. (1).

C. notatus Boddaert. The females reach puberty at 200 g. and up. The mean embryo count was 2.2, mode 2, range 1–4, with equal distribution in both horns. Most pregnancies were found from April to June. Of the males 50 per cent were fertile at 113 ± .7 g., and 95 per cent at 257 ± 1.0 g. (1). In Java the breeding season is in the wet monsoon (5).

C. prevosti Desmarest. Females are usually lactating during the first three months of the year (6).

C. tenuis Horsefield. Of the males 50 per cent were fertile at 50.2 ± .3 g., and 95 per cent at 88.3 ± .3 g. Embryo counts have ranged from 2 to 4 (1).

C. vittatus Raffles. This squirrel has 3 to 4 young to the litter (6).

1. Harrison, J. L. PZS., 125: 445–460, 1955.
2. Ho, H. J. Contr. Biol. Lab. Sci., China, Zool. Ser., 10: 245–287, 1933–5.
3. Wade, P. JM., 39: 429–433, 1958.
4. Carter, T. D. AM., 82: 95–114, 1943–4.
5. Dammerman, K. W. Treubia, 13: 429–470, 1931.
6. Banks, E. J. Malay Br. Royal Asiatic Soc., 9(2): 1–139, 1931.

Rhinosciurus laticaudatus Müller and Schlegel

In Malaya females with 1 and 2 embryos have been found (1).

1. Harrison, J. L. PZS., 125: 445–460, 1955.

Lariscus insignis Cuvier

In Malaya a female with 2 embryos was found (1).

1. Harrison, J. L. PZS., 125: 445–460, 1955.

Xerus

AFRICAN GROUND SQUIRREL

Xerus erythropus Geoffroy. This squirrel has 3 to 4 young to the litter (1).

X. inauris Zimmermann. This ground squirrel breeds twice a year and has from 3 to 6 young at a time. Puberty is reached at 6 months of age (2).

X. rutilis Cretzchmar. A female with 2 fetuses was found at the end of January (3).

1. Dekeyser, P. L. Les Mammifères de l'Afrique Noire Française. np., 1956.
2. Powell, W. Rodents. Dept. Public Health, South Africa, 1925.
3. Loveridge, A. J. East Africa and Uganda Nat. Hist. Soc., 17: 39–69, 1922.

Marmota marmota L.

MARMOT

This European marmot breeds about April. From the end of August onwards the testes are small, and the interstitial cells are laden with yellow-brown pigment. In March there is great testis growth, and spermatozoa may be found at the end of the month. The interstitial tissue grows and loses its pigment. At the end of April spermatogenesis ceases, but the interstitial tissue does not regress till July, when the testes return to the abdomen (1). Heat lasts for a half day (2). One female gave birth to 7 young after a gestation period of 33 to 34 days (3).

1. Courrier, R. AB., 37: 173–334, 1927.
2. Müller-Using, D. Zeit. Jagdwiss., 3: 24–28, 1957.
3. Psenner, H. Säugetierk, Mitt., 5: 4–10, 1957.

Marmota monax L.

WOODCHUCK, GROUND HOG

This North American marmot breeds at 2 years of age. Only 10 to 25 per cent breed earlier. The breeding season is from late February through March; most litters are born in mid-April and practically all by May 10 (1). The mean date of birth in south central Pennsylvania is April 10. In most years 75 to 80 per cent of the mature females are fertile. The mean litter size is about 3.9 to 4.0 (2), or 4.6 (1), but the resorption rate is about 15 per cent (2). In Canada the first litter is born about May 13, and the average number of young in newly emerged litters was 3.75 (3). There is practically no difference in the number of embryos in each horn of the uterus (4). The gestation period is 31 to 32 days (1, 5).

The interstitial cells of the ovary gradually enlarge during winter; after hibernation they become 3 to 4 times their original diameter. The maximal increase, due mainly to an accumulation of lipid and secretion granules, is found in females which do not become pregnant until late in the breeding season. The cells retrogress when corpora lutea are present. The latter persist for many weeks after parturition, but all have retrogressed by September. In the fall a number of fairly large graafian follicles may be observed (6). Young females have many polyovular follicles in their ovaries (7).

The testes show growth similar to that of the ovaries, but the changes are more sudden. The interstitial cells are at a minimum during late summer and fall, beginning to decrease in July when the testes retire to the abdomen. Free spermatozoa are present by the end of March, and regression sets in at the end of April. There is a new cycle of activity beginning in May, and the testes remain active for another two months. During the quiescent period the interstitial cells are laden with pigment, which disappears during the breeding season when these cells are at a maximum. The changes are similar to those observed in the European *M. marmota* (8).

The seminal vesicles have relatively coarse tubules, the bulbar gland is small, the penis is short, and the testes are abdominal most of the year (9).

1. Grizzell, R. A. AMN., 53: 257–293, 1955.
2. Snyder, R. J., and J. J. Christian. AR., 132: 509, 1958.
3. de Vos, A., and D. I. Gillespie. Canadian Field Nat., 74: 130–145, 1960.
4. Hamilton, Jr., W. J. Ann. Carnegie Mus., 23: 85–178, 1934.
5. Hoyt, S. F. JM., 33: 388–389, 1955.
6. Rasmussen, A. T. E., 2: 353–404, 1918.
7. Christian, J. J. JM., 31: 196, 1950.
8. Rasmussen, A. T. AJA., 22: 475–515, 1917.
9. Mossman, H. W., L. W. Lawlah, and J. A. Bradley. AJA., 51: 89–155, 1932.

Marmota bobak Müller. This Himalayan marmot mates after it emerges from hibernation. The gestation period is about 40 days; from 1 to 7 young are born (1).

M. caligata Eschscholtz. In the Yukon this marmot has its young late in the spring. Five is a usual number (2). The mating season is April and May, and the litter size, 4 or 5 (3). The age at first breeding is probably 2 years (4).

M. camtschatica Pallas. The litters are cast 1 or 2 weeks after the females emerge from hibernation. The litter varies from 3 to 11, and 5 is the usual number (5).

M. flaviventris Audubon and Bachman. This North American marmot mates in March (3) or earlier. The young are born about March 15 in the Snake River region and April 15 in the Upper Okanagan. Embryo counts have ranged from 3 to 6, with 4 to 5 as the average (6). There might be another breeding season in the fall (7).

M. sibirica Radde. This marmot hibernates from September and October to April, but spermatogenesis probably continues during this period. The female genitalia were in normal condition but no spermatozoa were observed in the vaginal smears (8).

1. Heptner, W. G., L. G. Morosowa-Turowa, and W. I. Zalkin. Die Säugetiere in der Schutzwaldzone. Berlin, 1956.
2. Rand, A. L. Canada Nat. Mus., Bull. 100, 1945.
3. Bailey, V. NAF., 55, 1936.
4. Cowan, I. McT., and C. J. Guiguet. British Columbia Provincial Mus., Handbook 11, 1956.
5. Kapitonov, V. I. ZZ., 39: 448–457, 1960.
6. Couch, L. K. The Murrelet, 11: 3–6, 1930.
7. Barnes, C. T. Utah Mammals. Salt Lake City, Utah, 1927.
8. Rjabov, N. I. ZZ., 27: 245–256, 1948.

Cynomys leucurus Merriam

PRAIRIE DOG

This prairie dog is sexually active at 1 year (1). According to one account the mating season is from the end of March into April. It must be short, because the reduction in testis size is very abrupt in April (2). Another account gives the end of April as the season of birth and the last week of January as the beginning of the mating season, which lasts for 2 or 3 weeks (3). The uteri have two cervices and, after copulation, a huge plug forms in them. There is a closure membrane across the vagina in the nonbreeding season (1). The ovum is 90 μ in diameter (4). The gestation period is 28 to 32 days. The litter size varies from 2 to 10, mode 5 and 6, and mean $5.5 \pm .2$ (1). In October and November the scrotum becomes pigmented and by January the testes have descended (3).

1. Stockard, A. H. Papers of Michigan Acad. Sci., Arts and Letters, 11: 471–479, 1929.
2. Stockard, A. H. JM., 10: 209–212, 1929.
3. Smith, R. E. Univ. Kansas Mus. Nat. Hist., Misc. Publ. 16, 1958.
4. Stockard, A. H. Papers of Michigan Acad. Sci., Arts and Letters, 22: 671–689, 1936.

Cynomys ludovicianus Ord

PRAIRIE DOG

Males and females of this species are anestrous from June to October. The females begin to experience heats in December, and the vaginal smears become cornified in late January and early February. The testes begin to grow in October; spermatozoa are present in January. Mating is in late January or early February and the young are born early in March after a gestation of 30 to 35 days (1). The reproductive tract is involuted from March to October (2). The litter size varies from 2 to 10, mode 5, mean 5.0 ± .2 (3). In South Dakota 66 litters averaged 4.9 (4). In the London zoo, where births were registered from March to June, the average was 3.7 and the mode was 2 (5).

According to another account the females are monestrous but the vaginal changes that accompany estrus are slow. The vulva is closed in the nonbreeding season. It enlarges and becomes pigmented late in January and begins to diminish by March 1. When it is possible to take vaginal smears they consist of nucleated epithelial cells with a few leucocytes. This stage lasts for 3 or 4 days. Then cornified cells gradually appear; this stage of mixed cell types lasts for from 1 to 3 days. The smear then changes rapidly to one with all cornified cells, but it is not the cheesy mass that characterizes the next stage, which soon appears and lasts for 1 to 2 days. Leucocytes follow and are present along with cornified cells for 1 to 2 days. Then the smear consists of nucleated epithelial cells and leucocytes, a condition that lasts until the next breeding season (6).

In the ruptured follicle there are prominent ingrowths of theca interna and theca externa which are well vascularized. There is no hemorrhage into the cavity. At the same time the endometrium thickens and the number of glands increases, owing to mitoses in the surface epithelium which grows inward. The mature corpus luteum contains many cells that display an intense basophilia. Placental scars persist for a long time. Changes in light or in temperature did not modify the season of breeding (6).

1. Anthony, A., and D. Foreman. Physiol. Zool., 24: 242–248, 1951.
2. Anthony, A. J. Morphol., 93: 331–369, 1953.
3. Wade, O. JM., 9: 149–151, 1928.

4. King, J. A. Univ. Michigan Contrib. Lab. Vert. Biol., 67, 1955.
5. Zuckerman, S. PZS., 122: 827–950, 1952–3.
6. Foreman, D. AR., 142: 391–405, 1962.

Cynomys gunnisoni Baird. The young of this prairie dog are born from mid-April to the end of the month (1) or into May (2). The litter size varies rather widely. Up to 8 young may be born at a time. One average of placental scars was 3.9 (3). An embryo count gave 4.8 per litter (1).

C. mexicanus Merriam. A female with 3 embryos has been found in March (4).

1. Aldous, S. E. JM., 16: 129–131, 1935.
2. Holdenried, R., and H. B. Morlan. AMN., 55: 369–381, 1956.
3. Longhurst, W. JM., 25: 24–36, 1944.
4. Baker, R. H. Univ. Kansas Mus. Nat. Hist., Publ. 9: 125–335, 1956.

Citellus beecheyi Richardson

CALIFORNIA GROUND SQUIRREL

The California ground squirrel is apparently polyestrous since in southern California it breeds at any time of the year. In northern California the breeding season is from February to mid-April. The testes decrease in size in May, and by June they are in the abdomen. They are found to be enlarging again in mid-October and are in the scrotum by mid-November. Probably each female has but one litter a year. The number of embryos varies from 4 to 15, and a very extensive series of counts gave a mean of 8.2 (1). Another gave a spread of 4 to 11, and 7.2 as the mean. In February the average litter size was 6.9, in March 7.3, in April 7.5, and in May 6.8. In the uplands birth is later than in the lowlands. The gestation period is probably 30 days (2). The young mature by the year following their birth and make up nearly half the breeding population (3).

1. Evans, F. C., and R. Holdenried. JM., 24: 231–260, 1943.
2. McCoy, G. W. United States Public Health Rpts., 27: 1068–1072, 1912.
3. Fitch, H. S. AMN., 39: 513–596, 1948.

Citellus columbianus Ord

COLUMBIAN GROUND SQUIRREL

This ground squirrel, which inhabits an area in British Columbia, Alberta, Washington, Idaho, and Montana, begins to breed about March 15 to 20 (1). They mate soon after they emerge from hibernation. The young are born early in May at low altitudes and may be as much as a month later in higher elevations. One litter a year is usual (2). Heat lasts for 2 to 3 days if the female is unbred, and it returns in 14 to 15 days, continuing this pattern into May. Gestation lasts 23 to 24 days, usually the latter (3). The number of young varies from 2 to 7, with 2 to 5 usually, and an average of 3.5 (1). Another count gave an average of 5.8 embryos (4).

1. Howell, A. H. NAF., 56, 1938.
2. Manville, R. H. JM., 40: 26–44, 1959.
3. Shaw, W. T. JM., 6: 106–113, 1925.
4. Couch, L. K. The Murrelet, 13: 25, 1932.

Citellus leucurus Merriam

ANTELOPE SQUIRREL

This squirrel breeds in early April. It may have a second litter in July (1). In Death Valley it breeds as early as the latter part of March, and young have been found on April 23 (2). The litter size is from 5 to 14, averaging about 9 (3).

1. Bailey, V. NAF., 53, 1931.
2. Grinnell, J. Proc. California Acad. Sci., Ser. 4, 23: 115–169, 1935–47.
3. Grinnell, J., and J. Dixon. California Commission of Horticulture Monthly Bull., 7: 597–708, 1918.

Citellus richardsoni Sabine

This ground squirrel mates in mid-April (1) and the young are born in mid-May (2). One litter a year is the rule; its size varies from 2 to 11, with a mode of 6 and 7. The gestation period is from 28 to 32 days (2), but a parturition following an observed mating gave 17 to 18 days if the female were not already pregnant (3).

1. Seton, E. T. Life Histories of Northern Animals. New York, 1909.
2. Howell, A. H. NAF., 56, 1938.
3. Denniston, R. H. JM., 38: 414–416, 1957.

Citellus suslicus Güldenstaedt

SUSLIK

This ground squirrel has one litter a year. The mating season is immediately after hibernation and it lasts for 15 to 20 days (1). In the Lublin region heat occurs before mid-April and pregnancy lasts until May or June (2). The litter size is usually 4 to 7, but as many as 12 may be born. The gestation period lasts for 22 to 26 days (1). In Byelorussia 80 per cent of the mature females participate in reproduction. They emerge from hibernation at the end of March and breed at once (3).

1. Heptner, W. G., L. G. Morosowa-Turowa, and W. I. Zalkin. Die Säugetiere in der Schutzwaldzone. Berlin, 1956.
2. Surdacki, S. Acta Theriol., 2: 203–234, 1958.
3. Petrovsky, Y. T. ZZ., 40: 736–748, 1961.

Citellus townsendii Bachman

In the State of Washington this ground squirrel mates at the end of January and the breeding season extends over two months. The young are born from mid-February to the end of March. One litter is produced during the year; the gestation period lasts for at least 23 days. The litter size is up to 10, with 4.75 as the average (1). A compilation of available data gave 4 to 16 as the range of embryo counts, mode 8 and 10, mean $9.5 \pm .2$. Fecundity probably varies from year to year.

1. Svihla, A. The Murrelet, 20: 6–10, 1939.

Citellus tridecemlineatus Mitchill

THIRTEEN-LINED GROUND SQUIRREL

This common ground squirrel inhabits central North America. It is one of the small group of animals known to ovulate only after coitus, but, from their behavior in the field, others of this genus probably fall into this group. Owing to the difficulty of keeping it in good reproductive health in the laboratory there has been a certain amount of confusion in the record of this species, but the main points of interest are now clearly emerging.

The ovaries begin to hypertrophy about January 1, while the females are hibernating. Upon their emergence, about the third week in April, these organs are fully developed and the females are ready to reproduce. The accessory organs follow the ovaries closely in their development. Older animals emerge in a more advanced state of reproductive activity than do younger ones. No effects upon the female have been observed when light, temperature, or diet have been altered in, or before, hibernation. The state of hibernation is said to be a prerequisite for reproduction in the female, though not in the male (1). Anestrum lasts from July to April (2).

The unbred female experiences persistent heat lasting 2 to 4 weeks; these periods recur irregularly at intervals of 2 to 4 weeks. Ovulation does not occur in the absence of copulation, but with copulation it takes place after an interval of 8 to 12 hours (2). The act of copulation lasts for 10 minutes (3). The litter size is 5 to 13, usually 6 to 10 (4), and the period of gestation 28 days (5). There is no postparturition heat (2). In the wild one litter a year is born.

In Manitoba the females breed first at one year old. They probably have a second litter during the season. A hot, dry spring advances mating. Most of the young are born late in May or early in June. Embryo counts vary from 4 to 12, with a mean of 8.1 ± .1 and a mode of 8 (6).

HISTOLOGY OF THE FEMALE TRACT

During anestrum the ovaries are inactive except that there is a good deal of follicular atresia (5). Some of the corpora lutea from the previous breeding season can be seen, but they are in a state of involution. When the females emerge from hibernation, the ovaries weigh 3 to 4 times their anestrous weight. The diameter of mature follicles at the time of heat is about 0.7 mm. During the breeding season the germinal epithelium produces new ova at all times in the cycle. The follicular cavity develops as a split between the granulosa cells at one side of the follicle. The theca externa and interna cannot be distinguished until the follicle is highly cellular (7).

The vulva enlarges considerably during heat. The vagina is small and is closed during anestrum. It enlarges in January and opens in February, remaining open throughout the breeding season, until May or June. It continues large for some months following parturition (5). During anestrum the vaginal epithelium consists of a few layers of small epithelial cells; in proestrum their number increases. The smear in estrus consists of cornified cells only, but in metestrum leucocytes and cornified and small cuboidal cells are found (8).

During anestrum the uterine muscle is thin and inactive; there is a great growth in proestrum and during heat (2,5). The cells are low columnar or cubical in anestrum and high columnar in heat (5). Glands are scanty and nonsecreting in anestrum, and their lumens are small or absent; in proestrum new glands appear by invagination of the epithelium (2). During heat they are coiled and are secreting (5). At ovulation and during metestrum the epithelial cells decrease in size, as they are actively secreting, while leucocytes appear in the stroma and make their way between the cells (2).

The oviduct is ciliated, with cells 12 to 18 μ in height during anestrum. At the time of heat they have increased to 25 to 30 μ, but again drop to 17 μ at 18 to 24 hours postovulation, after which there is a slight increase to 21 μ (2).

The pituitary gland undergoes seasonal variations in weight and in cell types. Basophils are larger and more numerous during the breeding season, and the granules are also large and numerous. These cells dedifferentiate to chromophobes during the summer. There are no sex differences in cell population. Gonadectomy increases the proportion of basophils, but this effect is prevented or abolished by the injection of gonadic hormones (1, 9).

The adrenal cortex, mainly the reticular zone, hypertrophies during the breeding season and also at other times when the gonads are stimulated by gonadotrophes (1).

PHYSIOLOGY OF THE FEMALE TRACT

Removal of the ovaries during pregnancy results in resorption of embryos or in abortion. Injection of corpus luteum extracts prevents resorption (10). Removal of the uterus during the cycle has no effect upon the ovaries (11). Gonadectomy during anestrum results in a slightly greater degree of atrophy of the accessory organs, so they must secrete a little hormone during this time. The ovaries respond to the injection of anterior pituitary extracts at any time of the year, and estrogens are effective upon the accessory organs of the ovariectomized female at all times (1).

The anterior pituitary is high in gonadotrophes during the breeding season, but they cannot be detected during anestrum. However, gonadectomy increases the potency whenever it is performed (1).

Injection of 100 R.U. of estrogen twice a week for 12 weeks and then three times a week for 4 weeks induced metaplastic changes in the uterine cornua and glands and in the urethra (12).

THE MALE

The males emerge from hibernation about 2 weeks before the females; the testes at this time are large and scrotal and already contain spermatozoa. By July 1 they have declined in size and have returned to the scrotum. Spermatozoa are no longer present, and the accessory organs have regressed (1). In the laboratory spermatozoa can be found in the adult testis by December; in younger animals they appear in January or February. If breeding males are kept at 4° C. through the spring and summer, they do not undergo sexual retrogression (13).

The pituitary of the male is high in gonadotrophic hormones during the breeding season, when spermatozoa are in the testis, and is extremely low during anestrum. Bilateral castration and low temperatures prevent the fall in potency. In prepuberal males the level is low until a few weeks before spermatozoa appear (13).

Large doses of male urine, P.U., or P.M.S. injected into hypophysectomized males all cause testis enlargement and spermatogenesis. The dose of P.M.S. was 20 to 40 R.U. daily, of Antuitrin S (P.U.) it was 25 to 100 R.U. daily. The shortest time required to produce spermatozoa was 19 days, the optimum length of treatment 30 days (14).

In the normal anestrous male pituitary implants cause descent of the testes in the anestrous period (15), but daily injections of 5 R.U. of Antuitrin S (P.U.) for 12 to 16 days caused enlargement of the testes, spermatogenesis, and development of the accessory organs (16). Spermatogenesis in the ground squirrel is more readily induced by gonadotrophes than it is in any other animal known. Gonadotrophes in doses as low as 2.5 R.U. daily caused some stimulation, even early in anestrum. Bull testis extract at the rate of 16 to 12.5 B.U. daily and male urine at the rate of 6 to 15 B.U. had similar effects, inducing spermatogenesis, but not until late November. The amount of androsterone needed was 1.5 mg. daily (17).

A gonad-stimulating substance is present in the blood serum of the male during the breeding season (18).

The injection of estrogens inhibits the growth of the testes; in castrates it causes growth of connective tissue and muscle in the accessory organs and cornification of the glandular tissue (18).

The anatomy of the male accessories has been studied in detail. The seminal vesicles are large and open separately into the prostatic urethra. The bulbar gland is enormous. The urethra is very broad and muscular with about 13 valvelike folds. The glans penis is broad with a flattened end. It bears hooks which are directed backwards from the os penis and which perforate the epithelium of the glans (19).

1. Moore, C. R., G. F. Simmons, L. J. Wells, M. Zalesky, and W. O. Nelson. AR., 60: 279–289, 1934.
2. Foster, M. A. AJA., 54: 487–511, 1934.
3. Wade, O. JM., 8: 269–276, 1927.
4. Howell, A. H. NAF., 56: 1938.
5. Johnson, G. E., M. A. Foster, and R. M. Coco. Trans. Kansas Acad. Sci., 36: 250–269, 1933.
6. Criddle, S. Canadian Field Nat., 53: 1–6, 1939.
7. Pliske, E. C. J. Morphol., 83: 263–287, 1938.
8. Coco, R. M. Proc. Louisiana Acad. Sci., 6: 83–84, 1942.
9. Hoffman, R. A., and M. X. Zarrow. AR., 131: 727–736, 1958.
10. Johnson, G. E., and J. S. Challans. E., 16: 278–284, 1932.
11. Drips, D. AJA., 25: 117–184, 1919.
12. Wells, L. J., and M. D. Overholser. AR., 78: 43–57, 1940.
13. Wells, L. J. E., 22: 588–594, 1938.
14. Wells, L. J., and M. D. Overholser. AR., 72: 231–247, 1938.
15. Johnson, G. E., E. L. Gann, M. A. Foster, and R. M. Coco. E., 18: 86–96, 1934.
16. Baker, B. L., and G. E. Johnson. E., 20: 219–223, 1936.
17. Wells, L. J., and C. R. Moore. AR., 66: 181–200, 1936.
18. Wells, L. J. AR., 62: 409–447, 1935.
19. Mossman, H. W., L. W. Lawlah, and J. A. Bradley. AJA., 51: 89–155, 1932.

Citellus undulatus Pallas

ARCTIC SUSLIK

This circumpolar species breeds in the early spring (1) and the young are born before mid-June (2) or a little later. In Arctic Alaska the season is very short and all are breeding in the last two weeks of May. The testes are developing in May and are abdominal in late June (3), when they decrease in size (4). Spermatozoa are present by May 20 (3). The gestation period is about 25 days and the litter size averages 7.8 (5). Another account gives a litter size from 5 to 12, mode 6, and mean 7.5 (6). At the height of the breeding season the seminiferous tubules measure 150 μ in diameter (3).

1. Rausch, R. The Murrelet, 34: 18–26, 1953.
2. Rand, A. L. Canada Nat. Mus., Bull. 99, 1945.
3. Mitchell, O. G. JM., 40: 45–53, 1959.
4. Bee, J. W., and E. R. Hall. Univ. Kansas Mus. Nat. Hist., Misc. Publ. 8, 1956.
5. Mayer, W. V., and E. T. Roche. Growth, 18: 53–69, 1954.
6. Macpherson, A. H., and T. H. Manning. Canada Nat. Mus., Bull. 161, 1959.

Citellus armatus Kennicott. In Yellowstone Park the young are born during the second half of May (1). No pregnant females were found in July or August and the males had no spermatozoa in their testes (2).

C. beldingi Merriam. One litter is born each year, from April 10 to 20 (3). In the subspecies, *C. b. oregonus,* which lives at high altitudes, the young are born in June (4). The litter size varies from 4 to 12 with 8 as the average (3).

C. citellus L. This suslik has a gestation period of 25 days (5).

C. eversmanni Brandt. The young are born as the females emerge from hibernation, i.e., about the second week of May (6). In the Altai Mountains there are indications of a fairly long breeding season but it is over by July (7).

C. franklini Sabine. This species breeds in the last week of April and the young are born early in July. One litter a year is the rule. The number of embryos varies from 4 to 10, with 8 as the mode and mean (8). According to another account the time of birth is May and June (4).

C. fulvus Lichtenstein. This suslik mates immediately after it emerges from hibernation. The gestation period is about 1 month and the litter size varies from 4 to 13, with 5 to 9 as the most frequent range and 6 the mean (9).

C. harrisii Audubon and Bachman. This species of Arizona and New Mexico mates from mid-January to mid-March. Four records each gave 6 embryos (4).

C. interpres Merriam. Lactating females have been found in March (10).

C. lateralis Say. The mantled ground squirrel has one short annual breeding season with a mating period of about 4 weeks (11). This begins in April (12) and the young are born in June or July (3). Embryo counts have varied from 2 to 8, mode 5, mean 5.5 ± .2.

C. major Pallas. This suslik mates immediately on emerging from hibernation (9).

C. mexicanus Erxleben. In Texas this species breeds at the end of March. The testes descend in March and April but begin to reascend by the beginning of May. One female with 19 embryos has been found (13), and another with 5 in June (10).

C. mohavensis Merriam. This species breeds from early to mid-March. One female was found with 6 embryos (3).

C. nelsoni Merriam. In California the young are born in the period of green feed, i.e., earlier than April 16. There is no evidence of another litter in the dry season (14).

C. pygmaeus Pallas. This suslik comes in heat only in the spring soon after hibernation. In Stavropol matings are over by May 15. From March to May the ascorbic acid content of the ovary averages 61.5 mg. per cent (ranges 20 to 98 mg.). In June and July it averages 141.2 mg. per cent (range 95 to 205) (15).

C. relictus Kashkarov. Some females reproduce at one year old. The breeding season is a long one (16).

C. spilosoma Brandt. This ground squirrel of western North America has two litters a year, the second one about August (17). The litter size varies from 5 to 8, with 6 and 7 the mode. Embryo counts in June and July have given 4.6 as the mean and 4 the mode (10).

C. tereticaudus Baird. The young are born in March or April but the females may have two litters a year (3). The litter size varies from 3 to 12. According to another account there is only one litter (18).

C. variegatus Erxleben. The Mexican rock squirrel mates from May to June and again in August or September. It apparently has two litters a year. The litter size is from 5 to 7 (19). In New Mexico a female with 5 embryos has been found in April (20).

C. washingtoni Howell. This species mates early in February. The litter size varies from 5 to 11, with 8.0 the average (21).

1. Wade, O. The Murrelet, 12: 76–78, 1931.
2. Negus, C., and J. S. Findley. JM., 40: 371–381, 1959.
3. Howell, A. H. NAF., 56, 1938.
4. Bailey, V. NAF., 55, 1936.
5. Kenneth, J. K. Gestation Periods. Edinburgh, 1943.
6. Allen, J. A. AM., 19: 101–184, 1903.
7. Hollister, N. Proc. United States Nat. Mus., 45: 507–532, 1913.
8. Sowls, L. K. JM., 29: 113–137, 1948.
9. Heptner, W. G., L. G. Morosowa-Turowa, and W. I. Zalkin. Die Säugetiere in der Schutzwaldzone. Berlin, 1956.
10. Baker, R. H. Univ. Kansas Mus. Nat. Hist., Publ. 9: 125–335, 1956.
11. Tevis, L. AMN., 53: 71–81, 1955.
12. Mullaby, D. P. JM., 34: 65–73, 1953.
13. Edwards, R. L. JM., 27: 105–115, 1946.
14. Hawbecker, A. C. JM., 28: 115–125, 1947.
15. Kratinov, A. G., A. M. Poljakova, E. A. Torbina, and A. T. Skirina. Bjull. eksp. Biol. Med., 22(1): 59–62, 1946.
16. Stroganova, A. S., and L. Chu. Trud. Zool. Inst. Akad. Nauk SSSR., 29: 81–100, 1961.
17. Davis, W. B., and J. L. Robertson, Jr. JM., 25: 254–273, 1944.
18. Stephens, F. California Mammals. San Diego, Calif., 1906.
19. Bailey, V. NAF., 53: 1931.
20. Holdenried, R., and H. B. Morlan. AMN., 55: 369–381, 1956.
21. Scheffer, T. H. JM., 22: 270–279, 1941.

Tamias striatus L.

EASTERN CHIPMUNK

The chipmunk is probably polyestrous; it breeds from March onward, and young born late in the previous season may mate and produce young in July and August. The number of young is 3 to 5, and the gestation period, 31 days (1). In some districts the litter size may be higher, as an average of 5.02 has been reported (2). Puberty is reached at the age of 2½ to 3 months (3), but in the area of Wisconsin around Madison females born in April or May often breed in July while their male litter mates do not mature until the following March or April (4). The vagina is swollen in March. A female that killed her young had a new heat about a week later. The testes are descended by the second week of February (5). In New York they remain large until early August, then gradually reduce and re-enter the body cavity by the end of September or October (6). In Wisconsin the testes are already abdominal by late July (7).

1. Hamilton, Jr., W. J. The Mammals of Eastern United States. Ithaca, N.Y., 1943.
2. Condrin, J. M. JM., 17: 231–235, 1936.
3. Burt, W. H. Univ. Michigan, Misc. Publ. 45, 1940.
4. Mossman, H. W., R. A. Hoffman, and C. M. Kirkpatrick. AJA., 97: 257–301, 1955.
5. Allen, E. G. New York State Mus., Bull. 314, 1938.
6. Yerger, R. W. AMN., 53: 312–323, 1955.
7. Panuska, J. A., and N. J. Wade. JM., 38: 192–196, 1957.

Eutamias amoenus Allen

CHIPMUNK

This western chipmunk has one short annual breeding season and the young reproduce in the season following that of their birth. The mating period, which lasts about 4 weeks (1), is in April; the young are born in late May or early June. In one season 84 per cent of the adult females were pregnant or lactating (2). Embryo counts have ranged from 4 to 8. One record gave an average of 5.8 (2), another 5.0 (1). A male with spermatozoa was found in August (3).

1. Tevis, L. AMN., 53: 71–78, 1955.
2. Broadbooks, H. E. Univ. Michigan Mus. Zool., Misc. Publ. 103: 5–42, 1958.
3. Negus, N. C., and J. S. Findley. JM., 40: 371–381, 1959.

Eutamias minimus Bachman

LITTLE NORTHERN CHIPMUNK

This chipmunk has its young before mid-June (1). No sexual activity was found in July or August (2). However, if the young are killed when only a few days old the mother mates again and a second litter may be born in late June. In captivity they may have 3 litters in the season, the last late in September (3). The litter size varies from 3 to 7, with 5 as the mode and a mean of 5.4 ± .14 (3, 4).

1. Hall, E. R. Mammals of Nevada. Berkeley, Calif., 1946.

2. Negus, N. C., and J. S. Findley. JM., 40: 371–381, 1959.
3. Criddle, S. Canadian Field Nat., 57: 81–86, 1943.
4. Anderson, S. Univ. Kansas Mus. Nat. Hist., Publ. 9: 405–414, 1959.

Eutamias sibiricus Laxmann

This species breeds early in the year, in late April or early May, at the time of emergence from hibernation (1). The young are born at the end of May or early in June (2). The gestation period is from 35 to 40 days and the number of young varies from 3 to 12, with 4 or 5 as the usual number (1). The mature graafian follicles measure 270 to 340 μ and the ovum 71 to 88 μ in diameter within the vitellus and zona pellucida. Few polyovular follicles are to be found (3).

1. Heptner, W. G., L. G. Morosowa-Turowa, and W. I. Zalkin. Die Säugetiere in der Schutzwaldzone. Berlin, 1956.
2. Suigirevskaya, E. M. ZZ., 41: 1395–1401, 1962.
3. Kang, Y. S. Univ. Seoul Coll. Theseon Sci. Nat., 4: 169–176, 1956.

Eutamias bulleri Allen. This Mexican species has been found lactating in late July and early August (1).

E. cinericollis Allen. This chipmunk probably has one litter a year (2). The young, 3 to 5 in number, are born in June and July (3).

E. dorsalis Baird. This species has two litters a year (2). Birth of young, from 2 to 8, has been recorded in July (3) and lactating females in early April (1) and June (4).

E. merriami Allen. The young are born in May or early June (3).

E. quadrimaculatus Gray. This species has one short annual breeding season; the young reproduce in the season following that of their birth. The mating period lasts about 4 weeks (5), and the young are born in May after a gestation lasting 31 days (6).

E. quadrivattus Say. In New Mexico this chipmunk breeds early; lactating females have been caught in April. At this time the testes are large and in the scrotum (7). In favored districts two litters a year may be produced, with the first mating in April. From 2 to 6 young at a time are usual (2).

E. speciosus Merriam. This chipmunk has one short annual breeding season with a mating period that lasts about 4 weeks. The young reproduce in the season following their birth. Four and 5 embryos have been reported (5).

E. townsendii Bachman. This species has one short annual breeding season lasting about 4 weeks (5). It occurs in March and April and the young are born in May or June (7). Embryo counts from 3 to 6 have been recorded, with 5 the usual. The young breed in the season following their birth (5).

E. umbrinus Allen. Lactating females have been found in July. In one of these there were 7 placental scars (8).

1. Baker, R. H. Univ. Kansas Mus. Nat. Hist., Publ. 9: 125–335, 1956.
2. Bailey, V. NAF., 53: 1931.
3. Mearns, E. A. United States Nat. Mus., Bull. 56, 1907.
4. Rand, A. L. Canada Nat. Mus., Bull. 99, 1945.
5. Tevis, L. AMN., 53: 71–78, 1955.
6. Ross, R. C. JM., 11: 76–78, 1930.
7. Bailey, V. NAF., 55: 1936.
8. Negus, N. C., and J. S. Findley. JM., 40: 371–381, 1959.

PETAURISTINAE

Petaurista

FLYING SQUIRREL

Petaurista magnificus Hodgson. This Indian squirrel breeds in the rainy season and has a single young (1).

P. petaurista Pallas. This squirrel appears to breed all the year. A single young is the rule (2).

P. volans L. The Russian flying squirrel has one or two litters, each of 2 to 4 young, in summer (3).

1. Blanford, W. T. The Fauna of British India. Mammals. London, 1888–91.
2. Banks, E. J. Malay Br. Royal Asiatic Soc., 9(2): 1–139, 1931.
3. Heptner, W. G., L. G. Morosowa-Turowa, and W. I. Zalkin. Die Säugetiere in der Schutzwaldzone. Berlin, 1956.

Glaucomys sabrinus Shaw

FLYING SQUIRREL

This flying squirrel probably has two litters a year. The first breeding season is in late winter and the young are born late in March or to early

May. The second season is in midsummer (1). In the northern part of its range one litter a year, born late in May or early in June, is usual (2). The gestation period is one month and 3 to 6 young, usually 4, is the rule (3).

1. Hamilton, Jr., W. J. Mammals of Eastern United States. Ithaca, N.Y., 1943.
2. Cowan, I. McT. JM., 17: 58–60, 1936.
3. Rand, A. L. Canada Nat. Mus., Bull. 100, 1945.

Glaucomys volans L.

FLYING SQUIRREL

The North American flying squirrel may breed several times a year, probably usually twice. Mating begins in late February and early March, and it has been observed again in July. The testes descend in late January or early February and are retracted in September. The vagina is closed by a membrane which opens during heat. At this time the vulva swells to five times its former size, and the vaginal epithelium cornifies. In an isolated female the vulva remained in a swollen state for 5 days, subsided for 11 days, and then again became swollen. She did not have another heat until the following year. The period of gestation is 40 days, and the litter size 1 to 4, usually 3 to 4, with a mean of 3.1 (1). A compilation of available litter sizes gave a mean of 3.4 ± .1. There seems to be considerable embryo wastage (2).

The male genital system is large in proportion to the body size. The testes are very large and thick, the seminal vesicles and bulbar glands are large, the prostate and Cowper's glands are very large. The penis has a very long, saw-edged baculum 20 mm. in length (3).

1. Sollberger, D. E. JM., 24: 163–173, 1943.
2. Uhlig, H. G. JM., 37: 295, 1956.
3. Mossman, H. W., L. W. Lawlah, and J. A. Bradley. AJA., 51: 89–155, 1932.

Iomys horsfieldi Waterhouse

Records of a female with 1 embryo (1) and of another with 2 (2) have been found.

1. Davis, D. D. Fieldiana Zool., 39: 119–147, 1958.
2. Harrison, J. L., and B. L. Lim. Bull. Raffles Mus., 23: 300–309, 1950.

GEOMYIDAE

GEOMYINAE

Geomys bursarius Shaw

POCKET GOPHER

This pocket gopher of North America probably has two litters a year with a breeding season that produces young from February to August. Embryos have first been seen in females collected on February 14. Corpus-luteum counts show that the season is at its height from February to July. It then falls off to October, and no corpora lutea are present from November to January. There is no relationship between the position of the testes and the breeding season, but the rate of spermatogenesis is low from July to October (1). In northeast Colorado one litter a year is usual and most females are pregnant in April and May, when the young are born. In May 86 per cent were either pregnant or lactating. In July 3 per cent of females were pregnant, and in August 5 per cent (2).

The number of young in nest varies from 1 to 4, with a mean of 2.7. Embryo counts have been from 1 to 5, with 2.7 average. Placental scars averaged 2.5 (1). Another compilation gave 1 to 6 as the litter range with a mean of 3.8 and a mode of 4 (3). In the Colorado series the litter size was 1 to 8, with 68 per cent carrying 3 or 4 embryos. Seven per cent of the females had reabsorbing embryos (2).

Puberty is reached two months after the end of the reproductive season, i.e., at age 3 months (1).

The oviduct is somewhat peculiar; the isthmus is very sharply and regularly coiled, forming a conical mass with its apex at the junction with the ampulla. It is very muscular. The ovaries are not encapsulated. The uterine horns are relatively short and thick, and are fused externally for 1.5 cm. above the two cervices (4). The theca interna is very thick, and the type of cells suggests that it is an endocrine gland. It reaches its greatest development during proestrum and heat. After ovulation it rapidly degenerates and takes no part in the formation of the corpus luteum, though it does form interstitial cells. It seems to be the only source of the latter (5).

The males produce spermatozoa at approximately 8 months old and they come into breeding condition about 2 months before the females. The testes enlarge rapidly in December and January; most adult males are fertile from January to June. Regression sets in during August and there are no spermatozoa from that month to November. The seminal vesicles and epididymides are largest in March (2). In castrated and normal males the injection of ovarian extracts (relaxin) causes the symphysis pubis to be reabsorbed. Testis grafts into ovariectomized females prevent resorption of the calcuim in the pubic bones (6).

1. Wood, J. E. JM., 30: 36–44, 1949.
2. Vaughan, T. A. JM., 43: 1–13, 1962.
3. Scheffer, T. H. Kansas Agric. Exp. Sta., Bull. 152, 1908.
4. Mossman, H. W., and F. L. Hisaw. AJA., 66: 367–391, 1940.
5. Mossman, H. W. AJA., 61: 289–319, 1937.
6. Hisaw, F. L. J. Exp. Zool., 42: 411–441, 1925.

Geomys pinetis Rafinesque

In north central Florida the females of this species breed all the year but there are spring and summer peaks in activity. Puberty is reached at about 6 months. Of 19 pregnant females 9 had 1 and 10 had 2 embryos. Combined data gave a mean of 1.5 ± .1 embryos. Fetal resorption accounts for at least 10 per cent of embryos. Placental scars averaged 1.73 ± .09. There are at least 2 litters a year (1).

During estrus, pregnancy, and just after parturition the vulva is open or lightly sealed with mucus. Copulation plugs may be found in March and April. The corpora lutea are small reddish and yellowish swellings. At the time of heat the uterus is noticeably turgid and swollen and is conspicuously vascularized. Placental scars persist for some time and overlap, as some females may be found with two sets. Most males are in breeding condition in January through June. All adult males are fertile from February through September and some in other months of the year (1).

1. Wing, E. S. JM., 41: 35–43, 1960.

Geomys arenarius Merriam. This species probably has a long breeding season. Females pregnant in June and August have contained 3 or 4 embryos (1).

G. tuza Barton. This species has one litter a year, with 3 to 4 young (2).

The sperm head measures 4.5 μ (range 4.0 to 5.0) $\times$ 2.2 μ (1.5 to 2.5); the midpiece width is 1.1 μ (1.0 to 1.2); and the tail length is 106.0 μ (95.0 to 110.0) (3).

1. Jones, J. K., and M. R. Lee. Southwestern Nat., 7: 77–78, 1962.
2. Hamilton, Jr., W. J. The Mammals of Eastern United States. Ithaca, N.Y., 1943.
3. Hirth, H. F. J. Morphol., 106: 77–83, 1960.

Thomomys bottae Eydoux and Gervais

POCKET GOPHER

In California this pocket gopher breeds throughout the year, but from August to January few pregnant females are found. Probably each female has 2 litters a year (1), and a post-partum heat is probable as a nursing female with embryos has been recorded (2). By November young females begin to show enlargement of their genital tracts and in January and February the vaginas of many of them are open. One litter a year seems to be usual; occasionally there are 2 (3). In Mexico embryos have been found in January, July, and December (4).

A series of 220 females gave an embryo distribution from 3 to 13, with 6 as the mode and 6.2 $\pm$ .1 the average (5). The embryo count increases with the size of the female, with a mean of 4.4 $\pm$ .6 at 80 to 99 g., and 6.0 $\pm$ .3 at 140 to 159 g. The embryo count tends to decrease throughout pregnancy. It is 5.8 $\pm$ .2 for small ones, 5.2 $\pm$ .3 for medium ones, and 5.4 $\pm$ .4 for large ones (1). One account gives about 30 days for the gestation period (3) and another, 19 days in two cases following observed copulations (6).

At all seasons the testis weight is very variable but it tends to be greatest in November and least from June to October. Spermatogenesis occurs in waves that seem unconnected with the time of year, and it never ceases completely. The accessory organs and interstitial cells also exhibit cycles. In the castrate male 0.5 mg. of testosterone propionate restores the accessory organs to functional condition (7). A copulation plug follows normal coitus in the wild gopher (3).

1. Miller, M. A. JM., 27: 335–358, 1946.
2. Burt, W. H. The Murrelet, 14: 42, 1933.
3. Howard, W. E., and H. E. Childs, Jr. Hilgardia, 29: 277–358, 1959.

4. Baker, R. H. Univ. Kansas Mus. Nat. Hist., Publ. 9: 125–335, 1956.
5. Bond, R. M. JM., 27: 172–174, 1946.
6. Schramm, P. JM., 42: 167–170, 1961.
7. Gunther, W. C. AMN., 55: 1–40, 1956.

Thomomys talpoides Richardson

POCKET GOPHER

The breeding season is in February and March (1), and the females may be polyestrous (2). The vulva swells during this season and the closure membrane opens. In Montana this occurs from mid-April to mid-May (3). The young are born in May or June (3), but at least two litters may be produced in quick succession during the year (4). There may be some variation, as some investigators give one litter a year as the rule (5). In Colorado the females show rapid enlargement of the uterus at 6 to 7 months when a pelvic gap is formed. The uterine horns reach their maximum size in May. The first pregnant females are found late in March and the last in June. The first parturient females are found in mid-April (6).

In one series the litter size, embryos, or placental scars numbered 4.4 ± .1 (33), but the number born alive is usually much smaller than the embryo count (4). Another count gave 2.9 ± .3 (5). Placental scars remain in evidence for about 2 months after parturition (3). In the Colorado series the embryo counts varied markedly with the locality. The range was from 6.4 ± .18 at Livermore to 4.4 ± .21 at Grand Mesa. In the Livermore series 4 per cent of embryos were resorbing (6).

In the nonbreeding season the testes are abdominal. As they enlarge with the onset of the season they descend into the inguinal canal, but only the caudal epididymis is in the scrotum. The epididymis is greatly enlarged at this time; the other accessory organs parallel the testes in their development. The latter are largest in March and April, and motile spermatozoa may be found from early March on. In northwest Montana they are present 2 weeks to a month earlier than in the Bozeman district (3). Retrogression begins in May (1). In the young males the testes increase rapidly in November and December. Just before the commencement of the breeding season the testes of both young and adults increase rapidly and before the increase in the accessory organs. Following the breeding season the testes decrease in size and become soft, flabby, and dark colored (6).

1. Scheffer, T. H. JM., 19: 220–234, 1938.
2. Bailey, V. NAF., 55: 1936.
3. Tryon, C. A. Montana Agric. Exp. Sta., Bull. 448, 1947.
4. Wight, H. M. JM., 11: 40–48, 1930.
5. Criddle, S. JM., 11: 265–280, 1930.
6. Hanson, R. M. JM., 41: 323–335, 1960.

Thomomys bulbivorus Richardson. This gopher breeds from March to July and has one litter a year of 5 to 9 young (1). One embryo count gave a mean of 4.2 (2).

T. monticola Allen. This gopher is polyestrous throughout the spring and summer (3).

T. townsendii Bachman. This species is polyestrous through the spring and summer. The litter size varies from 3 to 10, with 6 or 7 usually and 6.8 ± .4 the mean (4).

T. umbrinus Richardson. One January record gave a female with 2 embryos (5). There is also another record of the same number (6).

1. Wight, H. M. JM., 11: 40–48, 1930.
2. Scheffer, T. H. JM., 19: 220–234, 1938.
3. Bailey, V. NAF., 55, 1936.
4. Horn, E. E. JM., 4: 37–39, 1923.
5. Baker, R. H. Univ. Kansas Mus. Nat. Hist., Publ. 9: 125–335, 1956.
6. Davis, W. B. JM., 25: 370–403, 1944.

Cratogeomys

POCKET GOPHER

Cratogeomys castanops Baird. In Mexico this species has two distinct breeding seasons, from December to March and from June to August. Lactation has been observed in January, June, and July. The embryo count is 1 to 3, mean 1.8 (1).

C. fulvescens Merriam. This gopher probably has one litter a year, born in May or June (2).

C. merriami Thomas. This gopher has been reported from Mexico with 2 embryos in August and with half-grown young in the same month (2).

C. perotensis Merriam. This species probably has one litter a year, born in May or June (2).

1. Baker, R. H. Univ. Kansas Mus. Nat. Hist., Publ. 9: 125–335, 1956.
2. Davis, W. B. JM., 25: 370–403, 1944.

Heterogeomys hispidus Le Conte

POCKET GOPHER

This species probably has an extended season, and the females breed in their first year of life. In one instance 2 embryos were recorded (1).

1. Davis, W. B. JM., 25: 370–403, 1944.

HETEROMYIDAE

PEROGNATHINAE

Perognathus lordi Gray

POCKET MOUSE

Pregnant females have been found mostly in July, a few in May, June, and August. They do not appear to be pregnant in the other months. The gestation period probably lies between 21 and 28 days and there may be two pregnancies a year. The litter size, according to embryo counts, is from 2 to 8, with a mode of 5 and a mean of 5.2 ± .1 (1).

1. Scheffer, T. H. United States Dept. Agric., Tech. Bull. 608, 1938.

Perognathus baileyi Merriam. This pocket mouse of southwestern North America breeds in the spring and summer, from April to August. The number pregnant decreases in June and early July but is higher in August. The modal embryo number is 4 (1).

P. fasciatus Wied. This species breeds in the spring and summer (2), and probably has one litter a year (3). The 4 to 6 young are born after a gestation of about 4 weeks (4).

P. flavescens Merriam. Three records each of 4 embryos have been made in July and August (3).

P. flavus Baird. This species is probably polyestrous (5), and in New Mexico pregnant females have been found from February to October, with most from April to June (6). Three to 6 young are born as a rule (5).

P. formosus Merriam. In Utah this mouse has a restricted season from early April, or perhaps sooner, to early July, when atrophy of the testes and epididymides begins. Some females may have two litters in the season. The litter size is about 5 (7). In Death Valley 2 pregnant females, each with 3 embryos, have been taken in April (8).

P. hispidus Baird. This species has several litters of 4 to 7 young in the season (4).

P. longimembris Coues. In Utah this mouse has a restricted season from early April or sooner to early July, when atrophy of the testes and epididymides sets in. Some females may have 2 litters in the season, each with about 5 young (7).

P. merriami Allen. There is a June record with 4 embryos (9).

P. nelsoni Merriam. Pregnancies have been recorded from March to July. Embryo numbers have varied from 2 to 5, mean 2.8 (9).

P. parvus Peale. This mouse is probably polyestrous (10) and breeds from early May to August (11). The number of young varies from 3 to 8, usually 4 or 5, mean 5.1 ± .2 (11), or 5.2 (12).

P. penicillatus Woodhouse. This mouse breeds in the spring and summer, from May to September. The breeding intensity is less in June and early July but it recovers in August. The embryo count is from 2 to 4, mean 3.4 (1).

1. Reynolds, H. G., and H. S. Haskell. JM., 30: 150–156, 1949.
2. Warren, E. R. The Mammals of Colorado. Stillwater, Okla., 1942.
3. Bailey, V. NAF., 49, 1926.
4. Hall, E. R., and K. R. Kelson. The Mammals of North America. New York, 1959.
5. Bailey, V. NAF., 53, 1931.
6. Holdenried, R., and H. B. Morlan. AMN., 55: 369–381, 1956.
7. Duke, K. L. JM., 38: 207–210, 1957.
8. Grinnell, J. Proc. California Acad. Sci., Ser. 4, 23: 115–169, 1935–47.
9. Baker, R. H. Univ. Kansas Mus. Nat. Hist., Publ. 9: 125–335, 1956.
10. Dice, L. R. JM., 1: 10–22, 1919.
11. Scheffer, T. H. JM., 11: 466–469, 1929.
12. Scheffer, T. H. The Murrelet, 14: 51–54, 1933.

Microdipodops

PYGMY KANGAROO MOUSE

Microdipodops megacephalus Merriam. This mouse is probably polyestrous, at least until September, but most young are born in May and June (1). The number of embryos varies from 1 to 7, mean 3.9, and mode 4 (2).

M. pallidus Merriam. This mouse has from 1 to 6 embryos to a litter, with mean 3.9, and mode 4 (2).

1. Hall, E. R., and J. M. Linsdale. JM., 10: 298–305, 1929.
2. Hall, E. R., and K. R. Kelson. The Mammals of North America. New York, 1959.

DIPODOMYINAE

Dipodomys heermannii Le Conte

KANGAROO RAT

This rat of western North America has a prolonged season. The females breed at between 30 and 40 g. In the males the testes are abdominal until 50 g. weight is reached. The rats are probably less than 2 months old at this time and thus may produce at least 2 litters during the year in which they are born. The season extends from March to October but its intensity is low from August on. The vagina is closed during the nonbreeding season. At the time of heat it opens and the area surrounding the clitoris swells. The testes are within the abdomen in November and December, but they enlarge and descend in January or February (1).

The litter size is usually 3, and an average of 2.6 has been found (1). Another series gave 2.65 as the mean, with larger litters, 3.1, in spring, compared with 2.1 in the rest of the year (2, 3).

1. Fitch, H. S. JM., 29: 5–35, 1948.
2. Dale, F. H. AMN., 22: 703–731, 1939.
3. Stewart, G. R., and A. I. Roest. JM., 41: 126–129, 1960.

Dipodomys merriami Mearns

KANGAROO RAT

This species is polyestrous (1), breeding in the spring and fall (2). The early-born young probably reproduce in the year of their birth (3). At the time of heat the vulva swells and a bloody mucous discharge often appears. It remains open for 6 (3 to 11) days and then rapidly closes. These cycles recur every 25 (13 to 45) days. The gestation period lies between 17 and 23 days (4). The litter size varies from 2 to 4, averaging 3.1 (5).

The testes become scrotal and much enlarged in February and remain so until the end of September, but some males with scrotal testes may be found at any time of the year. Subadult males, probably born in March or April, have scrotal testes in July and August. The pregnancy peaks are in May and September; there are no pregnancies in winter. In a large series the embryo count averaged 2.02 ($n = 133$, range 1 to 3). There was not much seasonal difference (6).

1. Bailey, V. NAF., 53, 1956.
2. Duke, K. L. JM., 25: 155–160, 1944.
3. Hoffmeister, D. F. AMN., 55: 257–288, 1956.
4. Chew, R. M. JM., 39: 597–598, 1958.
5. Alcorn, J. R. JM., 22: 88–89, 1941.
6. Reynolds, H. G. JM., 48–58, 1960.

Dipodomys ordii Woodhouse

KANGAROO RAT

In central Utah this rat breeds from early January to March, has an anestrous period from late June until late August, and then breeds again from early September through October, when another anestrous period follows. In Nevada pregnancies have been found in May and June (1). Ovulation is spontaneous (2). Early-born young reproduce later in the year (3). The litter size varies from 2 to 4, mean 3, with birth after a gestation of 29 to 30 days (4). Another count puts the embryo number at

3.6, with 2 to 5 as the range (5). In New Mexico 2.4 was the average embryo count, and the main breeding season was from late February to early June, with April 1 the modal date for births (6); but pregnancies have been recorded for every month except December. There is a second peak of births in September (7).

In proestrum the vulva swells and it may remain enlarged for long periods, but ovulation is followed by rapid involution. The epithelium of the clitorine urethra is greatly thickened during anestrum. It is thin and atrophic in the periods of greatest ovarian activity. Ovariectomy causes involution of the vulva and thickening of the clitorine epithelium. Progesterone inhibits the effects of estrogens when they are given (8).

1. Duke, K. L. JM., 25: 155–160, 1944.
2. Pfeiffer, E. W. JM., 37: 449–450, 1956.
3. McCulloch, C. Y., and J. M. Inglis. JM., 42: 337–344, 1961.
4. Day, B. N., H. J. Egoscue, and A. M. Woodbury. Science, 124: 485–486, 1956.
5. Hoffmeister, D. F. AMN., 55: 257–288, 1956.
6. Johnston, R. F. Southwest Nat., 1: 190–193, 1956.
7. Holdenried, R., and H. B. Morlan. AMN., 55: 369–381,1956.
8. Pfeiffer, E. W. JM., 41: 43–48, 1960.

Dipodomys spectabilis Merriam

KANGAROO RAT

In New Mexico this rat breeds from December into May, with most pregnancies in March and weaning of young by August (1, 2). The first young of the season are born in January and up to 3 litters a year may be produced (2). A copulation plug is formed at mating. The litter size varies from 1 to 3 or 4, mean 1.8 ± .06, mode 2 (3). Another account gave a litter of 2.75 as the average (2).

1. Holdenried, R., and H. B. Morlan. AMN., 55: 369–381, 1956.
2. Holdenried, R. JM., 38: 330–350, 1957.
3. Vorhies, C. T., and W. P. Taylor. United States Dept. Agric., Bull. 1091, 1922.

Dipodomys deserti Stephens. In Death Valley this rat has been found with small embryos during April; lactating specimens have also been found in that month. One or 2 seems to be the usual litter size (1). Another count gave mean 3.3, mode 3. The gestation period was 29 to 32 days. A copulation plug is formed (2).

D. ingens Merriam. The giant kangaroo rat breeds from January to May. Embryo counts of 5, usually, and of 6 have been recorded (3).

D. microps Merriam. In central Utah this species breeds from early January to March; it has an anestrous period from late June to late August and breeds again in early September through October; another anestrous period follows (4). The litter size is from 2 to 4.

D. nitratoides Merriam. This rat breeds most of the year; 2 is the average litter size (5).

1. Grinnell, J. Proc. California Acad. Sci., Ser. 4, 23: 115–169, 1935–47.
2. Butterworth, B. B. JM., 42: 413–414, 1961.
3. Grinnell, J. JM., 13: 305–320, 1932.
4. Duke, K. L. JM., 25: 155–160, 1944.
5. Culbertson, A. E. JM., 27: 189–203, 1946.

HETEROMYINAE

Liomys

SPINY POCKET MOUSE

Liomys pictus Thomas. This mouse of southwestern North America breeds at any time of year, but usually in the spring or early summer. From 3 to 5 young, usually 4, is the rule (1).

L. salvini Thomas. This species breeds all the year and usually has 4 young (2). There are records of January (3), February, and March pregnancies with 2 and 3 embryos (4).

1. Hall, E. R., and K. R. Kelson. The Mammals of North America. New York, 1959.
2. Goodwin, G. G. AM., 87: 271–474, 1946.
3. Burt, W. H., and R. A. Stirton. Univ. Mich. Mus. Zool., Misc. Publ. 117, 1961.

Heteromys

Heteromys desmarestianus Gray. Three or 4 young seem to be usual (1, 2). The testes are active in January (3), and pregnant females have been found in July (2).

H. gaumeri Allen and Chapman. One record has been found of the capture in October of a female with 2 embryos.

H. oresterus Harris. The number of young varies from 3 to 5, but 4 is usual (5).

1. Murie, A. Univ. Michigan Mus. Zool., Misc. Publ. 26, 1935.
2. Kuns, M. L., and R. E. Tashian. JM., 35: 100–103, 1954.
3. Enders, R. K. HCMZ., 78: 385–502, 1935.
4. Hatt, R. T. JM., 19: 333–337, 1938.
5. Goodwin, G. G. AM., 87: 271–474, 1946.

CASTORIDAE

Castor canadensis Kuhl

BEAVER

There is little knowledge of the breeding habits of the beaver, mainly because of its aquatic habitat. The females do not breed until they are 2 years old, and then not all of them do (1). Mating usually occurs in January or February and the young are born in April or May (2). However, in some districts the time of breeding must be more irregular. In Texas nearly full-term fetuses have been found in February (3), and in the state of Washington births have been recorded in July and September (4). In the London zoo births have been distributed from May through September, with a peak in June (5). The number of young born is usually 3 to 4 but varies from 1 to 6. In one set of data placental scars averaged 3.4 and fetuses, 2.9 (6); in another, placental scars averaged 4.0 and embryo counts, 3.6 (6). One account gives the mean embryo number as 3.9, mode 3 and 5 (1), and another gives a litter size of 4.1 (8). The London zoo birth average was 1.8 per litter (5).

Gestation has been given as about 3 months but a pair seen copulating gave a probable period of 128 days (9). The sex ratio of fetuses is approximately equal, 45 males to 47 females (10, 11). The corpus luteum is 6 to 8 mm. in diameter. The corpora albicantia persist at least until the next breeding season (12).

In the male the uterus masculinus is prominent and very variable in size. The epididymis and glandular epithelia are least developed when the seminal vesicles weigh most (13).

1. Benson, S. B. Univ. Michigan Mus. Zool. Occ. Papers, 335, 1936.
2. Seton, E. T. Life Histories of Northern Animals. New York, 1909.
3. Miller, F. W. JM., 29: 419, 1948.
4. Guenther, S. E. JM., 29: 419–420, 1948.
5. Zuckerman, S. PZS., 122: 827–950, 1952–3.
6. Osborn, D. J. JM., 34: 27–44, 1953.
7. Hogdon, K. W. JWM., 13: 412–414, 1949.
8. Sanderson, G. C. Proc. Iowa Acad. Sci., 60: 746–753, 1953.
9. Bradt, G. W. JM., 20: 486–489, 1939.
10. Bradt, G. W. JM., 19: 139–162, 1938.
11. Tryon, C. A. JM., 27: 396–397, 1946.
12. Provost, E. E. JWM., 26: 272–278, 1962.
13. Conaway, C. H. JM., 39: 97–108, 1958.

Castor fiber L.

BEAVER

The Old World beaver mates from January to March and the young, numbering 3 or 4, are born in April or May after a gestation lasting 105 to 107 days. The females have one litter a year (1). The females reach puberty in their second year and most can be fecundated at 32 to 34 months old. In captivity births were from late April through July, with most in May and June. The litter size varied from 1 to 4, with a pronounced mode at 3 and a mean of 2.94. Ten per cent of the adult female population were not reproducing (2). Near Archangelsk 30 to 60 per cent of the 3-year-old and older females were reproducing and 10 per cent of the 2-year-olds (3).

The maximum development of the testes and accessory glands occurs in March. Retrogression begins in April and continues to June (2).

1. Heptner, W. G., L. G. Morosowa-Turowa, and W. I. Zalkin. Die Säugetiere in der Schutzwaldzone. Berlin, 1956.
2. Formicheva, N. I. Biull. Mosk. Obshch. Ispyt. Priody Otd. Biol., 64(3): 5–15, 1959.
3. Porovschchikov, V. Y. ZZ., 40: 466–467, 1961.

ANOMALURIDAE

ANOMALURINAE

Anomalurus

Anomalurus beecrofti Fraser. This West African squirrel has been recorded with a single embryo (1).

A. fraseri Waterhouse. This squirrel was found pregnant in the northern Cameroons in June and July only. It has 1 young (2). The penis is surmounted by a cartilaginous spear, and the vagina of the pregnant female is filled with a hard plug (2). In the Belgian Congo pregnant females have been found in February and June so that the species may have two seasons (3).

A. pelii Temminck. This squirrel has 2 litters a year, one in September. Two to 3 young are usually born (4).

1. Dekayser, P. L. Les Mammifères de l'Afrique Noire Française. np., 1956.
2. Sanderson, I. T. TZS., 24: 623–725, 1938–40.
3. Frechkop, S. EPNU., 1944.
4. Adams, W. H. PZS., 234–236, 1894.

PEDETIDAE

Pedetes

SPRING HARE

Pedetes cafer Pallas. In South West Africa this hare breeds once a year, in April (1). Usually a single young is born; twins are very rare (2).

P. surdaster Thomas. This species always has a single young (3).

1. Shortridge, G. C. The Mammals of South West Africa. London, 1934.
2. van der Horst, C. J. South African Biol. Soc., Pamphl. 8: 47, 1935.
3. Hollister, N. Smithsonian Inst., Bull. 99, 1918.

CRICETIDAE

CRICETINAE

Oryzomys palustris Harlan

RICE RAT

The rice rat of the eastern United States reaches puberty when it is about 50 days old. It is polyestrous and breeds from February to November (1). There is a post-partum heat, and 9 litters may be produced in a season (2). The cycle length averages 7.6 ± .2 days, with a range from 6 to 9 days; the duration of heat is usually 24 hours or less (3). The litter size is variable. One account gives an average of 3.0 (1), another an average of 5.0 ± .2 and a range from 4 to 6 (3), but young females have smaller litters than older ones (1). The duration of gestation is variously given as 25 (1) or from 21 to 24 days (4).

The vaginal smears resemble those of the rat. In proestrum nucleated epithelial cells increase and leucocytes decrease. During heat large, flat, cornified epithelial cells appear, then leucocytes and oval nucleated epithelial cells with scattered clumps of cornified cells (3). A bursa encloses the ovary, which contains much interstitial tissue. A thecal gland around the follicle is prominent. The preovulatory follicle has a diameter of 0.8 mm., and the corpus luteum, which is solid by 48 to 72 hours after ovulation, reaches its maximum diameter of 1.05 to 1.10 mm. at 72 to 96 hours. Before ovulation the theca layer folds into the lumen of the follicle (3).

The spermatozoon head length is 5.5 μ (5.0 to 6.0) × 2.1 μ (1.9 to 2.6); midpiece width, 0.9 μ (0.7 to 1.0); tail length, 73.9 μ (68.5 to 80.0) (5).

1. Svihla, A. JM., 12: 238–242, 1931.
2. Hamilton, Jr., W. J. JM., 30: 257–260, 1949.
3. Conaway, C. H. JM., 35: 263–267, 1954.
4. Steward, J. S. In A. N. Worden and W. Lane-Petter, eds., The UFAW Handbook on the Care and Management of Laboratory Animals. London, 1957.
5. Hirth, H. F. J. Morphol., 106: 77–83, 1960.

Oryzomys alfaroi Allen. This Central American rat has been recorded as pregnant in March and April (1), and the number of embryos has varied from 4 (1) to 7 (2).

O. caliginosus Tomes. This rat has no definite breeding season. From 3 to 7 young, 4 or 5 usually, are born at a time (3).

O. couesi Alston. Pregnancies have been reported in February, September, and November (1). The number of embryos varies from 1 to 6, with 4 and 5 most frequently and 3.8, the mean (1, 2).

O. devius Bangs. Embryo counts of 3 and 4 have been recorded (4).

O. fulvescens Saussure. A female with 3 fetuses has been reported in July (5) and others with 3 in December and January (6).

O. pyrrorhinus Wied. Pregnant females have been taken in August. The number of embryos is from 1 to 6, usually 5 (7).

O. talamancae Allen. Three and 4 embryos have been recorded for pregnant females taken in January and July. The species is probably polyestrous (5).

O. (Oecomys) bicolor Tomes. This rat has 2 to 4 young at a time. It probably breeds all year round (8).

O. (O) endersi Goldman. A February record gives 3 embryos (5).

1. Felten, H. SB., 39: 1–10, 1958.
2. Murie, A. Univ. Michigan Mus. Zool., Misc. Publ. 26, 1935.
3. Goodwin, G. G. AM., 87: 271–474, 1946.
4. Enders, R. K. Personal communication, 1945.
5. Enders, R. K. HCMZ., 78: 385–502, 1935.
6. Burt, W. A., and R. A. Stirton. Univ. Michigan Mus. Zool., Misc. Publ. 117, 1961.
7. Moojen, J. Bol. Mus. Nac. Rio de Janeiro, Zool., N5, 1943.
8. Hershkovitz, P. Proc. United States Nat. Mus., 110: 513–568, 1959.

Reithrodontomys humilis Audubon and Bachman

HARVEST MOUSE

This mouse breeds all year-round and there is a post-partum heat. The gestation period is 24 days or less. The mean litter size is 2.2, range 1 to 3, and the sex ratio at birth is even. The vulva opens first between 4 and 7 weeks after birth and breeding begins at 11 to 20 weeks of age. The testes enlarge at 7 to 8 weeks (1). Another record gives 2 to 5 as the range in litter size (2).

1. Layne, J. N. Florida Mus. Biol. Sci., Bull. 4: 61–82, 1959.
2. Brimley, C. S. JM., 4: 263–264, 1923.

Reithrodontomys megalotis Baird

HARVEST MOUSE

This mouse is widespread in North America. The female is polyestrous and breeds all the year round, but mostly from April to October, with a tendency to reduced breeding at midsummer (1). The duration of gestation is 23 to 24 days; the litter size at birth is 1 to 7, mean 2.6 ± .2. The earliest breeding age is 4 months 8 days (2). Embryo counts have ranged from 1 to 7, with a mean of 4.1 ± .3. There is a post-partum heat (3). In Southern California there is an increase in sexual activity in the fall (4).

1. Smith, C. F. JM., 17: 274–278, 1936.
2. Svihla, R. D. JM., 12: 363–365, 1931.
3. Egoscue, H. J. JM., 39: 306, 1958.
4. Dunmire, W. W. JM., 42: 489–493, 1961.

Reithrodontomys creper Bangs. For Panama two records, each of 2 embryos, have been made (1).

R. fulvescens Allen. In Mexico this mouse breeds throughout the year except on the high plateau. Four or 5 young are usually born (2).

R. gracilis Allen and Chapman. Records give 2 and 4 embryos for June and July (3).

R. mexicanus Saussure. This mouse probably breeds all the year (2). Embryo counts have varied from 2 to 5 (1).

R. montanus Baird. Puberty is reached at age 2 months. Breeding is probably the year-round as young have been recorded for most months. The litter size varies from 2 to 5, mean 3.0 ± .2, and the gestation lasts 21 days (4).

1. Enders, R. K. Personal communication, 1945.
2. Dalquest, W. W. Mammals of the Mexican State of San Luis Potosi. Baton Rouge, La., 1953.
3. Anderson, S., and J. K. Jones, Jr. Univ. Kansas Mus. Nat. Hist., Publ. 9: 519–529, 1960.
4. Leraas, H. J. JM., 19: 441–444, 1938.

Tylomys watsoni Thomas

There is a record of 3 embryos for this Central American mouse (1).

1. Enders, R. K. Personal communication, 1945.

Ototylomys phyllotis Merriam

Embryos have been found in July (1) and October (2) and in all the other months except May and June (3). The litter size is usually 2 but singles have been recorded (3).

1. Kuns, M. L., and R. E. Tashiam. JM., 35: 100–103, 1954.
2. Hatt, R. T. JM., 19: 333, 1938.
3. Burt, W. H., and R. A. Stirton. Univ. Michigan Mus. Zool., Misc. Publ. 117, 1961.

Nyctomys sumichrasti Saussure

NIGHT MOUSE

In the London zoo this Mexican mouse has had young in February and October (1).

1. Zuckerman, S. PZS., 122: 827–950, 1952–3.

Peromyscus boylii Baird

WHITE-FOOTED MOUSE

This mouse, which lives in western North America, is polyestrous (1). In lower districts it breeds the year-round but high on the Coast Range the young are not born until May (2). The age at first heat is 50.9 ± 1.9 days (3). In the Sierra Nevada there is a peak of breeding in May and June and another from August to September. The females have a post-partum heat. The litter size varies little with weight and time of year. Corpus luteum counts averaged 3.3 ± .1, and embryos 3.1 ± .1 (4). The number varies from 2 to 4, mode 3.

1. Bailey, V. NAF., 53, 1931.
2. Mearns, E. A. United States Nat. Mus., Bull. 56, 1907.
3. Clark, F. H. JM., 19: 230–234, 1938.
4. Jameson, E. W. JM., 34: 44–58, 1953.

Peromyscus californicus Gambel

This mouse is polyestrous and has no post-partum heat, thus differing from most other species of the genus (1). It breeds from March to September, with a scattering in other months. The testes weight increases in March and decreases in October; the accessory organs follow the same cycle. The average litter size, based on fetuses at half their prenatal growth, is 1.9, range 1 to 3, mode 2. An average of 3.25 litters is produced each season (2). The mean gestation period is 23.6, range 21 to 25 days (1).

The sperm head measures 5.5 μ (5.0–6.0) × 3.2 μ (2.9–4.0); the midpiece width 0.8 μ (0.6–0.9); and the tail length 73.0 μ (65.5–85.0) (3).

1. Svihla, A. Univ. Michigan Mus. Zool., Misc. Publ. 24, 1932.
2. McCabe, T. T., and B. D. Blanchard. Three Species of Peromyscus. Santa Barbara, Calif., 1950.
3. Hirth, H. F. J. Morphol., 106: 77–83, 1960.

Peromyscus eremicus Baird

DESERT PEROMYSCUS

This mouse is polyestrous but it experiences no heats immediately after parturition (1). In the laboratory the average age of females at the birth of their first litters was 10 months but some had begun to breed at 3 to 4 months, though they did not become pregnant at that age (2). The mean age at first heat is 39.2 ± 1.5 days (3). The mice breed the year-round in Mexico (4), but not in the mountains during winter (5). The mean litter size in the laboratory was 2.42 ± .05 and the sex proportion of 41 litters was

52.9 per cent males. The number of young increased to the sixth litter and then decreased (2). Compilation of data on embryo numbers gave 3.0 ± .2 as the average.

1. Svihla, A. Univ. Michigan Mus. Zool., Misc. Publ. 24, 1932.
2. Davis, D. E., and D. J. Davis. JM., 28: 181–183, 1947.
3. Clark, F. H. JM., 19: 230–234, 1938.
4. Baker, R. H. Univ. Kansas Mus. Nat. Hist., Publ. 9: 125–335, 1956.
5. Mearns, E. A. United States Nat. Mus., Bull. 56, 1907.

Peromyscus floridanus Chapman

DEER MOUSE

In captivity this mouse breeds all the year. The litter size has varied from 1 to 3 (1). The sperm head measures 5.4 μ (4.9–6.0) × 2.7 μ (2.5–3.0); the midpiece width is 0.8 μ (0.6–0.9) and the tail length 74.0 μ (70.1–78.7) (2).

1. Dice, L. R. JM., 35: 260, 1954.
2. Hirth, H. F. J. Morphol., 106: 77–83, 1960.

Peromyscus gossypinus Le Conte

COTTON MOUSE

In Florida pregnancies have been found at all times of the year except in June and July, with but few in August. The greatest number are found in November to January. The testes tend to retrogress from April to August (1).

In the laboratory the females are polyestrous, with a post-partum heat. The cycle averages 5.26 days with the range from 3.5 to 10 days. Of this, proestrum occupies 21 hours (8–30); estrus, 24 hours (12–36); metestrum, 29.5 hours (23-40); and diestrum, 49.5 hours (24–74). The vagina is usually open at the time of heat but not invariably so. The vaginal smear varies with the condition, but exclusively cornified smears are not often found; there are usually a few nucleated cells and leucocytes to be seen. Metestrous smears are characterized by a decrease in cornified cells and a rapid increase in leucocytes, but very few nucleated cells are present. The gestation period in the nonlactating female is 22.9 days. In one nursing it was 30 days. Births take place mostly in the morning. The number of embryos averages 3.9 ± .16, mode 4, range 2 to 6. For the number of young born the range was 1 to 7, mean 3.7 ± .14, mode 4. The sex proportion of 240 young was 49.6 per cent males (1).

Spermatozoa are found in the seminiferous tubules at 40 days and in the epididymis at 45 days. The vagina opens at 43 days; the earliest observed was at 36 days (1). The sperm head measures 5.6 μ (5.1–5.9) × 3.0 μ (2.8–3.1); the midpiece width 0.9 μ (0.8–1.1); and the tail length 72.8 μ (65.0–77.4) (2).

1. Pournelle, G. H. JM., 33: 1–20, 1952.
2. Hirth, H. F. J. Morphol., 106: 77–83, 1960.

Peromyscus leucopus Rafinesque

WOOD MOUSE

This mouse, widespread through North America, is polyestrous and breeds from early April to the beginning of October in the northern part of its range (1). In the south it breeds at all times (2), but reproduction tends to fall off in July and August (3). In the laboratory males are fecund throughout the year but females do not breed in winter (4). Puberty in the female is reached at 46.2 ± 3.2 days (5), or in 50 per cent of individuals at 15 g. weight (6). Another report gives the 50 per cent level at 79 to 85 mm. body length for the males and 74 to 88 mm. for the females, depending on the year (7). The estrous cycle lasts 4 to 5 days, and the vaginal smears are typical, like those of the rat and mouse (8). The litter size is 5.0 ± .1 embryos, range 3 to 7, in Ontario (1). In Michigan the number born was 4.1 ± .1 (9). There appear to be considerable differences; in New Jersey, for instance, the averages for September have varied from 4.4 to 4.9 (7). Four seems to be the modal number of embryos in the wild state.

There is a heat immediately after parturition, and the duration of gestation in the nonlactating female averages 23.2 days. In the lactating female gestation may be prolonged by as many as 14 days (9). Of 133 sexed embryos 51.8 per cent were males (6).

This species has been the subject of research on the influence of light upon the periodicity of breeding. In Michigan most litters are born in May and June, with few in July, and reproduction is most frequent again from late August to October. None are born from December to February. With 1 foot-candle of light to prolong the length of day to 18 hours during the short-day season the mice breed at that time as well as they do in the usual breeding season. Lowered temperature (4° to 6° C.) has no adverse effect if the light is maintained. Intense light is not an advantage. If the mice are blinded, fertility is not affected, but they do not experience cyclical reproduction. Continued darkness lowers reproduction, and it becomes noncyclical in type (10).

The sperm head measures 5.5 μ (5.0–6.0) × 3.0 μ (2.5–3.1); the midpiece width 0.9 μ (0.6–1.1); and the tail length 74.3 μ (68.1–80.4) (11).

1. Coventry, A. P. JM., 18: 489–496, 1937.
2. Mearns, E. A. United States Nat. Mus., Bull. 56, 1907.
3. Burt, W. H. Univ. Michigan Mus. Zool., Misc. Publ. 45, 1940.
4. Howard, W. E. JM., 31: 319–321, 1950.
5. Clark, F. H. JM., 19: 230–234, 1938.
6. Jackson, W. B. Ecol. Monogr., 22: 259–281, 1952.
7. Davis, D. E. JM., 37: 513–516, 1956.
8. Osgood, Jr., F. L. Personal communication, 1945.
9. Svihla, A. Univ. Michigan Mus. Zool., Misc. Publ. 24, 1932.
10. Whitaker, W. L. J. Exp. Zool., 83: 33–60, 1940.
11. Hirth, H. F. J. Morphol., 106: 77–83, 1960.

Peromyscus maniculatus Wagner

DEER MOUSE

This mouse is widespread throughout North America. The female is polyestrous and tends to breed the year-round (1), but the intensity of reproduction varies considerably with the climate and the year. In Ontario breeding ceases for the winter about the beginning of September (2), and in the Yukon no pregnancies were found in June or July (3). In the Sierra Nevada breeding is most intense from April to May, with some breeding the year-round (4). At elevations of 4,500 ft. there were two main seasons in fall and spring, with sporadic breeding in the winter and very little in summer. At 9,800 ft. breeding was in spring and summer, with some in winter. At 12,400 ft. it was mainly in late summer and early fall. At lower elevations the mice bred for a longer period and produced more litters and three times as many young as at high altitudes but at the high altitudes the number of young per litter was slightly larger (5). In New Mexico pregnancies were found at higher elevations from July to September, but lower down December to June was the main season (6).

The age at first heat is 48.7 ± 1.2 days (7), and the first litters are usually born at 10 weeks (8). Puberty has also been placed at from 32 to 35 days, at a weight of 13 g., and in the males at from 40 to 45 days, at 15 to 16 g. (4). The estrous cycle lasts 4 to 5 days and the changes in the vaginal smears are like those of the rat and mouse (9, 10). The female experiences heat immediately after parturition. The duration of gestation in the nonlactating female is 23.5 days, with a range from 22 to 27 days; lactation increases the length up to 10 days (11).

The average number of embryos has been given as 5.4 ± .2, range 2 to 8 (2). A compilation of other figures, which relate to embryo counts in several subspecies, gives 5.0 ± .1. The modal number is 5 and the range 2 to 9. One series of births gave 4.04 ± .04 (11). The degree of wastage during pregnancy may be gauged by the following figures for *P. m. gracilis:* ova shed (i.e., corpora lutea), 5.9 ± .2; implantations, 5.5 ± .1; and live embryos, 5.3 ± .1. For *P. m. bairdii* the corresponding figures were: ovulations, 5.8 ± .1; implantations, 5.3 ± .1; and live embryos, 5.1 ± .1 (12). In the laboratory the litter size has varied from 2 to 7, mode 5, and mean 5.0, with 4.0 litters per season (8). Another account gives the average number of ovulations in January as 3.1, rising to 4.8 by June, falling in the summer and rising again in October (4). In one account the average litter size was 4.5, range 1 to 9; the number of litters per year was 10.4, mean 7 to 12; and the number of young per year, 41.3, range 15 to 63 (13).

The limiting factor in the length of the estrous cycle is usually the length of diestrum. Proestrum lasts 20.7 hours (15 to 24); estrus 26.2 hours (24 to 27; $n = 27$). Ovulation is spontaneous and it occurs near the middle or end of the cornified vaginal stage. In diestrum the vaginal epithelium is 5 to 6 cells deep with leucocytes in the superficial cell rows. In proestrum it is 4 or 5 rows of lightly staining cells below the 10 to 13 rows of epithelial cells. At estrus the superficial layer consists of 10 to 12 rows of granular cornified cells upon 10 to 13 rows of normal epithelial cells. In early metestrum there are 9 to 12 rows of epithelial cells with no leucocytes. Later the leucocytes come in. The uterus is greatly distended and thin-walled in proestrum and early estrus (10).

1. Hamilton, Jr., W. J. The Mammals of Eastern United States. Ithaca, N.Y., 1943.
2. Coventry, A. P. JM., 18: 489–496, 1937.
3. Rand, A. L. Canada Nat. Mus., Bull. 99, 1945.
4. Jameson, E. W. JM., 34: 44–58, 1953.
5. Dunmire, W. W. Ecol., 41: 174–182, 1960.
6. Holdenried, R., and H. B. Morlan. AMN., 55: 369–381, 1956.
7. Clark, F. H. JM., 19: 230–234, 1938.
8. McCabe, T. T., and B. D. Blanchard. Three Species of Peromyscus. Santa Barbara, Calif., 1950.
9. Osgood, Jr., F. L. Personal communication, 1945.
10. Clark, F. H. Univ. Michigan Contr. Lab. Vert. Genet., 1: 1–7, 1936.
11. Svihla, A. Univ. Michigan Mus. Zool., Misc. Publ. 24, 1932.
12. Beer, J. R., C. F. MacLeod, and L. D. Frenzel. JM., 38: 392–402, 1957.
13. Egoscue, H. J. JM., 41: 99–110, 1960.

Peromyscus polionotus Wagner

The age at first heat is 29.6 .5 days (1), and one record was found of a female with 5 embryos in October. Young have been recorded in April (2).

The spermatozoa measure as follows: head, 5.4 μ (5.0–5.9) × 3.1 μ (2.9–3.5); midpiece width, 1.4 μ (1.2–2.0); tail length, 72.0 μ (68.2–77.4) (3).

1. Clark, F. H. JM., 19: 230–234, 1938.
2. De Witt Ivey, R. JM., 30: 157–162, 1949.
3. Hirth, H. F. J. Morphol., 106: 77–83, 1960.

Peromyscus truei Schufeldt

PIÑON MOUSE

This mouse is polyestrous and breeds throughout the year, but mostly from April to June (1) or September (2). Few young are born from October to February (1). The testis weight rises in March and diminishes through September; the accessory organs follow the same cycle. The vagina opens at about 10 weeks (3), but the age at first heat has been given as 50.1 ± 2.5 days (4). The litter size varies from 1 to 5, mode 3 and 4, mean 3.4, based on counts made at half the fetal growth. The females average 3.4 litters per season (4). In another series the mean litter size at birth was 2.8 ± .2, mode 2, considerably less than the usual number for the genus (5), but frequent in desert species. Embryo counts in the field have varied from 3 to 6 for a limited series. A single gestation in a lactating mouse lasted 40 days (5). One investigation gives the mean litter size as 3.6, range 1 to 6; the number of litters per year as 5.8, range 2 to 9; and the number of young per female per year as 15.6, range 3 to 26 (6).

1. Holdenried, R., and H. B. Morlan. AMN., 55: 369–381, 1956.
2. Baker, R. H. Univ. Kansas Mus. Nat. Hist., Publ. 9: 125–335, 1956.
3. McCabe, T. T., and B. D. Blanchard. Three Species of Peromyscus. Santa Barbara, Calif., 1950.
4. Clark, F. H. JM., 19: 230–234, 1938.
5. Svihla, A. Univ. Michigan Mus. Zool., Misc. Publ. 24, 1932.
6. Egoscue, H. J. JM., 41: 99–110, 1960.

Peromyscus banderanus Allen. One female with 3 embryos has been reported for August (1).

P. crinitus Merriam. At low altitudes this mouse breeds in spring only. At higher elevations there is a split season with reproduction in spring and late summer or early fall and a midwinter and summer drop (2). The litter size is from 2 to 5 (3, 4).

P. difficilis Allen. The number of young averages 3. Both embryos and young have been reported for July and August (1), and a lactating female in April (5).

P. megalops Merriam. This mouse has an extended season. One record gives 3 embryos (1).

P. melanophrys Osgood. The yucca mouse breeds until late in the year. Two females with 3 embryos each have been taken (6).

P. melanotis Allen and Chapman. The breeding season is long. The embryo count varies from 1 to 5, with 3.7 the mean (1). Another count gave 3.8, range 1 to 5, mode 5 (5).

P. mexicanus Saussure. This species has a long breeding season. Lactating females have been found in February and September, pregnancies in March and April, and fertile males from January to August (7). It probably breeds all year. Two to 4 embryos have been counted (6).

P. nasutus Allen. This mouse has a long season; males with large epididymides have been found from January to August and a pregnant female in August (8). The litter size varies from 3 to 6, but it is usually 4 (9).

P. nudipes Allen. This Central American species has 2 to 3 embryos at a time (10).

P. nuttalli Harlan. This mouse is polyestrous, breeding from February to October, and has a post-partum heat (11). The number of young has varied from 2 to 4, usually 3. The sperm head measures 4.2 μ (3.9–5.0) $\times$ 3.0 μ (2.7–3.1); midpiece width, 2.0 μ (1.4–2.2); and tail length, 63.5 μ (60.0–70.0) (12).

P. pectoralis Osgood. Embryos have been found from March to December, and lactating females in March. Reproduction may not occur in January and February as none of 18 specimens taken in January was pregnant (5). Three embryos is the usual count (6).

P. yucatanicus Allen and Chapman. There is one record of 2 embryos in October (13).

1. Davis, W. B. JM., 25: 370–403, 1944.
2. Dunmire, W. W. JM., 42: 489–493, 1961.
3. Bailey, V. NAF., 55, 1936.
4. Hall, E. R., and K. R. Kelson. The Mammals of North America. New York, 1959.
5. Baker, R. H. Univ. Kansas Mus. Nat. Hist., Publ., 9: 125–335, 1956.
6. Dalquest, W. W. Mammals of the Mexican State of San Luis Potosi. Baton Rouge, La., 1953.
7. Felten, H. SB., 39: 133–144, 1958.
8. Holdenried, R., and H. B. Morlan. AMN., 55: 369–381, 1956.
9. Bailey, V. NAF., 53, 1931.
10. Enders, R. K. Personal communication, 1945.
11. Goodpasture, W. W., and D. F. Hoffmeister. JM., 35: 16–27, 1954.
12. Hirth, H. F. J. Morphol., 106: 77–83, 1960.
13. Hatt, R. T. JM., 19: 333–337, 1938.

Calomyscus bailwardi Thomas

This vole has been recorded as having 3 to 5 young born in the period from March to June. Some females have 2 litters during the year (1).

1. Gambarian, P. P., and B. A. Martirosyan. ZZ., 39: 1408–1413, 1960.

Baiomys

PYGMY MOUSE

Baiomys musculus Merriam. This Central American mouse breeds throughout the year. Combined embryo counts and newborn young average 2.9, range 1 to 4 (1). Another series of embryo counts gave 2.0, range 1 to 3 (2).

B. taylori Thomas. This mouse is polyestrous (3) and breeds the year-round (4). Puberty is reached at about 44 days; the gestation lasts for 20 days and the litter size is 2.7 ± .1, range 1 to 5 (3). Another count gave 2.5, range 1 to 4 (1).

1. Packard, R. L. Univ. Kansas Mus. Nat. Hist., Publ. 9: 579–670, 1960.
2. Hall, E. R., and K. R. Kelson. The Mammals of North America. New York, 1959.
3. Blair, W. F. JM., 22: 378–383, 1941.
4. Dalquest, W. W. The Mammals of the Mexican State of San Luis Potosi. Baton Rouge, La., 1953.

Onychomys leucogaster Wied

GRASSHOPPER MOUSE

This central and western North American mouse is polyestrous from April to September (1, 2), but most litters are born from February to August. There is a post-partum heat (3). The females reach puberty in about 90 days (2) and the young, in number 2 to 6, usually 4, are born after a gestation of 33 days, or longer if the female is lactating (4). Another count gives a litter size of 3.6, range 1 to 6. The number of litters per year averages 5.1, range 1 to 10, and the number of young per year, 18.3, range 2 to 27. The gestation period in the nonlactating female was 29 to 32 days, and in the lactating female 32 to 38 days (3). In this account most females are said to breed first in the year following that of birth (3).

1. Warren, E. R. The Mammals of Colorado. Norman, Okla., 1942.
2. Bailey, V. NAF., 53, 1931.
3. Egoscue, H. J. JM., 41: 99–110, 1960.

Onychomys torridus Coues. In Mexico this mouse breeds from April to December and the number of young varies from 3 to 5 (1). In Death Valley young have been seen in mid-April (2).

1. Bailey, V., and C. C. Sperry. United States Dept. Agric., Tech. Bull. 145, 1929.
2. Grinnell, J. Proc. California Acad. Sci., Ser. 4, 23: 115–169, 1935–47.

Akodon

Akodon amoenus Thomas. The males of this South American mouse come into breeding condition towards the end of July, but no pregnant females were found at that time (1).

A. andinus Osgood. Breeding males were found in July and December (1).

A. berlepschii Thomas. Testis development begins in July and the breeding season follows immediately (1).

A. bolivensis Meyen. No pregnancies were found in July, but of 27 males, 2 were capable of breeding. In December the males had enlarged testes and a female with 5 embryos was taken (1).

A. jelskii Thomas. This mouse was not breeding in July or August. Breeding probably begins in October (1). The litter size is usually 4 (2).

1. Pearson, O. P. HCMZ., 106: 117–174, 1951–2.
2. Dorst, J. Mammalia, 22: 546–565, 1958.

Zygodontomys cherriei Allen

CANE RAT

Pregnant or nursing females were found in Panama in the months of November, January, March, and April. The litter size varied between 2 and 4 (1).

1. Enders, R. K. HCMZ., 78: 385–502, 1935.

Scotinomys teguina Alston

Four specimens of this Central American species each had 2 embryos (1). A lactating female has been found in April (2).

1. Enders, R. K. Personal communication, 1945.
2. Felten, H. SB., 39: 133–144, 1958.

Hesperomys

Hesperomys laucha Desmarest. In Venezuela a summer record gives 4 pregnant females, one with 3 embryos and 3 with 4 each, and a nest with 3 recent young (1).

H. sorella Thomas. A litter of 5 is usual (2).

1. Butterworth, B. B. JM., 41: 517–518, 1960.
2. Dorst, J. Mammalia, 22: 546–565, 1958.

Phyllotis

LEAF-EARED MOUSE

Phyllotis boliviensis Waterhouse. Males of this South American mouse were found in breeding condition after mid-September. Pregnant females with 3 to 4 embryos were taken in October and December. Young were found in November (1).

P. darwinii Waterhouse. Breeding males were taken from late July onward and pregnant females from late July to December, after which no more catching was done. Four was the average embryo number (1).

P. pictus Thomas. Breeding probably begins in September or October (1), and the litter size is 4 (2).

P. sublimis Thomas. At Santa Rosa males were taken with large testes at the end of July, but at Caccachara not until the beginning of September (1).

1. Pearson, O. P. HCMZ., 106: 117–174, 1951–2.
2. Dorst, J. Mammalia, 22: 546–565, 1958.

Neotomys ebriosus Thomas

In Peru this mouse was found with large testes at the end of July (1).

1. Pearson, O. P. HCMZ., 106: 117–174, 1951–2.

Reithrodon typicus Waterhouse

PAMPAS GERBIL

In the London zoo this species has had young throughout the year. The number has varied from 1 to 4, with 2 as the mean and mode (1).

1. Zuckerman, S. PZS., 122: 827–950, 1952–3.

Holochilus sciureus Wagner

In Brazil pregnant females have been found in September. The number of young was 4 to 6, usually 5 (1). In British Guiana litters of 2 to 5 young, averaging 3.5, have been recorded (2).

1. Moojen, J. Bol. Mus. Nac. Rio de Janiero, Zool., N5, 1943.
2. Twigg, G. I. JM., 43: 369–374, 1962.

Sigmodon hispidus Say and Ord

COTTON RAT

This cotton rat of southern North America is polyestrous in the wild, breeding from early spring to late fall, or longer (1). In the laboratory it breeds all the year round. The testes descend usually at 20 to 30 days, varying from 10 to 50 or more days, and the vaginal introitus is established at 30 to 40 days, as a rule, with a variation from 10 to 50 or more days. Puberty is reached earlier in the spring than it is in the fall. Spermatozoa are produced at 40 to 50 days of age (2). No female weighing less than 60 g. was found to be pregnant or with placental scars (3). In Texas the testes increased in size from February to July and atrophied from July to November. The pregnancy rate was highest in May and June, while none were pregnant from October to February (4).

The vaginal smear is characteristic of the Muridae and Cricetidae. There is a reduction in the number of leucocytes and an increase in nucleated epithelial cells during proestrum, with cornified cells in estrus, and nucleated epithelial cells followed by leucocytes in metestrum. The lengths of the parts of the cycle are most variable, and the following ranges have been found (2):

	Average	*Range*
Proestrum	14 hours	12– 21 hours
Estrus	46 hours	21–123 hours
Metestrum	14 hours	9– 21 hours
Diestrum	116 hours	42–156 hours

Another account gives vaginal heat as lasting on the average 3.4 days, range 1 to 12 days, and a cycle length of 9 days, range from 4 to 20 days. The uterus is distended during heat, and ovulation occurs late in this period. The post-parturition heat lasts 18 to 24 hours, a shorter period than that of the normal cycle, and ovulation occurs 6½ to 12 hours after littering. A vaginal plug is formed at coitus. The eggs take about 60 hours in their passage through the oviduct, and remain for nearly half that time in the last third. The placental sign, erythrocytes in the vagina, is given on the tenth day of pregnancy (5).

The average litter size is 5.6, with a range from 2 to 10, and the sex ratio at birth is 51.7 per cent (5). Other mean litter sizes are 4.75, range 3 to 8 (1), and 7.4, range 5 to 10 (6). Apparently, litter size varies in different districts or on different feeds. For the subspecies *chiriquensis* Enders has records of 2 and 3 embryos, lower than the usual number for the species. The duration of gestation is 27 days, and suckling has little effect upon its length (5).

The sperm head measures 6.1 μ (5.9–7.0) × 3.2 μ (3.0–3.6); the midpiece 17.2 μ (16.0–20.1) × 0.9 μ (0.8–1.1); and the tail length 80.9 μ (77.1–85.4) (7).

1. Svihla, A. JM., 10: 352–353, 1929.
2. Clark, F. H. Univ. Michigan Contr. Lab. Vert. Genet., 1936.

3. Odum, E. P. JM., 36: 368–378, 1955.
4. Haines, H. Texas J. Sci., 13: 219–230, 1961.
5. Meyer, B. J., and R. K. Meyer. JM., 25: 107–129, 1944.
6. Rinker, G. C. Trans. Kansas Acad. Sci., 45: 376–378, 1942.
7. Hirth, H. F. J. Morphol., 106: 77–83, 1960.

Sigmodon bogotensis Allen. The estrous cycle lasts 6.2 ± .45 days. The duration of heat is less than 1 day. Proestrum lasts 1.4 days; metestrum is less than 1 day; diestrum, 2.6 days; and the interestrous period, 5.4 days. The vaginal smear resembles that of *Sigmodon hispidus*. The variability in the cycle is from 4 to 8 days (1).

S. minimus Mearns. This species has been recorded as pregnant with 3 and 4 embryos in May (2).

S. ochrognathus Bailey. Two females caught in April each had 2 embryos. Four in October had 6, 7, 8, and 9, respectively (3).

1. Tamsitt, J. R., and D. Valdivieso. Mammalia, 26: 161–166, 1962.
2. Bailey, V. NAF., 53, 1931.
3. Baker, R. H. Univ. Kansas Mus. Nat. Hist., Publ. 9: 125–335, 1956.

Neotomodon alstoni Merriam

VOLCANO MOUSE

The breeding season stretches at least from June to September. The number of young averages 3.3, with a range from 2 to 5 (1). Embryo counts, varying from 2 to 5, gave 3.5 ± .2 as the average. Four was the mode (2).

1. Davis, W. B. JM., 25: 370–403, 1944.
2. Davis, W. B., and L. A. Follansbee. JM., 26: 401–411, 1945.

Neotoma albigula Hartley

WOOD RAT

This rat is polyestrous (1) and breeds all the year, though most pregnancies are found from March to May and the fewest from September to December

(2). The young may breed in the year of their birth, and some females may have two or more litters a year (3). The litter size is from 2 to 3, mean 2.2 ± .1, or, in another series, 1.95 (4), while the gestation period is less than 30 days (1).

1. Feldman, H. W. JM., 16: 300–303, 1935.
2. Holdenried, R., and H. B. Morlan. AMN., 55: 369–381, 1956.
3. Finley, Jr., R. B. Univ. Kansas Mus. Nat. Hist., Publ. 10: 213–552, 1958.
4. Vorhies, C. T., and W. P. Taylor. Arizona Agric. Exp. Sta., Tech. Bull. 86, 1922.

Neotoma cinerea Ord

WOOD RAT

This rat breeds from spring to early summer (1). The male season probably ends in August, and males of the first year do not reach puberty until their first winter (2). One litter a year is usual in the north of their range but probably more are produced in the south (1). The litter size has varied from 1 to 6, mode 4, and mean 4.0 ± .3.

In the laboratory they breed in the year following birth and are seasonally polyestrous. Births are from February to August, but most are from March to May. There is a post-partum heat. Mating usually takes place from 1:00 to 4:00 P.M. The gestation period is 27 to 31½ days long; lactation does not extend it. The litters have averaged 3.5, with mode 3 (3).

1. Warren, E. R. JM., 7: 97–101, 1926.
2. Finley, Jr., R. B. Univ. Kansas Mus. Nat. Hist., Publ. 10: 213–552, 1958.
3. Egoscue, H. E. JM., 43: 328–337, 1962.

Neotoma floridana Ord

WOOD RAT

This rat breeds from February to August and exceptionally into September (1), and is polyestrous (2), but only one litter a year is born (3). The testes enlarge late in January and are scrotal by February. In June most are abdominal, but they descend again in late July or early August (1). In the female puberty is reached at about 160 g. weight, or age 5 to 6 months, except for those that reach this age during the winter quiescent period. Early-born young may breed the same year (4, 5). The litter size varies from 1 to 6, with 3 as the mean; 2 to 4 are the more common litter and embryo numbers. The gestation period is probably between 31 and 36 days (6). In captivity 3.1 has been the average litter size and 32 days the gestation period (5).

The estrous cycle lasts for 3 to 8 days, usually 4 to 6. The vaginal smear has been divided into stages as follows:

Stage 1. Mostly large oval epithelial cells with some vacuolated. The vaginal epithelium is thick and contains leucocytes. Some desquamation is occurring. Many mucous cells are present and mucus may be found in the lumen. Ovulation occurs during this stage, which lasts from 12 to 30 hours.

Stage 2. Mostly vacuolated epithelial cells with pycnotic nuclei; many leucocytes. The epithelium is medium thick. New corpora lutea are present and the ova are in the oviducts.

Stage 3. Many leucocytes; pycnotic vacuolated cells. The smear becomes scarce late in this stage and the vaginal epithelium thin.

Stage 4. Diestrum. The smear is scanty, with degenerating cells. This stage lasts from 1 to 2 days (7).

During the heat period the external genitalia are enlarged and discolored; the vaginal orifice is prominent and gaping. The uterine horns are firm, enlarged, and pinkish (4).

The sperm head measures 8.1 μ (7.9–8.9) × 1.5 μ (1.0–1.8); the midpiece width 0.9 μ (0.8–1.0); and the tail length 130.0 μ (120.0–135.1) (8).

1. Rainey, D. G. Univ. Kansas Mus. Nat. Hist., Publ. 8: 536–646, 1954–6.
2. Dice, L. R. JM., 4: 107–112, 1923.
3. Bailey, J. W. The Mammals of Virginia. Richmond, Va., 1946.
4. Fitch, H. S., and D. G. Rainey. Univ. Kansas Mus. Nat. Hist., Publ. 8: 499–533, 1954–6.
5. Schwentker, V. In A. N. Worden and W. Lane-Petter, eds., The UFAW Handbook on the Care and Management of Laboratory Animals. London, 1957.
6. Poole, E. L. JM., 21: 249–270, 1940.
7. Chapman, A. O. Univ. Kansas, Sci. Bull. 34: 267–283, 1951–2.
8. Hirth, H. F. J. Morphol., 106: 77–83, 1960.

Neotoma fuscipes Baird

WOOD RAT

This wood rat has a wide range in the mountains and valleys of the western United States, and the differences in climate which it encounters may explain some of the discrepancies reported in its reproduction.

The females are polyestrous (1) and pregnancies have been reported from February to April or May (2). In California the testes increase late in September and become scrotal; in May they cease to be active and the accessory organs regress (3). Puberty is reached at about 240 g. February is the usual breeding season for adults; the young of the previous season are later in breeding (3). No vaginal plug is produced after mating. In two cases the duration of gestation was 33 days (4). The number of embryos, compiled from several sources, is 2.4 ± .1, mode 2 and 3, range 1 to 4.

In the female the vulva, when resting, is imperforate and the lips are small and pale. At heat they become inflamed and pink. The spermatozoa are about 128 μ long (3).

1. Warren, E. R. JM., 7: 97–101, 1926.
2. English, P. F. JM., 4: 1–9, 1923.
3. Linsdale, J. M., and L. P. Tevis, Jr. The Dusky-footed Wood Rat. Berkeley, Calif., 1951.
4. Wood, F. D. JM., 16: 105–109, 1935.

Neotoma lepida Thomas

WOOD RAT

In the laboratory this rat is polyestrous all year and the females breed at 2 to 3 months old (1). In the wild most females breed in the year following birth (2). Some females experience a post-partum heat; in others it is delayed until weaning. At heat the vulva is swollen and bright pink. The gestation period is from 32 to 36 days and the litter size 1 to 5, mode 2, mean 2.3 ± .1 (1). In the wild, breeding is from February to June. More than one litter a year is produced (3); the litter size is 1 to 5, but usually 3. In the laboratory the average number of litters a year is 5.8, range 2 to 9; the litter size is 2.8, range 1 to 5. The average number of young per year is 15.6, range 3 to 26 (2).

1. Egoscue, H. J. JM., 38: 472–481, 1957.
2. Egoscue, H. J. JM., 41: 99–110, 1960.
3. Hall, E. R. Mammals of Nevada. Berkeley, Calif., 1946.

Neotoma micropus Baird

WOOD RAT

This rat is polyestrous from late April onward (1). The gestation period is less than 33 days and the litter size is usually 2, with but one litter a year (2). In New Mexico the breeding season lasts from March to August (3). Another record cites the probability of 2 or more litters a year. The early young of the year may have litters in the summer (4). Lactating females have been seen in March and young in March and November (5).

1. Bailey, V. NAF., 53, 1931.
2. Feldman, H. W. JM., 16: 300–303, 1935.
3. Holdenried, R., and H. B. Morlan. AMN., 55: 369–381, 1956.
4. Finley, R. B. Univ. Kansas Mus. Nat. Hist., Publ. 10: 213–552, 1958.
5. Baker, R. H. Univ. Kansas Mus. Nat. Hist., Publ. 9: 125–335, 1956.

Neotoma goldmani Merriam. There is a March record of two females each with a single embryo (1).

N. mexicana Baird. This species has an extended season (2) and is probably polyestrous (3), with at least 2 litters a year. The litter size is from 1 to 4, usually 2 (4). A compilation gives an average of 2.25 young per litter.

N. (Teanopus) phenax Merriam. For this Mexican rat there is a February record of a female with 2 embryos (5).

1. Baker, R. H. Univ. Kansas Mus. Nat. Hist., Publ. 9: 125–335, 1956.
2. Davis, W. B. JM., 25: 370–403, 1944.
3. Bailey, V. NAF., 53, 1931.
4. Finley, Jr., R. B. Univ. Kansas Mus. Nat. Hist., Publ. 10: 213–552, 1958.
5. Burt, W. H. Univ. Michigan Mus. Zool., Misc. Publ. 39, 1938.

Rheomys thomasi Dickey

There is a February record of a female with a single 35mm. embryo (1).

1. Burt, W. H., and R. A. Stirton. Univ. Michigan Mus. Zool., Misc. Publ. 117, 1961.

Cricetus cricetus L.
HAMSTER

The gestation period for this species is 20 to 22 days (1), and the litter size ranges from 4 to 18, but is usually from 6 to 12 (2). The testes are active in March and resting in September (3).

1. Kenneth, J. H. Gestation Periods. Edinburgh, 1943.
3. Petzsch, H. Kleintier. u. Pelztier., 12(1): 1–83, 1936.
3. Kayser, C., and M. Aron. CRSB., 129: 225–226, 1938.

Cricetulus barabensis Pallas
CHINESE HAMSTER

The estrous cycle of the Chinese hamster lasts for 4 days, with an average of 4.3 ± .08 days, and there is no restricted breeding season. The vaginal smear resembles that of the rat and mouse, but the stages are more definite. Epithelial cells are present for 2 to 3 days, nucleated epithelial cells only for less than 1 day, and cornified cells for 1½ to 2 days, with ovulation toward the end of this stage. The proestrous stage lasts ½ day, estrus 1½ days, and diestrum or metestrum 2½ days. Ovulation is spontaneous, and the corpus luteum suffers a precipitous decline after 2 days of diestrum, in contrast to the condition in the rat and mouse. The mature follicle measures 0.45 mm. in diameter (1).

Females reach puberty at 8 to 12 weeks. In general, those born in the spring mature earlier than those born in the autumn. The gestation period is 20.5 days. In a large series the litter size varied from 1 to 10, with a mean of 4.6 ± .02 and a mode of 5 (2).

The ovaries resemble those of the rat and mouse. They are bean shaped and are completely enclosed within a fibrous capsule. The oviduct is long and much coiled. The uterus is bicornuate, and the horns fuse just above the cervix. The clitoris is large and it is perforated by the urethra (1).

The proestrous vaginal orifice is pink to red, smooth, and moist. At estrus the area is lighter, but the clitoris forms a darker red spot. In metestrum the swelling is still apparent but the prominent ridge about the orifice when it is spread laterally has disappeared. In diestrum the lumen is tightly closed; the clitoris is dry, scaly, and lightly tinted. The placental sign is evident in the vaginal smear on the fourteenth day of pregnancy (2).

During diestrum the uterine mucosa has a low-columnar epithelium, leucocytes are present in the stroma, and the glands are inactive. In proestrum the epithelium is the same as in diestrum, but no leucocytes can be found and the glands are secreting. In estrus the epithelial cells are tall and apparently pseudostratified. The glands are emptying, and, toward the end of this stage, leucocytes make their way into the stroma (1).

In some specimens sent from China a series of large glandular bodies about 1.5 to 2.0 mm. in diameter were present in the uterus on the mesometrial side. From 8 to 10 were found in each cornu, and as they closely resemble corpora lutea it is believed that they may have a glandular function (1).

1. Parkes, A. S. PRS., 108B: 138–147, 1931.
2. Yerganian, G. J. Nat. Cancer Inst., 20: 705–721, 1958.

Cricetulus triton de Winton

In the laboratory this hamster breeds from February to September and has 4 litters a season. Puberty is reached at 2 months. The litter size is from 6 to 7, and the gestation period lasts 17 to 18 days (1). In the London zoo litters have been from 1 to 5, mean 3 (2). The testes are enormous, up to one-fifth of the body weight (3).

1. Tupikova, N. V., and S. M. Kulogin. ZZ., 31: 476–478, 1952.
2. Zuckerman, S. PZS., 122: 827–950, 1952–3.
3. Loukashkin, A. S. JM., 25: 170–177, 1944.

Cricetulus eversmanni Brandt. This hamster usually has 2 litters of 4 to 6 young during the summer (1).

C. migratorius Pallas. This mouse breeds all the year and the litter size, mean 6.5, varies little with the season. There is a tendency for more embryos to be carried in the left uterus (2). In the south part of its range 2 litters a year, sometimes more, are usual. They breed at about a year old and have 3 to 10 young, usually 6, after a gestation of 11 to 13 days (1).

1. Heptner, W. G., L. G. Morosowa-Turowa, and W. I. Zalkin. Die Säugetiere in der Schutzwaldzone. Berlin, 1956.
2. Semenov, M. Y. ZZ., 40: 1743–1745, 1961.

Mesocricetus auratus Waterhouse

GOLDEN HAMSTER

The Syrian hamster first comes in heat at 7 to 8 weeks of age (1) or at a weight of 60 g. The vaginal introitus is established at 10 to 15 days and puberty sets in at 33 to 38 days. Sexual cycles cease when the females are 400 to 500 days old (2). The first postovulatory vaginal smear may be detected at 28 to 37 days, when the hamsters weigh 40 to 60 g. Three of 14 first cycles were found to be nonovulatory. The females are polyestrous and come in heat at any time of year, though pregnancies are infrequent from October to March. The failure in reproduction seems to be due to the male rather than the female, and better reproduction can be obtained during the winter by exposing the animals to an increased amount of light. The estrous cycle lasts for 4 days with little variation, but a mated female that does not become pregnant experiences pseudopregnancy, which lasts 7 to 13 days, usually 9 to 10 days, mean 9.6 ± .2 days (1, 4, 5). The cycle length averages 95 hours, of which estrus comprises 27.4 hours (6). Mating is nocturnal and usually occurs after 10 P.M. (1) or 8 P.M. (7), but it is also frequent before 9 A.M. (8). Ovulation is usually a few hours later. The bicornuate copulation plug is extruded rather soon, as a rule. There is no postparturient heat or ovulation, and, if the female is lactating, heat does not occur during this period.

The litter size varies from 1 to 12 (9), with 6 and 7 as the modal size (1). One set of records gave 6.3 as the mean, with a rise in number to the third and fourth litters, then a slight reduction for the fifth (10). Few females have more than 3 or 4 litters; subsequently they abort or do not become pregnant although they may come in heat regularly. The duration of pregnancy is 16 days (9), though it may last as long as 19 days (11). In another investigation the mean gestation length for month-old females was 373 hours. It rose to 402 hours with 14-month-old females (12). The secondary sex ratio was 54.7 per cent males (1). Blastocysts sexed at 80 to 90 hours after mating by the metaphase chromosomes gave a sex ratio of 64.3 per cent males (13).

THE FEMALE

The ovaries are small and completely encapsulated. The ovum is 60 to 63 μ in diameter and the ripe follicle, 690 to 780 μ (9). Across the zona pellucida the diameter of the ovum is 72 μ (14). At heat the theca interna becomes very congested, and 10 to 13 follicles usually rupture at this time. The corpora lutea develop rapidly, and those of the normal cycle decline precipitously after about 3 days, in contrast to the condition in rats and mice. In this type of corpus luteum, which appears to be nonfunctional, the lutein cells do not increase much in size. In the pseudopregnant corpus luteum the lutein cells are larger and vascularization is much greater. Regression sets in at 7 days. There are differences in the sizes of the various types of corpora lutea:

Size of corpus luteum of cycle	700 μ
Size of corpus luteum of pseudopregnancy	820–860 μ
Size of corpus luteum of pregnancy	900–1,000 μ

Regression of the corpus luteum of pregnancy is rapid after parturition (9).

The vagina may be divided into two parts: that comprising 1 cm. from the vulva, and the upper portion. The lower portion is continually growing cornified cells, which fill two lateral ventral pouches. If the region is squeezed, these cells may be extruded as two leaflike processes at any time of the cycle. Accordingly, the vaginal smear is not a good indication of the reproductive state. The epithelium of the upper vagina contains many mucous cells which are shed in large numbers after heat. In this region there are remarkable epithelial growths during heat. These consist of epithelial villi with 3 to 8 layers of cornified cells covered with irregular foliations of epithelial cells with oval nuclei. After ovulation these masses are shed and polymorphs invade the region. After mating the desquamated area regenerates and becomes covered with columnar cells which mucify (9).

The vaginal smear has received considerable attention. No fewer than nine complete accounts have been found (2, 6, 7, 15–20). Some workers have found that the smears have little reliability; others consider that the smear carefully taken from the upper vagina is a reliable indication of the reproductive state. One account (18) gives the changes as follows:

Stage I. *Proestrum.* Sudden appearance of nonnucleated scales, some nucleated epithelial cells, no leucocytes. The corpus luteum of the cycle is degenerating. This stage lasts from 3 to 6 P.M. of day 1.

Stage II. *Estrus.* Nucleated epithelial cells, columnar and oval, also elongated ones. Many are vacuolated and granular. A few nonnucleated scales. This from the early evening of day 1 into the morning of day 2. By 7 or 8 A.M. of day 2 the rupture points of the follicles are closed. The ovum is in the upper part of the oviduct, but its cumulus is still adherent.

Stage III. *Metestrum.* Leucocytes. The number of epithelial cells gradually decreases. By 7 or 8 A.M. of day 3 the corpora lutea are well established.

Stage IV. *Metestrum.* Fewer leucocytes and nucleated epithelial cells. Most of the latter are pycnotic and degenerating. By the morning of day 4 the ova are in the lower part of the oviduct and the cumulus has disappeared.

The evening before heat and on the day of heat the smear is sticky, changing to waxy (8). All the ova are not shed together, but the most favored time is 1 A.M. Fertile matings may occur as early as 6 P.M. of day 1 and as late as 9:30 A.M. of day 2 (18). In this account the day of proestrum is regarded as day 1. Inseminations made at the time of ovulation resulted in 97 per cent fertilized ova. By 6 hours the number had dropped to 60 per cent. Following this the reduction was very rapid—only 9 per cent at 9 hours and none at 18 hours. After the fertilizing capacity had been completely lost, 75 per cent of the ova extruded their second polar bodies and formed pronuclei, but cleavage was very rare (21). Under controlled light conditions most ovulations occurred from 8 to 9 hours after the onset of heat (22). Copulation neither hastened nor delayed the time.

The ova are not capable of being fertilized until a few hours after they have been shed (23). This might, however, be due to the need for capacitation of the spermatozoa. They are in the mid and caudal third of the ampulla at 2 to 10 hours after ovulation (23). The spermatozoa reach the oviducts in less than 30 to 60 minutes. Two to 4 hours are required for capacitation. The male is capable of mating 50 times in an hour, but there are no spermatozoa in the first 5 copulations (24). The spermatozoa retain their fertilizing power in the female tract for about 7 hours. There is a precipitous decline after 5 hours. The ova remain fertilizable for a few hours only (25).

The oviduct opens in a papilla at the tip of the uterine horn. There is no sphincter, and the utero-tubal junction is closed by absorption of fluid by the mucosal connective tissue. The tubular and uterine epithelia show cyclical changes resembling those in the rat and mouse. During heat and early metestrum many leucocytes are in the mucosa of the tubal opening (26).

Ovariectomy between 11 and 13 days of pregnancy results in abortion. Progesterone alone fails to continue the gestation, but if estrogen is given simultaneously it goes on to term. If the uterus is removed from the eighth to the thirteenth day of pregnancy, estrous cycles are resumed, but if the embryos only are removed and the placentae retained, cycles are inhibited (27).

If estrogens are injected into the spayed female, completely cornified vaginal smears are produced, a condition never found in the normal female (but see above for conditions in the upper vagina). Progesterone has very little effect upon uterine histology (19).

THE MALE

At 5½ weeks of age most seminiferous tubules contain spermatozoa but they have not yet appeared in the epididymis. At 21 months senile changes are evident in the testes. There is a breakdown of the cytoplasm of the germ cells. They disappear or become cystlike. The interstitial cells become fewer. Later, the tubules fill with connective tissue but some may be found with spermatozoa (28).

In the male the pituitary gland contains more basophils than acidophils. This is the reverse of the condition in the female (29). In the female neutrophils and basophils appear in May. Acidophils are most active in June. Their number is depressed in summer but it increases in the fall (30).

1. Bruce, H. M., and E. Hindle. PZS., 361–366, 1934.
2. Kupperman, H. S., R. B. Greenblatt, and L. Q. Hair. AR., 88: 441–442, 1944.
3. Shnider, S. M., and I. M. Thompson. Trans. Roy. Soc. Canada, 46(Proc.): 158, 1952.
4. Kent, G. C., and G. Atkins. PSEBM., 101: 106–107, 1959.
5. White, G. V. S. J. Tennessee Acad. Sci., 24: 216–219, 1949.
6. Kent, G. C., and R. A. Smith. AR., 92: 263–270, 1945.

7. Sheehan, J. F., and J. A. Bruner . Turtox News, 23: 65–78, 1945.
8. Ward, M. Personal communication, 1945.
9. Deanesly, R. PZS., 108A: 31–37, 1938.
10. Foote, C. L. AR., 120: 760–761, 1954.
11. Ben Menahem, H. Arch. Inst. Pasteur, Algiers, 12: 403–407, 1934.
12. Soderwall, A. L., H. A. Kent, Jr., C. L. Turbyfill, and A. L. Britenbaker. J. Gerontol., 15: 246–248, 1960.
13. Sundell, G. J. Embryol. Exp. Morphol., 10: 58–63, 1962.
14. Knigge, K. M., and J. H. Leathem. AR., 124: 679–707, 1956.
15. Mello, M. I. Rev. Brasil Biol., 9: 433–438, 1949.
16. Roig, C. E. Rev. Soc. Argentina Biol., 36: 363–368, 1960.
17. Comeaux, H. J. A. J. Tennessee, Acad. Sci., 24: 272–284, 1949.
18. Ward, M. AR., 94: 139–161, 1946.
19. Peczenik, O. JE., 3: 157–167, 1942–4.
20. Kent, G. C., and J. P. Mixner. PSEBM., 59: 251–253, 1945.
21. Yanagimachi, R., and M. C. Chang. J. Exp. Biol., 148: 185–204, 1961.
22. Harvey, E. B., R. Yanagimachi, and M. C. Chang. J. Exp. Zool., 146: 231–236, 1961.
23. Strauss, F. J. Embryol. Exp. Morphol., 4: 42–56, 1956.
24. Chang, M. C., and D. Sheaffer. J. Hered., 48: 107–109, 1957.
25. Bentley, A. J., and A. L. Soderwall. J. Gerontol., 8: 373–374, 1953.
26. Bögle, B. RSZ., 66: 211–227, 1959.
27. Klein, M. PRS., 125B: 348–364, 1938.
28. Spagnoli, H. B., and H. A. Charipper. AR., 121: 117–139, 1955.
29. Serber, B. J. AR., 131: 172–191, 1958.
30. Mogler, R. K. Zeit. f. Morphol. Okol. Tiere, 47: 267–308, 1958.

Mesocricetus raddei Nehring. The mean litter size in this hamster is 13.5. Litters tend to be smaller in summer (1).

1. Semenov, M. Y. ZZ., 40: 1743–1745, 1961.

Mystromys albicaudatus A. Smith

This species bears a litter of 2 to 5 young after a gestation of about 37 days (1).

1. Jameson, H. L. Ann. Mag. Nat. Hist., 8(14): 455–474, 1909.

NESOMYINAE

Macrotarsomys bastardi Milne-Edwards and Grandidier

This Madagascar gerbil probably breeds in the rainy season. A female with 2 embryos was captured (1). It breeds all year-round, has 2 to 3 young, usually 2, after a gestation of about 24 days (2).

1. Webb, C. S. The Odyssey of an Animal Collector. New York, 1954.
2. Letellier, F., and F. Petter. Mammalia, 26: 132–133, 1962.

MICROTINAE

Dicrostonyx groenlandicus Traill

COLLARED LEMMING

In captivity this lemming has an extended season, at least from March to September (1). In the wild, the rule seems to be one or two litters, according to climatic conditions. These are born at approximately an interval of one month, at the end of June and of July (2). Puberty is reached in the female at 25 to 30 days old and the vagina becomes perforate at from 25 to 27 days (3). The gestation period is 19 to 21 days, with some delay in implantation if the female is lactating (4). The litter size has ranged from 1 to 7, with 2 to 5 the most common numbers. This series gave an average of $3.4 \pm .2$, and the litter sequence had little influence on the number (4). In the wild, 3 to 5 appears to be usual (5), but litters up to 9 have been recorded (6).

1. Degerböl, M., and U. Möhl-Hansen. Medd. om Grönland, 131(11): 1–40, 1943.
2. Kolthoff, G. Kungl. Svenska Vetansk. Akad. Handlingar, 36(9), 1903.
3. Hansen, R. M. Arctic, 10: 105–117, 1957.
4. Manning, T. H. Arctic, 7: 36–47, 1954.
5. Smith, D. A., and J. B. Foster. JM., 38: 98–115, 1957.
6. Barkalow, Jr., F. S. J. Elisha Mitchell Sci. Soc., 68: 199–205, 1952.

Dicrostonyx rubricatus Richardson

In northeastern Siberia this lemming breeds throughout the year (1), as it does also on Southampton Island (2). In the Hudson Bay region the breeding season is over by mid-August. A post-partum heat is probable (3). The gestation period is from 21 to 22 days (4). The litter size is variable, from 2 to 11, with 3 to 8 the most usual range.

1. Allen, J. A., and N. G. Buxton. AM., 19: 101–184, 1903.
2. Sutton, G. M., and W. J. Hamilton, Jr. Mem. Carnegie Mus., 12(2): 1–111, 1932.
3. Preble, E. A. NAF., 22, 1902.
4. Morrison, P. R., F. A. Ryser, and R. L. Strecker. JM., 35: 376–386, 1954.

Dicrostonyx torquatus Pallas

More than one litter a year is usual. The litter size of young born in the wild was 3.9 (2 to 6), and in captivity 2.6 (1 to 4). The gestation period was 17 days or a little more (1). In continuous warmth, i.e., at 75° F., litters are smaller than they are in the wild. In cold conditions, 15°–22° F., ovarian atrophy occurs but spermatogenesis continues (2).

1. Quay, W. B., and J. F. Quay. Säugetierk. Mitt., 4: 174–180, 1956.
2. Quay, W. B. JM., 41: 74–89, 1960.

Synaptomys cooperi Baird

BOG LEMMING

This lemming breeds from February to November in the United States (1). There is a spring peak in the breeding cycle, but some lemmings continue through the autumn. The male tract retrogresses by mid-November (2). In captivity breeding continues through the winter (2). The gestation period in the lactating females was 23 days and the mean litter size was 3.0, mode 3, range 1 to 5 (2). Another series gave mean 3.2 ± .2, range 1 to 4 or 5 (3).

The sperm head measures 7.7 μ (7.1–8.2) × 4.0 μ (3.9–4.1); the midpiece width 0.7 μ (0.6–0.8); and the tail length 92.0 μ (87.0–96.1) (4).

1. Hamilton, Jr., W. J. Mammals of Eastern United States. Ithaca, N.Y., 1943.
2. Connor, P. F. Michigan State Univ. Mus., Biol. Ser., 1: 165–248, 1959.
3. Coventry, A. P. JM., 18: 489–496, 1937.
4. Hirth, H. F. J. Morphol., 106: 77–83, 1960.

Synaptomys borealis Richardson. This lemming has been found pregnant in May (1) and August (2). Four seems to be the usual embryo number.

1. Munro, J. A. British Columbia Provincial Mus., Occ. Papers, 6, 1947.
2. Rand, A. L. Canada Nat. Mus., Bull. 99, 1945.

Lemmus trimucronatus Richardson

LEMMING

On Southampton Island young are born throughout the year; pregnant females may be found in January and the testes of males are swollen at all times (1). There tend to be two age groups of fetuses at all times of the year, according to one report, suggesting that some breed a little later than others (2). In northern Alaska 21 per cent of females were pregnant in June, 60 per cent in July, and only 5 per cent in September (2). The litter size is variable. In Alaska the June average was 8.1 and that for September, 4.7 (2). Another record gave 7.3, range 4 to 9 (3); another, 7.0, range 3 to 10 (4).

1. Sutton, G. M., and W. J. Hamilton, Jr. Mem. Carnegie Mus., 12(2): 1–111, 1932.
2. Bee, J. W., and E. R. Hall. Univ. Kansas Mus. Nat. Hist., Misc. Publ. 8, 1956.
3. Barkalow, Jr., F. S. J. Elisha Mitchell Sci. Soc., 68: 199–205, 1952.
4. Macpherson, A. H., and T. H. Manning. Canada Nat. Mus., Bull. 161, 1959.

Lemmus lemmus L. In Norway during an irruption year peaks of reproduction were observed from April to early May, mid-June to July, and from the end of August to September (1). Considering the interest shown in these irruption years it is surprising that so little is known about reproduction in lemmings.

1. Curry-Lindahl, K. JM., 43: 171–184, 1962.

Clethrionomys gapperi Vigors

RED-BACKED VOLE

This North American vole is polyestrous and breeds from late winter to late fall in the wild, but throughout the year in captivity (1). There is a post-partum heat (2). The males attain puberty at a length of 120 to 139 mm. (3). Copulation is very brief and several matings occur in a few minutes, after which the female does not desire more. The gestation lasts from 17 to 19 days (1). One series of counts gave the corpora lutea as $6.65 \pm .13$; implantations, $6.2 \pm .12$; and live embryos, $6.1 \pm .12$. The losses were fewer in the larger females (4). A series of embryos from New York State gave a mean of $4.3 \pm .08$; mode 4; range 1 to 7 (5). The litter size varies from 1 to 8 (2).

1. Svihla, A. Michigan Acad. Sci., Arts and Letters, Paper 11: 485–489, 1930.
2. Hamilton, Jr., W. J. JM., 30: 257–260, 1949.
3. Smith, D. A., and J. B. Foster. JM., 38: 98–115, 1957.
4. Beer, J. R., C. F. MacLeod, and L. D. Frenzel. JM., 38: 392–402, 1957.
5. Patric, E. F. JM., 43: 200–205, 1962.

Clethrionomys glareolus Schreber

BANK VOLE

This European vole is polyestrous, with a breeding season from mid-April to the beginning of October. Reproduction is at a maximum in June, and the females usually bear 4 to 5 litters a year (1). In northwestern Scotland breeding starts in mid-May, reaches a peak in June, and then declines rather abruptly (2). At the beginning of the season many females undergo a varying number of cycles, usually 3, before they become pregnant. Females born early in the season reproduce before its close (1), and the young female has her first litter at age 60 days (3). Puberty is at 35 days (4). The spontaneous ovulation is near the end of heat, as copulation occurs before the follicles rupture. A hard copulation plug is formed. There is a postparturient heat. Implantation is delayed during lactation, and in the nonpregnant female lactation anestrum occurs. A winter anestrum is the rule, and the vagina is closed throughout this period (1). The number of young born varies from 2 to 8 (5), average 3.8 ± .02 (6). The gestation period is from 17.5 to 18 days (7), or 20 to 30, averaging 25 days (3). This larger figure may include instances of lactation-delayed implantations.

Cyclical changes in the histology of the reproductive tract are similar to those described for the rat and mouse. The ovaries are surrounded by a closed capsule. The number of corpora lutea averaged 4.4, with an observed maximum of 12, and the embryo count was 4.1, with an observed maximum of 6. The number of ovulations increases to a maximum in June and then falls off. The ripe follicle measures 550 to 800 μ; the ovum, 70 μ; and the corpus luteum, 0.8 to 1.2 mm. The maturation spindle is formed before ovulation, but the first division does not occur until after the follicle has ruptured. The uteri open into the vagina by separate cervices and are distended during heat, at which time, also, intense vaginal cornification occurs (1).

In the male sexual activity begins in March. The mean winter weight of the testes is 40 mg., and the mean summer weight 682 mg. Males born early in the season mature before its end, but those born later mature at the beginning of the next year. Regression of the testes begins in August, and spermatogenesis has ceased by November. The seminal vesicles are comparatively large (8). The head length of the spermatozoon is 6.85 ± .02 μ, and the entire length, 86.7 ± .4 μ (9).

1. Brambell, F. W. R. TRS., 226B: 71–97, 1936.
2. Delany, M. J., and I. R. Bishop. PZS., 135: 409–422, 1960.
3. Heptner, W. G., L. G. Morosowa-Turowa, and W. I. Zalkin. Die Säugetiere in der Schutzwaldzone. Berlin, 1956.
4. Sviridenko, P. A. ZZ., 38: 756–766, 1959.
5. Didier, R., and P. Rode. Mammalia, 3: 111–121, 1939.
6. Barrett-Hamilton, G. E. H. A History of British Mammals. London, 1910.
7. Wrangel, H. V. Zeit. Säugetierk., 14: 52–93, 1939.
8. Rowlands, I. W. TRS., 226B: 99–120, 1936.
9. Friend, G. F. QJMS., 78: 419–443, 1936.

Clethrionomys rufocanus Sundevall

In Finnish Lapland this vole is very dependent upon climatic factors. Males born early in the season are fecund in their first year. The adult females have 3 litters per season, while some of the juveniles have 2. Puberty in the male begins at about 20 to 24 g. weight, or, for the older ones, at 35 to 39 g., and the figures for the females are similar. Births begin in mid-June and there are subsequent waves of births at mid-July and mid-August so that a post-partum heat must occur. Embryo counts range from 2 to 10, mean 6.0 ± .1, mode 6. Resorption is of the order of 1 per cent, so it is not an important factor in fecundity. The litter size varies with the size of the mother (1). In Japan the litter size is 6.4 ± .4, mean 5, range 5 to 8 (2).

1. Kalela, O. Suom. Tied. Toim. (Ann. Acad. Sci. Fenn.), A(4)34, 1957.
2. Imaizumi, Y., and M. Yoshiyuki, J. Mamm. Soc., Japan, 1: 67–70, 1957.

Clethrionomys rutilis Pallas

The young are born from early May to early September, and the young of the earliest litters of the year may breed in the year of their birth. This is at 94 per cent of their adult weight (1), but there is a series of sterile cycles first (2). The testes reduce in size in August (3). There is a post-partum heat (1). The litter size varies from 4 to 10. A compilation of the available data gives an average of 6.3 ± .2; mode 5 and 7. In northern Norway the mean number of corpora lutea was 9.2 and of embryos, 8.2 (2).

1. Manning, T. H. Canada Nat. Mus., Bull. 144, 1956.
2. Hoyte, H. M. D. J. Anim. Ecol., 24: 412–425, 1955.
3. Bee, J. W., and E. R. Hall. Univ. Kansas Mus. Nat. Hist., Misc. Publ. 8, 1956.

Clethrionomys occidentalis Merriam. This vole is polyestrous and breeds into the fall of the year (1), from May to October (2). The gestation period in one female was 18 days (1), and the litters have ranged in size from 2 to 4.

1. Svihla, A. The Murrelet, 12: 54, 1931.
2. McNab, J. A., and J. C. Dirks. JM., 22: 174–180, 1941.

Eothenomys

Eothenomys fidelis Hinton. This East Asiatic species has been twice recorded with 3 embryos (1).

E. kageus Imaizumi. This Japanese species has a mean litter size of 2.5 ± .25, range 1 to 3 (2).

E. melanogaster Milne-Edwards. One and 2 embryos have been recorded (1).

E. smithii Thomas. In Japan the litter size is 3.2 ± .2, range 2 to 4 (2).

1. Hinton, M. A. C. Monograph of the Voles and Lemmings (Microtinae), Living and Extinct. London, 1926.
2. Imaizumi, Y., and M. Yoshiyuki. J. Mamm. Soc., Japan, 1: 67–70, 1957.

Arvicola amphibius L.

WATER VOLE

This vole, which lives in Great Britain, is polyestrous, with a breeding season from the end of March to the end of September, but males are ready to reproduce in February. Animals born early in the year breed during the same season, but cycles in adult females at the beginning of the season may be infertile. There is a postparturient heat and a lactation anestrum. The average number of corpora lutea is 6.4, and of embryos 5.7 (1) or 5.6 (2). The ripe follicle is 700 μ in diameter. The corpus luteum reaches its maximum size, 1.7 mm., fairly late in pregnancy. Large follicles are found at all stages of pregnancy. The diameter of the ovum is 80 μ. The uterus is distended with fluid during heat (1).

1. Perry, J. S. PZS., 112A: 118–130, 1942.
2. Barrett-Hamilton, G. E. H. A History of British Mammals. London, 1910.

Arvicola terrestris L.

VOLE

This vole is polyestrous (1) with a long breeding season extending at least into October (2). The number of young is usually 6 to 8 but it may be as high as 14. In one series the mean litter size was 6.5. It was highest in spring and lowest in autumn; the summer and winter figures were intermediate (3). Several litters a year are produced and the young may reach puberty in their first summer (4).

1. Didier, R., and P. Rode. Mammalia, 3: 111–121, 1939.
2. Cantuel, P. Mammalia, 10: 140–144, 1946.
3. Semenov, M. Y. ZZ., 40: 1743–1745, 1961.
4. Heptner, W. G., L. G. Morosowa-Turowa, and W. I. Zalkin. Die Säugetiere in der Schutzwaldzone. Berlin, 1956.

Ondatra zibethica L.

MUSKRAT

The muskrat is polyestrous and the breeding season varies in length with the latitude and climate. In Manitoba young are rarely born before mid-May—mostly in the last week—or early June. After this first surge of births there is a second smaller one almost exactly a month later, then a lesser third and an almost negligible fourth. Young are very rarely born after August (1). In southwest Idaho the season is about the same length with a peak in early June (2). In Iowa the season is from April to August (3); in Maryland the muskrat breeds at any time except possibly in November and December, but most young are born from mid-April to mid-September (4). In Manitoba the testis size increases with the approach of warm weather, but this change is much more pronounced in wild males than in captive ones. By mid-August reduction has set in and the testes are half their maximum size, but some spermatozoa are present as late as the third week of August (1). Puberty occurs at age 5 months (5).

Views diverge as to the length of the cycle. In one account the modal length was 3 to 5 days, with variation from 2 to 22 days. The nucleated epithelial cells increase slightly in the vaginal smear during proestrum. At heat 90 percent of cells are cornified and 7 per cent are leucocytes (1). Another account gives a 29-day cycle, range 24 to 34 days. In diestrum the vaginal smear consists largely of leucocytes with some nucleated epithelial cells. At the time of heat the vulva swells and sheets of cornified cells are sloughed off. Ovulation occurs spontaneously during heat (6). Pregnancy probably does not occur during the first ovulatory cycle. The vagina closes, usually incompletely, in the latter part of October (6). Each female bears at least two litters a year; but mature ones may produce 4 or even more (5). The litter size averages 6.5; it does not vary much with the season (3), though where the early young of the year reproduce late in the season of their birth they seldom produce more than 3 or 4 young (5). In Manitoba the number of young born averaged 7.3; mode 6; range 4 to 10. The seasonal litter production was 2.7 to 2.8, and the total young about 20 to a female (7). The sex ratio at birth is about 58 per cent males (8). Gestation lasts approximately 28 days (7).

In the female gonadotrophic hormone is highest in the anterior pituitary in springtime and early summer, least in late summer and early winter. The reproductive organs of the females weigh most from March to June and least from July to February (9).

The testes and male accessory organs begin to develop in February, reaching their greatest weight in April to May. They are smallest from September to January (9). Spermatozoa may be found in the epididymis from April to October. Interstitial cells are scanty in the period of rest; they increase just after spermatogenesis begins and decrease after August (5). Placental scar counts should be used cautiously as evidence of fecundity as they persist for a year. The seasonal production of young appears to be about 15.5 by this count, and in the group studied about 3 per cent of mature females did not bear young (10).

The total sperm length is $67.7 \pm .4\ \mu$, and the head length $5.39 \pm .04\ \mu$ (11).

1. McLeod, J. A., and G. F. Bondar. Canadian J. Zool., 30: 243–253, 1952.
2. Reeves, H. M., and R. M. Williams. JM., 37: 494–500, 1956.
3. Errington, P. L. JWM., 5: 68–89, 1941.
4. Smith, F. R. United States Dept. Agric., Circ. 474, 1938.
5. Miegel, B. Zeit. Mikros.-Anat. Forsch., 58: 521–598, 1953.
6. Beer, J. R. JWM., 14: 151–156, 1950.
7. Olsen, P. F. JWM., 23: 40–53, 1959.
8. Errington, P. L. JM., 20: 465–478, 1939.
9. Beer, J. R., and R. K. Meyer. JM., 32: 173–191, 1951.
10. Beer, J. R., and W. Truax. JWM., 14: 323–331, 1950.
11. Friend, G. F. QJMS., 78: 419–443, 1936.

Prometheomys schaposchnikovi Satunin

In the Georgian Republic this vole mates in April and the young are born in May. It has 2 litters a year. Juveniles have litters of 3 to 4, mean 3.5, while adults have from 1 to 5, usually 4 or 5. The females reach puberty at 45–49 g. weight and reproduce in the year of their birth. Lactation lasts for 3 weeks (1).

1. Yatsenko, E. N. ZZ., 38: 916–919, 1959.

Punomys lemminus Osgood

In Peru males are in breeding condition from August onward. A pregnant female was taken in November with 2 fetuses, one in each horn, 23 mm. long. Each ovary contained an extremely large, red corpus luteum and several smaller, apparently functional, corpora lutea (1).

1. Pearson, O. P. HCMZ., 106: 117–174, 1951–2.

Neofiber alleni True

ROUND-TAILED MUSKRAT

This muskrat is believed to breed the year-round. The usual number of young is probably 3, born after a gestation of not less than 30 days (1).

The head of the spermatozoon measures 4.9 μ (4.2–5.1) × 3.0 μ (2.5–3.5); the midpiece 21.0 μ (19.1–22.0) × 0.8 μ (0.7–1.0); and the tail length 70.0 μ (60.0–74.0) (2).

1. Hamilton, Jr., W. J. JM., 37: 448–449, 1956.
2. Hirth, H. F. J. Morphol., 106: 77–83, 1960.

Phenacomys intermedius Merriam

LEMMING MOUSE

At Fort Churchill, Manitoba, this mouse first has its young in mid-June and continues into August or September. The females mostly reach puberty in their first summer, at 4 to 6 weeks old, when they breed, but they produce fewer litters than they do in their second season. The males are not fully mature in their first summer. The current summer females also have smaller litters, 3.8 young, compared with the older ones who average 5.9 young per litter. The overall litter size averages 4.8, range 2 to 8. The duration of pregnancy is 21 to 24 days (1). In western North America the mouse is polyestrous (2) from April to August and the number of young varies from 3 to 8, with 4 to 6 usually (3). Records from Kewatin give June and July counts of 5 and 6 embryos (4).

1. Foster, J. B. JM., 42: 181–198, 1961.
2. Warren, E. R. The Mammals of Colorado. Stillwater, Okla., 1942.
3. Rand, A. L. Canada Nat. Mus., Bull. 100, 1945.
4. Harper, F. Univ. Kansas Mus. Nat. Hist., Misc. Publ. 12, 1956.

Phenacomys longicaudus True

TREE MOUSE

This mouse is polyestrous throughout the year, but mainly from late spring to late summer (1). Ovulation is induced. The cycle is 5.9 days, range 1 to 21 days. Nonlactating females have a 27- to 28.5-day gestation. If they are lactating it may continue for 40 to 41 days (2). The number of young is usually from 1 to 3, mostly 2 (3). There is a post-partum heat (4).

1. Benson, S. B., and A. E. Borell. JM., 12: 226–233, 1931.
2. Hamilton, III, W. J. JM., 43: 486–504, 1962.
3. Howell, A. B. NAF., 48, 1926.
4. Ingles, L. G. Mammals of California and Its Coastal Waters. Stanford, Calif., 1954.

Phenacomys sylvicola Howell. There is one record of 4 young found in April (1).

1. Wight, H. M. JM., 6: 282–283, 1925.

Pitymys pinetorum Le Conte

PINE MOUSE

This mouse breeds from January to early October, with a peak in March and April. The testes are flaccid and small in late September and October; they begin to grow in December. At mating a yellowish, viscous vaginal plug is formed that lasts for 2 days or more. The gestation period is about 21 days (1). In Indiana, October and November embryos have been reported

(2). In Oklahoma the females may breed all the year (3). Compilation of embryo counts gives a range from 1 to 7, mean 2.7 ± .1, mode 2 to 3. Spermatozoa have been found in the epididymides of males from January to March and in August (4).

1. Benton, A. H. JM., 36: 52–62, 1955.
2. Lindsay, D. M. JM., 41: 253–262, 1960.
3. Glass, B. P. JM., 30: 72–73, 1949.
4. Layne, J. N. AMN., 60: 219–254, 1958.

Pitymys duodecimcostatus de Sélys-Longchamps. In France this mouse has been reported with 3 fetuses in April and May (1).

P. quasiter Coues. There is an August record of 1 embryo for this species (2) and an average of 2.2, with a range from 1 to 4 (3).

P. subterraneus de Sélys-Longchamps. This mouse breeds usually from April to September (4), but October embryos have been recorded. The litter size is from 1 to 4; mean 2.6; mode 3 (5).

1. Carpenter, C. J. Mammalia, 10: 92–93, 1946.
2. Davis, W. B. JM., 25: 370–403, 1944.
3. Hall, E. R., and K. R. Kelson. The Mammals of North America. New York, 1959.
4. Heptner, W. G., L. G. Morosowa-Turowa, and W. I. Zalkin. Die Säugetiere in der Schutzwaldzone. Berlin, 1956.
5. Grummt, W. Zool. Anz., 165: 129–144, 1960.

Microtus agrestis L.

FIELD MOUSE

The European field mouse is polyestrous, and in Great Britain the breeding season is from February and March to September and October. Breeding starts later, and finishes earlier, in the south than it does in the north of the range (1). By shortening the length of exposure to light from 15 hours to 9 hours daily, reproduction is almost prevented, but the female is more affected than the male (2). The former experiences a postparturient heat; the gestation period is 21 days; males are sexually mature at 6 to 8 weeks of age, and the females at 3 weeks (3). The average litter size in the laboratory is 3.73, and the sex ratio at birth is 50.89 ± 2.22 per cent males (4).

Wild field mice caught in Wales evidently reach puberty, both males and females, at between 12 and 20 g. Young of both sexes breed during their first season if they are born early enough. The winter anestrum is accompanied by closure of the vagina in the female and by cessation of spermatogenesis in the male. Spermatozoa appear about a month before any pregnant females are found. If the female does not become pregnant at the postparturient heat, lactation anestrum sets in. The vaginal and uterine changes are similar to those found in the house mouse. At copulation a hard vaginal plug is formed. The average number of corpora lutea found was 5.5 ± .2, standard deviation 1.5, and embryos 4.9 ± .2, standard deviation 1.37 (5).

If the mice are kept in mixed groups, continuous heat results and ovulation is coitus-induced (6). Also, if the females are not mated immediately after parturition they come in heat at 4-day intervals and may be impregnated at any of these heats. The ensuing pregnancy is of the same length as it is in the nonlactating female (7).

The ovum has an overall diameter of 96 μ, and, within the vitelline membrane, of 60 μ. The spermatozoon is 104 μ long; the head, 7 μ; and the midpiece, 27 μ. At the time of ovulation a mean number of 124 spermatozoa are available at the site. In the sample studied a mean number of 4.1 ova were shed by each female; 82 per cent of these were fertilized. The first cleavage took place 24 hours after coitus and the ova passed into the uterus 40 to 50 hours after coitus (8).

1. Baker, J. R., and R. M. Ranson. PRS., 113B: 486–495, 1933.
2. Baker, J. R., and R. M. Ranson. PRS., 110B: 313–322, 1932.
3. Ranson, R. M. J. Anim. Ecol., 3: 70–76, 1934.
4. Ranson, R. M. PZS., 111A: 45–57, 1941.
5. Brambell, F. W. R., and K. Hall. PZS., 109A: 133–138, 1939–40.
6. Chitty, H., and C. R. Austin. Nature, 179: 592–593, 1957.
7. Chitty, H. JE., 15: 279–283, 1957.
8. Austin, C. R. J. Anat., 91: 1–11, 1957.

Microtus arvalis Pallas

FIELD MOUSE

In the neighborhood of Paris this mouse is polyestrous and breeds from March to October. The reproductive tract is atrophied during the rest of the year. In warmer climates and in the laboratory reproduction continues in winter. In the wild the testes are large from February to October and the seminal vesicles from March to October. Spermatogenesis begins in March (1). Mature spermatozoa are found in the testis at 35 days of age and in the epididymis at 40 days, while mature follicles are to be found in the ovary at 25 days (2). Testis weight at the time of appearance of spermatozoa is about 200 mg. In Germany the earliest litters have been born from females 33 to 34 days old, but those born in September do not mate until the following spring (3). The first litter is usually born at from 60 to 65 days old (4).

Ovulation is spontaneous and occurs between 12 midnight and 3 A.M. Spermatozoa may be found in the oviduct within 15 to 20 minutes after mating. The young are usually born at night (1). The females experience a post-partum heat. Gestation lasts an average of 19 days, with a spread from 16 to 23 days (4). The number of young is usually 4 to 7, but it may be as many as 15 (4), or 53 per cent of the mother's body weight (3). Litters from immature females are smaller, averaging 4.4, or, in the laboratory, 3.8 (1). Usually a female produces 3 to 5 litters a year (5). One in the laboratory has produced in her lifetime 33 litters with 127 young (3).

The mature graafian follicle measures 560 $\mu \times$ 420 μ; the ovum, 50 $\mu \times$ 70 μ within the zona pellucida (6).

1. Delost, P. Arch. Anat. Micros. Morph. Exp., 44: 150–190, 1955.
2. Delost, P. Arch. Anat. Micros. Morph. Exp., 45: 11–47, 1956.
3. Frank, F. JWM., 21: 113–121, 1957.
4. Heptner, W. G., L. G. Morosowa-Turowa, and W. I. Zalkin. Die Säugetiere in der Schutzwaldzone. Berlin, 1956.
5. Dalimier, P. Mammalia, 19: 498–506, 1955.
6. Delost, P. Bull. Soc. Zool. France, 80: 207–222, 1955.

Microtus californicus Peale

CALIFORNIA MEADOW MOUSE

This mouse is polyestrous all year, with most litters in late winter and spring and a minor peak in the fall (1). The male reaches puberty at 40 g. weight, and the female at 30 to 35 g., when she is 29 days old, but there is considerable variation (2). Ovulation is coitus-induced; stimulation of the cervix with a glass rod does not initiate the reaction. Accordingly the vaginal smear has a prolonged estrous cornified stage. Ovulation follows mating in less than 15 hours. The mean litter size is 4.2, with a range from 1 to 9, but there is much seasonal variation and change with age. Pregnancy lasts for 21 days, and there is a post-partum heat (2). In one experiment the mean corpus luteum count was 9.2, the mean embryo count, 4.2 ± .1, indicating considerable wastage of eggs and embryos during pregnancy (1).

Large follicles measure 500 to 600 μ in diameter, and the ova 65 to 70 μ at ovulation. The maximal postcopulatory follicle size is 900 μ. Fertilization takes place in the upper oviduct; the first polar body is extruded from the ovum before ovulation but after coitus. The zona pellucida has usually been lost by the 2-cell stage, and the 16-cell stage is reached by 48 to 72 hours—an unusually rapid rate of development. The mature corpus luteum, formed from both granulosa and thecal elements, measures 1,400 μ, and it is almost completely solid by 48 to 72 hours. In pregnancy the size increases to 2,000 μ or more. Involution has begun by 3 days after the young are born. The corpora lutea are very numerous and all are of the same age; many of them are "accessory"—formed after those of ovulation. In the young female the first cycle is a sterile one. As in many other species, polyovular follicles are common in the ovaries of young females (1).

1. Greenwald, G. S. JM., 37: 213–222, 1956.
2. Greenwald, G. S. Univ. California Publl. Zool., 54: 421–446, 1957.

Microtus guentheri Danforth and Alston

In the field this Asiatic vole may breed the year-round in favorable circumstances, but late fall and winter are the usual seasons. In warmer districts it is almost completely sterile in summer. From May to July the testes are abdominal. Ovulation is coitus-induced. Puberty is reached in the laboratory at 25 days and older for the females and at 30 days and older for the males, but spontaneous copulation does not begin until 10 to 20 days later. The litter size varies from 3 to 14, or even 17. Fertility is maintained until death at 2½ years for the females and 3 years for the males, but maximum fertility is expressed in the period from 150 to 450 days. The litter size rises from 4.6 for the first litter to 5.6 for the sixth; then it gradually decreases. The average embryo number is higher in irrigated than in non-irrigated fields; at the height of the season it was 8.5 and 5.2 respectively (1).

In the vaginal smear cornified cells are present at estrus and at the time of induced ovulation. A massive invasion of leucocytes follows (2). Pregnancy urine has no effect upon the ovaries, while P.M.S. always causes follicle growth without luteinization. Anterior pituitary extracts produce both (3).

At birth the ovary contains 23,000 ova; at age 4 days, 54,000. Then the number falls to 14,000 by 27 days, and at 75 days it is 8000 (4).

1. Bodenheimer, F. S. Problems of Vole Populations in the Middle East. Jerusalem, 1949.
2. Bodenheimer, F. S., and F. Sulman. Ecol., 27: 215–216, 1946.
3. Zondek, B., and F. Sulman. PSEBM., 43: 86–88, 1940.
4. Bodenheimer, F. S., and W. Lasch. Stud. Biol. Hist. Jerusalem, 1: 9–23, 1957.

Microtus montanus Peale

This vole is polyestrous but it does not breed in winter (1), except perhaps in Utah (2) and other more southerly states. The females first mate at 21 days (2) and the males reach puberty at 33 to 45 days (3). There is a post-partum heat and gestation lasts 20.5 to 21 days. The litter size averages 4.5, range 3 to 6 (3); or 2 to 8, mode 5 and 6, mean 6.3 (4). The vulva is sealed in the nonbreeding season (3).

The adrenals of sexually active males are smaller than those of inactive males. In the females the reverse is found (5).

1. Orr, R. T. Mammals of Lake Tahoe. San Francisco, 1949.
2. Bailey, V. NAF., 55, 1936.
3. Seidel, D. R., and E. S. Booth. Walla Walla Coll., Dept. Biol. Sci., Publ. 29, 1960.
4. Hall, E. R. Mammals of Nevada. Berkeley, Calif., 1946.
5. McKeever, S. AR., 135: 1–5, 1959.

Microtus ochrogaster Wagner

This vole of midwestern North America is polyestrous and breeds year-round, but most pregnancies are found in August and September and fewest in December and January. Most females breed before they are 6 weeks old, although those born in October wait until they are 15 weeks old (1).

In Ohio the females breed at over 12.5 cm. body length, at a weight of 20.5 g. and over (2). The gestation period is 21 days, or a little less, and the mean litter size is $3.4 \pm .75$ (3). The average female has 4.1 litters a year. The mean litter size in this group was $3.2 \pm .24$, mode 3, range 1 to 6 (1). Another count gave 4.7 placental scars and 3.5 embryos and fetuses; this suggests considerable wastage, but the sample was small (4). Another group gave a highest embryo count in March, when it was 3.9, with an average for the year of 3.4. In February 90 per cent of males had spermatozoa in their testes, and in March 77 per cent of the females were reproducing (5).

The dimensions of the sperm head are 7.9 μ (7.8–8.3) $\times$ 4.9 μ (4.7–5.0); those of the midpiece, 21.8 μ (21.0–23.0) $\times$ 0.8 μ (0.7–0.9); and the tail length is 86.5 μ (89.0–91.1) (6).

1. Martin, E. P. Univ. Kansas Mus. Nat. Hist., Publ. 8: 361–416, 1954–6.
2. De Coursey, Jr., G. E. JM., 38: 44–52, 1957.
3. Fitch, H. S. Univ. Kansas Mus. Nat. Hist., Publ. 10: 129–161, 1957.
4. Layne, J. N. AMN., 60: 219–254, 1958.
5. Jamison, E. W. Univ. Kansas Mus. Nat. Hist., Bull. 1: 125–151, 1947.
6. Hirth, H. F. J. Morphol., 106: 77–83, 1960.

Microtus oeconomus Pallas

This northern vole has a restricted season that ends about the last week of August. Puberty in the females is reached at about 9 to 10 cm. head and body length (1). The first heat is at about 40 days of age (2). The species is polyestrous and ovulation is spontaneous. The cycle length varies from 6 to 11 days and the vaginal smear is similar to those of the rat and mouse (1). The average litter size (embryos) is 7.5 (3), or 5.1, range 4 to 8 (2). The females probably lose their litters if they are lactating (1). Gestation is 20 to 21 days, and there is a post-partum heat (4).

1. Hoyte, H. M. D. J. Anim. Ecol., 24: 412–425, 1955.
2. Frank, F., and K. Zimmermann. Zeit. f. Säugetierk, 21: 58–83, 1956.
3. Bee, J. W., and E. R. Hall. Univ. Kansas Mus. Nat. Hist., Misc. Publ. 8, 1956.
4. Linn, I. In A. N. Worden and W. Lane-Petter, eds., The UFAW Handbook on the Care and Management of Laboratory Animals. London, 1957.

Microtus oregoni Bachman

OREGON CREEPING MOUSE

This mouse is polyestrous and breeds from early March until mid-September, but occasionally later; fertile females have been taken until November 23. The vulva is closed during the winter anestrum. Puberty is reached in females at 22 to 24 days and in males at 34 to 38 days, but the age varies with the season. No pregnancies are found until the females are at least 42 days old. There is a sterile period between the first heat and conception. The number of embryos varies from 1 to 5, mean 3.1, in wild-caught females, and the mean number of births in captivity is 2.8. In each group 3 was the mode. The number increases with litter sequence. Pregnancy lasts 23½ to 25 days and there is no indication that lactation prolongs it. A post-partum estrus occurs frequently but not always. Among 78 implantations 49 were in the right horn of the uterus and 29 in the left. One female had 6 litters in 320 days during a season artificially prolonged by suitable light and heating conditions (1).

1. Cowan, I. McT., and M. G. Arsenault. Canadian J. Zool., 32: 198–208, 1954.

Microtus pennsylvanicus Ord

FIELD MOUSE

This field mouse is polyestrous and breeds all the year round in the laboratory. In the wild it tends not to breed in the winter except in the south or in warm conditions. The females reach puberty and begin to breed at 25 days of age; the males at 45 days. In captivity the begin to breed before they are half grown (1). At Churchill, Manitoba, where the breeding season is restricted from late April or early May to August, at which time the testes regress, puberty is reached in both sexes at about 146 mm. body length, i.e., in the season following that of their birth (2). In a Minnesota winter during which there was no snow cover, reproduction ceased from November to March. In a year with snow cover the mice reproduced throughout the year. Females over 110 mm. long produced litters at the rate of 10.7 a year. In this sample embryo counts averaged 5.7, range 1 to 11. The calculated mean litter size at birth was 5.2 and it increased with the size of the mother. In females, 90 to 94 mm. was the lower range of size for fecundity. Testes descended at 65 to 69 mm. and above, but the males were not usually fecund until they were over 100 mm. long (3). A female can have as many as 17 litters in one year (1). In a series taken in the wild the average number of embryos was 5.4., which represented 85 per cent of the eggs shed. Recognizable fetal resorption occurred to the extent of 3.6 per cent of the embryos, the modal number of which was 5. Combination of these data with another series (4) gives the following figures for fetal litter size: mean 5.5 ± .06, standard deviation 1.55, mode 5, range 1 to 11. Findings for another series were: corpora lutea 6.1 ± .1; implantations 5.7 ± .1; and live embroys 5.6 ± .1. The extent of loss decreased with increasing size of the mothers but there was no well-defined seasonal trend (5). The average litter size increases to the fifth litter and then declines (6).

The ovaries are enclosed in a capsule, and the corpora lutea are salmon-colored. The vulva perforates at the first heat and is closed between heats. The clitoris is prominent. Copulation is rapid, and a vaginal plug is formed which lasts for 2 days. Copulation will take place immediately after parturition, and suckling does not lengthen the consequent gestation, which lasts 21 days or a little less. Low temperature alone does not interfere with breeding in winter. (7).

During heat the vaginal epithelium cornifies, and the copulation plug is lined with cells of this type, which break away from the wall. Changes in the smear are typical of the rat and mouse, and leucocytes invade the vagina soon after the end of heat. Corpus luteum formation is typical, but the amount of liquor folliculi which is retained seems to be unusually large (8).

The male has a well-defined scrotum, but the testes are often partly retracted. The seminal vesicles are large and directed cephalad for two thirds of their length, then they turn ventrally and caudally. The prostate is large and compound, with well-marked lobes. The bulbo-urethral glands are small and somewhat pear-shaped. Preputial glands are present, but small (6).

1. Bailey, V. J. Agric. Res., 27: 523–536, 1924.
2. Smith, D. A., and J. B. Foster. JM., 38: 98–115, 1957.
3. Beer, J. R., and C. F. MacLeod. JM., 42: 483–489, 1961.
4. Goin, O. B. JM., 24: 212–223, 1943.
5. Beer, J. R., C. F. MacLeod, and L. D. Frenzel. JM., 38: 392–402, 1957.

6. Poiley, S. M. JM., 30: 317–318, 1949.
7. Hamilton, Jr., W. J. Cornell Univ. Agric. Exp. Sta., Mem. 237, 1941.
8. Asdell, S. A., and W. J. Hamilton, Jr. Unpublished work.

Microtus chrotorrhinus Miller. The rock vole breeds from early spring to midautumn (1). The litter size varies from 2 to 5, with a mean of 3.6 ± .2 embryos (2).

M. gregalis Pallas. This vole breeds during the warm parts of the year. The litter size is usually 7 to 9, but it may be as high as 15 (3).

M. incertus de Sélys Longchamps. This mouse breeds throughout the year. It usually has 2 or 3 young, range 1 to 4 (4).

M. longicaudus Merriam. In the Yukon 3 to 6 young are born in the warmer months (5). In New Mexico pregnancies have been noted in April and August (6), and the usual season is from May to October, at least. Embryo counts varied from 2 to 8, with 5 and 6 the mode, and 5.6 the mean (7).

M. mexicanus Saussure. This vole is polyestrous (8), and the season is at least from May to the end of August. The interval between litters is about 30 days and the average number of young is 3 (9), or 2.7, mode 3 (10).

M. miurus Osgood. In Alaska this species, the singing vole, ceases to breed in August or a little later. The testes are large before late June. The females reach puberty at a length of 131 mm. and the average litter size at the height of the season is 8.2 embryos (11).

M. montebelloi Milne-Edwards. In Japan the average litter was 5.0 ± .26, range 3 to 8 (12).

M. mordax Merriam. This Rocky Mountain meadow mouse is polyestrous and breeds from May to September (8). The litter size is 2 to 8, usually 4 to 5, mean 4.1 ± .4. The female becomes pregnant when she is less than half grown (13).

M. nanus Merriam. The dwarf meadow mouse is polyestrous; the litter size varies from 4 to 10 (8), but is usually 6 to 8 (14).

M. nivalis Martins. In France this species is polyestrous, bearing 2 or 3 litters a year; the number of young is usually 3 or 4 (15).

M. operarius Nelson. In the Yukon this species has 2 or more litters a year, averaging 6 young, born in the warmer months (5).

M. orcadensis Millais. In the laboratory the Orkney vole had a mean litter size of 2.7; the highest was 8. The number rises with age to 22 to 33 weeks and then falls. The average age at first litter was 148 ± 11 days and the median, 123 days (16).

M. richardsoni De Kay. The water vole has been recorded with 4 to 7 embryos (14). The mean is 5, born from late June to late September (17). The males are sexually active at 28 g. weight and the females at 34 g. (18).

M. sikkimensis Hodgson. This vole has been recorded as having 3 to 4 young (19).

M. socialis Pallas. This vole breeds from early spring into winter, but mostly from early spring to early summer and then in the fall. Up to 9 young are produced each 30 to 35 days, with 3 to 5 litters a year and a gestation of 19 to 20 days. The young are capable of breeding at 45 to 60 days old (3).

M. townsendii Bachman. The litter size averages 4, range 1 to 9, born from March to September (20).

M. xanthognathus Leach. In the Yukon this vole has up to 11 young to a litter (5).

M. (Pedomys) ludovicianus Bailey. This vole breeds from late March to October and has 2 to 5 young (1).

M. (Phaiomys) blythi Blanford. One record gives 6 young (19).

1. Hamilton, Jr., W. J. The Mammals of Eastern United States. Ithaca, N.Y., 1943.
2. Coventry, A. P. JM., 18: 489–496, 1937.
3. Heptner, W. G., L. G. Morosowa-Turowa, and W. I. Zalkin. Die Säugetiere in der Schutzwaldzone. Berlin, 1956.
4. Lataste, F. Act. Soc. Linn. Bordeaux, 40: 293–466, 1886.
5. Rand, A. L. Canada Nat. Mus., Bull. 100, 1945.
6. Holdenried, R., and H. B. Morlan. AMN., 55: 369–381, 1956.
7. Hall, E. R. Mammals of Nevada. Berkeley, Calif., 1946.
8. Bailey, V. NAF., 53, 1931.
9. Davis, W. B. JM., 25: 370–403, 1944.
10. Baker, R. H. Univ. Kansas Mus. Nat. Hist., Publ. 9: 125–335, 1956.
11. Bee, J. W., and E. R. Hall. Univ. Kansas Mus. Nat. Hist., Misc. Publ. 8, 1956.
12. Imaizumi, Y., and M. Yoshiyuki. J. Mamm. Soc., Japan, 1: 67–70, 1957.
13. Barnes, C. T. Utah Mammals. Salt Lake City, Utah, 1927.
14. Bailey, V. NAF., 55, 1936.
15. Didier, R., and P. Rode. Mammalia, 3: 19–37, 1939.
16. Leslie, P. H., J. S. Tener, M. Vizoso, and H. Chitty. PZS., 125: 115–125, 1955.
17. Cowan, I. McT., and C. F. Guinguet. British Columbia Provincial Mus., Handbook 11, 1956.
18. Negus, N. C., and J. S. Findley. JM., 40: 371–381, 1959.
19. Blanford, W. T. Fauna of British India. Mammals. London, 1888–91.
20. Orr, R. T. Mammals of Lake Tahoe. San Francisco, 1949.

Lagurus curtatus Cope

SAGEBRUSH VOLE

This vole breeds all the year round (1), even at an elevation of 9,800 ft. (2). It mates again almost immediately after a litter is born. The litter size varies from 1 to 11, with 6.4 the mean (3), or 6.3, with a mode of 6 and 7 (1). The gestation lasts from 24 to 26 days (3).

1. Johnson, M. L., C. W. Clanton, and J. Gerard. The Murrelet, 29: 44–47, 1948.
2. Dunmire, W. W. JM., 42: 489–493, 1961.
3. James, W. B., and E. S. Booth. Walla Walla Coll., Publ. 1: 23–43, 1952.

Lagurus lagurus Pallas

This vole breeds in the warm parts of the year, and even in winter (1). Puberty is reached in both sexes at about 20 to 25 days, but if there is a drought the summer-born young do not reach puberty until autumn (2). Up to 7 litters a year may be produced, usually 4 or 5, and the mean is 6, with a range from 2 to 11.The gestation lasts 20, range 15 to 23 days (1).

1. Heptner, W. G., L. G. Morosowa-Turowa, and W. I. Zalkin. Die Säugetiere in der Schutzwaldzone. Berlin, 1956.
2. Kryltsov, A. I. ZZ., 36: 1239–1250, 1957.

Ellobius talpinus Pallas

MOLE-LEMMING

This species is polyestrous and breeds all year. It begins to reproduce at 90 days old. The duration of pregnancy is 26 days and the interval between litters is 34 to 36 days. In the south the litter size averages 3.9, and in the north 3.2 (1).

1. Zubko, Y. P., and S. I. Ostryakov. ZZ., 40: 1577–1579, 1961.

GERBILLINAE

Gerbillus simoni Lataste

GERBIL

This gerbil from Algeria is polyestrous and breeds throughout the year. The female usually comes in heat during the afternoon or evening and is out of heat next morning. The cycle is said to last about 10 days. A vaginal plug is formed after copulation. Gestation lasts 20 to 21 days, usually 20. The mean litter size is 4.7, range 1 to 7. There is a post-partum heat. Puberty in the female is at about 2 months and in the male at 60 days or more (1). If the female is suckling and becomes pregnant, implantation of the embryos is delayed (2). More information is needed on this species since the records of a 10-day cycle were made when pseudopregnancy had not yet been recognized. The cycle in the unmated gerbil may be shorter. These remarks also apply to other species of gerbils investigated by Lataste.

1. Lataste, F. Act. Soc. Linn. Bordeaux, 40: 293–466, 1886.
2. Lataste, F. CRSB., 43: 21–31, 1891.

Gerbillus campestris Levaillant. This gerbil is polyestrous and has a cycle of about 10 days (1). Gestation is less than 28 days (2).

G. cheesmani Thomas. In the wild a specimen with 3 embryos is recorded for September (3).

G. dasyurus Wagner. This species has been reported twice with 4 embryos each, in February and October (4).

G. gerbillus Olivier. In the London zoo this species has had from 3 to 5 young in April, May and June (5).

G. pyramidium Geoffroy. This species is polyestrous and has a 10-day cycle (1). In the London zoo litters of 2 and 5 young, born in August, have been recorded (5). Gestation is 25 days (2).

G. quadrimaculatus Lataste. There is a record of a female with 4 young (6).

G. swalius Thomas and Hinton. The litter number varies from 2 to 6, and fetuses have been found in April, October, and December (7).

1. Lataste, F. Act. Soc. Linn. Bordeaux, 40: 293–466, 1886.
2. Petter, F. Mammalia, Special Number, 25: 1–222, 1961.
3. Harrison, D. L. Ann. Mag. Nat. Hist., 12(8): 897–910, 1955.

4. Hatt, R. T. The Mammals of Iraq. Univ. Michigan Mus. Zool., Misc. Publ. 106, 1959.
5. Zuckerman, S. PZS., 122: 827–950, 1952–3.
6. Anderson, J. The Mammals of Egypt. London, 1902.
7. Shortridge, G. C. The Mammals of South West Africa. London, 1934.

Tatera afra Gray

GERBIL

The females experience a complete anestrum from the beginning of April to the end of July; the breeding peaks are in September and January (1). Females that are born early in the season breed during the same season (1). The males reach puberty at a weight of 40 to 60 g., or at about 6 weeks old, and those born early in the season breed the same season, but those born in December wait until the following spring, when they are 5 or 6 months old. Older males cease to breed in November and December but young ones continue into January and early February. Spermatogenesis occurs in the testes that weigh 3 g. or more the pair. The testes are in the abdomen in the nonbreeding season (2).

The number of corpora lutea averages 4.1, with a range from 2 to 6, and a mode of 4 and 5, while embryo counts averaged 4.0, mode 4. The number of ovulations, and also the embryo count, rise from August to December and then decrease. The number is also related to body weight. There is a post-partum heat and a lactation anestrum, but breeding may recommence immediately after lactation has finished. Ovulation is spontaneous. At coitus a vaginal plug is formed which remains for a brief period. The ovary is enclosed in a capsule. The vagina is closed in immature and anestrous females and sometimes in pregnant and lactating ones. Delayed implantation occurs in the pregnant lactating female (1).

1. Measrock, V. PZS., 124: 631–658, 1954–5.
2. Allanson, M. PZS., 130: 373–396, 1958.

Tatera brantsi A. Smith

GERBIL

In the wild this South African gerbil breeds the year-round, but the greatest intensity is from March to June and, again, to a lesser extent, in September. Breeding is least in July and from October to December. In the latter half of the year the young females increase considerably in weight without attaining puberty, which is delayed until January. In the first half of the year breeding begins at 55 g. weight (1). For males the size at puberty during the period from January to August is 50 to 65 g., or a length of from 12 to 14 cm. From October to December the corresponding figures are 74 to 98 g. and 13.8 to 16.2 cm. The rainy season is in September and October. Spermatozoa are present the year-round in testes that weigh 2 g. or more for the pair (2).

Ovulation is spontaneous and the cycle lasts for 4 to 6 days, of which 16 to 18 hours is proestrum, 20 to 28 hours estrus, 24 hours metestrum, and 2 days the diestrum. The average litter size is 2.75, mode 3; and gestation lasts for 22.5 days. These figures refer to females kept in the laboratory. In wild ones corpus-luteum counts gave an ovulation rate of 2.94, mode 4, range 1 to 4. Embryos averaged 2.64, with 3 the most frequent number (1).

The ovary is encapsulated and the vagina is closed in the immature and anestrous females, and sometimes in pregnant and lactating ones. At the time of heat the vulva is pinkish and in folds; the uterus is distended with fluid. The proestrous vaginal smear is composed mainly of cornified cells, while leucocytes make their appearance in metestrum. After copulation a vaginal plug rarely forms, and when it does, it lasts only briefly. Blood is present in the vaginal smear of the pregnant female at 14 to 15 days. In laboratory specimens a post-partum heat was detected, and lactation anestrum followed if the females were not impregnated at that time. In wild specimens there was evidently a post-partum heat with an ensuing pregnancy. A short resting period intervenes between the end of lactation and a new heat. In the lactating pregnant female delayed implantation is usual (1).

1. Measrock, V. PZS., 124: 631–658, 1954–5.
2. Allanson, M. PZS., 130: 373–396, 1958.

Tatera indica Hardwicke

This Indian antelope rat has litters from the last week of September to the first week of March. The males reach puberty in from 12 to 14 weeks; their testes are active from July to the end of April and are regressed in May and June. The interstitial cells and accessory organs follow the same cycle (1). The number of embryos varies between 5 and 8 (2), mean 6.3. Intrauterine mortality accounts for 8 per cent of embryos. The females reach puberty at 8 to 9 weeks and have 3 or 4 litters a season.

1. Prasad, M. R. N. Acta Zool., 37: 87–122, 1956.
2. Prasad, M. R. N. J. Bombay Nat. Hist. Soc., 52: 184–189, 1954.
3. Prasad, M. R. N. Acta Zool., 42: 245–256, 1961.

Tatera dichrura Thomas. Young of this African species have been found in March, November, and December. On one occasion 6 was the count of nest young (1).

T. lobengulae De Winton. A pregnancy has been noted in March (2).

T. nigricauda Peters. Juveniles have been found in June (3).

T. nyasae Wroughton. Juveniles have been found in August and September (4).

T. schinzi Noack. This gerbil breeds all the year, but mostly in spring and summer. The litter size is from 2 to 8, usually 4 or 5 (5). Puberty is reached at 3 months old (6).

T. valida Bocage. In Angola this gerbil has been recorded with 5 young, born in April (7).

T. vicina Peters. In East Africa this gerbil has been found in November and December with 4 and 5 fetuses (8).

1. Hatt, R. T. AM., 76: 457–604, 1939–40.
2. Frechkop, S. EPNU., 1944.
3. Allen, G. M., B. Lawrence, and A. Loveridge. HCMZ., 79: 31–126, 1935–7.
4. Lawrence, B., and A. Loveridge. HCMZ., 110: 1–80, 1953.
5. Shortridge, G. C. The Mammals of South West Africa. London, 1934.
6. Powell, W. Rodents. Dept. Public Health, South Africa, 1925.
7. Hill, J. E. JM., 22: 81–85, 1941.
8. Heller, E. In G. C. Shortridge, The Mammals of South West Africa. London, 1934.

Taterillus emini Thomas

This gerbil has been reported with 3 and 4 newborn young in November and December (1).

1. Hatt, R. T. AM., 76: 457–604, 1939–40.

Desmodillus auricularis A. Smith

SHORT-EARED GERBIL

This gerbil breeds at 3 months old and has 4 to 6 young about 4 times a year (1).

1. Powell, W. South Africa Dept. Public Health, Pamphl. 321, 1932.

Pachyuromys duprasi Lataste

BOUBIÉDA

This North African gerbil is polyestrous. After copulation a vaginal plug is formed with cornified tissue around it. The gestation period is 19 to 22 days, usually 20. The number of young is 3 to 6, mean 3.4, mode 3. Puberty in the female occurs at about 2 months old. Lataste's supplemental data suggest a cycle of 5 or 6 days. Matings are usually at night and births at midday (1). In the London zoo births occur the year-round and the litter size has varied from 1 to 9, with 5 the average (2).

1. Lataste, F. Act. Soc. Linn. Bordeaux, 40: 293–466, 1886.
2. Zuckerman, S. PZS., 122: 827–950, 1952–3.

Meriones longifrons Lataste

JIRD

This jird is polyestrous, breeding throughout the year, with a cycle of about 10 days. The gestation period is about 21 days, with delayed implantation if the female is suckling (1). Puberty in the female is at 2 months (2). In the London zoo births have occurred all year-round. The litters have varied from 1 to 8, mean 3.7, mode 4 (3), which compares with Lataste's average of 4.5 (2).

1. Lataste, F. Mem. Soc. Biol., 43: 21, 1891.
2. Lataste, F. Act. Soc. Linn. Bordeaux, 40: 293–466, 1886.
3. Zuckerman, S. PZS., 122: 827–950, 1955.

Meriones lybicus Lichtenstein

JIRD

This jird is polyestrous, breeding throughout the year, with a cycle of about 10 days. The gestation is about 20 to 25 days, with delayed implantation in lactating females (1, 2). There is a post-partum heat. Immature young have been collected from April to June; also males with enlarged testes, and pregnant females, so the breeding season is long. Embryo counts of 4, 6 and 7 have been made (3).

1. Lataste, F. Mem. Soc. Biol., 43: 21, 1891.
2. Lataste, F. Act. Soc. Linn. Bordeaux, 40: 293–466, 1886.
3. Harrison, D. L. Ann. Mag. Nat. Hist., 12(8): 897–910, 1955.

Meriones hurrianae Jerdon. The young of the desert jird are born from August to November and the litter size is from 3 to 5 (1, 2).

M. meridianus Pallas. In this Asiatic species breeding begins in March and continues through September. Three or 4 litters a year with 4 to 7 young are usual. Lactating females may be pregnant (3). In the northwestern Caspian region it breeds 3 to 5 months longer than it does in the Volga-Ural sands. Its most intensive season is at the end of summer (4).

M. persicus Blanford. In captivity this Asiatic species made its first matings at age 98 days. Gestation lasted for 22 days (5). In a state of nature it bred at the end of winter (6).

M. tamaricinus Pallas. This species breeds 3 to 5 months longer in the northwestern Caspian region than in the Volga-Ural sands (4). The gestation period is 25 to 29 days (7).

M. tristrami Thomas. The gestation period is less than 24 days (6).

M. vinogradovi Heptner. In the wild this species has 2 litters a year but in captivity it has up to 5. They do not breed in their first year. The gestation period lasts 21.5 to 23 days (6).

1. Prakash, I. JM., 41: 386–389, 1960.
2. Prakash, I. Mammalia, 26: 311–331, 1962.
3. Heptner, W. G., L. G. Morosowa-Turowa, and W. I. Zalkin. Die Säugetiere in der Schutzwaldzone. Berlin, 1956.
4. Pavlov, A. N. ZZ., 38: 1876–1885, 1959.
5. Eibl-Eibesfeldt, I. Zeit. Tierpsychol., 8: 400–423, 1951.
6. Petter, F. Mammalia, Special Number, 25: 1–222, 1961.
7. Rauch, H. G. Zeit. Säugetierk., 22: 218–240, 1957.

Psammomys obesus Cretzchmar

In the London zoo, births have occurred at all times of the year (1). From 3 to 7 young are born after a gestation of 25 days (2).

1. Zuckerman, S. PZS., 122: 827–950, 1952–3.
2. Dekeyser, P. L. Les Mammiféres de l'Afrique Noire Française. np., 1956.

Rhombomys opimus Lichtenstein

This species breeds from March to June. The gestation lasts 22 to 25 days, and two litters of 6 young each are usually produced in the season (1). In captivity a post-partum heat has been observed with pregnancy lasting for 29 to 34 days. In the absence of lactation it was 24 days. The placental scars resorb 3 months after the litter is cast (2).

1. Petter, F. Mammalia, Special Number, 25: 1–222, 1961.
2. Leontyeva, M. N. ZZ., 40: 1874–1882, 1961.

SPALACIDAE

Spalax kirgisorum Nehring

MOLE RAT

In Israel this rat breeds in December and January and pregnant females can be found only from January to March. Sixty-eight per cent are born from January 15 to February 15, but there is a second peak of births in the second half of March. The gestation period is above 28 days. The litter size varies from 1 to 9, with a very pronounced mode at 3 and 4 (1).

1. Nevo, E. Mammalia, 25: 127–144, 1961.

Spalax microphthalamus Güldenstaedt. The blind mouse breeds early in the year. It has 1 or 2 annual litters of 2 to 4 young (1). The litters are born in late February or early March. They consist of 4 to 5, occasionally 3, young (2).

1. Heptner, W. G., L. G. Morosowa-Turowa, and W. I. Zalkin. Die Säugetiere in der Schutzwaldzone. Berlin, 1956.
2. Samarsky, S. L. ZZ., 41: 1583–1584, 1962.

RHIZOMYIDAE

Rhizomys

BAMBOO RAT

Rhizomys pruinosus Blyth. This rat has 3 to 4 young to the litter (1).
R. vestitus Milne-Edwards. This rat has 3 to 4 young to the litter (2).

1. Blanford, W. T. Fauna of British India. Mammals. London, 1888–91.
2. Ho, H. J. Contrib. Biol. Lab. Sci. Soc. China, Zool. Ser., 11: 123–164, 1935.

MURIDAE

MURINAE

Vandeleuria oleracea Bennett

TREE MOUSE

This Indian mouse has 3 to 4 young to the litter (1).

1. Blanford, W. T. Fauna of British India. Mammals. London, 1888–91.

Micromys minutus Pallas

LITTLE HARVEST MOUSE

This mouse reaches puberty at 45 to 59 days old, and the young of early spring may produce a litter in August (1). The males reach puberty at 5 g. or up (2), or at 35 to 49 days of age (3). The females are not in breeding condition in March (2). They breed in the summer months (3). The number of young averages 3.8, mode 4, range 2 to 6, born after a gestation of 21 days. The females reach puberty at 31 to 32 days and bear their first litter at 52 days (3). In Japan the average litter size in the wild was 5.3, and in captivity 2.3. Most litters were born in the spring. (4).

The length of the spermatozoon is 63.9 ± .8 μ, and that of the sperm head, 5.67 ± .04 μ (5).

1. Kubik, J. Ann. Univ. Mariae Curie-Sklodowska, Sect. C, Biol., 7: 449–495, 1952.
2. Rowe, F. P. PZS., 131: 320–323, 1958.
3. Frank, F. Zeit. Säugetierk., 22: 1–44, 1957.
4. Shiraishi, S. J. Mamm. Soc. Japan, 1: 121–127, 1959.
5. Friend, G. F. QJMS., 78: 320–323, 1958.

Apodemus sylvaticus L.

FIELD MOUSE

This mouse is polyestrous and breeds from March to October (1). The young breed at 80 to 90 days old (2). On St. Kilda the males become fecund in February and March and the females become pregnant from April to June. Some females may reproduce in the year of their birth (3). On Fair Island these mice breed from late February to early September, but most intensively from early June to late August (4). The litter size is from 1 to 12, with 5 or 6 the usual number (2). One series of young gave 5 as the mode, mean 4.8 ± .2, range 1 to 9 (5). Another gave a mean of 5.3, with somewhat smaller litters in winter, and a strong tendency for more embryos to be carried in the right uterus (6). The gestation period lasts from 23 to 29 days (7). There is a post-partum heat (8). In December and January the testes are abdominal and the accessory organs are small. In March the testes have hypertrophied, but not appreciably the accessory organs (1). The testes can be made functional in winter by good feeding (9). The length of the spermatozoon is $132.8 \pm .5$ μ, and of the sperm head, $9.83 \pm .04$ μ (10).

1. Raynaud, A. CRSB., 144: 938–940, 1950.
2. Heptner, W. G., L. G. Morosowa-Turowa, and W. I. Zalkin. Die Säugetiere in der Schutzwaldzone. Berlin, 1956.
3. Boyd, J. M. PZS., 133: 47–65, 1959–60.
4. Delany, M. J., and P. E. Davis. PZS., 136: 439–452, 1961.
5. Barrett-Hamilton, G. E. H. A History of British Mammals. London, 1910.
6. Semenov, M. Y. ZZ., 40: 1743–1745, 1961.
7. Kenneth, J. H. Gestation Periods. Edinburgh, 1943.
8. Laver, —. Reference lost.
9. Raynaud, A. CRSB., 144: 945–948, 1950.
10. Friend, G. F. QJMS., 78: 419–443, 1936.

Apodemus agrarius Pallas. This species has 3 to 5 litters a year with 5 to 6 young to a litter. The young reach puberty at 2½ months old (1). In Korea breeding is probably over for the season by October (2).

A. flavicollis Melchior. This mouse has 2 to 4 litters a year, usually with 6 young apiece, but the number may be as high as 12. The young females have litters at about 3 months old and there is an interval of from 49 to 60 days between litters (1). In Iraq a female with 5 embryos was found in November (3). The sperm head is $8.78 \pm .04$ μ long and the whole spermatozoon measures $125.4 \pm .5$ μ (4).

A. mystacinus Danford and Alston. In Bulgaria this mouse reproduces from March to October. The modal litter size is 4, range 2 to 9 (5).

A. speciosus Temminck. In central Korea the females were in breeding condition in August but the males not until the end of the month (2).

1. Heptner, W. G., L. G. Morosowa-Turowa, and W. I. Zalkin. Die Säugetiere in der Schutzwaldzone. Berlin, 1956.
2. Jones, J. K., and A. A. Barber. JM., 38: 377–392, 1957.
3. Hatt, R. T. The Mammals of Iraq. Univ. Michigan Mus. Zool., Publ. 106, 1959.
4. Friend, G. F. QJMS., 78: 419–443, 1936.
5. Pechev, T. Mammalia, 26: 293–310, 1962.

Thamnomys surdaster Thomas and Wroughton

This species breeds mainly late in the rains and early in the dry season, from September to April (1). The number of young varies from 1 to 5, with 2 or 3 the commonest. Litters are regularly born at 5- or 6-week intervals, but those that were born in captivity failed to breed (2).

1. Pirlot, P. L. Ann. Mus. Congo Belge Tervueren, Ser. Zool. Ser. Nouv. en 4°, 1: 41–46, 1954.
2. Rodhain, J. Ann. Mus. Congo Belge Tervueren, Ser. Zool. Ser. Nouv. en 4°, 1: 74–76, 1954.

Mesembriomys gouldi Gray

JERBOA RAT

In North Queensland a female with 2 young has been reported (1). This is the usual litter size (2).

1. Le Souef, A. S., and H. Burrell. The Wild Animals of Australasia. London, 1926.
2. Troughton, E. Furred Animals of Australia. New York, 1947.

Oenomys hypoxanthus Pucheran

This African rat breeds from August to April (1, 2), or in the dry season. The litter size varies from 1 to 5, mean 2.7, mode 3 (2).

1. Sanderson, I. T. TZS., 24: 623–725, 1938–40.
2. Hatt, R. T. AM., 76: 457–604, 1939–40.

Dasymys incomtus Sundevall

WATER RAT

This rat has an extended season, from June to October in South West

Africa (1). In Central Africa pregnancies have been noted in January (2) and April (3).

1. Shortridge, G. C. The Mammals of South West Africa. London, 1934.
2. Hatt, R. T. AM., 76: 457–604, 1939–40.
3. Hill, J. E. JM., 22: 81–85, 1941.

Arvicanthus

STRIPED MOUSE

Arvicanthus abyssinicus Rüppell. There is a record of 6 fetuses in June (1).

A. niloticus Desmarest. In the London zoo this mouse has had 2 and 4 young in February and June (2).

1. Allen, G. M., and A. Loveridge. HCMZ., 75: 47–140, 1933–4.
2. Zuckerman, S. PZS., 122: 827–950, 1952–3.

Hadromys humei Thomas

BUSH MOUSE

There is a September pregnancy record (1).

1. Roonwal, M. L. Trans. Nat. Inst. Sci., India, 3: 67–122, 1949.

Golunda ellioti Gray

COFFEE RAT

In Ceylon this rat usually has 3 young to the litter, born early in the year (1).

1. Phillips, W. W. A. Ceylon J. Sci., 14B: 209–293, 1927–8.

Pelomys

CREEK RAT

Pelomys fallax Peters. This rat has been found pregnant in March and May (1).

P. frater Thomas. In Angola this rat has 2 or 3 young born about April (2).

1. Frechkop, S. EPNU., 1944.
2. Hill, J. E. JM., 22: 81–85, 1941.

Lemniscomys

STRIPED MOUSE

Lemniscomys barbara L. In the London zoo litters have been born throughout the year. They ranged from 3 to 6 and averaged 4 (1). In the wild, young have been found in November, December, and February (2).

L. griselda Thomas. This species is polyestrous and has 5 to 12 young according to one account (3), or 2 to 4, born in April and May, according to another (4).

L. striatus L. Young or fetuses have been recorded from November to January (5), and from August to October (2). The litters have varied from 2 to 5, with 4 both mean and mode.

1. Zuckerman, S. PZS., 122: 827–950, 1952–3.
2. Hatt, R. T. AM., 76: 457–604, 1939–40.
3. FitzSimons, F. W. The Natural History of South Africa. London, 1919.
4. Hill, J. E. JM., 22: 81–85, 1941.
5. Allen, G. M., and A. Loveridge. HCMZ., 89: 147–214, 1941–2.

Rhabdomys pumilio Sparrman

STRIPED RAT

This rat is polyestrous and breeds from September to April (1). Puberty is at 3 months old (2). The number of young is usually from 3 to 8 and it breeds 4 times a year (2).

1. FitzSimons, F. W. The Natural History of South Africa. London, 1909.
2. Powell, W. Rodents. Dept. Public Health, South Africa, 1925.

Hybomys univattatus Peters

BACK-STRIPED MOUSE

From 1 to 3 fetuses have been found in females taken in January, September, and October (1).

1. Hatt, R. T. AM., 76: 457–604, 1939–40.

Millardia meltada Gray

This rat has 6 to 8 young at a time (1).

1. Blanford, W. T. Fauna of British India. Mammals. London, 1888–91.

Aethomys

BUSH RAT

Aethomys chrysophilus De Winton. In South West Africa this rat has been found with 2 to 4 fetuses in October and November (1).

A. kaiseri Noack. This rat has been found in December with 3 embryos (2).

A. thomasi De Winton. This rat breeds fairly continuously throughout the year. One litter of 3 was reported (3).

1. Shortridge, G. C. The Mammals of South West Africa. London, 1934.
2. Hatt, R. T. AM., 76: 457–604, 1939–40.
3. Hill, J. E., and T. D. Carter. AM., 78: 1–211, 1941.

Thallomys

TREE RAT

Thallomys damarensis De Winton. This rat has been found with 3 to 4 young in September and October (1).

T. namaquensis A. Smith. This rat breeds all the year round and has 2 to 5 young at a time (1). The gestation period may be 22 days or more (2).

T. nigricauda Thomas. This rat has been reported to breed from September to December and also in April. The number of young is from 2 to 4 (1).

1. Shortridge, G. C. The Mammals of South West Africa. London, 1934.
2. Meester, J. JM., 39: 302–304, 1958.

Rattus assimilis Gould

This Australian rat breeds all year, but mostly from March to May, in the rainy season. In captivity the vagina opens at age 46 days (35 to 57), at a weight of 56 g. (37 to 73). The males reach puberty at 2½ months, at 75 g. weight. The litter size averages 3.8, mode 4, range 1 to 7. The mean number of litters a year was 3.6, i.e., 13.7 young per female. The vaginal cycle is 4.5 days, either 4 or 5 days. In the proestrous smear nucleated epithelial cells are present, and the lips of the vagina are slightly swollen. This stage lasts for 10 to 14 hours. During estrus the smear consists of cornified cells and the vaginal lips are swollen. The stage lasts for 24 hours. In metestrum cornified cells and leucocytes are present and the lips of the vagina are no longer swollen. This stage lasts for less than 10 hours and the females mate early in it. The maximum distention of the uterus is during estrus. The gestation period lasts 22 to 24 days, with 22.8 the average, and with 60 per cent at 23 days. Eighty per cent of births occur between 2 P.M. and 8 P.M. Stimulation of the cervix uteri at the time of heat causes pseudopregnant activation of the corpora lutea. This period lasts for 15 to 17 days. There is a post-partum heat, usually on the first or second day after parturition, and delayed implantation of the embryos. If the female does not become pregnant at this time a lactation anestrum sets in (1).

The diameter of the mature follicle is 0.8 to 0.9 mm. The mean number of corpora lutea of ovulation is 7, range 3 to 11. The weight of a single mature testis is 2.4 g. (2.0 to 3.5); of the epididymis, 0.55 g. (0.4 to 0.76); and of a seminal vesicle, 1 g. (0.6 to 2.4) (1).

1. Taylor, J. M. Univ. California Publ. Zool., 60: 1–66, 1961.

Rattus conatus Thomas

The males of this Queensland rat reach puberty at 9 to 10 weeks old, when they weigh 65 to 70 g. The females experience their first ovulation at the same age, at a weight of 50 to 55 g., but the proportion of pregnant rats is low at an earlier age than 5 months. The vagina opens at 37 days old, range 33 to 45 days, when the body weight is 35 g. They breed all year and the mean litter size is 6 (1), or, in the rainy season, 7, varying from 4 to 11 (2). They are born after a gestation of 21 to 22 days (1).

The vaginal smear changes are similar to those of *R. norvegicus*. The corpora lutea retrogress more rapidly than in that species, so that fewer may be found in the ovaries. Usually only 2 to 3 sets are present. Functional corpora lutea measure 1 mm. in diameter, but at term they measure nearly 2 mm. The mature follicle has a diameter of 0.79 to 0.84 mm. (1).

1. McDougall, W. A. Queensland J. Agric. Sci., 3: 1–43, 1946.
2. Troughton, E. Furred Animals of Australia. New York, 1947.

Rattus norvegicus Berkenhout

NORWAY RAT

The wild brown rat breeds throughout the year in most situations. In Baltimore there was no difference in the seasonal pregnancy rate of 40.3 per cent for large rats, but for small ones pregnancy was more frequent in spring and fall; the average rate for this group was 10.8 per cent. The annual litter rate varied from 2.2 for small rats to 8.2 for large ones. The mean embryo count was 9.3 ± 2.3 per litter. The 50 per cent point for initiation of ovulation was at 153 g. weight and 144 mm. length. The 50 per cent point for vaginal perforation was 102 g. and 143 mm. (1). The 50 per cent point for testis descent was 105 ± 45 g., and spermatozoa were present in 50 per cent of males at 200 ± 70 g. weight. There was little change with season, and most of the seasonal variation in breeding was due to the females (2). In farm rats 50 per cent of males were with scrotal testes at 136 g. and 50 per cent of females had perforate vaginas at 88 g. For city rats the corresponding weights were 119 g. for males and 105 g. for females. Few city rats were found pregnant at less than 200 g. weight (3). Polyovular follicles were found in 12 per cent of rats, and cases of more embryos than corpora lutea in 16.8 per cent. The number of sets of corpora lutea is usually fewer than in the laboratory rat. Embryonic mortality accounts for 15.8 per cent of ovulations (4).

In England there are bursts of intensive breeding at intervals, and the season is at its height in March to June. The heavier the females the more young are born, and the pregnancy rate is also higher. The gestation is probably prolonged if the female is lactating (5). Judging by the presence of corpora lutea, 50 per cent of females have reached puberty at 125 to 135 g., and 92 per cent at 155 to 195 g. The average embryo count is 9.0, mode 8 and 10, and it is correlated with the empty weight of the mother. Fertility in country districts is lowest in the summer; both pregnancy rate and embryo counts were adversely affected. In winter, November to February, the pregnancy rate was 27.0 per cent in rats dwelling in grain ricks and 2.3 per cent in those not dwelling in ricks. (6).

The remainder of this article refers to the laboratory rat, usually the albino.

The rat is polyestrous all the year round, and ovulation is spontaneous, occurring near the end of heat. Heat lasts about 20 hours, and the cycle from 4 to 6 days. The corpus luteum formed after rupture of the follicle is physiologically inactive unless the cervix uteri has been stimulated mechanically or by coitus. The interval between heats thus represents proestrum, and its length is determined by the time required to ripen new follicles. When the cervix is stimulated, prolactin is apparently released from the anterior pituitary, enabling the corpus luteum to secrete progesterone. This continues for about 14 days, during which time the rat is "pseudopregnant." New follicles do not ripen, heat is not experienced, and the uterus undergoes various changes, such as growth of the glands and sensitization of the endometrium to trauma. Subsidence of the corpora lutea is followed by the ripening of new follicles, heat, and ovulation.

The vaginal epithelium undergoes well-marked changes during the cycle.

Heat is characterized by marked cornification and the disappearance of leucocytes. At the end of heat the cornified layer is sloughed off, and an invasion of leucocytes occurs. The vaginal smear is, therefore, an excellent indicator of the stage of the cycle. The uterine lumen is distended with fluid at heat, but otherwise uterine changes are not marked in either type of cycle. The duration of gestation is about 21 days, and the average litter size 7 to 9. Quantitative hormonal relationships have been worked out in considerable detail.

THE ESTROUS CYCLE

In the young rat the vagina is a cord of cells without a lumen. As the first ovulation approaches, the central cells separate, forming a tube closed at the vaginal orifice by a thin membrane. This ruptures at, or a little before, the first ovulation; hence, the establishment of the opening is a good indication of the arrival of puberty. Opening is not, however, entirely dependent on sexual activity as it occurs in the absence of the ovaries at about the usual time in the Long-Evans strain (7). In the Cornell Nutrition strain ovariectomy delays the opening for about 14 days (8).

TABLE 1. Age and Weight of Rats at Time of Vaginal Opening

STRAIN	AGE, DAYS	WEIGHT, GRAMS
Columbia Univ., specially fed (9)	41.1 ± .45	127.0 ± 1.6
Wistar (10)	42.0 ± .7	87.8 ± 1.9
Hooded (10)	46.5 ± 1.3	91.0 ± 2.6
Brown Univ. (11)	49.4 ± .26	———
McCarrison (10)	50.6 ± 1.7	96.8 ± 2.1
Cornell Nutrition (8)	60.8 ± .6	120.8 ± 3.0
Long-Evans (7)	76.5	———

The average age at puberty varies considerably with the strain of rats and with the rate of growth. Rats from small litters and those which are specially fed reach this age earlier than others (9). The mean weight at this time is about 52.5 per cent of the mature weight, but, as the coefficient of variation for weight is greater than that for age, the latter is more closely related to the attainment of puberty (8). However, the body length, mean, 166.0 ± .6 mm., is still less variable (9). Another series of data gives 148 to

150 mm. as the body length at puberty or first appearance of corpora lutea, and it seems that this is the less variable measurement (12). In an inbred gray strain puberty was reached at 36 days, and in a similar white one at 41.6 days (13).

The opening of the vagina and ovulation are practically simultaneous in 46 per cent of rats, and they have occurred within 10 days of each other in 80 per cent. If males are with the females all the time, the first fertile copulation occurs, on the average, about 16 days later than vaginal opening in the Long-Evans strain (7), and 6 days later in the Cornell Nutrition strain (8).

The modal length of the estrous cycle in the Long-Evans strain is 4 to 5 days, with 82 per cent of all cases falling between 4 and 6 days. The average of all cases 8 days and less in length was 4.8 days, and this may be taken as the mean since periods longer than 8 days may include some rats one of whose heat periods has been missed (7). The Stanford colony gave a mode of 4 days for rats of all ages, though the mean for older rats (above 283 days old) was about 1 day longer than that for young ones because of the greater irregularity of the former (14). The Brown University colony gave a mean of 4.4 days, with 87 per cent between 4 and 6 days. Occasional split heats were observed (15). In South Africa the modal cycle length for albino rats was 4 days, with 80 per cent from 3 to 5 days (16). It is thus generally agreed that the modal length of the cycle is 4 days and the mean about 4.8 days.

The length of the pseudopregnant cycle after electrical stimulation of the cervix is 14.03 days (17), and after vasectomized copulation, 14.5 days (18). The difference is probably not significant. The mode in the first report cited was 12 to 13 days. In the Long-Evans strain the modal length was 12 to 14 days (7). Pseudopregnancy is also induced by the application of silver nitrate solution to the nasal mucosa during heat (19). Anesthetization of the nasal mucosa and also ablation of the sphenopalatine ganglia have a similar effect (20). Bilateral abdominal sympathectomy does not affect the normal cycle (21) or the induction of pseudopregnancy after infertile copulation, but it does prevent pseudopregnancy from following mechanical or electrical stimulation of the cervix. In the latter, therefore, the sympathetic chain is involved (22).

The duration of heat has been determined by hourly copulation tests. This method gave 13.7 ± .12 hours, with a range of 1 to 28 hours and a standard deviation of 4.55 hours. Eighty per cent were between 9 and 20 hours. Copulation early in heat shortens the period slightly (15). Another determination gave a distribution from 12 to 18 hours (23). Heat usually begins between 7 and 8 P.M., with 75 per cent beginning between 4 P.M. and 10 P.M. and less than 1 per cent between 3 A.M. and 11 A.M. (15, 23). The average duration of the first heat is slightly less than that of later heats, the mean being 9.1 hours. It usually begins between 7 P.M. and 5 A.M. (11). In the normal heat period the female is most receptive during the first 3 hours (23). The usual time of ovulation is 8 to 11 hours after the beginning of heat (24, 25). The relation of ovulation to the vaginal smear is considered later.

As the rat is a more or less nocturnal breeder, the influence of light upon the reproductive processes is of interest. Rats kept from birth or from 21 days of age in continuous light experienced the vaginal opening at an average of 6 days earlier than controls, while those kept in continuous darkness were 10 days late. In light the heat period was longer, and in darkness the interval between heats was longer, than usual (26). Reversal of the hours of light and darkness, but with their relative lengths retained, caused a shift of the onset of heat to the new time of darkness. Establishment of an 8-hour rhythm of light and darkness had no effect, and exposure to constant light increased the length of the heat period and the duration of vaginal cornification (27, 28). Blindness and continuous darkness did not affect the cycle (28).

The removal of active corpora lutea during pseudopregnancy results in a return of heat in 4 days' time. If it is done late in pregnancy, abortion sometimes occurs, more often resorption of the embryos, and occasionally the pregnancy is not interrupted (29). Ovariectomy, apparently a less severe operation than removal of corpora lutea by electrocautery, results in abortion if it is done before 13 days of pregnancy; if it is done later, the pregnancy continues (30).

Parturition usually occurs in the afternoon, least frequently from midnight to 6 A.M. It takes from 10 to 70 minutes. The postparturition heat begins after a mean interval of 18.5 hours (range 4 to 36 hours). The time of its commencement is somewhat irregular, but it is related to the time of parturition in such a way as to bring it between 6 P.M. and midnight, the usual time for heat to begin. The average length of this heat period is 10 hours. Occasionally ovulation occurs without heat after parturition (31).

There is little difference in the relative activity of the two ovaries. A count of corpora lutea gave 51 per cent of ovulations from the left ovary (12).

The results of insemination at different times in the cycle are given in Table 2 (32, 33). This work sets the maximum fertile life of the spermatozoa in the female tract at 14 hours, and the life of the ovum at about 12 hours. But as the ova become old they are less able to give rise to normal embryos.

TABLE 2. Relation of Time of Insemination to Fertility and Fecundity

TIME OF INSEMINATION	PREGNANCIES, PER CENT	NORMAL PREGNANCIES, LITTER SIZE	ABNORMAL PREGNANCIES, PER CENT
More than 14 hours before ovulation	0	0	0
12–14 hours before ovulation	20	3.5	0
10–12 hours before ovulation	50	3.5	0
Near ovulation time	83	6.7	11
6 hours after ovulation	47	4.6	48
10 hours after ovulation	22	1.8	79
12 hours after ovulation	4	0	100
18–20 hours after ovulation	0	0	0

The rate of passage of the ovum through the oviduct appears to be rather rapid until the last tenth of the distance remains to be traversed, when it is much slower. The ovum takes 3 days or a little more to reach the uterus (34, 7).

The mean litter size varies with the strain. Stock albino rats at the Wistar Institute averaged 6.1 (35); Wistar rats in Edinburgh, 9.26 ± .18 (36); albinos in South Africa, 7.4 (16); Cornell Nutrition strain, 6.22 ± .08 (37); wild rats brought into the laboratory, 10 to 11 (38). The second litter is the largest, and the litter size decreases sharply after the tenth litter (35, 39). The litter size tends to be slightly smaller in fall and winter (16). In a sample of 35 rats that had 8 or more litters the maximum litter size was reached at the third litter (40). Unilateral ovariectomy is followed by the production of the number of ova usual for the rat from the remaining ovary, so that the litter size falls only from 8.6 to 8.1, while the number of corpora lutea increases from 10.3 to 10.6. No migration of ova across the abdominal cavity occurs; the ova could not migrate through the uteri, as the latter open to the vagina by two cervices. In these experiments, therefore, the pregnant uterus had twice the normal number of young in it with little increase in the amount of fetal atrophy (41, 42, 43). In the normal rat loss of ova and fetal atrophy account for about one third of the eggs shed (7).

A comparison has been made of the effects of breeding rats at the earliest possible time, i.e., at the time of vaginal opening, at 100 days of age (normal time), and at 280 days old. In all these groups lactation for 21 days was allowed before the rats were rebred. A fourth group was not allowed to lactate: the young were removed immediately after parturition and the mother was put in a cage with a male. In all these groups the second litter was the largest; the "bred late" rats had the smallest litters throughout their lives; and the "bred early" rats had larger initial litters than the normals but ceased reproduction rather earlier than usual. The "nonlactating" rats had consistently larger litters after the first, and, though they had more litters, they tended to break down at an earlier age. The size of the first litter, and the shortness of the interval between the first and second litters in the lactating rats, was a fairly reliable indication of their lifetime ability to reproduce (39). The menopause, or cessation of breeding, was found to be a process that might occur at any age or not at all; no definite period for cessation, as in man, was found (25). When it does occur it is characterized by persistent estrus and a final phase of anestrum (44).

The sex proportion at birth in Wistar stock is 51.4 per cent males (35). This seems to be characteristic, since the same stock in Edinburgh has given a proportion of 51.2 per cent males (36). South African albino stock gave fewer males than females, i.e., 43.3 per cent males (16), and the Cornell Nutrition stock, 49.4 per cent males (39). A more recent estimate also gave 51.4 per cent males (45).

The gestation period is 22 days with very little variation. A series in which the time was accurately obtained by the use of obstetrical cages gave 90 per cent from 21½ to 22 days, and the average of all gestations was 21.8 days (7).

If the mother is suckling 6 or more young, implantation in a new pregnancy is delayed, and the pregnancy is consequently prolonged (46). In the absence of pregnancy lactation causes the corpora lutea of the postparturition ovulation to persist. Removal of the young is followed by heat 4 days later (7).

HISTOLOGY OF THE FEMALE TRACT

OVARY. The ovary of the immature rat less than 15 days old contains ova surrounded by granulosa cells. These loosen and antra develop at about 15 days. At this time the percentage of vesicular follicles is about 12. It increases up to 38 days, then falls until just before puberty, when there is a sudden rapid increase (37). The fall may be due to the large amount of follicular atrophy. The number of binovular and triovular follicles is very small, about 0.05 per cent (38), in contrast to their relative frequency in the mouse. The number of ova of all sizes in the two ovaries at birth is about 35,000; at 100 days it has fallen to 8,000; at 400 to 600 days to 5,000; and at 950 days to about 2,000. At any one time degenerating ova account for 10 per cent of all those present (6). It is now known that the age of the female is related to the proportion of polyovular follicles: the younger she is the more she has. One report gives the proportion of females with polyovular follicles as 12 per cent (49).

Comparatively few mitoses are found in follicles below 20 μ in diameter, indicating a slow rate of growth up to this size (50). Follicles that are destined to rupture begin to grow at the previous heat of the 4-day cycle. The growth rate is fairly constant until just before heat, when there is an 8-fold acceleration. Follicles that will not rupture but that are destined to become atretic can be picked out at the third day of the cycle (25). The greatest number of mitoses in the granulosa cells is found early between heats, and the volume of the theca interna is greatest just before the heat in which the follicles will rupture (50).

The first sign of impending ovulation is a slight indipping of the theca interna. As this occurs the first polar body is extruded, and some of the outermost granulosa cells have minute lipid deposits within them. At this time the follicle is about 0.9 mm. in diameter (7). The theca interna thins out at the point of rupture. When ovulation occurs, the follicle collapses; some of the granulosa cells and a little liquor folliculi are retained. A small crater forms at the rupture point, but it soon heals over, and the follicle again becomes somewhat distended by the secretion of more fluid (25). Ovulation takes place at 3 A.M. and copulation helps to standardize the time. By 4 hours after ovulation the follicle wall has begun to organize, especially the theca interna; there is no activity in the granulosa until 6 to 8 hours. The stigma is closed about 12 hours after rupture and the central cavity begins to fill with fluid. Invading fibroblasts are in the central cavity at 20 to 22 hours and development is complete by 50 to 54 hours (51). At 24 hours after ovulation a layer of lutein cells has been formed, and at 48 hours these cells are well developed. They surround a small core of connective tissue, which occasionally contains extravasated blood. The corpora lutea reach their maximum size, about 0.8 mm. in diameter, after 3 days, and retrogression sets in immediately (52), though they may be easily distinguished for three cycles (7). The corpora lutea of pregnancy last throughout gestation and retrogress slowly, and those of pseudopregnancy continue for 12 days and then slowly retrogress. The most apparent sign of retrogression is a reduction in the amount of lipid in the lutein cells (7). It has been stated that the mean maximum diameter of the corpus luteum of the normal cycle is 1.77 mm., of that of pseudopregnancy, 1.52 mm., and of that of lactation, 1.59 mm. The corpus luteum of pregnancy increases slowly to 1.45 mm. by the eleventh day; then there is a rapid increase to 2.0 mm., at which size it remains until parturition. In the delayed implantation caused by lactation the secondary growth of the corpus luteum is also delayed. The increase can be brought about by the injection of estrogens, probably supplied normally by the placenta (53).

In the ripe follicles alkaline phosphatase, ascorbic acid, and lipids are limited to the theca interna. These compounds may be found in the granulosa when atresia sets in and they are also present in the corpus luteum (54). Phospholipids are present in the theca interna of both growing and ripe follicles. They are also present in corpora lutea and reach their maximum in mature corpora lutea (55).

VAGINA. Vaginal changes in the rat are definitely related to the estrous cycle and are exceptionally clear-cut, so much so that the vaginal smear is an excellent means of determining the stage of the cycle. During metestrum the vaginal wall is moist and pinkish, but in estrus it is dry, white, and lusterless (7). These changes are associated with the cornification of the surface layers during heat and the extensive desquamation at the end of this period. It may be noted that these color changes are the opposite of those occurring in the domestic ungulates, but in these species not cornification but mucification is the rule in estrus. During the interval the epithelium is 4 to 7 cells deep with little squamous transformation of the surface layers and few mitoses in the basal layer. Leucocytic infiltration is always present. Toward the end of this period, i.e., during proestrum, mitoses are much more frequent and the epithelium increases to 8 to 9 layers. The surface cells swell and become a characteristic surface layer, while, with continued growth, those beneath become cornified. These two layers are gradually sloughed off during heat, and at the end of this period a heavy invasion of leucocytes appears. This is followed by the "interval" type of epithelium.

Along with these changes in the vaginal epithelium there are well-marked changes in the type of vaginal smear. In the interval nucleated epithelial cells and leucocytes, with an occasional cornified cell, are formed. As heat approaches, the leucocytes disappear and nucleated epithelial cells increase in number and are gradually replaced by cornified nonnucleated cells. The beginning of heat (sexual receptivity) usually occurs when the smear contains 75 per cent nucleated and 25 per cent cornified cells. This is followed by a second stage in which cornified cells only are present. Gradually "pavement" cells come in. These are flat, nucleated epithelial cells. During the latter part of the cornified stage the smear becomes very abundant and cheesy in texture. Heat may end at any time during the cheesy stage. It is followed by the appearance of large numbers of leucocytes and the virtual disappearance of cornified cells. Two schemes denoting the stages of the vaginal smear of the rat have been given, and they are summarized in Table 3.

TABLE 3. Vaginal Smears in the Rat

LONG AND EVANS (7)	YOUNG, BOLING, AND BLANDAU (56)
STAGE I Small, round, nucleated cells only. Duration about 12 hours PROESTRUM	STAGE I From small, round, nucleated cells only, to 75 per cent cornified cells. Beginning of heat is when 25 per cent of the cells are cornified. Duration 8–11 hours
STAGE II Cornified cells only. Mating mostly in this stage ESTRUS	STAGE II From cornified cells only with ovulation early in this stage, to cornified 25 per cent, pavement nucleated cells 75 per cent
STAGE III Late cornified stage, abundant cheesy smear. Mating no longer allowed as a rule. Stages II and III last about 27 hours	STAGE III Pavement epithelial cells only
STAGE IV Cornified cells and leucocytes. Duration about 6 hours METESTRUM	STAGE IV Pavement cells and leucocytes
STAGE V Epithelial cells and leucocytes. Duration about 57 hours DIESTRUM	

Using Shorr's technique for vaginal smears it was found that the number of acidophils rises in proestrum and then declines. In late metestrum basophils predominate, and these cells become vacuolated in diestrum. The complete cellular cycle is not established until 30 days after the first opening of the vagina (57). Cyclical changes may be detected in the vaginal smears of adult ovariectomized rats but they are not as clear as in the normal animal. Acidophil nucleated cells largely replace the cornified cells. Leucocytes are never wholly absent (58). In the normal vaginal epithelium alkaline phosphatase is most plentiful at proestrum; it declines during heat and is least in metestrum and early diestrum (59).

The length of proestrum is 13.5 hours; of estrus, 25.3 hours; of metestrum, 13.3 hours; and of diestrum, 56.5 hours (60). Another account divides the cycle as follows: early heat, 18 hours; heat, 25 hours; late heat, 6 hours; early diestrum, 24 hours; diestrum, 28 hours; and late diestrum, 7 hours (61).

UTERUS. The chief characteristic of the uterine changes is the marked distension with fluid that occurs during estrus. In diestrum the uterus is always slender, with a slitlike lumen. It is lined by a simple columnar epithelium, and the glands are scanty. During vaginal stage I, or proestrum, there are marked vascular engorgement and an accumulation of fluid in the lumen, which distends to 5 mm. in diameter, in contrast to the diestrous diameter of 1.6 to 2.5 mm. The distension converts the columnar epithelium into a cuboidal one. Leucocytes are usually absent from the epithelium. These changes are continued in the early part of stage II, but at about the end of this stage the distension disappears, and the fluid is lost by drainage through the cervices. With the reduction in distension the uterine wall becomes flaccid, and the epithelium reverts to the columnar type, but the cells show signs of vacuolar "degeneration." During stage III the vacuolization increases, but regeneration is apparent in stages IV and V (7). Degenerated ova from the previous heat remain in the uterus and are washed out with the fluid in the uterus during heat (62).

During pseudopregnancy—a period in which the vagina remains in stage V—the main change is that more mitotic activity is found in the uterine epithelium at about 7 to 8 days after heat; the stroma also shows growth which precedes that of the epithelium by a day or two. The glandular changes at any time are but slight (63).

During the normal cycle the endometrium does not produce deciduomata after traumatization, but they may be produced in pseudopregnancy from the fourth to the seventh day (7, 63). This reaction is due to the secretion of progesterone by the corpora lutea and affords evidence that this does not occur, either at all or as much, in the normal cycle as in the long cycle. On the other hand, during lactation, when the corpora lutea become active and persist, the endometrium remains sensitized from the fourth to the seventeenth day from delivery. After this time it is no longer sensitive (7, 64).

OVIDUCT. The distal folds of the oviduct are distended by fluid at the time of ovulation, and during the 12 hours that the distension lasts the eggs are in this part of the tube (7). When the eggs are shed, they are surrounded by granulosa cells which are dispersed by the first spermatozoa to reach them. The average size of the eggs is 71 × 76 μ (65). The tubo-uterine junction is 0.3 mm. from the tip of the uterus on the anterior side, not on the mesenteric side as in most mammals. The mucosal margin is slightly raised, and the last part of the tube is surrounded by a strong muscle (66).

The level of phospholipids is high in the mucosa during estrus and metestrum. The cells are secreting in diestrum (67). The epithelium is relatively eosinophilic. Only the cells of the fimbriated end contain lipids. The isthmus cells contain akaline phosphatase. No changes in the amounts of these substances could be correlated with the cycle (54).

PITUITARY. The cells of the anterior pituitary in relation to the cycle have been studied thoroughly, and the work has been summarized in a paper which also contains many original observations. It was found that the eosinophil cells do not vary in relative numbers during the normal cycle, but the granules are more distinct when the females are in heat. There is, however, a drop in the proportion of these cells during the lutein phases of pseudopregnancy and pregnancy. The basophilic cells become degranulated during diestrum (68).

Changes in the adrenal cortex in relation to the cycle have been described. The cells of the zona fascicularis and of the cell nests near the medulla are enlarged during heat, and their lipid content increases (69). Castration or ovariectomy of immature rats causes the gland to hypertrophy; if the operation is performed after puberty, however, atrophy results in females but not in males (70).

PHYSIOLOGY OF THE FEMALE TRACT

The ovary of the mature rat contains 173 ± 6.5 follicles per ovary. Luteal and interstitial cells contain little alkaline phosphatase (71). If the female was mated at 5 to 10 hours before ovulation the proportion of ovulated eggs lost before the end of gestation was 16.7 per cent. Most were lost as nonfertilized or lost after implantation. If coitus was delayed until 8 hours after ovulation there was a 68 per cent loss; of this, 27.1 per cent were not fertilized, 14.6 per cent were abnormal, 7.5 per cent were lost before implantation and 18.8 per cent after implantation. The mean duration of fertile life of the rat egg is 13 hours (72). During the fertilization process 47 per cent of the ova are penetrated by the spermatozoa during the first hour, 54 per cent by the end of the second hour, and 90 per cent by the end of the third hour (73). Another account gives a fertile life for the ovum of 10 to 15 hours (74). Ovulation occurs between midnight and 4 A.M. The whole process takes about an hour. Capacitation of the spermatozoa takes about 2 hours, and of the egg 2 to 4 hours (75). The maturation of the ovum has been studied. The germinal vesicle appears before the onset of heat; the chromatin mass at 1 to 2 hours after its onset; the prespindle and spindle stages of the first metaphase at 3 to 4 hours; early telophase at 3 to 5 hours; late telophase at 4 to 5 hours; dissolution of the telophase spindle at 6 to 7 hours; second metaphase 7 to 8 hours; and ovulation at 9 to 10 hours after the onset of estrus (76).

The application of X-rays to the ovaries of the immature rat in doses of 540 r produces precocious puberty, while 830 r, or more, produce castrate atrophy of the accessory organs (77). In the adult female the castration dose is approximately 3,000 r, but the corpus luteum is entirely resistant to radiation (78), and in some cases the rats still experience estrous cycles (79).

In general, the thyroid gland has little relationship to reproduction, but it has been reported that the feeding of 0.25 to 0.3 g. of thyroid substance daily to females for 3 to 5 days in the cooler parts of the year considerably increases the litter size. If it is done during hot weather, or if the dose is increased, adverse effects result (80).

It is well known that the thymus gland retrogresses at the time of puberty, but not in the gonadectomized animal. The injection of pituitary extract or of estrogens accelerates, or brings about, this change (81).

Pseudopregnancy is induced by the sensitization of the corpora lutea of the cycle when the cervix uteri is stimulated. It has already been mentioned that the reaction depends upon the integrity of the abdominal sympathetic chain. If the stimulation of the cervix is done by a single electric shock during late proestrum or estrus, 86 per cent become pseudopregnant at once. On the first day of diestrum 54 per cent react at once, and 34 per cent have one more cycle. On the second day 13 per cent react at once, and 54 per cent have one more cycle. Stimulation during pseudopregnancy does not prolong this period (82). Anesthesia of the animal during heat reduces the response considerably (83). Pseudopregnancy is also evoked by injecting prolactin; hence stimulation of the cervix is believed to cause the release of this hormone, which activates and maintains the corpus luteum (84). The length of pseudopregnancy, if it is brought about by electrical stimulation of the cervix, is 13.0 days. If it is accompanied by extensive bilateral deciduomata it lasts for 19.9 days. When the corpora lutea are thus activated there is an increase in the cholesterol content of this organ but it lasts only 2 days (85).

The pH of the vaginal contents during the normal cycle is as follows:

Beginning of proestrum	pH 5.4
During estrus	pH 4.2
In diestrum	pH 6.1

In the ovariectomized rat the pH is 7.0; if 8 I.U. of estrogen are injected, it falls to 4.1 (86). These reactions are more acid than those recorded for most species.

In diestrum the amount of fluid in the uterus weighs 0.025 g., but in estrus 0.5 g. is present. Removal of the fluid greatly delays the passage of spermatozoa through the tract (87). The pH varies with the stage of heat. The following figures may be quoted (88):

Stage I	pH 7.29
Stage II	pH 7.31
Stage III	pH 7.69
Stage IV (diestrum)	pH 7.43

At the time of heat the pH in the uteri is 7.74 (7.43–8.31); in the dilated ampulla, 8.04 (7.31–8.53); and in the periovular sacs, 8.05 (7.66–8.41) (89).

It has been mentioned that ovariectomy does not always interrupt pregnancy if it is done during the last third of that period. If the embryos are removed, the placentae are retained for the normal time, and paraffin pellets also remain in the uterus if placentae are present, either in the same or the opposite uterus (90). If all fetuses but one are removed, pregnancy is continued when the ovaries are removed provided that the placentae are left *in situ,* but the birth mechanism is impaired. It would seem that part of the corpus-luteum function is taken over by the placenta (91). If the pregnant female is ovariectomized at the ninth day all the fetuses are lost within 2 days. If the operation is performed on the seventeenth day 60 per cent are retained until term. After ovariectomy the fetuses show pressure symptoms. If the females were given 5 mg. progesterone daily, gestation was prolonged to 12 days. For further prolongation a daily injection of 10 mg. was needed (92).

The mean contraction rate of excised uterine muscle is lower (31 per hour) in estrus than in diestrum (52 per hour) (93) according to one paper, but according to another the reverse relationship holds. Contractions tend to begin at the tubal end (94). Anaerobic glycolysis and O_2 consumption are greatest at proestrum and lowest at the end of heat, but aerobic glycolysis shows the opposite relationship (95). Adrenalin inhibits contractions in all stages. The minimal effective dose *in situ* is 0.0005 mg. per kg. of body weight (96).

The tonus of the cervix uteri is low in diestrum. About 4.3 (1–8) hours before the onset of heat it is sufficiently contracted to retain fluid within the uterus. This condition is maintained for 8.5 (5–12) hours after onset of heat. The average duration of this contracted state is 12.8 (7–17) hours. The cervix begins to relax about 4 (0–11) hours before the end of sexual receptivity. No spermatozoa from a male with ligated vesicular and coagulating glands were able to penetrate the contracted cervix (97). In diestrum much mucus and numerous leucocytes are within the lumen. This is in contrast to the condition in metestrum. At estrus some cornification of the lower segment is evident. The upper segment has a coating of columnar epithelium (98).

If the ovarian bursa is opened, fecundity is reduced (99). Probably this is due to the chance given the eggs to escape, as it will be recalled that the upper third of the oviduct is distended with fluid at the time when the eggs have been shed (7).

In the female rat creatine is absent from the urine only during diestrum; it is also absent if the rats are ovariectomized, and it reappears if estrogens are injected. It is absent in the mature male and present in the castrate, it is increased in amount following the injection of estrogens, and it disappears if androgens are injected (100). These results have been questioned (101). Choline esterase is higher in mature females than in young ones or in males. It falls after ovariectomy and is not restored by estrogens, but estrogen and progesterone together cause it to rise to normal, or higher (102).

There is a small quantity of estrogen in pregnancy urine but not in the placentae (103).

It is known that a neurohumoral agent is released from the hypothalamus that causes the release of L.H. by the pituitary gland and thus produces ovulation. This agent begins to be released about 2 P.M. of the day preceding ovulation. Pituitary stimulation is complete in 50 per cent of rats before 4 to 5 P.M. (104). The stimulation lasts for half an hour (105). A uterine factor is involved in corpus-luteum involution, for if the sacral spinal cord is severed it prevents involution of the corpora lutea (106). The injection of 1.5 mg. of progesterone on the first day of diestrum causes a 4-day cycle to become a 5-day one by delaying release of the neurohumoral agent. In a 5-day cycle the same amount given on the third day of diestrum accelerates ovulation to 4 days (107).

The pituitary of the rat is relatively rich in gonadotrophic hormones, and particularly so in follicle-stimulating hormone (108). The amount of gonadotrophic hormones present is high at birth, rises until the twenty-first day, then falls a little until puberty, when the fall becomes abrupt. The amount remains at a low level until old age, when it rises slowly. Up to puberty the content of the female pituitary is greater than that of the male, but afterwards their relative amounts are reversed; this is due not only to the fall in the content in the female but to a rise in that of the male (109–111). Ovariectomy causes the pituitary to become about 20 per cent more potent; it increases rapidly for the first 6 weeks after the operation, then the rate of increase becomes slower (112, 113). The increase is mainly in F.S.H. content, and slightly in L.H. Injection of estrogen into the ovariectomized female lowers the L.H. content to below normal level and causes the F.S.H. to fall to just above the normal content (114). The level of gonadotrophic hormones in the anterior pituitary shows no variation with the cycle. In the blood, F.S.H. is highest in proestrum and estrus. There are two peaks in the blood, one in early heat and the other in early diestrum. In the pituitary, L.H. is highest late in estrus. In the blood there are two peaks, one in proestrum and the other late in heat (115). The weight of the pituitary increases markedly a short while before parturition. Lactogenic hormone increased in the first two thirds of pregnancy. The peak of .051 I.U./mg. was reached on day 16; then it decreased and was lowest, at .004 I.U./mg., on day 21 (116). The prolactin content of the pituitary of the female is about 0.28 bird units per mg. in the immature animal. It rises to 0.48 B.U. after maturity. During pregnancy it falls slightly, but in lactation there is a marked rise. After ovariectomy it falls gradually, reaching 0.18 B.U., which is practically the level in the normal male. The injection of estrogens or of progesterone raises the level to about 0.7 B.U. Treatment with 200 I.U. of estrogen daily for 20 days is more effective than higher doses, which tend to depress the prolactin content (117). Prolactin is responsible for the exhibition of maternal behavior, i.e., the retrieving and nursing instincts (118).

The level of estrogens necessary to induce the fully cornified vaginal smear has been accepted as the "Rat Unit" of estrogenic activity. Tests must be made under standard conditions on at least 20 mature ovariectomized females. The International Unit (I.U.) is approximately one third of the R.U. and is defined by the League of Nations Commission on Biological Standardization as the specific estrus-producing activity contained in 0.1 γ (= 0.0001 mg.) of a standard preparation of estrone (119). The potencies of various steroid hormones have been determined in terms of Rat Units and are on record (120).

The properties of estrogen in causing vaginal and uterine changes and also copulatory responses can be demonstrated in the immature rat. The latter is relatively insensitive until 30 days after birth, when the effective amounts are 10 R.U. of estrogen followed by 0.4 mg. of progesterone in a single injection (121), or 2 R.U. of estrogen daily for 3 to 4 days (122). Some differences in the response to various estrogens have been observed; estrone is more potent in causing cornification and uterine hypertrophy, and estriol is more potent in opening the vaginal introitus (123, 124).

In mature ovariectomized rats threshold doses of estrogen do not produce continuous heat even when they are injected daily. After the normal length of estrus the females cease to be in heat. With daily doses of 3.75 to 10.0 I.U. (1 to 3 R.U.) a cyclical rhythm can be set up in many rats with a cycle length of 4.8 ± .1 days. If the pituitary, also, is removed, the mean length is the same, but with the additional removal of the adrenals cycles are suppressed, either anestrum or continuous heat ensuing. The cycles may be renewed in some cases if cortin is injected (125). Desoxycorticosterone and progesterone have the same effect (126).

Sexual receptivity is produced more effectively by the injection of progesterone along with estrogen than by estrogen alone. Thus, while 20 R.U. of estrogen alone produced receptivity in only 10 per cent of ovariectomized females and 100 R.U. in 90 per cent, 10 R.U. of estrogen with 0.4 I.U. of progesterone was 100 per cent effective (127).

Continued injections of estrogens into normal females are injurious to the ovaries; daily injections of 0.25 to 2.0 R.U. produce a state of continuous anestrum or of continuous heat. This is believed to be due to an effect of the estrogens upon the pituitary gland (128).

The injection of estrogens into the immature rat causes hypertrophy of the uterus. The maximal increase is brought about by 0.5 γ estradiol (6 R.U.); 0.075 γ increases the weight about 50 per cent; and 0.025 γ (0.3 R.U.) is without effect (129). In the ovariectomized female 0.001 mg. of estradiol propionate together with 0.1 mg. of progesterone produces maximal estrous distention of the uterus (130). Under the same conditions 0.25 γ of estradiol benzoate daily injected subcutaneously maintains the uterine weight (131).

It is now believed that estrogens cause the persistence of the corpus luteum. In the rat a minimum of 50 I.U. estrogen daily is needed to produce this effect, but the corpora lutea so produced do not last beyond 20 days, their normal duration in pregnancy (132).

The level of progesterone needed to maintain pregnancy in the ovariectomized rat is about 2 Rab.U. daily; 1 Rab.U. daily maintained 3 pregnancies in 8; below 1 unit the injections were ineffective (133). This hormone, if injected into the normal pregnant rat, prolongs gestation for from 30 to 150 hours, but after 70 hours the fetuses are nonviable (134). In pseudopregnant rats after ovariectomy 1.5 Rab.U. given over 4 to 8 days were not sufficient to maintain the decidual reaction, but 3.0 Rab.U. and above were effective. Estrogen inhibits this effect if it is given in fairly large amounts, but it augments the reaction in small doses. Thus, 0.15 γ of estradiol augments 1.5 Rab.U. of progesterone, but 3 γ inhibits the effect of 3 Rab.U., and 9 γ inhibits 6 Rab.U. but 0.6 γ augments them (135). Progesterone will not prolong the life of deciduomata indefinitely (136).

Pregnancy is maintained after the removal of all but 2 corpora lutea on the eighth day of pregnancy. It was partially maintained in two-thirds of rats in which all but 1 were removed on the fifteenth day (137). The corpora lutea must produce at least 0.5 mg. of progesterone daily to neutralize the normally occurring estrogens of pregnancy (138). During the last third of pregnancy the rat ovary contains 7.9 μg/g. of tissue, and the blood 0.07 μg/g. The 4-pregnen-20α-ol-3-one content of the ovary is 10.1 μg/g. and of blood 0.03 μg/g. (139).

A very large amount of work, in which the rat has been used as the experimental animal, has been done upon the effects of gonadotrophic hormones and especially upon pituitary-gonadic relationships. Since the object of this book is to bring together as far as possible the major quantitative data, much has been neglected unless it seemed to throw light upon the physiology of the species. We now give some data relating to the major reactions produced by these hormones.

The ovary of the immature rat will not respond to the injection of gonadotrophic hormones by the growth of follicles before 21 days, and apparently a well-developed granulosa is necessary for this to occur (140). F.S.H. alone will not cause the ovaries to develop beyond a weight of 45 mg., and L.H. alone will not cause any increase at all; but if F.S.H. is present to cause follicular development, L.H. is able to act and will produce a large increase in weight (141). International and Rat standards have been adopted in some cases. These are measured by the amount of substance needed to double the weight of immature ovaries under certain standard conditions. At present there is some confusion as the method of testing varies with the laboratory. One of the most pressing needs is the adoption of standard methods of assay and their general application.

It is possible to obtain superovulation in immature rats by carefully regulating the dose, and in these animals as many as 23 young may be born, and up to 33 embryos have been found. The maximum prolificacy has been obtained when the rats were treated just before the normal time of puberty (142). Sixteen R.U. of P.M.S. is about the optimum dose for this purpose; a greater amount causes too much luteinization (143). With F.S.H. an average of 17 implantation sites was produced against a normal of 10 for mature rats. Ten to fifteen R.U. was the optimum dose, and the number of young born was almost never in excess of normal, and the average was 6.4 against 9.0 for normals (144). These rats were slightly younger than those used in the P.M.S. work. In mature rats superovulation is not produced, but the injection of 12 R.U. of P.M.S. in metestrum does increase the average litter size if the rats are bred at the next heat period, but not beyond the normal extreme (145).

The injection of 2 R.U. of estrogen daily for 30 days or more impairs the gonadotrophic function of immature rats (146), and 10 R.U. daily for 5 days degranulates the pituitary basophils and, to a lesser extent, the eosinophils (147).

Castration and ovariectomy cause changes in the anterior pituitary. The most marked of these is the production of highly vacuolated (signet-ring) basophil cells. These changes may be prevented in the female by a minimum of 2 B.U. of androgens daily, but the male requires 10 B.U. (148). Estrogens prevented changes when given daily in doses of 0.03 R.U. to females and of 0.4 R.U. to males (149). Progesterone does not prevent them in doses up to 3 mg. (150). Ovariectomy increases the gonadotrophic potency of the anterior pituitary. After 5 days the latter is nearly 4 times as potent as the normal; at 60 days, 31 times; and eventually, 52 times (151). These figures relate to the gland as a whole. Castration increases the I.C.S.H. of the anterior pituitary about fourfold in 3 months. Estradiol decreases it but not to the level of the intact males. Estrogen enlarges the pituitaries of intact and castrated males equally, but the degree of enlargement in the intact female is less than it is in the ovariectomized female (152).

The minimum doses of hormones that will produce pseudopregnancy appear to be as follows: estrogen, 200 I.U. daily (153); testosterone, 2.5 to 5.0 mg. given over 10 days (154); testosterone propionate, 0.1 mg. daily for 16 days (155); progesterone, 1 mg. daily for 9 to 11 days (156). They act by modifying the anterior pituitary function.

Large acidophilic cells appear in the anterior pituitary during pregnancy. These cells increase in numbers from 3 days to 12 days, after which there is no further increase, but they persist throughout pregnancy and lactation. They also appear in pseudopregnancy and last about 12 days (157). Hypophysectomy before the eleventh day of pregnancy is followed by resorption of the fetuses; when performed at from 11 to 20 days of pregnancy, normal young are born after a prolonged gestation; but if the operation is performed on the twenty-first day, normal parturition follows (158). In another experiment hypophysectomy on the twelfth day was followed with normal delivery at term in 25 per cent. The others prolonged pregnancy for 1 to 3 days and the young were delivered late, or resorbed or died (159). Pregnant mare's serum will not maintain pregnancy in early hypophysectomized rats, but 20 I.U. daily of prolactin does so in most cases (160).

Large doses of estrogens injected before the nineteenth day of pregnancy cause the death of the fetuses, but not after this time (161). Injection of 500 γ of testosterone propionate daily during late pregnancy delays parturition. If injected early it prevents, or modifies, fetal development (162). Injections made before the sixteenth day produce modifications in the accessory sex organs of female fetuses. Administration of 750 to 3,300 I.U. causes vasa deferentia to be produced in all female young, but intermediate doses above 300 I.U. cause them to appear in some of the fetuses. Androsterone is just as effective on an I.U. basis, whereas androstene-dione is not quite so effective, The order of sensitivity of the tract in increasing difficulty of modification is as follows:

Ventral prostate
Posterior prostate
Coagulating gland
Seminal vesicles
Vas deferens and epididymis

The Wolffian duct of the left side is more sensitive than that of the right (163). Similar injections of estrogens produce effects in male embryos. With injections of 1.0 mg. of estradiol dipropionate slight hypospadias was produced in male fetuses; with 2.0 mg. there was a change in the position of the gonad and the degree of hypospadias increased, or the external genitalia were modified toward the female. From 3.0 mg. upwards the injections caused the external genitalia to be female. No effects were observed upon the testes, but there was a relative inhibition of the epididymis, vas deferens, and prostate. The Mullerian ducts remained vestigial (164).

A summary of the relationships between body weight and the weights of parts of the reproductive tract, together with much data on prenatal growth, has been published in convenient form (165).

The water content of the uterine tissue decreases at the time of heat but that of the oviducts does not (166). The placental scar is not a good criterion of the number of young born (167).

THE MALE

In male rats in the University of Chicago colony sperm heads appear in the testes at 33 to 35 days of age. Treatment with anterior pituitary preparations did not hasten their appearance by as much as 2 to 3 days, but it did accelerate the development of the accessory tract (168).

Spermatozoa may be found in the testes first at age 56 days. They enter the epididymis on the seventy-second day of life (169). According to one account the whole of the process of spermatogenesis takes a little less than 48 days (170). Another gives a 4-day interval between mitoses and about 16 days for the whole process (171). The length of the spermatozoon is $190.0 \pm .5\ \mu$, and of the head $12.05 \pm .05\ \mu$ (172). The length of the seminiferous tubules in a single testis is 20 meters (173).

The plug formed in the vagina during copulation has two prongs which fit into the cervices. The vulval end is pointed in contrast to that of the mouse, in which species it is blunt. It gradually disappears by disintegration (174). The semen is, in part, injected directly in the uteri, as spermatozoa appear at the apices within 2 minutes after copulation (175). By 15 minutes after coitus they are in the uterine segment of the oviduct in 42 per cent of cases, and in the ovarian segment in 21 per cent. In 30 minutes the corresponding figures are 88 and 62 per cent. After an hour has elapsed spermatozoa can be found throughout the tract in all cases. Only a small proportion of those which are deposited in the female reach the infundibulum (176).

The copulation plug appears to be necessary for the effective insemination of the female, in spite of the belief that some semen is injected directly into the uterus. Removal of the vesiculae seminales and the coagulating gland, a portion of the prostate, was followed by infertility in the majority of instances. In some cases where the operation was apparently incomplete, spermatozoa were found in the uteri only when a plug had been formed (177).

Spermatozoa can survive in the uterus up to 12 hours, in the vagina for 14½ hours, and in the oviducts for 16 to 17 hours (178). Those which are stored in the epididymis, with the vasa efferentia ligated, remain capable of motility for 42 days, and of fertility for 21 days (179). If the testes are anchored in the body cavity, sperm survival is not affected, but castration reduces their life to 14 days (180). Testes that are anchored in the abdominal cavity cease sperm production within 5 days, but there is no diminution in the output of hormones, at least up to 60 days (181).

The pH of various parts of the male tract has been given as follows (182):

Testes	pH 7.2 -7.4
Epididymis	pH 6.5 -6.6
Prostate	pH 7.14
Seminal vesicles	pH 6.32-6.34

The effects of the failure to produce testosterone after castration are made apparent by a degranulation of seminal vesicle cells in 2 to 3 days (183), and by similar damage to the prostate cells within 5 days (184). The weight of the seminal vesicles decreases to about one fifth in 5 weeks, but subsequently a partial recovery ensues (185). This may possibly be related to the secretion of androgens by the adrenal cortex, as this organ is able partially to maintain the weight of the ventral prostate in the castrated rat; the effect, however, is said to cease after they are 31 days old (186). The vas deferens does not have spontaneous motility but acquires it after castration (187). Androgens restore the immotility, but estrogens do not (188).

The ventrocaudal part of the scrotal sac is reddish yellow in color and is wrinkled. These are secondary sexual characters as they are lost after castration (189). After this operation the scrotum loses weight and its muscle contracts with much less readiness than formerly (185).

In the Long-Evans strain the prepuce begins to cornify at 31 days of age, cleavage to two surfaces begins at 45 days, and separation is complete at 52 days. If the rats are castrated at 26 days of age, 0.012 mg. of testosterone propionate daily causes separation in 3 weeks, and larger doses shorten the time (190). If rats are castrated at birth, the minimal effective daily dose of testosterone needed to grow the accessory organs is: to 20 days, 0.03 mg.; at 30 days, 0.025 mg.; and from 40 to 60 days, 0.005 mg. After this time the amount needed increases rapidly to 0.025 mg. (191). In adult life the amounts needed daily to maintain various structures are given below:

		SEMINAL VESICLES	
	PROSTATE	WEIGHT	GRANULES
Androsterone (192)	1 mg.	> 2 mg.	
Testosterone (193)	0.03-0.05 mg.	0.2 mg.	0.6 -0.7 mg.
Testosterone propionate (193)		< .05 mg.	0.15 mg.

The amount of testis tissue needed to maintain normal prostate and seminal vesicles is 0.07 g., or 3.1 per cent of the total testis weight (194). It may be inferred that the testes are capable of secreting about 20 mg. of testosterone daily if the fragment of testis continues to secrete at the normal rate, an ample margin of safety. According to another account the amount of testosterone propionate that will maintain the accessory organs in castrated rats of 250 to 300 g. weight is 50 μg daily injected in oil (195). In the hypophysectomized male 45 μg of testosterone propionate maintained the accessory organs at nearly normal weights. The testis weight increased and spermatogenesis was continued (196).

In the immature rat the daily injection of 0.05 mg. of testosterone propionate prevents normal testicular growth (197), but higher doses, i.e., 30 mg. a week for 4 weeks, stimulate the growth—unless they are continued for a longer period, when inhibition results (198). In immature hypophysectomized rats 2 mg. daily induces sperm formation. Smaller doses are ineffective though they stimulate growth of the accessory organs (199).

The anterior pituitary of the castrate is twice as potent in gonadotrophic hormone as is that of the normal (112). The potency is mainly in F.S.H., and it continues so to the twentieth day; from then to 9 months more L.H. is produced, and after that time F.S.H. again (200).

The injection of 1 R.U. of P.M.S. daily in immature rats caused the testis tubules to develop, but no hastening of spermatogenesis was observed. The interstitial tissue also developed. High doses caused tubule damage (201). The daily injection of 0.5 M.U. of P.M.S. maintained the testis weights of hypophysectomized rats if the injections were started at once, but did not maintain fertility; 1.0 M.U. daily was necessary to do this. If treatment was delayed, 5 M.U. were needed to maintain testis weights, but fertility was lost. Pregnancy urine was less effective in immediate treatment, but more so in delayed treatment (202). The daily injection of 10 to 25 R.U. of pregnancy urine into normal immature rats did not hasten puberty, but, injected into immature hypophysectomized rats, it caused development of the interstitial tissue and accessory organs, though regression set in after about 30 days (203).

The prolactin content of the pituitary of the male is about 0.16 B.U. per mg. of tissue. It varies little in the cryptorchid, but falls to about 60 per cent in the castrate (116).

1. Davis, D. E., and O. Hall. Physiol. Zool., 24: 9–20, 1951.
2. Davis, D. E., and O. Hall. Physiol. Zool., 21: 272–282, 1948.
3. Davis, D. E. AR., 99: 575–576, 1947.

4. Dall, O., and D. E. Davis. Texas Rpt. Biol. Med., 8: 564–582, 1951.
5. Perry, J. S. PZS., 115: 19–46, 1945–6.
6. Leslie, P. H., U. M. Venables, and L. S. V. Venables. PZS., 122: 187–238, 1952–3.
7. Long, J. A., and H. M. Evans. Univ. California Mem., 6: 1–148, 1922.
8. Bogart, R., G. Sperling, L. L. Barnes, and S. A. Asdell. AJP., 128: 355–371, 1940.
9. Engle, E. T., R. C. Crafts, and C. E. Zeithaml. PSEBM., 37: 427–432, 1937.
10. Andersen, D. H. J. Physiol., 74: 49–64, 1932.
11. Blandau, R. J. AR., 86: 197–214, 1943.
12. Arai, H. AJA., 27: 405–462, 1920.
13. Cole, H. H., and R. B. Casady. E., 41: 119–126, 1947.
14. Slonaker, J. R. AJP., 68: 294–315, 1924.
15. Blandau, R. J., J. L. Boling, and W. C. Young. AR., 79: 453–463, 1941.
16. Murray, G. N. Onderstepoort J. Vet. Sci. Anim. Ind., 16: 331–539, 1941.
17. Shelesnyak, M. C. AR., 49: 179–183, 1941.
18. Slonaker, J. R. AJP., 89: 406–416.
19. Rosen, S., and M. C. Shelesnyak. PSEBM., 36: 832–834, 1937.
20. Shelesnyak, M. C., S. Rosen, and L. R. Zacharias. PSEBM., 45: 449–451, 1940.
21. Herren, R. Y., and H. O. Haterius. AJP., 100: 533–536, 1932.
22. Haterius, H. O. AJP., 103: 97–103, 1933.
23. Cooley, C. L. and J. R. Slonaker. AJP., 72: 595–613, 1925.
24. Blandau, R. J., and A. L. Soderwall. AR., 79: 419–431, 1941.
25. Boling, J. L., R. J. Blandau, A. L. Soderwall, and W. C. Young. AR., 79: 313–331, 1941.
26. Fiske, V. M. E., 29: 187–196, 1941.
27. Hemmingsen, A. M., and N. B. Krarup. Kgl. Danske Vid. Selskab, Biol. Med., 13 (7): 1937.
28. Browman, L. G. AR., 67 (suppl.): 107, 1936.
29. McKeown, T., and S. Zuckerman. PRS., 124B: 464–475, 1938.
30. Marshall, F. H. A., and W. A. Jolly. TRS., 198B: 99–141, 1905.
31. Blandau, R. J., and A. L. Soderwall. AR., 81: 419–431, 1941.
32. Soderwall, A. L., and R. J. Blandau. J. Exp. Zool., 88: 55–64, 1941.
33. Blandau, R. J., and E. S. Jordan. AJA., 68: 275–291, 1941.
34. Huber, G. C. J. Morphol., 26: 247–358, 1915.
35. King, H. D. AR., 27: 337–366, 1924.
36. Hain, A. M. AR., 59: 383–391, 1934.
37. Babcock, M. J., R. Bogart, G. Sperling, and S. A. Asdell. J. Agric. Res., 50: 847–854, 1940.
38. Miller, N. Am. Nat., 45: 623–635, 1911.
39. Asdell, S. A., R. Bogart, and G. Sperling. Cornell Univ. Agric. Exp. Sta., Mem. 238, 1941.
40. Ingram, D. L., A. M. Mandl, and S. Zuckerman. JE., 17: 280–285, 1958.
41. Bunde, C. A., and A. A. Hellbaum. Proc. Oklahoma Acad. Sci., 19: 23–25, 1939.
42. Arai, H. AJA., 28: 59–79, 1920–1.
43. Slonaker, J. R. AJP., 81: 620–627, 1927.
44. Bloch, S., and E. Flury. Gynaecologia, 147: 414–438, 1959.
45. Kidwell, J. F., and H. J. Weeth. J. Hered., 49: 303–304, 1959.

46. Krebhiel, R. H. AR., 81: 43–65, 1941.
47. Lane, C. E. AR., 61: 141–153, 1935.
48. Lane, C. E. AR., 71: 243–247, 1938.
49. Davis, D. E., and O. Hall. AR., 107: 187–192, 1950.
50. Lane, C. E., and F. R. Davis. AR., 73: 429–442, 1939.
51. Pederson, E. S. AJA., 88: 397–427, 1951.
52. Boling, J. L. AR., 82: 131–145, 1942.
53. Weichert, C. K., and A. W. Shurgast. AR., 83: 321–334, 1942.
54. Deane, H. W. AJA., 91: 363–393, 1952.
55. Buno, W. Compt. Rend. Assn. Anat., 43: 842–847, 1956.
56. Young, W. C., J. L. Boling, and R. J. Blandau. AR., 80: 37–45, 1941.
57. Jaworski, Z. Ann. d'Endocrinol., 11: 361–388, 1950.
58. Mandl, A. M. J. Exp. Biol., 28: 585–592, 1951.
59. Ring, J. R. AR., 107: 121–131, 1950.
60. Okigaki, T. Japanese J. Genet., 34: 15–22, 1959.
61. Mandl, A. M. J. Exp. Biol., 28: 576–584, 1951.
62. Blandau, R. J. AR., 87: 17–26, 1943.
63. Allen, W. M. AR., 48: 65–103, 1931.
64. Lyon, R. A., and W. M. Allen. AJP., 122: 624–626, 1938.
65. Gilchrist, F., and G. Pincus. AR., 54: 275–287, 1932.
66. Andersen, D. H. AJA., 42: 255–305, 1928.
67. Gerard, G., and H. Poletti. Arch. Soc. Biol. Montevideo, 23: 74–82, 1958.
68. Wolfe, J. M., and R. Cleveland. AR., 55: 233–249, 1933.
69. Andersen, D. H., and H. S. Kennedy. J. Physiol., 76: 247–259, 1932.
70. Andersen, D. H., and H. S. Kennedy. J. Physiol., 79: 1–30, 1933.
71. Ford, D. H., and A. Hirschman. AR., 121: 531–547, 1955.
72. Braden, A. W. H. Fertility and Sterility, 10: 285–298, 1959.
73. Odor, D. L., and R. J. Blandau. AJA., 89: 29–62, 1951.
74. Braden, A. W. H., and C. R. Austin. Science, 120: 610–611, 1954.
75. Austin, C. R., and A. W. H. Braden. Australian J. Biol. Sci., 7: 179–194, 1954.
76. Odor, D. L. AJA., 97: 461–491, 1955.
77. Mandel, J. AR., 61: 295–309, 1935.
78. Fels, E. Strahlentherapie, 54: 279–293, 1935.
79. Levine, W. T., and E. Witschi. PSEBM., 30: 1152–1153, 1933.
80. Kraatz, C. P. PSEBM., 40: 499–502, 1939.
81. Chiodi, H. CRSB., 129: 1258–1259, 1938.
82. Greep, R., and F. L. Hisaw. PSEBM., 39: 359–360, 1938.
83. Meyer, R. K., S. L. Leonard, and F. L. Hisaw. PSEBM., 27: 340–342, 1929.
84. Evans, H. M., M. E. Simpson, and W. R. Lyons. PSEBM., 46: 586–590, 1941
85. Dawson, A. B., and J. T. Velardo. AJA., 97: 303–329, 1955.
86. Beilly, J. S. E., 25: 275–277, 1939.
87. Warren, M. R. AJP., 122: 602–608, 1938.
88. Hall, B. V. Physiol. Zool., 9: 471–497, 1936.
89. Blandau, R. J., L. Jensen, and R. Rumery. Fertility and Sterility, 9: 207–214, 1958.
90. Kirsch, R. E. AJP., 122: 86–93, 1938.
91. Haterius, H. O. AJP., 114: 399–406, 1936.

92. Alexander, D. P., J. F. D. Frazer, and J. Lee. J. Physiol., 130: 148–155, 1955.
93. Blair, E. AJP., 65: 223–228, 1923.
94. Clark, A. J., H. H. Knaus, and A. S. Parkes. J. Pharmacol. Exp. Therap., 26: 359–369, 1925.
95. Kerly, M. Biochem. J., 31: 1544–1552, 1937.
96. Knaus, H. H., and A. J. Clark. J. Pharmacol. Exp. Therap., 26: 347–358, 1925.
97. Blandau, R. J. AJA., 77: 253–272, 1945.
98. Hamilton, C. E. AR., 97: 47–62, 1947.
99. Kelly, G. L. AR., 73: 401–405, 1939.
100. Kun, H., and O. Peczenik. Pflüger's Arch. ges. Physiol., 236: 471–480, 1935.
101. Allison, J. B., and S. L. Leonard. AJP., 132: 185–192, 1941.
102. Berkhäuser, H., and E. A. Zeller. Helv. chim. Acta, 23: 1460–1464, 1940.
103. D'Amour, F. E., D. Funk, and M. B. Glendenning. PSEBM., 35: 26–27, 1936.
104. Everett, J. W., and C. H. Sawyer. E., 45: 581–595, 1949.
105. Everett, J. W. E., 59: 580–585, 1956.
106. Hill, R. T., and M. Alpert. E., 69: 1105–1108, 1961.
107. Everett, J. W. E., 43: 389–405, 1948.
108. Witschi, E. E., 27: 437–446, 1940.
109. Clark, H. M. AR., 61: 175–192, 1935.
110. Williams, M. McQ. PSEBM., 32: 1051–1052, 1935.
111. Lauson, H. D., J. B. Golden, and E. L. Sevringhaus. AJP., 125: 396–404, 1939.
112. Clark, H. M. AR., 61: 193–202, 1935.
113. Evans, H. M., and M. E. Simpson. AJP., 89: 371–374, 1929.
114. Leonard, S. L. E., 21: 330–334, 1937.
115. Soliman, F. A., and H. Nasr. Nature, 194: 154–155, 1962.
116. Reece, R. P., and C. W. Turner. Missouri Agric. Exp. Sta., Res. Bull. 266, 1937.
117. Grosvenor, C. E., and C. W. Turner. E., 66: 96–99, 1960.
118. Riddle, O. E., L. Lahr, and R. W. Bates. PSEBM., 32: 730–734, 1935.
119. League of Nations, Bulletin of the Health Organization, 10: 80–82, 1942–3.
120. Allen, E., ed., Sex and Internal Secretions. Baltimore, 1939.
121. Wilson, L. G., and W. C. Young. E., 29: 779–783, 1941.
122. Leonard, S. L., R. K. Meyer, and F. L. Hisaw. E., 15: 17–24, 1931.
123. Dorfman, R. I., T. F. Gallagher, and F. C. Koch. E., 19: 33–41, 1935.
124. Meyer, R. K., L. C. Miller, and T. R. Cartland. J. Biol. Chem., 112: 597–604, 1936.
125. Bourne, G., and S. Zuckerman. JE., 2: 268–310, 1940–1.
126. Del Castillo, E. F., and G. di Paola. E., 30: 48–53, 1942.
127. Boling, J. L., and R. J. Blandau. E., 25: 359–364, 1939.
128. Button, L. L., and C. I. Miller. PSEBM., 34: 835–839, 1936.
129. Varangot, J. CRSB., 131: 1027–1028, 1939.
130. Williams, M. F. AJA., 83: 247–307, 1948.
131. Paesi, F. J. A., and E. M. van Soest. Acta Endocrinol., 16: 88–99, 1954.
132. Donahue, J. K. E., 23: 521–523, 1938.
133. Rothchild, I., and R. K. Meyer. PSEBM., 44: 402–404, 1940.
134. Nelson, W. O., J. J. Pfiffner, and H. O. Haterius. AJP., 690–695, 1930.
135. Rothchild, I., R. K. Meyer, and M. A. Spielman. AJP., 128: 213–224, 1940.

136. Selye, H., A. Borduas, and G. Masson. AR., 82: 199–208, 1942.
137. Kelsey, R. C., and R. C. Meyer. PSEBM., 75: 736–739, 1950.
138. Freire, J. R. C. Ann. Acad. Brasil Cienc., 27: 79–82, 1955.
139. Wiest, W. G. E., 65: 825–830, 1959.
140. Smith, P. E., E. T. Engle, and H. H. Tyndale. PSEBM., 31: 744, 1934.
141. Fevold, H. L., and F. L. Hisaw. AJP., 109: 655–665, 1934.
142. Cole, H. H. Science, 91: 436–437, 1940.
143. Cole, H. H. AJA., 59: 299–331, 1936.
144. Evans, H. M., and M. E. Simpson. E., 27: 305–398, 1940.
145. Cole, H. H. AJP., 119: 704–712, 1937.
146. Meyer, R. K., S. L. Leonard, F. L. Hisaw, and S. J. Martin. PSEBM., 27: 702–704, 1929–30.
147. Wolfe, J. M., and C. S. Chadwick. E., 20: 503–510, 1936.
148. Nelson, W. O., and T. F. Gallagher. AR., 64: 129–145, 1935.
149. Hohlweg, W., and M. Dohrn. Wien. Arch. f. Inn. Med., 21: 337, 1931.
150. Cutuly, E. E., 29: 695–701, 1941.
151. Lauson, H. D., J. B. Golden, and E. L. Sevringhaus. E., 25: 47–51, 1939.
152. Paesi, F. J. A., and S. E. de Jongh. Acta Physiol. et Pharmacol. Neerl., 7: 277–288, 1958.
153. Nelson, W. O., and S. W. Pickette. AR., 84 (suppl.): 72, 1942.
154. Fluhmann, C. F., and G. L. Laqueur. PSEBM., 54: 223–225, 1943.
155. Freed, S. C., J. P. Greenhill, and S. Soskin. PSEMB., 39: 440–442, 1938.
156. McKeown, T., and S. Zuckerman. PRS., 124B: 362–368, 1937.
157. Haterius, H. O. AR., 54: 343–353, 1932.
158. Pencharz, R. I., and J. A. Long. AJA., 53: 117–139, 1933.
159. Fortgang, A., and M. E. Simpson. PSEBM., 84: 663–666, 1953.
160. Cutuly, E. PSEBM., 48: 315–318, 1941.
161. D'Amour, F. E., and C. Dumont. Quart. J. Exp. Physiol., 26: 215–224, 1937.
162. Hamilton, J. B., and J. M. Wolfe. AR., 70: 433–440, 1938.
163. Greene, R. R., M. W. Burrill, and A. C. Ivy. AJA., 65: 415–469, 1939.
164. Greene, R. R., M. W. Burrill, and A. C. Ivy. AJA., 67: 305–345, 1940.
165. Donaldson, H. H. The Rat: Data and Reference Tables. Mem. Wistar Inst., 6, 1924.
166. Odor, D. L., and R. J. Blandau. PSEBM., 70: 540–543, 1949.
167. Davis, D. E., and J. T. Emlen, Jr. JWM., 12: 162–166, 1948.
168. Moore, C. R. AR., 59: 63–88, 1936.
169. Reid, B. L. Australian J. Zool., 7: 22–38, 1959.
170. Clermont, Y., C. P. Le Blond, and B. Messier. Arch. Anat. Micros. Morphol. Exp., 48: 37–55, 1959.
171. Roosen-Runge, E. C. AJA., 88: 163–176, 1951.
172. Friend, G. F. QJMS., 78: 419–443, 1936.
173. Bascom, K. F., and H. L. Osterud. AR., 31: 159–169, 1925.
174. Parkes, A. S. PRS., 199B: 151–170, 1926.
175. Hartman, C. G., and J. Ball. PSEBM., 28: 312–314, 1930.
176. Blandau, R. J., and W. L. Money. AR., 90: 255–260, 1944.
177. Blandau, R. J. AR., 91 (suppl.): 4–5, 1945.

178. White, W. E. J. Physiol., 79: 230–233, 1933.
179. White, W. E. PRS., 113B: 544–550, 1933.
180. Moore, C. R. J. Exp. Zool., 50: 455–494, 1928.
181. Jeffries, M. E. AR., 48: 131–139, 1931.
182. Lanz, T. van. Pflüger's Arch. ges. Physiol., 222: 181–214, 1929.
183. Moore, C. R., W. Hughes, and T. F. Gallagher. AJA., 45: 109–135, 1930.
184. Moore, C. R., D. Price, and T. F. Gallagher. AJA., 45: 71–107, 1930.
185. Tyrrell, W. P., F. N. Andrews, and M. R. Zelle. E., 31: 379–383, 1942.
186. Burrell, M. W., and R. R. Greene. E., 26: 645–650, 1940.
187. Martin, T., and J. R. doValle. CRSB., 127: 464–466, 1938.
188. Martin, T., and J. R. doValle. CRSB., 127: 1385–1388, 1938.
189. Hamilton, J. B. PSEBM., 35: 386–387, 1936.
190. Lyons, W. R., I. Berlin, and S. Friedlander. E., 31: 659–663, 1942.
191. Hooker, C. W. E., 30: 77–84, 1942.
192. Callow, R. K., and R. Deanesly. Biochem. J., 29: 1424–1445, 1935.
193. Moore, C. R., and D. Price. AR., 71: 59–78, 1938.
194. Hansen, I. B. AR., 60 (suppl.): 56, 1934.
195. Cavazos, L. F., and R. M. Melampy. Iowa State Coll. J. Sci., 31: 19–24, 1956.
196. Dvoskin, S. AR., 99: 329–351, 1947.
197. Biddulph, C. AR., 73: 447–463, 1939.
198. Shay, H., J. Gershon-Cohen, K. E. Paschkis, and S. S. Fels. E., 28: 485–494, 1941.
199. Cutuly, E., E. C. Cutuly, and R. D. McCullogh. PSEBM., 38: 818–823, 1938.
200. Hellbaum, A. A., and R. O. Greep. AJA., 67: 287–304, 1940.
201. Cole, H. H. AJA., 59: 299–331, 1936.
202. Liu, S. H., and R. L. Noble. JE., 1: 7–14, 1939.
203. Smith, P. E., and S. L. Leonard. AR., 58: 145–173, 1934.

Rattus rattus L.

BLACK RAT

Most of the published work on this species has been devoted to studies of its rate of increase in different environments. The females are polyestrous and in most districts breed throughout the year. In Texas most young are born in January to March and fewest in July and August (1). In London the lowest point is reached in February, and the highest from March to October, with a possible peak in September (2). In Cyprus no pregnant females were found in January and very few in February. The principal seasons were March to April and September. Embryo counts varied from 5.2 to 8.0, with the higher figures from more northerly countries. It was high in those living in ships. In the port of London the amount of fetal atrophy was 1.9 per female; for ship dwellers it was 1.5 (2).

Puberty is reached at 70 g. weight, at the age of 1.5 to 2.5 months (3), but in many areas it appears to be much later. One account from a northern climate gives 6 months for males and 7 months for females (4). Figures for several Malayan subspecies show that 50 per cent of males were fertile, judging by the presence of spermatozoa in the testes, at weights varying from 57 to 76 g. The weights for the 95 per cent fertility level were from 88 to 201 g. In this investigation it was also found that different subspecies had different embryo counts, but much of the variation seemed to depend on the environment. The numbers varied from 2.4 to 6.0, and they varied with the size of the rat. The distribution between right and left uteri was even except for *R. r. jalorensis* in which the right uterus carried a significantly higher proportion (5). In the Ceylonese subspecies, *R. r. kandianus,* puberty was reached in females at 5 to 6 months and the usual litter size was 3 to 4 (6).

Gestation lasts for 21 days and there is a post-partum heat. The average litter size in a small sample in captivity was 5.6 (7).

The sperm head measures 11.0 μ (9.0–12.1) × 1.8 μ (1.5–2.1); the midpiece width, 1.0 μ (0.7–1.2); and the tail length, 147.0 μ (130.3–160.8) (8).

1. Davis, D. E. JM., 28: 241–244, 1947.
2. Watson, J. S. PZS., 120: 1–12, 1950–1.
3. Gomez, J. C. AMN., 63: 177–193, 1960.
4. Jirsik, J. Säugetierk. Mitt., 2: 21–28, 1954.
5. Harrison, J. L. PZS., 121: 673–694, 1951–2.
6. Phillips, W. W. A. Ceylon J. Sci., 14B: 209–293, 1927–8.
7. Kelway, P., and H. V. Thompson. In A. N. Worden and W. Lane-Petter, eds., The UFAW Handbook on the Care and Management of Laboratory Animals. London, 1957.
8. Hirth, H. F. J. Morphol., 106: 77–83, 1960.

Rattus (*Mastomys*) *coucha* A. Smith

MULTIMAMMATE RAT

A subspecies (*R. c. erythroleucus* Temminck) of the multimammate rat from Sierra Leone has been investigated. It is polyestrous and breeds at all times of the year, but most females are found pregnant in October and

November, at the end of the rainy season. At any other time of the year some females may be in anestrum, but the males breed at any time. At puberty the males weight 45 g., and the females 40 g., compared with their respective mature weights of 105 g. and 83 g. Ovulation is spontaneous, and there is a postparturient heat. If the female does not become pregnant at this time, a lactation anestrum sets in. A vaginal plug is formed after copulation. The uterus is filled with fluid during heat, but not at the postparturient heat. The vaginal epithelium becomes intensely mucified during proestrum. During heat it becomes cornified below the layer of mucous cells, and in metestrum both layers are sloughed off. The mean litter size is 11.5 by embryo counts, with a range from 7 to 17. The number of corpora lutea is 12.1, with a range from 5 to 19 (1).

In the laboratory another subspecies, *R. c. natalensis* A. Smith, gave a mean weight at vaginal perforation of 35.7 ± 1.8 g., when the age was 76 ± 6 days. The range of ages was from 56 to 115 days. At the onset of first heat the weight was 39 ± 1.4 g., and the age 104 ± 9 days, while the age range was from 64 to 145 days. The greater the weaning weight, the earlier was the first heat. The mean interval between heats was 8.8 ± .4 days, with a mode of 6 days. Anestrous periods were fairly lengthy. Proestrum lasted 1.9 ± .15 days, and metestrum for about the same period. Pseudopregnancy could not be induced by stimulating the cervix. A postpartum heat was the rule and the length of gestation was 23.1 ± .7 days. Erythrocytes could be detected in the vaginal smear at 15.4 ± 3.2 days after conception. According to this account there is possibly delayed implantation during lactation, but no lactation anestrum (2). The secondary sex ratio was 51.9 per cent males ($n = 3{,}928$), and the average litter size 7.3 young (3).

The vaginal smear has been described as follows:

Stage I. Proestrum. Fibrous-looking mucus enclosing bunches of proliferating epithelial cells. In later stages large epithelial cells increase. These have degenerating excentric nuclei. Some cornified cells are present.

Stage II. Estrus. Cornified cells.

Stage III. Early metestrum. Leucocytes enter the smear.

Stage IV. Late metestrum. Leucocytes, degenerating cornified cells, and a few proliferating epithelial cells (1).

Fifty per cent of the males have reached puberty at about 40 g. and the testis weight at maturity is 97 ± .02 mg. In this colony vaginal perforation was at about 44 g. (4).

In the north Cameroons pregnant females were found only from October to December (5), but in the Belgian Congo these rats bred during the dry season, from April to September (6). In the London zoo litters have been recorded throughout the second half of the year (7). In Tanganyika adult males are in breeding condition all the year. Breeding falls off after July and August and there is little between October and April, i.e., at the end of the rainy season and beginning of the dry season. In 10 pregnant females the range of embryos was 3 to 16, mean 11.2 (8).

1. Brambell, F. W. R., and D. H. S. Davis. PZS., 111B: 1–11, 1941.
2. Johnston, H. L., and W. D. Oliff. PZS., 124: 605–613, 1954–5.
3. Oliff, W. D. J. Anim. Ecol., 22: 217–226, 1953.
4. Pirlot, P. L. Mammalia, 21: 385–395, 1957.
5. Sanderson, I. T. TZS., 24: 623–725, 1938–40.
6. Pirlot, P. L. Ann. Mus. Congo Belge, Tervueren, Ser. Zool., Nouv. Ser. in 4°, 1: 41–46, 1954.
7. Zuckerman, S. PZS., 122: 827–950, 1952–3.
8. Chapman, B. M., R. F. Chapman, and I. A. D. Robertson. PZS., 133: 1–9, 1959–60.

Rattus bowersi Anderson. In Malaya 50 per cent of the males are fertile at 166.5 ± 4.9 g., and 95 per cent at 338.1 ± 1.4 g. Embryo counts of 2 and 5 have been recorded (1).

R. canus Miller. In Malaya 50 per cent of males are fertile at 67.4 ± .5 g., and 95 per cent at 145.4 ± .5 g. The females reach puberty at 120 g. and over. The mean embryo count was 3.0, range 2 to 6, mode 2 (1).

R. cremoriventer Miller. In Malaya 50 per cent of males are fertile at 19.0 ± .2 g., and 95 per cent at 61.2 ± .2 g. The females reach puberty at 50 g. and over. The embryo count has ranged from 2 to 5, mode 3, mean 3.7 (1).

R. exulans Peale. In Malaya 50 per cent of males are fertile at 22.2 ± .3 g., and 95 per cent at 24 g. (2). The embryo count varied from 1 to 8, mean 4.3, and mode 4 (1). In Selangor the rat breeds all year, and at any time 25 to 30 per cent of the mature females are pregnant (3). In New Caledonia limited numbers give a season from November on and from April on, i.e., in spring and fall (4).

R. hawaiiensis Stone. This rat has a very uniform season throughout the year. There is a post-partum heat, and gestation is not less than 21 days. The mean litter size is 3 to 4. Fetal counts have averaged 3.8 (5).

R. mülleri Jentink. In Malaya 50 per cent of males are fertile at 172 ± .6 g., and 95 per cent at 378.7 ± .9 g. The females reach puberty at 280 g. upward. Most pregnancies are found from July to December. The mean embryo count is 3.8, mode 3, and range 1–9. They are equally distributed in the uteri (1).

R. rajah Thomas. For the subspecies, *R. r. surifer,* a report gives 50 per cent of males fertile at 67.5 ± .3 g., and 95 per cent at 145.1 ± .4 g. Females reached puberty at 150 g. and up. The mean embryo count was 3.3, range 2 to 5, mode 3, equally distributed between the uteri (1). For the subspecies, *R. r. pellex,* 50 per cent fertility for males was at 68.9 ± .8 g., and 95 per cent at 160.7 ± .9 g., with female puberty at 150 g. upwards (1).

R. sabanus Thomas. This Malayan rat breeds in Selangor the year round, and 11 to 17 per cent were found pregnant at one time (3). However, most pregnancies are found from July to September. The females attain puberty at 20 g. and up. The embryo count varied from 1 to 7, mean 3.1, mode 2 and 3. The embryos are equally distributed between the uteri (1).

R. whiteheadi Thomas. In Malaya this rat breeds throughout the year but most pregnancies are found from October to March. Fifty per cent of males are fertile at 35.6 ± .1 g., and 95 per cent at 44.7 ± .1 g. Puberty in the female is attained at 35 g. and up. Embryos are equally distributed between the uteri and number from 1 to 6, mean 3.0, mode 3 (1). In Selangor 20 to 29 per cent of the adult females examined were pregnant (3).

R. (*Hylomyscus*) *alleni* Thomas and Wroughton. A specimen has been recorded with 3 embryos (6).

R. (*H.*) *carillus* Thomas. Lactating females of this African rat have been found in May (7).

R. (*Ochromys*) *woosnami* Schwann. In South West Africa this rat has been found in November with 2 fetuses (8).

R. (*Praomys*) *morio* Trouessart. Pregnancies have been reported from the Cameroons in January and February, each with 3 embryos (9).

R. (*P.*) *taitae* Heller. A lactating female of this African rat has been reported in April (10).

R. (*P.*) *tullbergi* Thomas. In the Belgian Congo this rat bred towards the end of the rains and early in the dry season, from September to March (11). There are reports of females in December with 3 and 6 embryos and of one in January with 3 (6). Two records each gave 2 nestlings (12, 13).

R. (*Stochomys*) *longicaudatus* Tullberg. A female of this African rat was found in August with 2 embryos, both in the left uterus. A lactating female was found in December (6).

1. Harrison, J. L. PZS., 125: 445–460, 1955.
2. Harrison, J. L. PZS., 121: 673–694, 1951–2.
3. Harrison, J. L. Raffles Mus. Bull., 24: 109–131, 1952.

4. Nicholson, A. J., and D. W. Warner. JM., 34: 168–179, 1953.
5. Svihla, A. The Murrelet, 17: 3–14, 1936.
6. Hatt, R. T. AM., 76: 457–604, 1939–40.
7. Hill, J. E., and T. D. Carter. AM., 78: 1–211, 1941.
8. Shortridge, G. C. The Mammals of South West Africa. London, 1934.
9. Eisentraut, M. Zool. Jahrb., Syst., 85: 619–672, 1957.
10. Allen, G. M., B. Lawrence, and A. Loveridge. HCMZ., 79: 31–126, 1935–7.
11. Pirlot, P. L. Ann. Mus. Congo Belge, Tervueren, Ser. Zool., Nouv. Ser. in 4°, 1: 41–46, 1954.
12. Allen, G. M., and A. Loveridge. HCMZ., 89: 147–214, 1941–2.
13. Frechkop, S. EPNA., 1938.

Melomys cervinipes Gould

TREE RAT

Embryo and litter counts of this Queensland rat give a range from 1 to 4, mean and mode 3 (1, 2). Breeding ceases in the dry season (3).

Puberty is at age 7 months (4).

1. McDougall, W. A. Queensland J. Agric. Sci., 3: 1–43, 1946.
2. Tate, G. H. H. AM., 102: 199–203, 1953.
3. Troughton, E. Furred Animals of Australia. New York, 1947.
4. Davies, S. J. J. F. J. Roy. Soc. Western Australia, 43: 63–66, 1960.

Uromys caudimaculatus Krefft

This Australasian mouse has been reported with 4 young (1).

1. Tate, G. H. H. AM., 102: 199–203, 1953.

Coelomys bicolor Thomas

This Ceylonese rat has been reported in February with 3 embryos (1).

1. Phillips, W. W. A. Ceylon J. Sci., 14B: 209–293, 1927–8.

Malacomys longipes Milne-Edwards

In the north Cameroons this mouse breeds in the early rains, from March to June (1). A January record of a female with 3 embryos has been made (2).

1. Sanderson, I. T. TZS., 24: 623–725, 1938–40.
2. Hatt, R. T. AM., 76: 457–604, 1939–40.

Mus bactrianus Blyth

PERSIAN HOUSE MOUSE

This Asiatic mouse when kept in the laboratory is polyestrous and breeds all the year around. It experiences a postparturient heat. The mean age at which the female gives birth to her first litter is 177 ± 6.2 days. Gestation lasts 20 days if the female is not suckling. The mean litter size is 4.4 ± .09 (1).

1. Green, C. V. JM., 13: 45–47, 1932.

Mus musculus L.

HOUSE MOUSE

The house mouse is polyestrous all the year round, and in all respects its reproductive cycles closely resemble those of the rat. As the cycle has been described in detail under that species, reference should be made to it there.

THE WILD MOUSE

In urban buildings 22 per cent of the female mice are pregnant at one time and in grain ricks, 40.6 per cent. Similarly, the number of litters per year is 5.5 in buildings and 10.2 in grain ricks; the number of young, 30.9 and 57.2 respectively. Puberty in the females is reached between 8 and 12 g., and in the males between 10 and 15 g. There was no marked seasonal trend, but the weight did vary with the environment. There is a slight tendency for more embryos to be carried in the right uterus (1). Mice living in grain ricks in England had a live embryo count of 5.66 ± .07. The average number of resorbing embryos per female was 1.9, and 16.6 per cent of females displayed some amount of resorption (2).

In Mississippi the pregnancy rate was 46.8 per cent of mature females and the average embryo count was 4.8 ± .09. There was little difference by months. Breeding tended to be a little lower from December to March and in July. A body length of 2½ inches denoted sexual maturity (3). In Malaya 50 per cent of males were fertile at 9.0 ± .01 g., and 95 per cent at 12.1 ± .01 g. Females reached puberty at 8 g. and up. The mean embryo count was 4.3, mode 4 and 5, range 1 to 7, equally distributed between the uteri (4). In Ceylon the mouse has 3 to 5 litters a year, usually with 5 or 6 young. They breed throughout the year (5).

In Britain confined populations have a higher proportion of nonfecund females than wild-caught ones. When the former were allowed to disperse in a large pen the reproduction rate increased (6). A low-density population had embryo counts from 6 to 23 and a very high-density one from 5 to 11 (2).

The subspecies, *M. m. molossinus* Temminck, of Japan breeds throughout the year but the pregnancy rate is highest in the fall, less in spring and summer, and least in winter. Spermatozoa may be found in the epididymis all year. Testis weight and accessory-gland activity are highest in the fall and least in winter. The count of normal fetuses was 5.6 ± .28. Atrophic fetuses amounted to 0.3 per female. The highest fetus count, 6.6, was found in the fall; the number of atrophic fetuses was then at its lowest (7). The average litter size at birth was 5.4 ± .09 (8). The cycle length, by vaginal smears which are typically murine, was 6.0 ± .08 days, mode 4 and 5. Proestrum lasted 0.7 days; estrus 2.3; metestrum 0.6; and diestrum 2.4 (9). Copulation usually occurs between proestrum and estrus. The placental sign was found at 8 to 13 days, mean 11.2 ± .13. Most parturitions occur round midnight. The gestation period in the nonsuckling female lasted 20.24 ± .12 days; it was higher in the suckling female (10). The vagina opened at 64.2 ± .5 (35–76) days, and mature follicles were present in the ovaries at 65 days. Reproduction declined at 600 days. The testes descended at 25.4 ± .4 days, range 16 to 45 days. Spermatozoa were present in the epididymis at 40 days (11).

THE LABORATORY MOUSE

The age of establishment of the vaginal orifice is somewhat variable. One investigation gave a spread from 28 to 49 days, a mean of 35.5 ± .3 days, and a standard deviation of 4.75 days. In this investigation the weight at opening gave a slightly closer fit than age. The mean weight was 13.15 ± .08 g. (12). Temperature has some effect on the time of opening. In mice raised at 60 to 64° F. cycles began at 30.8 days, while in those raised at 80 to 92° F. they began at 34.2 days (13). There was no difference between mice kept in total darkness, in darkness with 1 hour of light daily, with red light, or kept normally (14). The vagina opens 1.4 days before the onset of the first heat (15). In a strain of albino mice the females reached puberty at 27.91 ± .26 days (16).

The length of the estrous cycle varies from 3 to 9 days, with an average at 4.5 days, and a mode of 4 and 6 days. However, the length appears to vary with the color of the mouse. Brown mice have the longest cycles, a modal length of 5 and 6 days; grays and yellows, 5 days; blacks, 4 days; and albino "browns," 4 days. Most of the prolongation was in the duration of heat (17). A small decrease in environmental temperature has no marked effect upon the duration of cycles, but a reduction to 0° C. increases the interval from 4.02 ± .14 to 11.16 ± .59 days. After one week at the lower temperature the cycle is restored to its normal length (8). The onset of heat, as judged by willingness to mate, usually occurs between 10 P.M. and 1 A.M. The spontaneous ovulation is usually between 12 midnight and 2 to 3 A.M., and the spread is from 11:30 P.M. to 4:40 A.M. It occurs, therefore, about 2 to 3 hours from the onset of heat, but the interval is variable. If mice are submitted to a reversal of the time of light, and are also subjected to 16 to 17 hours of light, the onset of the first heat is accelerated and they mate at the time customary under usual light conditions, i.e., in the dark hours (19). Mating has no effect on the time of ovulation. The first polar body is nearly always extruded between the onset of heat and ovulation. At the time of first acceptance of the male the vaginal smear consists of from 21 to 95 per cent cornified cells (20). The conception rate is highest from June to August and lowest from October to January. The litter size and sex ratio are relatively constant throughout the year. For over 20,000 deliveries the litter size, including stillbirths, was 5.6 (21).

The mouse requires a sterile copulation or stimulation of the cervix uteri to activate the corpus luteum in the absence of fertile copulation, and then she displays the phenomenon of pseudopregnancy which lasts for 10 to 12 days (22).

Heat and ovulation occur within 24 hours after parturition. If the female is allowed a sterile copulation at that time, the ensuing pseudopregnancy is prolonged to 21 to 29 days if she is suckling. In nonsuckling animals and in those suckling only 1 or 2 young this prolongation is not observed (22).

The ova take 3 days to traverse the oviduct and remain for 1 day in the last portion of the tube (23).

The litter size in mice has been the subject of many papers (11, 12). The means reported vary from 4.5 to 7.4. The second litter is the largest, and there is a steady decrease afterwards so that the sixth litter averages less than the first (12). Fertility rises slightly through the year to July and September. Litters during lactation tend to be 0.3 lower than those born when the mother is not lactating (13). Counts of corpora lutea, embryos, and young born have averaged 8.4, 6.35, and 5.86, respectively, showing the considerable loss that occurs from ovulation to birth, due to loss of eggs, failure of implantation, and fetal resorption. The embryos are distributed about equally between the two uteri, but there are too many occurring all on one side, because of pathology of one oviduct (14). If uniovular twins occur they must be infrequent, less than 0.7 per cent (15). The sex ratio is about 52 per cent males (16), and the duration of gestation is 19 days, varying from 18 to 20. If the female is suckling at the time, implantation is delayed, and gestation is prolonged, on the average, by 21 hours for each suckling young (17). Most births take pace from 4 P.M. to 4 A.M., and the fewest from 4 A.M. to 8 A.M. The female remains in seclusion an average of 4 hours and 40 minutes, during which time the litter is born (18).

Twenty-four females were paired monogamously and kept with their mates for a year. During this time the number of pregnancies was 197 and the young weaned numbered 1,149. Another 24 were caged 4 to a cage with a male and were then kept from the males before the birth of their litters and until after weaning. They had 110 pregnancies and a total of 559 young weaned. There was no difference in mortality of breeding females; the litter size, weight of young at weaning, and the sex ratio at weaning were significantly affected by the number of young weaned (32).

For lactating mice the mean gestation period was 28.8 ± 1.3 days. In this strain the mean litter size was 9.6 for the first litter (33). At the post-partum heat the females ovulate between 8 P.M. and 8 A.M. on the night following delivery, irrespective of the time of delivery. Most cyclic females ovulate between 11:30 P.M. and 1:30 A.M. (34). In a count of corpora lutea per litter order the number rose to the third litter and then remained fairly constant. It was not influenced by genetic differences between strains (35). In 106 litters without prenatal loss, a count which was designed to estimate the primary sex ratio, the proportion was 50.1 per cent males (36). The maximum frequency of the commencement of birth was between 1 and 4 A.M. The time can be changed by altering the light rhythm (37). In one strain the sex proportion of stillborn mice was 47.3 ± 2.2 per cent males, the same as that of living sibs. In 14-day old fetuses it was 55.9 ± 3.9 per cent males (38).

HISTOLOGY OF THE FEMALE TRACT

OVARY. At every heat period an average of 400 to 500 young ova in each ovary differentiate, but most are destined for atresia. Semispaying does not increase the number differentiating, but it does increase the number that mature, thus giving the normal number of ovulations from one ovary instead of from two (39). The graafian follicles measure about 0.4 mm. in diameter 2 days before heat, when they begin their rapid ripening growth. They measure about 0.7 mm. when ovulation occurs (40). After rupture of the follicle the granulosa cells mostly remain and hypertrophy to form the lutein cells. The spindle cells of the theca interna and externa divide by mitosis to form connective-tissue strands. Theca interna cells do not enter into the formation of lutein cells, but polygonal cells are formed from them, which, however, degenerate within 60 hours of ovulation (41).

New ova are produced from the germinal epithelium, and most mitoses are found in this layer immediately following ovulation (42). The injection of estrogens during the diestrous period causes mitoses to occur during that time, when normally there are few (43). Follicular atresia varies in a cyclical manner; it is highest the day after heat and lowest on the next day (44). A count made in 100 ovaries revealed 2 polynuclear ova, 16 binovular and 2 triovular follicles, and 13 anovular follicles, which were in an advanced stage of atresia (45). The average diameter of the ovum is 78.4 μ or, with the zona pellucida, 95.4 μ (46). The distribution of ova shed from the ovaries, i.e., the relative ovulation rate, is random (47).

VAGINA, etc. The vagina and vaginal smears are similar to those found in the rat. By the use of colchicine and estrogens in the ovariectomized mouse it has been shown that growth of the vaginal epithelium starts at the uterine end and proceeds toward the vulva (48).

The uterine changes are also similar to those in the rat. There is a heavy leucocytic infiltration of the endometrium just after heat (49).

The ciliated and nonciliated cells of the oviduct secrete and show maximal activity during heat. Eosinophil lipids are formed during diestrum and are secreted during estrus (50).

PHYSIOLOGY OF THE FEMALE TRACT

The ovary of the mouse seems to have remarkable powers of regeneration, since in 121 mice doubly ovariectomized with the removal of the capsules and parts of the oviducts all ceased to experience cycles at first, but later 11 of them came into heat. Of these, 8 were found to have ovarian tissue with follicles and corpora lutea (51). The injection of anterior pituitary extract did not increase the percentage of recovery (52). The role of the ovary in maintaining estrous cycles has been questioned, as it was found that X-ray sterilization before puberty, with oöcytes, follicles, and follicular tissue all apparently rendered nonfunctional, did not prevent most of the mice from having estrous cycles. The tubular genitalia remained normal (53). When the experiment was repeated with adult mice, the average length of cycles was increased to 6.0 to 6.6 days, but it became more variable (54). Perhaps these experiments should be considered in conjunction with the preceding in which regeneration after ovariectomy was recorded, though they have recently been used as evidence for the secretion of estrogens by tissues other than the granulosa cells.

The average time required for completion of ovulation is 5 hours, 4 of which may be taken up with capacitation of the ovum. The interval between ovulation and penetration by the spermatozoon is 5 hours. Capacitation of the spermatozoon requires about 1 hour. The spermatozoa may reach the ova in 15 minutes after mating but in the majority of females not until 1 hour has elapsed (55).

It is believed that the graafian follicles secrete sufficient estrogen to induce heat early in their growth, since ovariectomy 36 to 48 hours before heat does not prevent its occurrence at the usual time (56).

If 0.04 to 0.06 R.U. of estrogen is injected into mice, i.e., considerably less than the amount needed for cornification, the vaginal epithelium is converted into a mucus-secreting type (57). The same effect can be brought about if progesterone is injected together with a cornifying amount of estrogen (58). If a pituitary extract is injected during diestrum ovulation follows 11 to 13 hours later (34).

Deciduomata can be produced by traumatization of the endometrium during pseudopregnancy. The reaction is strongest at 3 to 4 days and slight at 5 to 6 days. They can also be evoked, but not so strongly, in a pseudopregnancy which is concurrent with lactation. They cannot be produced after X-ray sterilization, or after X-ray treatment and the injection of anterior pituitary extract to cause luteinization (59). Removal of the ovaries at any time during pregnancy causes resorption of the embryos or abortion (60, 61). Hypophysectomy at mid-pregnancy does not cause the pregnancy to end (62).

Pregnancy cells appear in the anterior pituitary of the mouse at the appropriate time. The reaction depends upon the ovary, as the injection of anterior pituitary extract can produce the effect, but only in the intact animal. It can also be produced in males by injection if ovaries have been transplanted into them. In other words the reaction is dependent upon the presence of corpora lutea (63). During pregnancy there is a leakage of red blood cells from the placenta, known as the placental sign, which can be detected in the vaginal smear from 9 to 10 days (64). Ovariectomy 1 to 12 hours after the vaginal plug has been formed locks the eggs in the oviduct. One injection of 200 R.U. of estrogen is sufficient to repair the damage to the tubes and allow the eggs to descend (65).

The vaginal pH varies with the cycle as follows: diestrum, 7.4; proestrum, 7.0–7.2; estrus, 6.8–7.0; metestrum, 7.8. For 2 to 3½ days after mating it remains constant at 7.6, from which one may infer that fertilization has occurred (66).

The pH of the uterine fluid at different stages of the cycle is as follows (67):

Stage	pH
Stage I	6.97
Stage II	7.13
Stage III	7.50
Stage IV (diestrum)	7.20

These figures are all about 0.2 units below the corresponding ones in the rat.

The uterine muscle is slightly more responsive to oxytocin during heat than between heats, and the response during pseudopregnancy is comparatively low, just as it is during pregnancy (68).

The Mouse Unit of estrogen is often used as a measure, but it varies with the laboratory from 0.2 to 1.0 of a Rat Unit. This variation may be caused either by differences in the method of testing or by strain differences. It should be valuable if several strains could be tested by a standardized technique and the results expressed in international units. If there is a great difference between strains, an inquiry into the mode of inheritance might yield valuable results. There is also a seasonal difference, since the response to 1 I.U. is three times as great in May as it is in November (69). If estrogen and progesterone are injected in sequence, sexual receptivity is more easily produced in the ovariectomized mouse than it is if estrogens alone are injected. The optimal amount of progesterone is 0.05 I.U., and the optimal interval is 48 hours (70). The pregnant mouse produces about 1.5 mg. (= 1.5 Rab.U.) of progesterone daily, according to the method of estrogen neutralization (71). The adult female's pituitary contains 0.25 B.U. of prolactin per mg. of tissue (72). The non-pregnant mouse excretes 1 to 2 I.U. of estrogen per month (73).

During pseudopregnancy the levels of estrogen and progesterone in the blood are low. Progesterone rises for 3 days and then remains high until the ninth day, when it decreases, reaching a low level at 12 days. Estrogen declined for 3 days and was absent until the ninth day when it rose but immediately declined. In pregnancy much the same pattern is followed, but on day 9 progesterone decreases abruptly and remains low. Estrogen was low until the eleventh day. Then it rose somewhat, and there was another slight increase immediately before parturition. In a more recent report free progesterone in the serum of the pregnant mouse rose to a peak of 8 progesterone equivalents per ml. at days 7 and 8, then it fell to day 13, rose again on day 15, and then fell steadily. The histology of the corpus luteum suggests that this organ ceases to be active about day 14 (75). In diestrum the free plasma progesterone is lowest, at 0.5 progesterone equivalents per ml. During proestrum and estrus it rises to 4 progesterone equivalents per ml., and then decreases through metestrum (76).

During pregnancy the level of gonadotrophic hormone in the pituitary gland is highest at 12½ to 15½ days. It is intermediate at 6½ days and least at 9½ and 18½ days. Most secretion is released early in pregnancy. The weight of the gland increases significantly during pregnancy (77).

The ovaries of immature mice do not respond to pituitary implants until the mice are 15 days of age (78). After this age, but while they are still immature, 2.5 I.U. of P.M.S., or 1.5 I.U. of P.U., are needed to cause ovulation. During pregnancy it may be induced by 0.7 I.U. of P.U., but in diestrum 1.0 I.U. is required (79). Hysterectomy reduces the response to threshold doses of chorionic gonadotrophin in 20-day-old mice (80). Prolactin injected into the female suspends cycles for 3 weeks; then a prolonged estrus of 4 to 8 days occurs in spite of continued injections (81). Three micrograms of estrone injected daily produces continued heat, as judged by the cornified vaginal smear, but by 17 days this reaction ceases (82).

The minimal dose of estrogen that will produce uterine distension is 100 M.U. (divided into 4 injections given over a period of 36 hours), but distension did not occur in all cases until the dose was raised to 400 M.U. At over 100 M.U. copulation was induced in 50 per cent of cases (83). The injection of 500 M.U. produced overstimulation of the uterus, and greatly distended glands, with a Swiss-cheese effect, were observed (84). In the normal cycle the injection of 7.5 R.U. of estrogen when the eggs are in the tube causes them to be retained or "tube-locked" (85). Similar effects are produced by testosterone propionate, 0.5 mg. of which causes some delay in the passage of ova, while 2.0 mg. completely prevents it (86). One to three M.U. of estrogen induces estrous vaginal smears in mice pregnant 2 to 6 days, with resorption of the embryos. The same effect with resorption or abortion is caused by the injection of 2 to 2.5 M.U. between 4 and 10 days. After 11 days 3.0 M.U. or more are needed (64). In the pregnant mouse ovariectomized at 14 to 15 days, 1.0 mg. of progesterone daily is needed to maintain the pregnancy (87). The daily amount of progesterone needed to sensitize the endometrium to produce deciduomata on traumatization lies between 0.25 and 0.5 mg. (88). In normal pseudopregnancy deciduomata produced on the third day persist for 5 days. If 1 mg. of progesterone daily is injected subcutaneously for a period beginning within 3 days of traumatization, they survive for 10 days, but not longer even though the injections are continued. The same effect can be produced in ovariectomized females. The additional injection of estrogens or of androgens do not affect the survival time (89).

In lactating mice with what should be delayed implantation, an injection of 0.03 of estradiol causes implantation to occur at the normal time. The injection of 5 or 10 I.U. of serum gonadotrophin on day 3 or 5 also results in normal implantation. In these experiments the litter size was standardized at 8 (33).

In females ovariectomized before implantation, 1 mg. of progesterone daily caused implantation. But with suckling mice ovariectomized on the sixth day, progesterone was without effect, although a single injection of 0.5 g. of estrogen brought it about. It was inferred that no estrogen is secreted by suckling mice as no follicles develop during this time (90).

In the diestrous mouse 8 R.U. of chorionic gonadotrophin, 2 to 5 R.U. of equine gonadotrophin, or 0.4 R.U. of sheep or pig pituitary caused ovulation in 50 per cent of the mice (91).

If a mated female is exposed within 4 days to a strange male, pregnancy or pseudopregnancy is blocked and the female returns to estrus within 3 to 4 days of exposure. The influence causing this blockage is olfactory (92).

The X-zone of fuchsinophil cells has been observed in the adrenal cortex. The injection of 0.4 mg. of testosterone propionate daily causes it to disappear or prevents its formation in adolescent male and females, adult females, and in adult castrated males (93).

Estrous cycles are suppressed by the daily injection of 20 μg. of testosterone and also of 200 μg. of progesterone (98).

The prenatal growth of the mouse has been studied in detail, and tables of growth have been prepared (95).

THE MALE

The testes develop rapidly to 60 days of age and then slow down. They are fully mature at 40 days. Spermatozoa are first found in the testes at 30 days old (96). The length of seminiferous tubules in one testes is 2 meters (97). The sperm head measures 7.9 μ (7.5–8.3) × 3.2 μ (2.6–3.4); the mid-piece, 18.4 μ (16.0–21.1) × 1.3 μ (0.8–1.8); and the tail length, 115.0 μ (110.0–125.4) (98). Another report gives the sperm length as 125.5 ± .5, and the sperm-head length 8.27 ± .04 (99).

Study of the growth of the seminal vesicles indicates that testicular androgens equivalent to 65 μ of testerone propionate are secreted daily between the ages of 30 and 50 days (100). Spermatogenesis takes 34½ days (101).

In the developing male the seminal vesicles increase in weight most rapidly at a body weight from 19 to 24 g. If postpuberal castration is performed, they lose half their weight in 7 days and then gradually sink to a level of less than 10 per cent of their maximal weight (102). The accessory glands are maintained in castrated males if ovaries are implanted into the ears and the mice are then kept at a temperature of 22° C. If the environmental temperature is 33° C., they are not maintained. The inference has been drawn that androgens are secreted by ovaries maintained at the lower temperature (103).

The fertile life of spermatozoa in the male tract is about 10 to 14 days, since fertility is retained for that period after spermatogenesis is halted by X-raying the testes (104). Spermatozoa enter the oviducts 1¼ hours after copulation and meet the ova at the end of 2 hours. In the oviducts spermatozoa retain their fertilizing ability for 6 hours and their motility for 13½ hours. The presence of a copulation plug is not necessary for fertilization (105). The plug seals better than does that of the rat. It extends up each cervical canal and into the vulva; the end is flat, not pointed as in the rat, and is usually flush with the vulva. The plug is removed by softening at the surface. It usually falls out after 18 to 24 hours, though it may persist for 36 hours to 2 days. The plug formed at a post-partum copulation is not so definite owing to a copious discharge of fluid and leucocytes. Under these circumstances the plug either does not form or breaks into a mass of debris (22).

Two milligrams of androsterone daily is just insufficient completely to maintain the prostate and seminal vesicles of castrated mice, which fact indicates that the mouse requires more than the rat (106). The minimum daily dose of testosterone that will bring the castrated mouse into a condition in which ejaculation can occur when pernostone and yohimbine are injected is about 40 γ (107).

The injection of crystalline androgens prevents the degeneration of the seminiferous tubules in hypophysectomized males for at least 23 days, but it does not prevent the degeneration of the interstitial cells. Androstanedione was most effective in a dose of 0.4 mg. daily (108). Injection of larger quantities is harmful in the normal adult mouse. Two milligrams of testosterone daily for 20 days produced moderate tubal atrophy and more complete atrophy of the interstitial cells. Five to ten milligrams daily did not damage the seminiferous tubules, but atrophy of the interstitial cells occurred. The injection of 2 mg. of progesterone also caused atrophy of the interstitial cells (109). Estrogens in quantity are also harmful to the male; 40 R.U. daily decreased the weight of the prostate (also of the uterus in the female) and caused the urine retention and bladder dilatation in both the intact and the castrated male. Metaplasia of the columnar epithelium into stratified squamous epithelium occurred, and it was more marked in the anterior than in the posterior prostatic lobe. The effect was greater in castrates than in intact mice. Injection of P.U. and of androgens prevented the metaplasia but not the retention of urine (110).

1. Laurie, E. M. O. PRS., 133B: 248–281, 1946.
2. Southwick, C. H. PZS., 131: 163–175, 1958.
3. Smith, W. W. JM., 35: 509–515, 1954.

4. Harrison, J. L. PZS., 125: 445–460, 1955.
5. Phillips, W. W. A. Ceylon J. Sci., 14B: 209–293, 1927–8.
6. Crowcroft, P., and F. P. Rowe. PZS., 131: 357–365, 1958.
7. Hamajima, F. Kyushu Univ. Fac. Agric. Sci., Bull. 19: 193–113, 1961–2.
8. Hiraiwa, Y. K., and F. Hamajima. Kyushu Univ. Fac. Agric. Sci., Bull. 18: 167–173, 1960–61.
9. Hamajima, F. Kyushu Univ. Fac. Agric. Sci., Bull. 19: 115–124, 1961–2.
10. Hiraiwa, Y. K., and F. Hamajima. Kyushu Univ. Fac. Agric. Sci., Bull. 18: 161–165, 1960–1.
11. Hiraiwa, Y. K., and F. Hamajima. Kyushu Univ. Fac. Agric. Sci., Bull. 18: 181–186, 1960–61.
12. Engle, E. T., and J. Rosasco. AR., 36: 383–388, 1927.
13. Ogle, C., and C. A. Mills. AJP., 105: 76–77, 1933.
14. Kirchhoff, H. Arch. f. Gyn., 163: 141–185, 1937.
15. Mirskaia, L., and F. A. E. Crew. Proc. Roy. Soc. Edinburgh, 50: 179–186, 1930.
16. Venge, O. Kon. Lantbr. Högskol Ann., Uppsala, 26: 251–268, 1960.
17. Allen, E. AJA., 32: 293–304, 1923.
18. Parkes, A. S., and F. W. R. Brambell. J. Physiol., 64: 388–392, 1928.
19. Gresson, R. A. Proc. Roy. Soc. Edinburgh, 60: 333–343, 1939.
20. Snell, G. D., E. Fekete, K. P. Hummel, and L. W. Law. AR., 76: 39–54, 1940.
21. Bluhm, A. Roux Arch. Entw. Mech. Organ., 141: 15–32, 1941.
22. Parkes, A. S. PRS., 100B: 151–170, 1926.
23. Smith, H. P. AR., 11: 407–410, 1916–7.
24. Watt, L. J. JM., 15: 185–189, 1934.
25. Wanke, S. Zeit. Tierzucht u. Zuchtungsbiol., 42: 269–280, 1938.
26. Parkes, A. S. Brit. J. Exp. Biol., 2: 21–31, 1924.
27. Danforth, C. H., and S. B. de Aberle. AJA., 41: 65–74, 1928.
28. Stevens, W. L. Ann. Eugenics, 8: 70–78, 1937.
29. Parkes, A. S. Brit. J. Exp. Biol., 4: 93–104, 1926–7.
30. Enzmann, E. V., N. R. Saphir, and G. Pincus. AR., 54: 325–341, 1932.
31. Merton, H. Proc. Roy. Soc. Edinburgh, 58: 80–97, 1937–8.
32. Bruce, H. M. J. Hyg., 45: 420–430, 1947.
33. Whitten, W. K. JE., 13: 1–6, 1955–6.
34. Rummer, M. AR., 99: 564–5, 1947.
35. MacDowell, E. C., and E. M. Lord. AR., 31: 41, 1925.
36. MacDowell, E. C., and E. M. Lord. AR., 31: 143–148, 1925.
37. Svorad, D., and V. Sachova. Physiol. Bohemosh., 8: 439–442, 1959.
38. Weir, J. A., H. Haubenstock, and S. L. Beck. J. Hered., 49: 217–222, 1958.
39. Allen, E. AJA., 31: 439–481, 1922–3.
40. Brambell, F. W. R. PRS., 193B: 258–272, 1928.
41. Togari, C. Aichi J. Exp. Med., 1(4): 1924.
42. Bullough, W. S., and H. F. Gibbs. Nature, 148: 439–440, 1941.
43. Bullough, W. S. Nature, 149: 271–272, 1942.
44. Engle, E. T. AJA., 39: 187–203, 1927.
45. Engle, E. T. AR., 35: 341–343, 1927.
46. Lewis, W. H., and E. S. Wright. CE., 459: 113–144, 1935.

47. Falconer, D. S., R. G. Edwards, R. E. Fowler, and R. C. Roberts. J. Reprod. and Fertility, 2: 418–437, 1961.
48. Allen, E., G. M. Smith, and W. U. Gardner. AJA., 61: 321–341, 1937.
49. Allen, E. AJA., 30: 297–371, 1922.
50. Henin, A. AB., 52: 97–115, 1941.
51. Parkes, A. S., U. Fielding, and F. W. R. Brambell. PRS., 101B: 328–354, 1927.
52. Hill, M., and A. S. Parkes. J. Anat., 65: 212–214, 1931.
53. Parkes, A. S. PRS., 100B: 172–199, 1926.
54. Parkes, A. S. PRS., 101B: 421–449, 1927.
55. Braden, A. W. H., and C. R. Austin. Australian J. Biol. Sci., 7: 552–565, 1954.
56. Brambell, F. W. R., and A. S. Parkes. Quart. J. Exp. Physiol., 18: 185–198, 1927–8.
57. Meyer, R. K., and W. M. Allen. Science, 75: 111–112, 1934.
58. Allen, W. M., and R. K. Meyer. AR., 61: 427–439, 1935.
59. Parkes, A. S. PRS., 104B: 183–188, 1929.
60. Harris, R. G. AR., 37: 83–93, 1927.
61. Parkes, A. S. J. Physiol., 65: 341–349, 1928.
62. Gardner, W. U., and E. Allen. AR., 83: 75–97, 1942.
63. Haterius, H. O., and H. A. Charipper. AR., 51: 84–101, 1931.
64. Parkes, A. S., and C. W. Bellerby. J. Physiol., 62: 145–155, 1926.
65. Whitney, R., and H. O. Burdick. E., 24: 45–49, 1939.
66. Lipkow, J. Zeit. Vergl. Physiol., 40: 593–609, 1958.
67. Hall, D. V. Physiol. Zool., 9: 471–497, 1936.
68. Robson, J. M. J. Physiol., 82: 105–112, 1934.
69. Duszynska, J. Nature, 142: 673–674, 1938.
70. Ring, J. R. E., 34: 269–275, 1944.
71. Robson, J. M. Quart. J. Exp. Physiol., 28: 195–205, 1938.
72. Reece, R. P., and C. W. Turner. Missouri Agric. Exp. Sta., Res. Bull. 266, 1937.
73. Karnofsky, D. A., I. T. Nathanson, and J. C. Aub. Cancer Res., 4: 772–778, 1944.
74. Atkinson, W. B., and C. W. Hooker. AR., 93: 75–92, 1945.
75. Forbes, T. R., and C. W. Hooker. E., 61: 281–286, 1957.
76. Guttenberg, I. E., 68: 1006–1009, 1961.
77. Ladman, A. J., and M. N. Runner. E., 53: 367–379, 1953.
78. Smith, P. E., and E. T. Engle. AJA., 40: 159–217, 1927.
79. Burdick, H. O., H. Watson, V. Ciampa, and T. Ciampa. E., 33: 1–15, 1943.
80. Palmer, A., and L. Fulton. Nature, 148: 596, 1941.
81. Dresel, I. Science, 82: 173, 1935.
82. Bishop, P. M. F., and T. McKeown. JE., 2: 339–342, 1940–1.
83. Marrian, G. F., and A. S. Parkes. J. Physiol., 69: 372–376, 1930.
84. Parkes, A. S. Lancet, 1: 485–486, 1935.
85. Whitney, R., and H. O. Burdick. E., 20: 643–647, 1936.
86. Burdick, H. O., B. B. Emerson, and R. Whitney. E., 26: 1081–1086, 1940.
87. Robson, J. M. Cold Spring Harbor Symposia, 5: 65, 1937.
88. Hooker, C. W. PSEBM., 46: 698–700, 1941.
89. Atkinson, W. B. AR., 88: 271–280, 1944.
90. Bloch, S. Gynaecologia, 148: 157–174, 1959.
91. Saunders, F. J. E., 40: 1–8, 1947.

92. Parkes, A. S., and H. M. Bruce. Science, 134: 1049–1054, 1961.
93. Starkey, W. F., and E. C. H. Schmidt. E., 23: 339–344, 1938.
94. Robson, J. M. J. Physiol., 92: 371–382, 1938.
95. MacDowell, E. C., E. Allen, and C. G. MacDowell. J. Gen. Physiol., 11: 57–70, 1927.
96. Onuma, H., and Y. Nishikawa. Bull. Nat. Inst. Agric. Sci. Chiba, G(10): 365–375, 1955.
97. Bascom, K. F., and H. L. Osterud. AR., 31: 159–169, 1925.
98. Hirth, H. F. J. Morphol., 106: 77–83, 1960.
99. Friend, G. F. QJMS., 78: 419–443, 1936.
100. Chai, C. K. AJP., 186: 463–467, 1956.
101. Oakberg, E. F. AJA., 99: 507–516, 1956.
102. Deanesly, R., and A. S. Parkes. J. Physiol., 78: 442–450, 1933.
103. Allen, E. Cold Spring Harbor Symposia, 5: 104–110, 1937.
104. Snell, G. D. J. Exp. Zool., 65: 421–441, 1933.
105. Merton, H. Proc. Roy. Soc. Edinburgh, 59: 207–218, 1938–9.
106. Callow, R. K., and R. Deanesly. Biochem. J., 29: 1424–1445, 1935.
107. Loewe, S. PSEBM., 37: 483–486, 1937.
108. Nelson, W. O., and C. E. Merckel. PSEBM., 38: 737–740, 1938.
109. Selye, H., and S. Friedman. E., 28: 129–140, 1941.
110. Weller, D., M. D. Overholser, and W. O. Nelson. AR., 65: 149–163, 1936.

Mus bellus Thomas. This African mouse has been recorded with young ranging from 3 to 8 in February (1), March, April (2), and August (3).

M. booduga Gray. In Ceylon this mouse probably breeds all year, but most young are found in March and April. The number is usually 6, but may be as many as 10 (4).

M. calliwaerti Thomas. In Africa lactating females have been found in July and August (5).

M. gratus Thomas and Wroughton. From Africa there come records of 3 to 5 young in December (6), January (7), and February (2).

M. minutoides A. Smith. In Northern Rhodesia a litter of 4 was born at the end of December (8).

M. musculoides Temminck. In Africa a litter of 4 young has been recorded in November (6). In the London zoo litters of 4 young have been born in October and November (9).

M. triton Thomas. Nestling young have been recorded in September (10) and December (7). In the first of these 2 was the number found.

1. Frechkop, S. EPNU., 1944.
2. Allen, G. M., and A. Loveridge. HCMZ., 89: 147–214, 1941–2.
3. Shortridge, G. C. The Mammals of South West Africa. London, 1936.
4. Phillips, W. W. A. Ceylon J. Sci. 14B: 209–293, 1927–8.

5. Hill, J. E., and T. D. Carter. AM., 78: 1–211, 1941.
6. Hatt, R. T. AM., 76: 457–604, 1939–40.
7. Allen, G. M., B. Lawrence, and A. Loveridge. HCMZ., 79: 31–126, 1935–7.
8. Ansell, W. F. H. JM., 41: 405, 1960.
9. Zuckerman, S. PZS., 122: 827–950, 1952–3.
10. Lawrence, B., and A. Loveridge. HCMZ., 110: 1–80, 1953.

Colomys goslingi Thomas and Wroughton

This wood rat has been recorded in April with 2 embryos and in July with 2 half-grown young (1).

1. Hatt, R. T. AM., 76: 457–604, 1930–40.

Lophuromys aquilus True

In the Belgian Congo this mouse bred at the end of the rains and early in the dry season, from December to April (1) or May (2). In East Africa embryo or fetuses have been recorded in January (3), February, July (4), August, and December (5). This suggests breeding for most, if not all, of the year. The litter size varies from 2 to 4.

1. Pirlot, P. L. Ann. Mus. Congo Belge, Tervueren, Ser. Zool., Nouv. Ser. en 4°, 1: 41–61, 1954.
2. Frechkop, S. EPNA., 1943.
3. Allen, G. M., and A. Loveridge. HCMZ., 75: 47–140, 1933–4.
4. Allen, G. M., and A. Loveridge. HCMZ., 89: 147–214, 1941–2.
5. Hatt, R. T. AM., 76: 457–604, 1939–40.

Lophuromys sikapusi Temminck. A female with 2 embryos has been recorded (1).

1. Eisentraut, M. Zool. Jahrb., Syst., 85: 619–672, 1957.

Notomys

Notomys gouldi Gould. This mouse has 4 young usually, but occasionally up to 6 (1).

N. mitchelli Ogilby. In the London zoo this Australian mouse has had litters of 2 or 3 young in December, February, and March (2). In the wild it usually has 4 young (1).

1. Troughton, E. Furred Animals of Australia. New York, 1947.
2. Zuckerman, S. PZS., 122: 827–950, 1952–3.

Acomys

SPINY MOUSE

Acomys caharinus Desmarest. In the London zoo this Egyptian mouse has had litters all the year. The number of young has varied from 2 to 11, with 4.5 the mean. A gestation period of 12 days has been suggested (1).

A. dimidiatus Cretzchmer. In Cyprus the breeding season is from April to October, and young born early in the season breed in the same year. The average embryo number is 2.1 (2).

A. hunteri de Winton. In the London zoo litters of 2 to 5 young have been recorded in March, May, July, and October (1).

A. ignitis Dollman. Pregnant females have been reported in January, April, and October. The number of young was 1 or 2 in each instance (3).

A. percivali Dollman. Females with 1 or 2 embryos have been recorded in July and October (3).

A. wilsoni Thomas. A female with 4 embryos has been recorded in July (3).

1. Zuckerman, S. PZS., 122: 827–950, 1952–3.
2. Watson, J. S. Colonial Res. Publ., London, 9: 1–66, 1951.
3. Heller, E. In G. C. Shortridge, The Mammals of South West Africa. London, 1934.

Bandicota bengalensis Gray and Hardwicke

INDIAN MOLE RAT

In the London zoo this rat has had litters of 1 and 7 young in August and October (1). The usual litter size is recorded as from 8 to 10 (2), but the subspecies, *B. b. kok,* is said to have litters of 2 as a rule (3).

1. Zuckerman, S. PZS., 122: 827–950, 1952–3.
2. Blanford, W. T. The Fauna of British India. Mammals. London, 1888–91.
3. Webb-Peploe, C. G. J. Bombay Nat. Hist. Soc., 46: 629–644, 1946–7.

Bandicota gracilis Nehring. In Ceylon this rat has 10 to 12 young at a time (1).

B. malabarica Shaw. This rat has no definite breeding season (1).

1. Phillips, W. W. A. Ceylon J. Sci., 14B: 209–293, 1927–8.

Nesokia indica Gray and Hardwicke

Pregnancies have been reported for late March. Late in April females at full term were found (1).

1. Hatt, R. T. The Mammals of Iraq. Univ. Michigan Mus. Zool., Misc. Publ. 106, 1959.

Beamys major Dollman

In Nyasaland this mouse breeds from November to May, i.e., in the wet and early dry seasons. All July males had retracted testes, which began to descend at the beginning of the wet season. The age at initial breeding was 9 to 13 months, and the length, 136 mm. The litter size varied from 4 to 7, and was usually 4 (1).

1. Hanney, P., and B. Morris. JM., 43: 238–248, 1962.

Saccostomus campestris Peters

POUCHED MOUSE

Embryos have been found in December (1) and July (2), while nestlings are present in February (3) and males with enlarged testes in October (4). These facts suggest a prolonged season. The number of young varies from 4 to 8.

1. Shortridge, G. C. The Mammals of South West Africa. London, 1934.
2. Heller, E. In (1).
3. Lawrence, B., and A. Loveridge. HCMZ., 110: 1–80, 1953.
4. Hill, J. E., and T. D. Carter. AM., 78: 1–211, 1941.

Cricetomys gambianus Waterhouse

GIANT RAT

This African rat has a prolonged season; embryos have been found in February (1), nursing young in April (2) and males with enlarged testes in August and December (3). The young, usually 4, are born after a gestation period of 42 days (4).

1. Allen, G. M., B. Lawrence, and A. Loveridge. HCMZ., 79: 31–126, 1935–7.
2. Allen, G. M., and A. Loveridge. HCMZ., 75: 47–140, 1933–4.
3. Hill, J. E., and T. D. Carter. AM., 78: 1–211, 1941.
4. Dekeyser, D. L. Les Mammifères de l'Afrique Noire Française. np., 1956.

DENDROMURINAE

Dendromus

TREE MOUSE

Dendromus acraeus Wroughton. This African mouse has been reported with 3 embryos in July (1) and August (2). Young have been seen in February (2).

D. arenarius Roberts. This mouse breeds during the winter months and has 3 to 4 young at a time (3).

D. insignis Thomas. Births of 3 to 4 young have occurred in February (4).

D. melanotis Smith. A female has been reported in June with 8 nest young (5).

D. messorius Thomas. Newborn young have been found in January (4), and fetuses in April and August (6). The number has been 3 or 4 on each occasion.

D. ochropus Osgood. A female with 5 embryos has been reported in January (1).

D. whytei Wroughton. Newborn young have been found in April and July. In the first litter there were 7, and in the second, 4 young (4).

1. Hollister, N. Smithsonian Inst., Bull. 99, 1918.
2. Frechkop, S. EPNU., 1944.
3. Roberts, A. Reference lost.
4. Allen, G. M., and A. Loveridge. HCMZ., 89: 147–214, 1941–2.
5. Allen, G. M., and A. Loveridge. HCMZ., 75: 47–140, 1933–4.
6. Hatt, R. T. AM., 76: 457–604, 1939–40.

Malacothrix typicus A. Smith

MOUSE GERBIL

This African mouse is probably polyestrous; it breeds 3 times a year, for the first time at age 3 months. Four to 5 is the usual litter size (1).

1. Powell, W. Rodents. Dept. Public Health, South Africa, 1925.

Petromyscus

PIGMY ROCK MOUSE

Petromyscus collinus Thomas and Hinton. In South West Africa a specimen with 2 fetuses was found in September (1).

P. shortridgei Thomas. Pregnant specimens have been taken in May, usually with 2 or 3 fetuses (1).

1. Shortridge, G. C. The Mammals of South West Africa. London, 1934.

Steatomys

FAT MOUSE

Steatomys opimus Pousargues. A specimen of this African mouse has been reported with 5 fetuses in November (1).

S. pratensis Peters. This mouse usually has 4 to 6 young (2). It has been recorded as pregnant in December (3).

1. Hatt, R. T. AM., 76: 457–604, 1939–40.
2. Shortridge, G. C. The Mammals of South West Africa. London, 1934.
3. Frechkop, S. EPNU., 1944.

OTOMYINAE

Otomys

Otomys anchietae Bocage. Lactating females and males with enlarged testes have been recorded in August (1).

O. denti Thomas. A pregnant specimen has been found in January (2).

O. irroratus Brants. This mouse is probably polyestrous, and breeds about 3 times a year, for the first time at age about 4 months. The litter size is usually 3 to 5 (3). In South West Africa it is reported as breeding from June to August, with 2 to 4 young, usually 3 and never more than 4 (4).

O. tropicalis Thomas. Suckling females have been reported in December and February (5), and a female with 2 small embryos in the right uterus in March (6). Two seems to be the usual litter size.

1. Hill, J. E., and T. D. Carter. AM., 78: 1–211, 1941.
2. Frechkop, S. EPNA., 1938.
3. Powell, W. Rodents. Dept. Public Health, South Africa, 1925.
4. Shortridge, G. C. The Mammals of South West Africa. London, 1934.
5. Allen, G. M., B. Lawrence, and A. Loveridge. HCMZ., 79: 31–126, 1935–7.
6. Hatt, R. T. AM., 76: 457–604, 1939–40.

Deomys ferrugineus Thomas

Two specimens of this mouse, each with 2 small embryos in the left uterus, have been reported for February (1).

1. Hatt, R. T. AM., 76: 457–604, 1939–40.

Paratomys

Paratomys brantsii A. Smith. This African mouse breeds 4 times a year beginning at age 3 months (1).

P. littledalei Thomas. A female with 3 fetuses has been reported in August, also newborn young in November (2).

1. Powell, W. Rodents. Dept. Public Health, South Africa, 1925.
2. Shortridge, G. C. The Mammals of South West Africa. London, 1934.

PHLOEOMYINAE

Pogonomys macrourus Milne-Edwards

In Papua this mouse ceases to breed in the latter half of the dry season (1).

1. Tate, G. H. H. AM., 72: 501–728, 1936–7.

Chiropodomys gliroides Blythe

TREE MOUSE

In Malaya 50 per cent of males are fertile at 7.2 ± .1 g., and 95 per cent at 165 ± .1 g., while the females reach puberty at 139 g. and upward. Most pregnancies are found from January to March. The embryo count is 1 to 3, mean 2.2, and mode 2, equally distributed in each uterine horn (1).

1. Harrison, J. L. PZS., 125: 445–460, 1955.

HYDROMYINAE

Hydromys chrysogaster Geoffroy

WATER RAT

The males reach puberty at 400 to 600 g. weight and the females at 425 g. Mating takes place in late winter and continues through the spring, i.e., from September to January. The females are anestrous in late summer, autumn, and early winter. One litter a year is normal and the litter size is 1 to 7, usually 4 to 5 (1).

1. McNally, J. Australian J. Zool., 8: 170–180, 1960.

GLIRIDAE

GLIRINAE

Glis glis L.

EDIBLE DORMOUSE

This dormouse begins the breeding season about 3 weeks after it emerges from hibernation. The young reach puberty at age 10 to 11 months. The number of young varies from 3 to 10, born after a gestation of 20 to 25 days (1). There is one litter a year (2).

1. Heptner, W. G., L. G. Morosowa-Turowa, and W. I. Zalkin. Die Säugetiere in der Schutzwaldzone. Berlin, 1956.
2. Freiherr von Vietinghoff-Riesch, A. Säugetierk. Mitt., 3: 113–121, 1955.

Muscardinus avellanarius L.

DORMOUSE

This dormouse is probably polyestrous, bearing litters from May to October (1). Possibly 2 is the usual number of litters (2). The young number from 2 to 7 (1), born after a gestation of 21 days (3).

1. Barrett-Hamilton, G. E. H. A History of British Mammals. London, 1910.
2. Heptner, W. G., L. G. Morosowa-Turowa, and W. I. Zalkin. Die Säugetiere in der Schutzwaldzone. Berlin, 1956.
3. Kenneth, J. H. Gestation Periods. Edinburgh, 1943.

Eliomys quercinus L.

DORMOUSE

This dormouse is probably polyestrous, with heats at intervals of about 10 days. After copulation a vaginal plug is formed. The gestation period is 22 days (1). The number of young is 3 to 6 and one litter a year is the rule; only very rarely is there a second (2). Follicular atresia is characterized by a transformation of granulosa cells into interstitial type cells (3).

1. Lataste, F. Act. Soc. Linn. Bordeaux, 40: 293–466, 1886.
2. Heptner, W. G., L. G. Morosowa-Turowa, and W. I. Zalkin. Die Säugetiere in der Schutzwaldzone. Berlin, 1956.
3. de Pinko, A. V. AR., 30: 211–220, 1925.

Dryomys nitedula Pallas

TREE DORMOUSE

According to one account this dormouse bears one litter a year consisting of 2 to 6 young, but usually 3 or 4 (1). According to another, 2 or 3 litters are produced, in spring, March to April; in midsummer, July; and in autumn, September to October. Three young is the usual number (2).

1. Heptner, W. G., L. G. Morosowa-Turowa, and W. I. Zalkin. Die Säugetiere in der Schutzwaldzone. Berlin, 1956.
2. Navo, E., and E. Amir. Bull. Res. Council, Israel, Sect. B (Zool.), 9B(4): 200–201, 1961.

GRAPHIURINAE

Graphiurus

AFRICAN DORMOUSE

Graphiurus christyi Dollman. This dormouse possibly breeds all year, as young have been recorded in January and September (1).

G. kelleni Reuvens. This species has been recorded with 2 (2) and 5 young (3).

G. lorraineus Dollman. Nest and half-grown young have been recorded for January, April, and November (1).

G. murinus Desmarest. This South African species has a short season about midsummer. Four or 5 is the usual litter size (4).

G. nanus de Winton. One record gives 4 young (4).

G. vulcanicus Lönnberg and Gyldenstolpe. This dormouse usually has 2 to 4 young. Newborn and nest young have been reported from January to March (5).

1. Hatt, R. T. AM., 76: 457–604, 1939–40.
2. Bradfield, R. D. Field Notes on S. W. African Mammals. London, nd.
3. Hill, J. E. JM., 22: 81–85, 1941.
4. FitzSimons, F. W. The Natural History of South Africa. London, 1919.
5. Frechkop, S. EPNA., 1943.

ZAPODIDAE

SICISTINAE

Sicista

BIRCH MOUSE

Sicista betulina Pallas. This mouse mates in May, and the young are of breeding age when they emerge from their first hibernation. The gestation period lasts from 4 to 5 weeks (1).

S. subtilis Pallas. This mouse has one litter a year. The young breed the same year (2). This probably means at the beginning of the following season.

1. Kubik, J. Ann. Univ. Mariae Curie-Sklodowska, Sect. C., Biol., 7: 1–63, 1952.
2. Flent, V. E. ZZ., 39: 942–946, 1960.

ZAPODINAE

Zapus hudsonius Zimmermann

JUMPING MOUSE

In Minnesota this mouse has 3 peaks of births—one in the latter part of June, one in mid- and late July, and a third in mid-August. It has 2 or 3 litters a year and a gestation period of 17½ days. There is a post-partum heat, and gestation is longer if the mouse is lactating. The young of the early litters probably breed during their first summer, so that they may produce young at the age of about 2 months (1). The number of embryos varies from 2 to 8, mean 5.4 (2).

In the Yukon 1 or 2 litters a year are born in spring or summer (3). In New York State the first litter is born in June, with a second in July or, more frequently, in September (4).

The sperm head measures 5.1 μ (4.9–5.5) × 2.9 μ (2.7–3.1); the midpiece width is 0.8 μ (0.6–0.9); and the tail length is 71.0 μ (69.1–75.6) (5).

1. Quimby, D. C. Ecol. Monogr., 21: 61–95, 1951.
2. Krutsch, P. H. Univ. Kansas Mus. Nat. Hist., Publ. 7: 351–472, 1952–5.
3. Rand, A. L. Canada Nat. Mus., Bull. 100, 1945.
4. Hamilton, Jr., W. J. AMN., 16: 187–200, 1935.
5. Hirth, H. F. J. Morphol., 196: 77–83, 1960.

Zapus princeps Allen. Pregnant specimens of this mouse may be found from late May to mid-July, and lactating ones till late August. The embryo count varies from 2 to 7, with an average of 5 (1, 2).

Z. trinotatus Rhoads. This mouse has one litter a year, with 4 to 8 young at a time (3).

1. Krutsch, P. H. Univ. Kansas Mus. Nat. Hist., Publ. 7: 351–472, 1952–5.
2. Negus, N. C., and J. S. Findley. JM., 40: 371–381, 1959.
3. Bailey, V. NAF., 55: 1936.

Napaeozapus insignis Miller

WOODLAND JUMPING MOUSE

The young of this mouse of eastern North America are born late in June or early in July, but sometimes a second litter is born in September (1). In New Hampshire pregnancies have been found from May to August (2). In captivity more than two litters a year may be produced, but the number of young per litter is smaller with the later litters. There is no post-partum heat. The mean litter size was 4.7 (3). In the wild the litter size varies from 3 to 6 and averages 4.5 embryos.

1. Hamilton, Jr., W. J. The Mammals of Eastern United States. Ithaca, N.Y., 1943.
2. Preble, N. A. JM., 37: 196–200, 1956.
3. Schwentker, V. In A. N. Worden and W. Lane-Petter, eds., The UFAW Handbook on the Care and Management of Laboratory Animals. London, 1957.

DIPODIDAE

DIPODINAE

Dipus sagitta Pallas

JERBOA

This Eurasian jerboa reproduces from April to August, but mainly from mid-May to mid-August. The females have 2 litters a year and the early young breed in the year of their birth (1). Mating begins in March, soon after the end of hibernation, and gestation lasts 25 to 30 days (2). Embryo counts have varied from 2 to 5, with a mean of $3.0 \pm .1$, and a pronounced mode of 3 (1).

1. Feniuk, B. K., and J. M. Kazantzeva. JM., 18: 409–426, 1937.
2. Heptner, W. G., L. G. Morosowa-Turowa, and W. I. Zalkin. Die Säugetiere in der Schutzwaldzone. Berlin, 1956.

Jaculus

AFRICAN JERBOA

Jaculus jaculus L. This jerboa has 3 to 4 young at a time (1). Another account places the number at 4 to 5 (2).

J. orientalis Erxleben. In the London zoo this Egyptian species has had 3 young in litters born in April and October. The gestation period is 42 days (3).

1. Dekeyser, P. L. Les Mammifères de l'Afrique Noire Française. np., 1956.
2. Petter, F. Mammalia, Special Number, 25: 1–222, 1961.
3. Zuckerman, S. PZS., 122: 827–950, 1952–3.

Scirtopoda telum Lichtenstein

JUMPING MOUSE

The young of this mouse are born in May or June. Three or 4 is the usual number (1).

1. Heptner, W. G., L. G. Morosowa-Turowa, and W. I. Zalkin. Die Säugetiere in der Schutzwaldzone. Berlin, 1956.

Allactaga

JERBOA

Allactaga elater Lichtenstein. This jerboa mates soon after the end of hibernation. The young, usually 5, are born in May (1).

A. major Kerr. This species mates soon after emergence from hibernation, and the litter of 3 to 6 young is born in May or June. Possibly second and third litters may be born in the season (1).

A. sibirica Forster. This jerboa mates in May and the young are born in June or July (1).

1. Heptner, W. G., L. G. Morosowa-Turowa, and W. I. Zalkin. Die Säugetiere in der Schutzwaldzone. Berlin, 1956.

Alactagulus pumilio Kerr

This jerboa has 2 litters a year of 3 to 6 young each (1).

1. Heptner, W. G., L. G. Morosowa-Turowa, and W. I. Zalkin. Die Säugetiere in der Schutzwaldzone. Berlin, 1956.

HYSTRICIDAE

HYSTRICINAE

Hystrix

OLD WORLD PORCUPINE

Hystrix africaeaustralis Peters. This porcupine has 2 litters a year and 1 to 4 young, usually 2 (1). The gestation period is given as 112 days (2).

H. brachyurus L. In the London zoo births have been from March to October, and the litter size is 1 to 2 young (3). In India the species breeds in the spring and the young are born when the crops begin to ripen (4).

H. cristata L. In the London zoo births have occurred throughout the year. One is the usual number of young, with twins once in 5 births (3). Gestation has been variously given as lasting 63 and 112 days (5).

H. hodgsoni Gray. The Himalayan porcupine breeds in the spring and usually has 2 young at a time (6).

H. leucura Sykes. In Ceylon this porcupine has no general breeding season but most young are born during the northeast monsoon. A single young is always produced (7). In the London zoo births have been distributed through the year and the number of young has been 1 or 2 (3).

H. subcristatus Swinhoe. This porcupine breeds in the spring and has 3 or more young at a time (8).

1. Haagner, A. K. South African Mammals. London, 1920.
2. Dekeyser, P. L. Les Mammifères de l'Afrique Noire Française. np., 1956.
3. Zuckerman, S. PZS., 122: 827–950, 1952–3.
4. Jerdon, J. C. The Mammals of India. Roorkee, 1867.
5. Kenneth, J. K. Gestation Periods. Edinburgh, 1943.

6. Blanford, W. T. The Fauna of British India. Mammals. London, 1888–91.
7. Phillips, W. W. A. Ceylon J. Sci., 14B: 209–293, 1927–8.
8. Ho, H. J. Contrib. Biol. Lab. Sci. Soc. China, Zool. Ser., 10: 245–287, 1933–5.

ATHERURINAE

Atherurus

Atherurus africanus Gray. In the wild this species has been recorded with single large and small embryos in December (1). In the London zoo a single young has been born in September (2). Births of single young occur throughout the year and the gestation period is from 100 to 110 days (2).

A. macrourus L. In the London zoo single young have been born in May and September (3).

1. Hatt, R. T. AM., 76: 457–604, 1939–40.
2. Rahm, U. Mammalia, 26: 1–9, 1962.
3. Zuckerman, S. PZS., 122: 827–950, 1952–3.

ERETHIZONTIDAE

ERETHIZONTINAE

Erethizon dorsatum L.

PORCUPINE

The common porcupine of North America mates during November. At this time the hair surrounding the mammary glands of the female temporarily acquires a cinnamon coloration. The number of young is nearly always 1; twins are very occasional. The period of gestation is 16 weeks (1).

The growth of the corpus luteum has been described in detail. The rupture points are relatively broad and the central cavity is wide; it contains many polymorphs and an eosinophilic albuminous precipitate. Many mitoses are found in the lutein cells. Some thecal cells differentiate into true lutein cells. The corpus luteum has solidified by the time the embryo is in the gill-slit stage. Corpora lutea of previous pregnancies are buried deep in the ovary and are highly pigmented. Many atretic follicles become "accessory" corpora lutea but these do not produce a progestational reaction (2).

The ovary is not enclosed by a bursa. At the junction of the two myometrial layers there is a good deal of edematous connective tissue, especially at the time of estrus and during pregnancy (2).

1. Struthers, P. H. JM., 9: 301–308, 1928.
2. Mossman, H. W., and I. Judas. AJA., 85: 259–260, 1953.

Erethizon epixanthum Brandt. This porcupine has a breeding season in September and October. The number of young is usually 1, but occasionally 2 may be born. The duration of gestation has been given as 9 months, a length that must be queried in view of the known length in *E. dorsatum* (1).

1. Bailey, V. NAF., 55, 1936.

Coendou

HAIRY TREE PORCUPINE

Coendou mexicanum Kerr. This porcupine breeds in February; it has 4 young after a gestation of 60 to 70 days (1).

C. villosus Cuvier. In the London zoo birth of a single young in July has been recorded (2).

1. Gaumer, G. F. Mamiferos de Yucatan. Mexico City, 1917.
2. Zuckerman, S. PZS., 122: 827–950, 1952–3.

CAVIIDAE

CAVIINAE

Cavia porcellus L.

GUINEA PIG

The guinea pig in the laboratory is polyestrous all the year round. Heat lasts for less than half a day, and the complete cycle is 16½ days. The cyclical changes in the female tract of the guinea pig have been studied in more detail than in any other mammal. The vaginal smear is a clear indication of the reproductive state, and its detailed investigation made possible the recent rapid developments in the fields of hormone isolation and physiological analysis. The number of young is usually 3 to 4, and the duration of gestation is about 67 to 68 days, a very long period for so small an animal. Ovulation is spontaneous and the corpus luteum of the cycle is functional, in contrast to the condition found in all the Muridae that have been investigated. The vagina is closed by a membrane, which opens spontaneously at heat and which thus provides a useful index of heat if it is read in conjunction with the vaginal smear.

THE ESTROUS CYCLE

The age of puberty in the guinea pig is from 55 to 70 days under normal conditions of management. If the food is richer than usual, causing more rapid growth, puberty may occur at from 45 to 60 days; slow growth on poor food delays its onset. Puberty occurs at the same time whether males are kept with the females or not (1). One series of records gave a mean age at first heat of 67.8 ± 2.0 days, with a standard deviation of 21.5 days, and a spread from 33 to 134 days. In this work it was found that the mean age at first rupture of the vaginal closure membrane was 58.2 days. The usual interval between first rupture and heat was from 0 to 4 days, and an appreciable number of the long intervals were the duration of a cycle or a multiple of it, suggesting that in some cases the earlier heats are not so intense as later ones (2).

The duration of the estrous cycle is usually given as 16½ days. One set of data gives a modal length of 16 days (3); another, 17 days with a mean of 17.7 ± 1.8 days (4); and another a modal length of 15 to 16 days with a mean of 16.34 ± .10 days, standard deviation of 1.89 days, and a spread of 13 to 25 days (5). A series in which the animals were observed at two-hour intervals instead of daily or twice daily gave a mean cycle length of 16 days, 6 hours. As heat begins most frequently in the evening, the distribution shows a double mode, at 16 and 17 days (6). Vasectomized copulation does not modify the cycle length (7, 6).

Proestrum, in which there is congestion and swelling of the external genitals and a slight serous discharge from the vagina, usually lasts from 1 to 1½ days, and heat or sexual receptivity lasts 6 to 11 hours in 90 per cent of all cases (1). Another investigation gave an average of 8.21 ± .07 hours, with a range from 1 to 18 hours, for the duration of heat (6). The vagina usually remains open about 4 days (4). Ovulation occurs usually at 10 hours after the beginning of heat or sexual receptivity. There is no relation between the length of heat and the number of follicles which rupture (8), but the number is highest in those females with the most intense heat, as judged by the frequency with which they are mounted (9). About 64 per cent of all heat periods begin between 6 P.M. and 6 A.M., but during the months of short daylight, i.e., October to December, the curve of distribution of the onset of heat shifts about 2 hours and heat tends to begin earlier in the afternoon (6). If the females are kept in the dark, they come in heat at any time; the tendency for heat to begin in the evening vanishes. Guinea pigs so kept do not change the length of the cycles nor the duration of heat (10).

There is little or no difference in the number of ovulations from the two ovaries, since 51 per cent of embryos were found in the left horn of the uterus (11). As internal migration of ova is held to be impossible on anatomical grounds (12), this must represent the proportion of ova shed from the ovary. The absence of internal migration has also been shown experimentally (13).

The removal of corpora lutea immediately after heat shortens the cycle to 11 days (14). This represents an exceptionally long time required for the ripening of new graafian follicles. Heat occurs immediately after parturition in about 64 per cent of females. It usually begins within 2 hours of the end of parturition but is short, lasting 3.5 hours instead of the normal average of 8.6 hours. It is always associated with ovulation, though more than half of those guinea pigs which did not come in heat had ovulated (15).

Insemination at different times has given a maximum life for the ovum of 30 hours, but within about 8 hours after ovulation the number of normal

TABLE 4. Effects of Insemination Time on Fertility

TIME OF INSEMINATION	PER CENT PREGNANT	AVERAGE LITTER SIZE	PERCENTAGE OF NORMAL PREGNANCIES
HOURS AFTER OVULATION			
Controls	83	2.6	88
8	67	1.7	66
14	56	1.6	27
20	31	1.3	10
26	7	0	0
32	0	0	0

pregnancies and of eggs fertilized drops markedly. The data are summarized in Table 4 (16). Pregnancy is said to be rare following a copulation in the first 3 or 4 hours of heat (1). The eggs arrive in the uterus about 3½ days after they are shed (17).

The average litter size for a very large number of individuals was 2.58 ± .006; the range was from 1 to 8; the standard deviation, 1.02; and the mode, 2 to 3. It is highest in spring and summer (18). The mode of the duration of gestation is 68 days, with a distribution of 58 to 72 days (16). Parturition occurs equally at any time of day (15). Litter size is more closely related to maternal weight than to age or parity (19). For a single young born the gestation lasted 70.5 days, and for litters of 6 it was 66.8 days (20). Another set of figures shows the same trend (21). Litters with an excess of males are carried for about half a day longer than those with an excess of females (22).

The sex proportion is 50.59 ± .19 males. Variations with litter size are not significant (18). Among fetuses of various ages the proportion of males is 55.9 ± 2.09 (11). There is no sign that monovular twins are produced (23).

HISTOLOGY OF THE FEMALE TRACT

OVARY. The cavity in some graafian follicles is established when the guinea pig is 4 to 7 days old, but the theca interna is not well differentiated at this time; mitoses are frequent throughout the ovary. At 18 days atresia has begun in many follicles (24). There is a wave of follicular growth during the cycle which begins 2 to 3 days after ovulation. The follicles grow at a constant rate until the beginning of heat, when rapid development sets in. Those which will rupture protrude from the surface of the follicle; the tunica albuginea is reduced so that the germinal epithelium and a much-thinned granulosa layer alone cover the follicle at the rupture point (25). The ovary of the guinea pig is remarkable for the extent and variety of forms of follicular atresia. The wave of follicular growth before heat produces in each ovary about 40 to 50 follicles with cavities, of which only about 2 are destined to rupture. Those which will not rupture can be detected about 2 days before heat by degenerative changes in the granulosa next the theca interna. The nucleus of the ovum in these follicles is active, and the chromosomes arrange themselves along the equator of the spindle, where they remain. In a follicle that will rupture they are able to separate, and the first polar body is produced before ovulation. In the degenerating follicles the granulosa cells break up and are usually removed by cytolysis, but occasionally by phagocytosis, leaving a cavity which is reduced by the proliferation of the theca interna. In the smaller of these follicles the theca interna proliferates rapidly and encloses the egg, which often fragments (7). This has been interpreted as a parthenogenic development (26, 27) but, in the opinion of the writer and of others (28, 29), the changes are definitely degenerative in nature. Unilateral ovariectomy doubles the number of ovulations from one ovary (9).

The ripe follicle is about 0.8 mm. in diameter, and the ovum measures about 65 μ (30). The outer diameter is 122.5 μ and tubal ova do not acquire an albuminous coat (31). Alkaline phosphatase may be found in the theca interna and the granulosa of the follicle (32). When the follicle ruptures there is little hemorrhage, but a small pool of follicular fluid is retained in the center of the developing corpus luteum for 2 to 3 days. The formation of the corpus luteum proceeds along classical lines, but the theca interna, perhaps, contributes to a greater degree than it does in other species. Fatty degeneration sets in at about the thirteenth day of the cycle, and the corpus luteum is reduced to a few strands of connective-tissue cells, which gradually merge with the stroma of the ovary. The mature corpus luteum is oval in shape, at right angles to the surface of the ovary, and it is very difficult to extirpate completely because of its depth and the high vascularity of the ovarian stroma (30, 7). The corpus luteum of the cycle grows until the twelfth day; that of pregnancy until the eighteenth or twentieth. It continues in the pregnant animal to day 35; then a very gradual decline sets in. The blastocyst attaches at 6 days (33).

A striking feature of the ovary of the guinea pig is the great frequency with which cysts of the rete ovarii develop (9).

VAGINA. During diestrum the vaginal epithelium is thin, consisting at first of a low stratified squamous type with a ragged appearance. After 8 to 10 days it becomes more regular and the lower layer is cuboidal or columnar, with long flat cells overlying it. As heat approaches, an intense proliferation takes place; the epithelium becomes several cells thick, with the superficial layers flattened. At the beginning of heat these cells cornify and tend to desquamate, a process which becomes more rapid as heat continues. Toward the end of heat the vaginal wall, which has become very congested and edematous, is invaded by large numbers of leucocytes, which penetrate through the epithelial layer and pass into the lumen. Desquamation is particularly severe during metestrum, but repair soon sets in and the epithelium is restored to the typical diestrous state. During pregnancy the epithelium changes to a high-columnar mucous type (34, 35). The vaginal smear changes during heat and metestrum. At the beginning leucocytes disappear; squamous epithelial cells are present, followed by cornified cells, then by small epithelial cells, and lastly by an invasion of leucocytes. A late stage of metestrum has been described (34) in which red blood cells are present; this has been denied by others (36), but in the writer's experience it occurs occasionally and is caused by bleeding from the uterus. An outline of the various schemes which have been suggested for classifying the stages of the cycle is given in Table 5. The smears in the postparturient heart are not typical (35).

TABLE 5. Vaginal Smears of the Guinea Pig

<table>
<tr><th>STOCKARD AND PAPANICOLAOU (34)</th><th>SELLE (36)</th><th>YOUNG (37)</th></tr>
<tr><td>STAGE I
a. Squamous epithelial cells staining gray with H. and E. Pycnotic nuclei; abundant mucus</td><td>STAGE I</td><td>STAGE Ia
Superficial cells only, 25 per cent come in heat</td></tr>
<tr><td rowspan="2">b. Cornified cells staining red with H. and E.</td><td>STAGE II</td><td rowspan="2">STAGE Ib
Early: less than 25 per cent cornified; Middle: 25–75 per cent cornified } 60 per cent in heat
Late: 75 per cent or more cornified. 15–20 per cent in heat
Ovulation in Stage Ib</td></tr>
<tr><td rowspan="2">STAGE III
Cornified and small epithelial cells, or epithelial cells only</td></tr>
<tr><td>STAGE II
Nucleated epithelial cells; smear abundant and cheesy</td><td>STAGE II</td></tr>
<tr><td>STAGE III
Liquefaction of cheesy mass; cells as last, with numerous leucocytes</td><td>STAGE IV</td><td>STAGE III</td></tr>
<tr><td>STAGE IV
Slight hemorrhage, not always present</td><td>Absent</td><td></td></tr>
</table>

UTERUS. During the vaginal stage Ia there is some growth of the uterine epithelium, the cells are tall and pseudostratified, and the surface tends to be irregular. The glands are not well developed, are inactive, and contain no mitoses. In stage Ib the mucous layer becomes very edematous, and the usual leucocytic invasion begins. Rare mitoses may be found in the glands. In stage II the epithelial cells tend to become vacuolated. In stage III vacuolization increases, the epithelium breaks, patches are shed, and the leucocytic invasion is well marked. At 1 to 2 days postestrum the uterine wall is still somewhat broken, but edema has subsided and repair is in progress; a few mitoses are present in the gland cells but not in the stroma, which is very vascular. The glands are swollen with secretion. At 3 days the glands are much more tortuous, the cells are swollen so as to obliterate the lumen which has hitherto been open, and mitoses are abundant. The epithelium is tall, with nuclei at the bases of the cells, the upper margins of which are level. Mitosis has ceased. The mucous layer is thicker. At 7 days the gland cells have secreted into the lumen and hence have shrunk in size. The epithelium is cubical to columnar, and mitoses are common in the stroma cells. At 10 days the glands are retrogressing and are less coiled; at 14 days the retrogression is still more apparent. Occasionally in metestrum one finds extravasation of blood, which gathers as hematoma below the epithelium and passes into the lumen of the uterus, and hence into the vaginal smear. In one guinea pig seen by the writer the hemorrhage was so severe that it killed her (38, 7).

Lipids may be found in the uterine epithelium for most of diestrum. The level is high on days 3 and 9 but they are absent on day 7 (39). There is a diffuse basophilia, most evident during diestrum. The glands are also basophilic at this time. Protein-bound disulfide groups are more in evidence in the surface epithelium than in the glands. The reaction for them is most intense during proestrum and estrus. Traces of glycogen may be found in the surface cells and also in the stroma during estrus and metestrum. Mucus is most in evidence in metestrum and least in late diestrum. Lipids are very abundant in the glands during estrus and metestrum. Alkaline phosphatase is most abundant in the glands during proestrum and estrus (40).

OVIDUCTS. The tubo-uterine junction is very complex. The isthmus is tortuous, with moderately firm muscular walls. The mucosa has 4 major primary folds with a minor primary fold in the furrow between each pair. The tube passes through the uterine wall at right angles to the uterus and opens into the cavity about 0.5 mm. from its tip. The entrance is guarded by mucosal lips which are glandular. A thick spincterlike muscle extends into these lips. The tube is further guarded, just before its extension into the uterus, by a papilla which projects into its lumen (41). Attempts to force fluid from the uterus into the oviducts failed; the uterus always ruptured first (42).

PITUITARY. Variations in the number of cells of different types in the anterior pituitary in relation to the cycle are not statistically significant. Granulation is highest at 6 to 8 days postestrum, then falls, rises again to the beginning of heat, and falls rapidly during heat. These remarks apply both to acidophils and basophils. The Golgi apparatus hypertrophies during the first half of the cycle and tends to fragment during the second half. The mitochondria tend to concentrate in the Golgi zone while it is intact, and to scatter when it is not. They tend to concentrate at the periphery of many cells in proestrum and estrus (43). The degranulation during heat has been observed by others (44). There are no special pregnancy cells, but both types of chromophils tend to degranulate at this time (43). The special type of castration cells formed after operation in many species is not found in the guinea pig of either sex (45, 46).

It is said that the thyroid gland undergoes cyclical proliferation which is connected with the estrous cycle. It is at a maximum at 4 to 8 days after heat and at a minimum at 8 to 10 days (47).

PHYSIOLOGY OF THE FEMALE TRACT

The estrous cycle is upset if the uterus is removed. This is due to the persistence of the corpus luteum, which may last for a considerable time (48, 49). Hysterectomy during heat or in gestation prevented the recurrence of cycles for at least 5 months. If the operation was performed before puberty, the first heat occurred at the normal time, but the corpora lutea persisted and no further heat periods ensued (50). Hysterectomy on the fifth day inhibited the cycle for at least 8 months. The corpora lutea persisted and contained progesterone. When the operation was performed at day 10 the corpora lutea persisted for the duration of a normal pregnancy. Hysterectomy at day 15 caused delay in estrus in some, i.e., in 70 per cent of the female (51). Ovariectomy before the twenty-sixth day of pregnancy has always resulted in abortion or resorption of the embryos, but after this time, occasionally, the pregnancy continues normally (52, 53). However, if ovariectomy is performed at 3, 4, or 5 days post coitum the ova implanted in 35 of 42 guinea pigs but the implantation sites subsequently regressed. The failure could be prevented by injecting 10 mg. of progesterone daily from day 11 (54). Under normal circumstances the corpus luteum is necessary for pregnancy until day 15 (55). In the writer's experience removal of corpora lutea is more severe in its results than is ovariectomy. This may be due to the disturbing effect of estrogens from new follicles in the first case (7). The vaginal closure membrane opens on the twenty-sixth and twenty-seventh days of pregnancy. This is regarded as indicating the time of change from ovarian to placental control for the remainder of pregnancy (56).

If the uterine mucosa is traumatized at any time from the third to the eighth or ninth day after heat, deciduomata are produced. The reaction depends upon the presence of the corpora lutea (57), and it has been elicited in other species and in ovariectomized animals given progesterone.

Contractions of the uterus are not so great, but are more frequent, during diestrum when the corpora lutea are active (58, 59). The frequency is highest at 4 days postestrum, and it slows to the rate found during heat by the tenth day. In the ovariectomized female the uterine muscle has the same slow rate of contraction, but the injection of 1 mg. of progesterone increases it beyond that found during the cycle, and 2 mg. still further increases the rate (59).

No estrogens have been found in the urine during diestrum, but when the female is in heat 250 R.U. per liter have been observed. During early pregnancy there is none; at 20 days, 250 R.U. per liter; and at 40 to 63 days, 500 R.U. per liter (60). There is a peak of 17-ketosteroid excretion at the time of estrus. It then amounts to 600 μg. in 24 hours. The excretion is reduced by ovariectomy but the response is gradual, suggesting that the source of these steroids may be the adrenals and not the ovaries. The level also rises during pregnancy (61, 62).

The gonadotrophe level of the pituitary is low, one-quarter (or less) that of the rat; ovariectomy increases the amount whereas estrogens depress it (63). Follicle stimulator is low, and it appears to be at its maximum just before heat (64). Luteinizer is at least ten times less than the level in the rat's pituitary (65).

The prolactin content of the pituitary is low; in immature females it is 220 B.U. per g., in mature females 520 B.U. per g. In lactating females and in males it is higher, and the injection of estrogens also raises it (66). The level is higher during heat than it is in diestrum (67).

It is very difficult to induce sexual receptivity in the spayed female by the injection of estrogens, but the injection of 40 I.U. of estrogen, followed 36 to 48 hours later by 0.2 I.U. of progesterone, invariably caused the females to be receptive to the male (68). In later experiments the use of 50 I.U. of progynon B (estradiol benzoate in oil) as the primer, with 0.1 I.U. of progesterone, was successful in 73 per cent of the animals. Lower doses were much less effective, but when the progesterone was increased to 0.4 I.U., the successes rose to 92 per cent. When the dose of progesterone was kept at 0.2 I.U. and the priming dose was varied, 20 I.U. of estrone was 33 per cent successful and the percentage rose gradually to 80 per cent with 50 I.U., after which there was not much gain (69). At 2 to 6 days of age the immature females are entirely unresponsive to these injections, but at 15 days there is some response, and a complete response is given at 30 days (70). Vaginal changes are abnormal when the ovariectomized female is injected with 175 I.U. of estradiol benzoate for 4 days. But if 0.2 I.U. of progesterone is given after the third injection, cornification becomes complete. Optimal results are obtained if the progesterone is preceded by injection of 350 to 400 I.U. of estradiol benzoate (71).

The spayed adult guinea pig requires 0.65 I.U. of progesterone daily for 6 days to sensitize the endometrium so that it will produce deciduomata. If the injections were preceded by estrogens, the reaction was slightly intensified. Continued progesterone injections caused the uterine mucosa to become refractory. Immature animals required 1 I.U. daily to produce sensitivity (72). In the normal animal a minimum of 100 to 160 R.U. of estrogen suppresses the power to give a decidual reaction (73).

The ovary of the guinea pig is very refractory to the injection of gonadotrophic hormones or to pituitary implants. This is largely due to the ease with which the theca interna of the follicles becomes luteinized. It is most difficult to grow normal follicles, but it can be done by the implantation of four guinea-pig pituitaries daily (7).

The injection of P.M.S. into immature females causes little increase in ovarian weight, but the uterus is greatly enlarged (74). Injection into adults 3 to 5 days before heat was expected caused a great growth of the ovaries. These organs increased with progressive increase in dosage to 350 per cent at 150 R.U., and to 400 per cent with 500 R.U. The increase was mainly due to theca luteinization and, at the higher doses, edema; but there was some granulosa luteinization, especially at the higher levels. The uterus increased in weight, with a maximum increase at 100 R.U., after which it became less with progressively higher doses. Heat occurred at the usual time, but ovulation was suppressed as there were no follicles in a condition to rupture (75). The injections caused an enlargement of the clitoris similar to that found when pregnancy urine is injected (76). The injection of 4 mg. of testosterone propionate daily prevents follicular growth and ovulation (77).

The uterus of the virgin or the ovariectomized guinea pig is inhibited by adrenalin. If the females are injected with estrogens the myometria of 50 per cent have motor reactions to adrenalin. With progesterone the reaction continues to be inhibitory (78).

The relaxation of the pelvic girdle normally found at parturition and the requirement of the corpus-luteum hormone, relaxin, in this process have been investigated (79). The prenatal growth of the guinea pig has been the subject of considerable study (11, 80).

THE MALE

The growth rates of the testes and accessory organs accelerate from 30 to 40 days of age. Electric ejaculation techniques show that some secretion is present at 21 days at the earliest, and most males respond at 30 days. The first spermatozoa appear at 50 days, and their production is fairly uniform by 70 days (81). If one testis is removed before puberty, hypertrophy occurs in a few cases. This is believed to be not compensatory hypertrophy but an acceleration of growth (82). The mean volume of ejaculate obtained by electrical stimulation was 0.76 ml. In 50 per cent of males spermatozoa may be obtained at 11 weeks old and in all at 18 weeks. The mean sperm count was 9,762,000 per milliliter. The mean body weight at first appearance of spermatozoa in the ejaculate was 611 g. Body weight was more closely correlated with the first appearance of spermatozoa than was age (83). The length of seminiferous tubules in a single testis was 40 meters (84).

Spermatozoa stored in the isolated epididymis retain their fertility for 20 to 35 days, and they remain motile for 59 days (85). If the epididymides are placed in the abdomen, motility lasts only 14 days. If they are scrotal but the testes are removed, motility ceases after 23 days (80). The injection of androgens increases both motility and survival time (87).

Electric ejaculation is a convenient means of obtaining semen for study. The male is lightly anesthetized. One electrode is placed through the skin on the back of the neck at the base of the skull. The other, a blunt one, is put in the mouth and a 33-volt alternating current is passed. The normal amount of semen thus produced is 1.5 to 3 g. There is no seasonal difference in the amount (88, 89).

Spermatozoa probably take 14 to 18 days to pass through the epididymis. If it is isolated from the testis, the time increases to 25 to 35 days (90). In the female they retain their fertility for 22 hours, but beyond 17 hours it is reduced (91).

A copulation plug is normally formed in the vagina after coitus. If the proximal prostatic lobe is removed, this no longer happens. The anatomy and histology of the prostate is described in detail in this paper (92), and the seminal vesicles, which are large and produce a jellylike secretion, have also been carefully investigated (92).

The accessory organs tend to increase in weight up to 600 g. of body weight. Castration causes them to decrease to 50 per cent of their former weight, a reduction which is usually complete by 30 days after castration (94). Castration also leads in the young guinea pig to arrested atrophy and subsequent hypertrophy of the thymus (95). Inositol is present in the seminal-vesicle secretion to the extent of 26 mg. per 100 ml. (96). Fructose measures over 100 mg. per 100 g., and critric acid 350 mg. per 100 g. Secretion begins at age 4 weeks and is fully active at 4 months. In the castrate 1 mg. of testosterone daily was sufficient to cause the glands to grow and produce secretion. This was in 28-day-old males; in older ones 2.5 mg. were required (97).

After hypophysectomy the testes decrease from 2.5 to 3.5 g. (the normal weight) to 0.5 g. This decrease takes 45 days, but spermatogenesis ceases at 35 days. Spermatozoa are found in the tubules to 56 days, though they are few after between 5 and 14 days. The rate of decrease of the accessory organs is at about the same rate as after castration (98). After hypophysectomy the injection of 3 mg. of testosterone propionate daily preserves spermatogenesis in most of the tubules (99).

In the normal male 2 mg. daily of androgens for 30 days causes atrophy of the testes, but this effect is prevented if an extract of horse pituitary (rich in F.S.H.) is also given. There are differences in the responses of the prostate and seminal vesicles to individual androgens (100).

The pituitary of the male contains twice as much gonadotrophe as that of the female. In the cryptorchid and castrate the content rises about 70 per cent (63).

1. Ishii, O. Biol. Bull., 38: 237–250, 1920.
2. Young, W. C., E. W. Dempsey, C. W. Hagquist, and J. L. Boling. J. Comp. Psychol., 27: 49–68, 1939.
3. Bacsich, P., and G. M. Wyburn. Proc. Roy. Soc., Edinburgh, 60: 33–39, 1939–40.
4. Nicol, T. Proc. Roy. Soc., Edinburgh, 53: 220–238, 1932–3.
5. Guttmacher, A. F., and S. A. Asdell. Unpublished data.
6. Young, W. C., E. W. Dempsey, and H. I. Myers. J. Comp. Psychol., 19: 313–335, 1935.
7. Asdell, S. A. Unpublished data.
8. Young, W. C., H. I. Myers, and E. W. Dempsey. AJP., 105: 393–398, 1933.

9. Young, W. C., E. W. Dempsey, H. I. Myers, and C. W. Hagquist. AJA., 63: 457–487, 1938.
10. Dempsey, E. W., H. I. Myers, W. C. Young, and D. B. Jennison. AJP., 109: 307–311, 1934.
11. Ibsen, H. L. J. Exp. Zool., 51: 51–94, 1928.
12. Kelly, G. L. AR., 40: 365–372, 1928.
13. Kinney, P. P. AR., 25(proc.): 137, 1923.
14. Dempsey, E. W. AJP., 120: 126–132, 1937.
15. Boling, J. L., R. J. Blandau, J. G. Wilson, and W. C. Young. PSEBM., 42: 128–132, 1939.
16. Blandau, R. J., and W. C. Young. AJA., 64: 303–329, 1939.
17. Lams, H. AB., 28: 229–323, 1913.
18. Haines, G. J. Agric. Res., 42: 123–164, 1931.
19. Eckstein, P., and T. McKeown. JE., 12: 115–119, 1955.
20. Goy, R. W., R. M. Hoar, and W. C. Young. AR., 128: 747–757, 1957.
21. McKeown, T., and B. MacMahon. JE., 13: 195–200, 1956.
22. McKeown, T., and B. MacMahon. JE., 13: 309–318, 1956.
23. Hagedoorn, A. L., and A. J. W. Hagedoorn. Genetica, 23: 315–328, 1943.
24. Loeb, L. J. Morphol., 22: 37–70, 1911.
25. Myers, H. I., W. C. Young, and E. W. Dempsey. AR., 65: 381–401, 1936.
26. Loeb, L. J. Am. Med. Assn., 56: 1327, 1911.
27. Courrier, R., and —. Oberling. Bull. Soc. Anat., Nov. 1923.
28. Clark, E. B. AR., 313–337, 1923.
29. Kampmeier, O. F. AJA., 43: 45–76, 1929.
30. Loeb, L. AA., 28: 102–106, 1906.
31. Squier, R. R. CE., 137: 223–250, 1932.
32. Corner, G. W. CE., 575: 1–8, 1948.
33. Rowlands, I. W. Ciba Foundation Colloquia in Ageing, 2: 69–83, 1956.
34. Stockard, C. R., and G. N. Papanicolaou. AJA., 22: 225–283, 1917.
35. Kelly, G. L. AJA., 43: 247–287, 1929.
36. Selle, R. M. AJA., 30: 429–449, 1922.
37. Young, W. C. AR., 67: 305–325, 1937.
38. Loeb, L. Biol. Bull., 27: 1–44, 1914.
39. Nicol, T., and R. S. Snell. J. Obst. Gyn. Brit. Emp., 61: 216–222, 1954.
40. Burgos, M. H., and G. B. Wislocki. E., 59: 93–118, 1956.
41. Andersen, D. H. AJA., 42: 255–305, 1928.
42. Kelly, G. L. AJA., 40: 373–383, 1927.
43. Hagquist, C. W. AR., 72: 211–229, 1938.
44. Chadwick, C. S. AJA., 60: 129–147, 1936.
45. Kirkman, H. AJA., 61: 233–287, 1937.
46. Severinghaus, A. E. AJP., 101: 309–315, 1932.
47. Chouke, K. S., and H. T. Blumenthal. E., 30: 511–515, 1942.
48. Loeb, L. PSEBM., 20: 441–443, 1923.
49. Loeb, L. AJP., 83: 202–224, 1927–8.
50. Herlant, M. CRSB., 114: 273–275, 1933.
51. Rowlands, I. W. J. Reprod. and Fertility, 2: 341–350, 1961.

52. Herrick, E. H. AR., 39: 193–200, 1938.
53. Benazzi, M. Arch. Sci. Biol., 18: 409–419, 1933.
54. Deanesly, R. Nature, 186: 327–328, 1960.
55. Courrier, R. Ann. Fac. Med., Montevideo, 35: 767–770, 1950.
56. Ford, D. H., R. C. Webster, and W. C. Young. AR., 109: 707–715, 1951.
57. Loeb, L. Surgery, Gyn. and Obst., 25: 300–315, 1917.
58. Ohkubo, Y., and S. Endoh. J. Japanese Soc. Vet. Sci., 17: 170–183, 1938.
59. Greig, K. A. AJP., 125: 547–550, 1939.
60. Schmidt, I. G. AR., 64: 255–266, 1936.
61. Charrolais, E. J., K. Ponse, and M. F. Jayle. Ann. d'Endocrinol., 18: 109–119, 1957.
62. Brooks, R. V., B. E. Clayton, and J. E. Hammart. JE., 20: 24–35, 1960.
63. Nelson, W. O. PSEBM., 32: 1605–1607, 1935.
64. Schmidt, I. G. E., 21: 461–468, 1937.
65. Lipschutz, A. CRSB., 108: 646–647, 1931.
66. Reece, R. P., and C. W. Turner. Missouri Agric. Exp. Sta., Res. Bull. 266, 1937.
67. Reece, R. P. PSEBM., 42: 54-56, 1939.
68. Dempsey, E. W., R. Hertz, and W. C. Young. AJP., 116: 201–209, 1936.
69. Collins, V. J., J. L. Boling, E. W. Dempsey, and W. C. Young. E., 23: 188–196, 1938.
70. Wilson, J. G., and W. C. Young. E., 29: 779–783, 1941.
71. Ford, D. H., and W. C. Young. E., 49: 795–804, 1951.
72. Blumenthal, H. T., and L. Loeb. Arch. Pathol., 34: 49–66, 1942.
73. Raynaud, R. Thesis, Algiers, 1934.
74. Rowlands, I. W. J. Physiol. 92: 8P, 1938.
75. Rogel, I. Thesis, Cornell University, 1945.
76. Falk, E. A., and G. N. Papanicolaou. AR., 64(suppl.): 16, 1935–6.
77. Boling, J. L., and J. B. Hamilton. AR., 1–15, 1939.
78. Balassa, G. Quart. J. Pharmacol., 14: 347–355, 1941.
79. Hisaw, F. L., M. X. Zarrow, W. L. Money, R. V. N. Talmage, and A. A. Abramowitz. E., 34: 122–134, 1944.
80. Draper, R. L. AR., 18: 369–392, 1920.
81. Sayles, E. D. Physiol. Zool., 12: 256–267, 1939.
82. Lipschutz, A., and E. Vinals. CRSB., 100: 984–985, 1929.
83. Freund, M. Fed. Proc., 19: 371, 1960.
84. Bascom, K. F., and H. L. Osterud. AR., 31: 159–169, 1925.
85. Young, W. C. J. Morphol. Physiol., 48: 475–491, 1929.
86. Moore, C. R. J. Exp. Zool., 50: 455–494, 1928.
87. Kabak, J. M., and A. L. Paduceva. Trud. Dinam. Razvit., 8: 82, 1934.
88. Moore, C. R., and T. F. Gallagher. AJA., 45: 39–69, 1930.
89. Dalziel, C. F., and C. L. Phillips. Am. J. Vet. Res., 9: 225–232, 1948.
90. Tothill, M. C., and W. C. Young. AR., 50: 95–107, 1931.
91. Soderwall, A. L., and W. C. Young. AR., 78: 19–29, 1940.
92. Engle, E. T. AR., 34: 75–90, 1926.
93. Warnock, A. W. AR., 25: 154–155, 1923.
94. Sayles, E. D. J. Exp. Zool., 90: 183–197, 1942.

95. Halnan, E. T., and F. H. A. Marshall. PRS., 88B: 68–89, 1914.
96. Hartree, E. F. Biochem. J., 66: 131–137, 1957.
97. Ortiz, E., D. Price, H. G. Williams-Ashman, and J. Banks. E., 59: 479–492, 1956.
98. Allanson, M., R. T. Hill, and M. K. McPhail. J. Exp. Biol., 12: 348–354, 1935.
99. Cutuly, E. PSEBM., 47: 290–292, 1941.
100. Bottomley, A. C., and S. J. Folley. J. Physiol., 94: 6P, 1938.

Galea musteloides Meyen

The males of this cavy are in breeding condition at the end of July and the testes are greatly enlarged from September to December. All females examined were pregnant during these months. They breed during their first year of life. The embryo count is from 1 to 4, with mean 2.0, but embryonic mortality is high (1).

1. Pearson, O. P. HCMZ., 106: 117–174, 1951–2.

Microcavia australis Geoffroy and D'Orbigny

This cavy has 2 young at a time (1).

1. Allen, J. A. Rpt. Princeton Univ. Exped. to Patagonia, 1896–7, 3: 1–210, 1905–11.

DOLICHOTINAE

PATAGONIAN HARE-CAVY

Dolichotis patagona Zimmermann. This cavy has produced young in the London zoo from March to October. Of 19 births, 1 was of triplets while singles were twice as frequent as twins (1).

D. salincola Burmeister. In the London zoo this species has given birth to young from March to July (1). In the wild a female with 2 embryos has been found in August (2).

1. Zuckerman, S. PZS., 122: 827–950, 1952–3.
2. Eisentraut, M. Zeit. f. Säugetierk., 8: 47–69, 1933.

HYDROCHOERIDAE

Hydrochoerus

CAPYBARA

Hydrochoerus hydrochoeris L. In the London zoo this species has given birth to 2 young in November after a gestation lasting between 119 and 126 days (1).

H. isthmius Goldman. In Panama this species experiences estrus at the end of February. The gestation period is between 104 and 111 days. Three young seems to be the usual number, but one killed at the end of August had 4 fetuses near term (2).

1. Zuckerman, S. PZS., 122: 827–950, 1952–3.
2. Trapido, H. JM., 30: 433, 1949.

DASYPROCTIDAE

CUNICULINAE

Cuniculus paca L.

SPOTTED CAVY

This species breeds early in the winter and the two young are born in the dry season, in winter and early spring (1). A female with a single embryo has been recorded for January (2).

1. Gaumer, G. F. Mamiferos de Yucatan. Mexico City, 1917.
2. Felten, H. SB., 38: 145–155, 1957.

DASYPROCTINAE

Dasyprocta

AGOUTI

Dasyprocta agouti L. In the London zoo this agouti has produced young throughout the year. One young is the rule with twins at 1 birth in 3 (1). The gestation period is about 104 days (2).

D. azarae Lichtenstein. The young are born at the end of the year (3).

D. cristata Desmarest. In the London zoo births have been recorded in January and November (1).

D. prymnolopha Wagler. In the London zoo young have been born throughout the year. Slightly more twins are produced than singles (1).

D. punctata Gray. In the wild this species has no fixed season (4). In the London zoo single young have been born in July and August (1). The Panama records give 2 as the usual number with a single record of 4. The bursa incompletely surrounds the ovary (4).

D. variegata Tschudi. In the London zoo single young have been born in May, June, and August (1).

1. Zuckerman, S. PZS., 122: 827–950, 1952–3.
2. Brown, C. E. JM., 17: 10–13, 1936.
3. Kühlhorn, F. Säugetierk. Mitt., 2: 66–72, 1954.
4. Enders, R. K. HCMZ., 78: 385–502, 1935.

CHINCHILLIDAE

Lagostomus maximus Desmarest

VISCACHA

The viscacha breeds in the Argentine during the months of March and April, while the young are born from July to August. Some breed in November and have their young in March. The first is the principal cycle as it represents about 64 per cent of the annual births (1). The gestation period is about 145 days (2). In the London zoo births have occurred all the year-round, and twins have been slightly more frequent than singles (3).

1. Llanos, A. C., and J. A. Crespo. Rev. de Investig. Agric., 6: 289–378, 1952.
2. Kenneth, J. H. Gestation Periods. Edinburgh, 1943.
3. Zuckerman, S. PZS., 122: 827–950, 1952–3.

Lagidium peruanum Meyen

MOUNTAIN VISCACHA

The males of this species breed first at 1 kg. weight, or at 7 months old. The females breed first at the same weight if it is reached in October, November, or December; if in August or September they wait until the later months. Almost no females are pregnant in August, September or early October. The vagina is closed except at the time of heat. The copulation plug is large. Ovulation occurs at about the time of copulation. One ovum is released, almost always from the right ovary. Pregnancy lasts about 3 months (1); it is nearly always in the right horn of the uterus (2). Occasionally there is a post-partum heat followed by pregnancy (2).

The testes grow rapidly between the sizes of 14 and 22 mm., and all that are larger than 18 mm. contain spermatozoa (2). After 6 weeks of pregnancy accessory corpora lutea are formed in the right ovary so that a dozen may be present at the end of gestation. The ovaries are partly enclosed in a bursa (2).

If the right ovary is removed, ova are shed from the left, and pregnancy is then in the left horn of the uterus (1).

1. Pearson, O. P. JM., 29: 345–374, 1948.
2. Pearson, O. P. AJA., 84: 143–173, 1949.

Chinchilla laniger Molina

CHINCHILLA

The chinchilla, a native of South America, breeds at any time of the year but is said to mate most readily in December and March (1), through according to one account most litters are born in September and January. Mating is nocturnal (2). A copulation plug is formed, which is shed entire, surrounded by the cornified layer of the vaginal wall. It hardens rapidly on exposure to the air. The estrous cycle is said by breeders to last 24 days and heat for 2 days (3). If it is correct, this is a surprisingly long cycle length, but it is similar to that of the nutria, which is rather closely related to the chinchilla. The period of gestation is 105 to 111 days (2), or 111 days, and 1 to 4 young are born at a time (4). The female experiences a postparturient heat 12 hours after she has given birth (5). The average number of corpora lutea is 4.3, with little difference between the ovaries. In one count the right ovary contained an average of 2.1 corpora lutea and the right horn of the uterus 1.4 embryos, while the left ovary contained 1.8 corpora lutea and the left horn 1.3 embryos. The sex proportion of 2,855 young was 54.7 per cent males (6).

The average adult testis weight was 2.6 g. (7). The average semen ejaculate measures 0.1 to 0.5 cc., and it contains from 20 to 200 million spermatozoa. These are capable of traveling at the rate of 1 inch in 4 minutes (8). The male pituitary weighs 3.46 mg./per cent of body weight and the corresponding figure for the pituitary of the female is 4.02 mg. (7).

1. U.S.D.I., Bureau of Biological Survey, Wildlife Leaflet BS-151, 1940.
2. Dennler, G. Deutsche Pelztierzucht, 14: 388–390, 1939.
3. Asdell, S. A. Unpublished work.
4. Brun, G. M. Am. Fur Breeder, 15: 20–26, 1942.
5. Metayer, —. Bull. Soc. Acclim. Française, 80: 235–236, 1933.
6. Hilleman, H. H., and D. Tibbitts. Fur Trade J., Canada, 34(6): 40, 42–44, 1957.
7. Roos, T. B., and R. M. Shackleford. AR., 123: 301–312, 1955.
8. Bullard, R. W., and L. R. Parkinson. Fur Trade J., Canada, 31(3): 23–27, 1953.

CAPROMYIDAE

Capromys pilorides Say

HUTIA

In the London zoo 4 births have occurred in March, May (twice), and August. The number of young was 1 and 3 (1). In the wild a female with 4 embryos has been reported (2).

1. Zuckerman, S. PZS., 122: 827–950, 1952–3.
2. Hall, E. R., and K. R. Kelson. The Mammals of North America. New York, 1959.

Myocaster coypus Molina

NUTRIA

The South American coypu is bred for its fur and is known commercially as the nutria. In several countries escapes have become naturalized and have flourished. The females are polyestrous, though to get them to breed through the winter the temperature must be kept at 60° F. (1). In Louisiana there is a breeding period in December and January and a secondary one in June and July (2). Puberty in the male is at age 4 to 5 months (3). In the female it is given as occurring at 1,800 to 2,200 g. weight, or at 5 months (4), but ovulations may not begin until 10 to 12 months, though vaginal smears show estrous changes at 6 months (5). Reports on cycle lengths, etc., have been extremely varied. The cycle in young animals is 24 to 27 days, rarely 14 to 17 days; in older ones it lasts 27 to 29 days. Heat lasts 2 to 4 days but it is rarely shown by females above the age of 3½ to 4 years (6). Another account gives 17 days as the length of the cycle in 8 to 10-month-old females and 19 days in multiparous females, with heat lasting for 1 day in the young, and 2, occasionally 3 to 4 days, in older ones (7). Yet another account gives the cycle length as varying from 5 to 28 days. One female experienced continuous estrus for periods up to 14 days long with 5- or 6-day intervals between them (8). These conflicts suggest that ovulation is coitus-induced. Vaginal smears resemble those of the rat but are less distinct in February to April than at other times (7). They indicate a cycle of 6.4 days and estrus of 1.7 days (9). There is a post-partum heat 48 hours after the young are born (10). In Louisiana the mean fetus count was 5.0, range 1 to 11, and 64 per cent at the mode of 4 and 5 (2). Another account gives 5.6 for primiparous females and 6.1 for older ones (11).

The diameter of the mature follicle is 0.94 to 1.8 mm., and that of the secondary oöcytes, 83 μ. The mature follicle does not protrude beyond the general surface of the ovary. Following ovulation a spherical corpus hemorrhagicum about 1.8 mm. in diameter is formed. Radially arranged cords of lutein cells extend centrally. There are no mitoses. The fully formed luteal cells measure 0.12 × 0.20 mm. Some are binucleate. The corpus luteum of pregnancy is red to pink; its size may rise to 3.2 mm. Accessory corpora lutea are found with luteinized theca interna, but in these no oöcytes were found (12).

During heat the cholesterol and ketosteroid levels in the ovaries rise (5). Mating in the first half of heat usually results in pregnancy, but mating in the second half seldom does (7). During the glandular proliferation of the uterus compact cellular nests proliferate towards one side of the existing glands and bud inward to produce the complex progestational pattern (13). The anterior pituitary contains very little L.H. (14). The ovary is not encapsulated (15). Spermatogenesis decreases when the males reach 4 to 5 years of age (3). At coitus a copulation plug is formed (12).

1. Federspiel, M. N. Am. Fur Breeder, 13: 13–20, 1941.
2. Adams, W. H. Proc. Louisiana Acad. Sci., 19: 28–41, 1956.
3. Pietrzyk-Walknowska, J. Folia Biol., 4: 22–34, 1956.
4. Laurie, E. M. O. J. Anim. Ecol., 15: 22–34 ,1956.
5. Konieczna, B. Folia Biol., 4: 139–150, 1956.
6. Kraetge, E. Deutsche Pelztierzucht, 12: 117–120, 1937.
7. Skowron-Cendrzak, A. Folia Biol., 4: 119–138, 1956.
8. Wilson, E. D., and A. A. Dewees. JM., 43: 362–364, 1962.
9. Bartha, T., and E. P. Gayer. Allattemyesztes, 7: 179–184, 1958.
10. Matthias, K. E. K. Am. Fur Breeder, 14: 18–20, 1941.
11. Ocetkiewicz, J., J. Kawinska, and J. Jarosz. Roczn. Nauk, 76B: 785–799, 1960.
12. Stanley, H. P., and H. H. Hilleman. J. Morphol., 106: 277–299, 1960.
13. Molina-Ahumada, J. B., and O. Orias. Rev. Soc. Argentina Biol., 18: 321–325, 1942.
14. Lipschutz, A., and C. Oviedo. CRSB., 118: 333–334, 1935.
15. Hilleman, H. A., A. I. Gaynor, and H. P. Stanley. AR., 130: 513–532, 1958.

OCTODONTIDAE

Octodon degus Molina

BUSH RAT

In the London zoo this South American rat breeds all year. Litters have ranged from 1 to 10, mean 4 (1).

1. Zuckerman, S. PZS., 122: 827–950, 1952–3.

Aconaemys fuscus Waterhouse

In early November this South American rodent was found with small young. A female with 2 very large fetuses was also taken at this time (1).

1. Osgood, W. H. Field Mus. Nat. Hist., Zool. Ser., 30: 1–268, 1943.

CTENOMYIDAE

Ctenomys

TUCO-TUCO

Ctenomys opimus Wagner. Pregnant females, each with 2 fetuses, were taken in September (1).

C. peruanus Sanborn and Pearson. A specimen with 5 embryos was found in December (1).

1. Pearson, O. P. HCMZ., 106: 117–174, 1951–2.

ABROCOMIDAE

Abrocoma cinerea Thomas

CHINCHILLA RAT

Pregnant and lactating females were found in December. Two had 2 fetuses each, one of them with 4 corpora lutea. This female was lactating (1).

1. Pearson, O. P. HCMZ., 106: 117–174, 1951–2.

ECHIMYIDAE

Proechimys cayennensis Desmarest

SPINY RAT

This Central and South American rat has been found with embryos from January to August. These numbered usually 2 to 3, but may be as high as 6 (1, 2).

1. Enders, R. K. HCMZ., 78: 385–502, 1935.
2. Goodwin, G. G. AM., 87: 271–474, 1946.

THRYONOMYIDAE

Thryonomys

CANE RAT

Thryonomys harrisoni Thomas and Wroughton. Females with 3 embryos have been found in February and March (1).

T. swindlerianus Temminck. This South African rat probably has a restricted season, as the young are born from June to August. The litter size varies from 2 to 4, usually 3 (2).

1. Hatt, R. T. AM., 76: 457–604, 1939–40.
2. Shortridge, G. C. The Mammals of South West Africa. London, 1934.

PETROMYIDAE

Petromys typicus A. Smith

DASSIE RAT

This South African rat breeds from November to February, in the hot weather. One or 2 appears to be the usual litter size (1).

1. Shortridge, G. C. The Mammals of South West Africa. London, 1934.

BATHYERGIDAE

Cryptomys

MOLE RAT

Cryptomys damarensis Ogilby. This South African rat probably has a fixed breeding season. Females with 5 fetuses each have been reported for April (1).

C. hottentotus Lesson. There is a single record of nestling young found in October (2).

C. mechowii Peters. This rat has 1 to 2 young born in January or February (3).

C. mellandi Thomas. Pregnant specimens have been found in February and August (4).

1. Shortridge, G. C. The Mammals of South West Africa. London, 1934.
2. Lawrence, B., and A. Loveridge. HCMZ., 110: 1–80, 1953.
3. Hill, J. E. JM., 22: 81–85, 1941.
4. Frechkop, S. EPNU., 1944.

Heliophobius argenteocinereus Peters

Pregnant specimens of this rat have been taken in January and March (1).

1. Frechkop, S. EPNU., 1944.

CTENODACTYLIDAE

Ctenodactylus gundi Rothman

GUNDI

This African rodent appears to have two breeding seasons, in January and April. The young are born in March and June, while anestrum lasts from July to December (1). In North Africa a female with 3 fetuses near term has been reported near the end of April (2).

1. de Lange, D. Zeit. Mikros. Anat. Forsch., 36: 488–496, 1934.
2. Lataste, F. Act. Soc. Linn. Bordeaux, 39: 129–299, 1885.

Cetacea

If we consider the difficulties in observing the breeding habits of whales it is amazing that so much information on the subject is available. We owe these data to the observers placed on factory ships by the International Whaling Commission and to the scientists of the *Discovery* expeditions who have done an immense amount of work. Far more is now known about whales than about porpoises and dolphins.

Whales appear to be polyestrous and to have fairly extended breeding seasons, but their habit of experiencing several ovulations for each pregnancy after the first is puzzling. Considering the great size of many species the gestation periods are remarkably short, and, as the young generally are very large in comparison to their mothers, the rate of fetal growth must be regarded as phenomenal. The habit of preservation of corpora lutea in the ovaries throughout life has been a boon to the observer.

PLATANISTIDAE

PLATANISTINAE

Platanista gangetica Lebeck

GANGES DOLPHIN

This dolphin has 1 young born from April to July after a gestation of 8 to 9 months (1).

1. Blanford, W. T. Fauna of British India. Mammals. London, 1888–91.

INIINAE

Inia geoffroyensis Gray

AMAZON DOLPHIN

This dolphin has 1 young at a time (1).

1. Cabrera, A., and J. Yepes. Historia Natural Ediar. Mamiferos Sud-Americanos. Buenos Aires, 1940.

ZIPHIIDAE

Mesoplodon

BEAKED WHALE

Mesoplodon gervaisi Deslongchamps. This whale has a bursa ovarii (1). *M. mirus* True. A single record gives one fetus for this whale (2).

1. Rankin, J. J. AR., 139: 379–386, 1961.
2. Brimley, H. H. JM., 24: 199–203, 1943.

Berardius bairdi Stejneger

BEAKED WHALE

Females of this northern whale reach puberty at 10 meters long. All ovulations are shown by the persistent corpora lutea (1).

1. Matsuura, Y. Zool. Mag., Tokyo, 54: 466–473, 1942.

Hyperoödon ampullatus Forster

BOTTLENOSE WHALE

The mating time of the bottlenose whale is April and May. In the Arctic the young are born before May and June. The female may be pregnant while lactating. One young is the rule, and the period of gestation is about 12 months (1).

1. Ohlin, A. Lund, Acta Univ., 29(2), 1892–93.

PHYSETERIDAE

PHYSETERINAE

Physeter catodon L.

SPERM WHALE

In the Azores region the males of the sperm whale are sexually mature at about 9.6 m. long. Above that length testis volume begins to increase materially. The females become mature within the range of 7.9 to 9.5 m. long, mean 8.8 m. If the ovaries weigh 0.5 kg. or more, the female is mature (1). In Japanese waters the males reach puberty at 13.1 m. (2), and all females of 10.8 m. and over are sexually mature (3). In the Southern Hemisphere 12 m. has been given as the probable length for puberty in the males and 9 to 9.5 m. in the females (4).

The fetuses fall into two groups, suggesting a limited breeding season and a gestation period of more than a year. In the Azores most births occur between July and August and the time of mating is probably from mid-February to mid-June, with a few as early as January and as late as July. The whole period of births is from May to November, with most from July to September (1). In Japanese waters mating is in March and birth from the end of July to the beginning of October after a gestation of about 17 months (2). In the Southern Hemisphere the breeding season is from August to

December, with October the time of most activity. Births occur from December to April, with most in February (4). The gestation period lasts about 16 months (1, 4). There is no long anestrum. Ovulations, probably about 4 to each season after the first, occur near the end of lactation (4). Since lactation lasts about 15 months this would give a pregnancy about every third year, a figure that agrees with the estimate for these whales in Japanese waters (5). One young at a time is the rule, but twins have been found in 0.66 per cent of pregnancies. Combined data from many sources give 50.57 per cent males as the sex proportion among fetuses (1).

The Azores data suggest a limited breeding season for the males, from March to May (1), but the material from the Southern Hemisphere does not, as some spermatogenesis and spermatozoa may be found all the year round (4).

The combined testis volume at puberty is about 3.6 l. or more (1). In one instance a single testis weighed, without epididymis, 6.4 kg. The internal diameter of the tubule in the tail of the epididymis was 2 mm., and of the ductus deferens 5 mm. The head length of the spermatozoon was $4.9 \pm .02\ \mu$, and the breadth, $2.7 \pm .01\ \mu$. The tail length was $40.6\ \mu$ (6). Spermatogenesis is general when the males reach 12.3 m. length, but 10.8 to 11.4 m. is the critical stage (3). Seventy-five per cent of testes are producing spermatozoa when the males are 12.6 to 13.0 m. long. They then weigh 1.5 kg. The left testis is the heavier in 56 per cent of cases. These figures are for whales taken in the Antarctic and they suggest an earlier maturity in the north (7).

The right ovary ovulates more frequently than does the left (8).

1. Clarke, R. *Discovery* Rpts., 28: 237–298, 1955–7.
2. Mizue, K., and H. Jimbo. Sci. Rpts. Whales Res. Inst., Japan, 3: 119–131, 1950.
3. Nishiwaki, M., and T. Hibiya. Sci. Rpts. Whales Res. Inst., Japan, 6: 153–165, 1951.
4. Matthews, L. H. *Discovery* Rpts., 17: 93–168, 1937–8.
5. Matsuura, Y. Zool. Mag., Tokyo, 48: 260–266, 1936.
6. Yamane, J. Zeit. f. Zucht., 34B: 105–109, 1936.
7. Nishiwaki, M. Sci. Rpts. Whales Res. Inst., Japan, 10: 143–149, 1955.
8. Chuzhakina, E. S. Trudy Inst. Okeanol. Akad. Nauk, 18: 95–99, 1955.

KOGIINAE

Kogia breviceps Blainville

PYGMY SPERM WHALE

This whale breeds late in the summer and the single young is born in the following spring after a gestation of about 9 months (1). In November a lactating female was taken; her uterus was enlarged, thick-walled, and spongy (2).

1. Allen, G. M. Field Mus. Nat. Hist., Zool. Ser., 27: 17–36, 1941.
2. Smalley, A. E. JM., 40: 452, 1959.

MONODONTIDAE

Delphinapterus leucas Pallas

WHITE WHALE

There is a record of a female with a single fetus, 19 inches long, taken in October (1). The time of birth varies. In the White Sea it occurs from the middle to the end of June; in the Barents Sea, early July; and in the Kara Sea at mid-July (2).

1. Soper, J. D. JM., 25: 221–254, 1944.
2. Belkovich, V. M. ZZ., 39: 1414–1422, 1960.

Monodon monoceros L.

NARWHAL

This whale of Arctic waters has no definite breeding season. One young is the rule (1).

1. Parsild, M. P. JM., 3: 8–13, 1922.

DELPHINIDAE

Delphinus delphis L.

COMMON OCEAN DOLPHIN

The age at puberty is 3 years (1). The young are still attached to the mother by the umbilical cord several days after parturition. At this time the placentae and corpora lutea are still present, and the latter show no signs of involution (2). The period of gestation is given as 276 days (3). The vaginal smear is helpful in diagnosis of estrus but it is not completely reliable (4).

1. Sleptzov, M. M. Bull. Soc. Nat. Moscow, Biol., N.S., 49: 50, 1940.
2. Khvatov, V. P. Bull. Biol. Med. Exp., U.R.S.S., 5: 27–88, 1938.
3. Kenneth, J. H. Gestation Periods. Edinburgh, 1943.
4. Sokolov, V. E. Trudy Soveshch. Ikhtiol. Komis. Akad. Nauk, S.S.S.R., 12: 68–71, 1961.

Grampus rectipinna Cope

PACIFIC KILLER WHALE

The young of this whale are born in spring and summer (1).

1. Scheffer, V. B., and J. W. Slipp. AMN., 39: 257–337, 1948.

Tursiops truncatus Montagu

BOTTLE-NOSED DOLPHIN

The females of this dolphin probably reach puberty at age 4 years (1). The time of mating and parturition is from February to May and the single young is born after a gestation of about 12 months. The young are weaned at about 18 months (2).

1. McBride, A. F., and H. Kritzler. JM., 32: 251–266, 1951.
2. Tavolga, M. C., and F. S. Essapian. Zoologica, 42: 11–31, 1957.

Lagenorhynchus obliquidens Gill

STRIPED DOLPHIN

According to one account mating is probably in the summer, and the young are born in the following spring or summer. There are records of a 12-cm. fetus in September and of a 37-cm. one in December. The single young is being suckled in July (1, 2). Another record mentions that copulations were observed in March and that a well-developed fetus was also found in that month (3).

1. Scheffer, V. B. AMN., 44: 750–758, 1950.
2. Gubertat, J. E. The Murrelet, 17: 56, 1936.
3. Wilke, F., T. Taniwaki, and N. Kuroda. JM., 34: 488–497, 1953.

Lagenorhynchus acutus Gray. A newborn young has been seen early in July (1).

1. van Utrecht, W. L. Mammalia, 23: 100–122, 1959.

Pseudorca crassidens Owen

FALSE KILLER WHALE

This whale is polyestrous except in winter, when it is in the anestrous condition. One ovum is shed at a time (1).

1. Comrie, L. C., and A. B. Adam. Trans. Roy. Soc. Edinburgh, 59: 521–531, 1938.

Globiocephala melaena Traill

BLACKFISH

The blackfish or caaing whale is probably polyestrous and ovulation is spontaneous. Thecal cones appear in the follicle but these are believed to be transient. The corpus luteum is yellow to pale ochre and the stigma is markedly herniated. The luteal cells are derived from the granulosa and many are multinuclear. Theca interna cells remain at the periphery of the corpus luteum. Since the corpora albicantia gradually shrink, they are not useful in determining the age of the blackfish. The luteinized cells are 30 μ in diameter (1). The gestation period is about 13 to 16 months. Puberty in the male is reached at age 13 years, and in the female at age 6 years (2).

1. Harrison, R. J. J. Anat., 83: 238–253, 1949.
2. Slijper, E. J. Whales. New York, 1962.

Globiocephala macrorhyncha Gray. Most of the young are born in the spring but a few not until fall (1).

G. scammoni Cope. The young of this blackfish are said to be born at any time of year (2). In captivity a pair has mated from February to April and again in October and November (3).

1. Starrett, A., and P. Starrett. JM., 36: 424–429, 1955.
2. Bailey, V. NAF., 55, 1936.
3. Brown, D. H. Zoologica, 47: 59–64, 1962.

PHOCAENIDAE

Phocaena phocoena L.

PORPOISE

Porpoise females breed at about 14 months old, when they are 150 cm. long and weigh 50 kg. They breed every year, most of them mating in July and August (1, 2). Pregnancy lasts about 10 to 11 months and lactation about 8 months. A single young is usual (1). A birth was recorded for early July (3). Another account gives the mating season as lasting from July to October with births from March to July (4).

1. Möhl-Hansen, U. Videnskab. Medd. Dansk Naturhist. Foren., Köbenhavn, 116: 369–396, 1954.
2. Scheffer, V. B., and J. W. Slipp. AMN., 39: 257–337, 1948.
3. van Utrecht, W. L. Mammalia, 23: 100–122, 1959.
4. Slijper, E. J. Whales. New York, 1962.

Phocaena dalli True. Pregnant females of Dall's porpoise have been captured in May (1).

1. Wilke, F., T. Taniwaki, and N. Kuroda. JM., 34: 488–497, 1953.

Neophocaena phocaenoides Cuvier

INDIAN PORPOISE

This porpoise breeds in October and has 1 young (1). The young are also said to be born in this month (2).

1. Ho, H. J. Contrib. Biol. Lab. Sci. Soc. China, Zool. Ser., 10: 245–287, 1933–5.
2. Blanford, W. T. Fauna of British India. Mammals. London, 1888–91.

RHACHIANECTIDAE

Rhachianectes gibbosus Erxleben

GRAY WHALE

Mating has been observed in April. The young are born in spring after a gestation of about a year (1). Another account gives the season for mating as December and birth in the same month. Lactation continues for 5 months, and puberty in each sex is reached at 4 years of age (2). Adult cows bear one calf at 2-year intervals (3).

1. Bailey, V. NAF., 55, 1936.
2. Slijper, E. J. Whales. New York, 1962.
3. Gilmore, R. M. United States Fish and Wildlife Service Special Sci. Rpt., Fish, 342: 1–30, 1960.

BALAENOPTERIDAE

Balaenoptera acutirostrata Lacepède

LITTLE PIKED WHALE

Puberty is reached at 7.5 m. long in the females, and in the males at 6.5 to 7 m. (1). Mating is most frequent in February and March, and a single young is born after a gestation of 10 months (2). In northeast Atlantic waters mating extends from January to the end of May (3). In the Pacific most of the mating is at the end of February and the beginning of March, with births at the end of December and the beginning of January. There is also a lesser period from August to September, and births follow in June and July. Most females bring forth a calf once a year, but some bear at 18-month intervals (1).

1. Omura, H., and H. Sakiura. Sci. Rpts. Whales Res. Inst. Japan, 11: 1–37, 1956.
2. Scattergood, L. W. The Murrelet, 30: 1–16, 1949.
3. Jonsgård, Å. Norsk Hvalfangst-tid., 40: 209–232, 1951.

Balaenoptera borealis Lesson

SEI WHALE

The males reach puberty at 13 to 14 m. long and the females at 14 m. (1, 2), when they are about 2½ years old (3). There are no indications of a male rutting season but in the Southern Hemisphere most pairing is from May to August and the season is at its height in July (1). In the Northern Hemisphere pairing takes place in January, for the most part, though there is a 3-month spread, and calving in November (4). The gestation period has been estimated as a little less than 11 months (4), and as 12 months, followed by a mean lactation period of about 5 months and then by anestrum until the next pairing season. Of 17 whales with a single corpus luteum 8 were pregnant and 4 had been pregnant. Thus, 70 per cent become pregnant at their first ovulation. Between pregnancies there are 3 to 5 ovulations. The diameter of the corpus luteum of pregnancy is about 8.5 cm. (1). Twins have been reported in 1.1 per cent of pregnancies (5).

1. Matthews, L. H. *Discovery* Rpts., 17: 183–290, 1937–8.
2. Mizue, K. Sci. Rpts. Whales Res. Inst. Japan, 3: 106–118, 1950.
3. Ruud, J. T. Hvalrådets Skrifter, 29, 1945.
4. Mizue, K., and H. Jimbo. Sci. Rpts. Whales Res. Inst. Japan, 3: 119–131, 1950.
5. Paulsen, H. Norsk Hvalfangst-tid., 12: 464, 1939.

Balaenoptera physalus L.

FINNER WHALE

In the Southern Hemisphere the males reach puberty at 19.2 m. long, and the females at 19.9 m. (1), when the whales are 3 years old (2). Another account gives 5 years as the average for both sexes (3), but there is reason to believe that the age at puberty has decreased with increased rate of growth following the diminution of numbers. The pregnancy rate has also been increasing (4). In the Northern Hemisphere the testes increase in size at 19 m. and the females reach puberty at 20.6 m. (5). In the Southern Hemisphere the testes become active in the early winter, roughly from April to June, when the breeding season begins. Gestation lasts for nearly a year (1), or for 12 months and 10 days (6). Pregnancy is rare during lactation (1) which lasts for 6 months (3). In the Northern Hemisphere matings take place during the months from November to March (7). The fetal sex proportion of about 13,000 specimens was 52.0 per cent males (1). Of 569 births, 11 were twins, or 1.9 per cent. Of these 11 cases, in one only was there a single corpus luteum. The tentative conclusion is reached that about 10 per cent of twins *may* be identical (8). In another investigation twins occurred 123 times and triplets, twice, in 13,186 pregnancies, a rate of 0.95 per cent (9). In 64 per cent of pregnancies the fetus has been on the right side (10).

The uterine endometrium at the time of estrus is 3 times as thick as it is during anestrum. The glands are numerous, closely packed, and greatly convoluted. The capillaries bulge into the uterine lumen and the mucosa is edematous (11). During pregnancy the anterior hypophysis is large and hyperemic. At this time it weighs about 34 g., in comparison with 18 g. in males and nonpregnant females (12). The anterior hypophysis increases in weight suddenly in males at 19.7 m. long and at 20.6 to 21 m. in females. Seventy-five per cent of the males are sexually mature at 19.0 m. The anterior hypophysis then weighs 12 g. In the female it weighs 20 g. (5). When mature the two testes weigh 5 kg. and up (5). The functional corpus luteum measures 12.9 cm. in diameter (13).

1. Mackintosh, N. A. *Discovery* Rpts., 22: 197–300, 1942–3.
2. Ruud, J. T. Hvalrådets Skrifter, 29, 1945.

3. Slijper, E. J. Whales. New York, 1962.
4. Laws, R. M. In E. D. Le Cren and M. W. Holdgate, eds., The Exploitation of Natural Animal Populations. Oxford, 1962.
5. Nishiwaki, M., and T. Oye. Sci. Rpts. Whales Res. Inst. Japan, 5, 1948.
6. Mizue, K., and H. Jimbo. Sci. Rpts. Whales Res. Inst. Japan, 3: 119–131, 1950.
7. Zenkovic, B. A. Dokl. Akad. Nauk, S.S.S.R., 2: 337–343, 1935.
8. Brinkmann, Jr., A. Hvalrådets Skrifter, 31: 5–38, 1948.
9. Paulsen, H. Norsk Hvalfangst-tid., 12: 464, 1939.
10. Slijper, E. J. Bijdr. Dierk., 28: 416–448, 1949.
11. Matthews, L. H. J. Anat., 82: 207–232, 1948.
12. Sverdrup, A., and K. Arnesen. Hvalrådets Skrifter, 36, 1952.
13. Mizue, K., and T. Murata. Sci. Rpts. Whales Res. Inst. Japan, pp. 73–131, n.d.

Megaptera novaeangliae Borowski

HUMPBACK WHALE

At puberty, as judged by the condition of the testes, the males of this species average 11.3 m. long, range 10.3 to 12.6 m. At this time the mean testis weight is 2 kg. Fertility is low for a year. In Western Australian waters the testis weight increases and spermatogenesis is more active in the winter months (1). The bulk of conceptions are from August to November, with a few from December to June (2), and gestation lasts nearly a year (3). At puberty the females are 11.9 m. long, range 11.0 to 13.4 m. They may breed immediately or delay for a year (4). The females are polyestrous and there are about 4 ovulations for each pregnancy after the first (2). In one account lactation is of 5 months' duration and is followed by a period of anestrum from December to February (2), but, according to another, it lasts for 10½ months and is not followed by anestrum (3). However, a post-partum ovulation may occur, though it is not known whether this is usual (4). Twins are born occasionally; one account gives the frequency as 0.4 per cent (5), and the sex proportion at birth is 51.4 per cent males (3). The sex proportion of 1,700 fetuses was given at 57.4 per cent males (6). About 86 per cent of mature females are pregnant (7).

The ripening follicles project from the surface of the ovary, and the corpus luteum is functional throughout pregnancy. After parturition, and after ovulation without an ensuing pregnancy, it becomes small and fibrous but it probably persists throughout life (2). No preference is shown in activity between the ovaries or uterine horns (3). The functional corpus luteum measures 12.6 cm. in diameter (8).

The adult testes weigh about 7 kg., and there is little difference in size between right and left. Branched seminiferous tubules are common. The length of the spermatozoon is 52.5 μ, range 32.2 to 64.4 μ. The penis length is very variable. In sexually mature males it ranges from 3.2 to 6.2 ft. long, with 3.5 ft. the average (1).

1. Chittleborough, R. G. Australian J. Marine Freshwater Res., 6: 1–29, 1955.
2. Matthews, L. H. *Discovery* Rpts., 17: 7–92, 1937–8.
3. Chittleborough, R. G. Australian J. Marine Freshwater Res., 5: 159–169, 1954.
4. Chittleborough, R. G. Australian J. Marine Freshwater Res., 6: 315–327, 1955.
5. Paulsen, H. Norsk Hvalfangst-tid., 12: 464, 1939.
6. Mackintosh, N. A. *Discovery* Rpts., 22: 197–300, 1942–3.
7. Omura, H. Sci. Rpts. Whales Res. Inst. Japan, 8: 81–102, 1953.
8. Mizue, K., and T. Murata. Sci. Rpts. Whales Res. Inst. Japan, pp. 73–131, n.d.

Sibbaldus musculus L.

BLUE WHALE

This whale reaches puberty in the male at 22.6 m. long, and in the female at 22.5 m. (1). This occurs in the females at age 3 to 7 years, usually 5, and in the males at 2 to 6 years, mostly 4 or 5 (2). In the Southern Hemisphere the testes become most active in the early winter months, roughly from April to June, and this is the usual breeding season. Gestation lasts for nearly a year and pregnancy during lactation is rare (1). Lactation is of about 7 months' duration, but it may be longer. Some females become pregnant every 2 years but an unknown proportion do so every 3 years (3). About 43 per cent of mature females caught are pregnant (4). The females are probably polyestrous, as corpora lutea accumulate at the first pregnancy at the rate of slightly less than 2 and then at the rate of slightly more than 2 each 2 years. The maximum number produced at one breeding season is about 4 (2). Twins, etc., account for about 0.74 per cent of all births, and one record of quadruplets has been made (5). Another account gives 0.68 per cent of twins (6). The sex proportion of about 10,000 fetuses was 52.7 per cent males (1). In a limited series 59.8 per cent of fetuses were in the left horn of the uterus (7, 8). The corpus luteum of pregnancy has a mean diameter of 12.7 cm. It has a very pronounced scar with a raised corona (3). Another account gives the diameter of the functional corpus luteum as 17.9 cm. (7). The ovary weighs 4 to 5 kg., and it contains about 500 cc. of liquor folliculi which has an estrogen value of about 2,000 M.U. per liter. The corpus luteum may weigh 4 kg., and, after preservation in a frozen state for a year, the tissue gave a progesterone value of 60 Rab.U. per kg. The anterior pituitary weighs 30 g. (9). The corpus luteum remains throughout life in a more or less degenerated state (10).

The anterior pituitary suddenly increases in weight at 23 m. long in the males and at 26.5 m. in the females. At the critical point, i.e., when 75 per cent are sexually mature, it weighs 17 g. in males and 23 g. in females (4).

1. Mackintosh, N. A. *Discovery* Rpts., 22: 197–300, 1942–3.
2. Ruud, J. I., Å. Jonsgård, and P. Ottestad. Hvalrådets Skrifter, 33: 1950.
3. Laurie, A. H. *Discovery* Rpts., 15: 223–284, 1937.
4. Nishiwaki, M., and T. Oye. Sci. Rpts. Whale Res. Inst. Japan, 5: 91–167, 1948.
5. Brinkmann, Jr., A. Hvalrådets Skrifter, 31: 1–38, 1948.
6. Paulsen, H. Norsk Hvalfangst-tid., 12: 464, 1939.
7. Mizue, K., and T. Murata. Sci. Rpts. Whales Res. Inst. Japan, pp. 73–131, n.d.
8. Slijper, E. J. Bijdr. Dierk., 28: 416–448, 1949.
9. Jacobsen, A. P. Nature, 136: 1029, 1935.
10. Peters, N. Zool. Anz., 127: 193–204, 1939.

BALAENIDAE

Eubalaena

RIGHT WHALE

Eubalaena australis Desmoulins. The southern right whale gives birth to its single young in June or July (1). It probably has 4 ovulations to a season after its first (2).

E. glacialis Borowski. The Atlantic right whale has its young in March after a gestation lasting about a year (3).

1. Shortridge, G. C. The Mammals of South West Africa. London, 1934.
2. Matthews, L. H. *Discovery* Rpts., 17: 169–182, 1937–8.
3. Millais, J. G. The Mammals of Great Britain and Ireland. London, 1906.

Balaena mysticetus L.

GREENLAND WHALE

This whale mates in February and March; the young are born in December after a gestation of 9 to 10 months. Lactation lasts for 12 months (1).

1. Slijper, E. J. Whales. New York, 1962.

Carnivora

AND

PINNIPEDIA

THE patterns of reproduction in Carnivora vary widely in different families. The Ursidae appear to be monestrous; the young are born in a very immature state and delayed implantation of the blastocysts is suspected. The Mustelidae have a restricted breeding season in which waves of follicles ripen. They seem to be very susceptible to variations in the intensity of light. Induced ovulation is the rule in several species and probably occurs in others. Some species have an excessively long gestation period, almost a year, and in these delayed implantation is the cause. The latent period may be shortened by exposure to greater amounts of daylight. Coitus-induced ovulation is found in some of the Viverridae. Most Canidae are monestrous, and they have a prolonged pseudopregnancy in the absence of pregnancy. Extravasation of blood in the uterus seems to be the exception rather than the rule. It has been found definitely only in the genus *Canis*. The Felidae appear to be seasonally polyestrous in temperate regions and completely polyestrous in tropical regions, but our knowledge of the smaller tropical species is inadequate for generalization. In the Procyonidae induced ovulation has been found.

In the Pinnipedia breeding is once a year, in spring soon after they have given birth to their young. This gives a gestation period of almost a year and it is associated with delayed implantation. In several species coitus-induced ovulation is suspected, but it is difficult to obtain rigid proof. Lactation is exceptionally short. The corpus luteum is versatile; it seems to be capable of taking on new life during lactation, and its activity seems to be in abeyance in the period before implantation of the embryos.

For several reasons the writer believes that coitus-induced ovulation is a more primitive mechanism than spontaneous ovulation. If so, then these two suborders show this trait to a remarkable degree. This should be taken into account in evolutionary sequences, if such are justified in this realm of taxonomy.

CANIDAE

CANINAE

Canis familiaris L.

DOG

The domestic dog is monestrous and usually has two heat periods in a year, in late winter or early spring and in the fall. Proestrum lasts for about 7 to 9 days and is characterized by a swelling of the vulva and a discharge of blood derived from the uterus. Estrus lasts about the same time, and about 1 day after its commencement the ova are shed. These are almost unique in that the first polar body is not shed, and fertilization is apparently not possible, until some time after they have left the graafian follicles. If the bitch does not become pregnant, the corpora usually persist for about 2 months, during which time a condition of pseudopregnancy exists. This period is followed by anestrum lasting about 2 months.

In the London zoo the dingo, wild dog of Australia, usually has its young between February and April, with a few in November. The litter size is 1 to 8, mode 4, and average 4.1. The gestation period is 63 days (1).

THE ESTROUS CYCLE

The bitch is usually said to have two heat periods a year, one in the early spring and one in the fall. Curves of births according to seasons made from data in kennel club records do not show this fact. The curves are almost unimodal, with peaks denoting conception very early in the spring. The true nature of the breeding season is obscured in these records by the breeders' habit of mating their dogs usually in spring and by the method of estimating dogs' ages, which makes it advantageous for show and racing purposes to have them born early in the year. However, a series of American records for several breeds shows a slight bimodality, with most conceptions in February and April (2). Another compilation from American Kennel Club records for cockers, great Danes, setters, and Pekingese showed no peaks of consequence (3). Similar records for greyhounds in England have peaks in January and March. It is not clear whether these are natural phenomena or the result of breeding customs (4). In kennels, Airedales show heats fairly evenly distributed through the year, but with most in July, and fewest from September to October. The interval between heats was mostly 7 or 8 months (5). The Basenji has only one heat a year, almost exclusively in the fall; 60 per cent were between September 20 and October 10, and all between August 15 and December 1 (6). When they were crossed with other breeds it was ascertained that the single annual season was due to a single recessive gene but the heterozygotes were rather variable (7). Puberty is reached at 6 to 8 months as a rule (8), although a good deal depends upon the amount of liberty allowed. Kenneled Airedales reached puberty at 15 to 24 months and mongrels by 16 months (5).

The onset of heat in the dog is gradual and is preceded by a long proestrum, which is marked by considerable swelling of the vulva and bleeding from the uterus. The length of proestrum varies somewhat widely. One series gives for greyhounds 7 to 22 days, as judged by erythrocytes in the vaginal smear (9); another, from the first appearance of a sanguineous discharge to first acceptance, gave 4 to 13 days with a mean of $9 \pm .5$ days (5). Heat in the greyhound lasts 7 to 9 days (9), and for a limited series of dogs of various breeds 4 to 13 days; mean, $9 \pm .5$ days (10). In Airedales bleeding lasted for 12 days, range 5 to 27 days; and estrus for 10 days, range 6 to 14 days. In foxhounds the corresponding figures were: bleeding, 5 days, range 2 to 7; estrus, 12 days, range 6 to 16 (5). There is often an overlapping of the external signs of the two periods. During the proestrum the male and female are interested in each other, but coitus is not usually allowed until the last day or two of bleeding. Ovulation is usually 1 to 3 days after first acceptance by the male (9, 10). The dog and fox differ from other mammals so far investigated in that the first polar body is not extruded from the ovum for some days after it has been shed. Accordingly the ova are apparently not ready for fertilization for some days after ovulation (11), which is spontaneous (12). Service after the bitch has been in heat for 5 days results in lowered fertility (5).

There is some uncertainty concerning the length of time during which the eggs may be fertilized after they have been shed. It is certain that they may survive for 4 days, and probably for twice this time. They travel quickly to the middle of the oviduct, taking a day or less, then they remain in this part, or in the portion nearest the uterus, for several days (10).

After ovulation corpora lutea are formed, which persist, in the absence of pregnancy, in a functional condition for at least 30 days. Then they gradually degenerate, but they may be detected in the ovaries at the time of the next heat. This persistence is accompanied by proliferative changes in the uterus and mammary glands, and the decline is marked, in many cases, by lactation and the formation of a nest, a usual reaction at the end of a true pregnancy (10, 13). The duration of this pseudopregnancy is somewhat variable and is usually given as about 2 months. Probably it is too gradual to set any definite limit.

There appears to be practically no precise information on the litter size of the dog and its variation with the breed. For the German sheep dog the mean litter size is $7.15 \pm .02$, with a range from 1 to 17. The sex proportion in this series, which is drawn from stud books, was $52.75 \pm .12$ per cent males. It tends to decrease with litter size. The material is peculiar in that there are too many combinations, exclusively or predominantly, of one sex (14). This may indicate that uniovular twins are frequent. In another study of the German sheep dog, in which the mean litter size and its range were the same as in the work just quoted, 83 per cent of the litters were between 4 and 10. Stillbirths were 2.29 per cent of the total, and their sex proportion was 62.25 per cent (15), which suggests that the sex proportion may be higher early in gestation than at birth, a condition which holds also in man and the pig. Yet another German sheep dog investigation gave a litter range from 1 to 15, mean $6.6 \pm .03$, mode 5 and 8, and a sex proportion of 51.5 per cent males (2). The litter size for the Hungarian sheep dog was 6.7, of which 51.2 per cent were males (16). The litter size decreased gradually in females over 3 years old. Litter size for Airedales has been given as $7.7 \pm .32$, and for foxhounds as $7.3 \pm .93$ (17). For the greyhound it was 6.5 (18), and for the border collie in Australia, 5.7 (19). It increases with the size of the bitch. The sex proportion for the schnauzer is given as 50.5 per cent males, and for the French bull, 50.4 per cent, but for bull terriers the proportion has been recorded as 55.9 per cent (20).

The gestation period is usually given as between 58 and 63 days (12, 8).

HISTOLOGY OF THE FEMALE TRACT

OVARY. At birth the ovaries contain 700,000 ova; at puberty the number has been reduced to 355,000; at age 5 years to 33,800; and at 10 years only 518 remain (21). At the beginning of the proestrum large numbers of small and medium-sized graafian follicles can be found in the cortical zone of the ovary, but most of these already show degenerative changes, and very few are destined to rupture. Those follicles which continue to grow have folds of granulosa cells with vascularized cores of theca interna, which are more complex than those found in any other mammal that has been investigated. As the bitch enters heat, these folds become more complex, but the theca interna cells show less than the usual amount of hypertrophy. Ovulation usually occurs on the first day of heat or acceptance, when the follicles are about 6 mm. in diameter, and there is little growth during proestrum. All rupture within a short time. At first the granulosa lutein cells are arranged in irregular columns resembling an open lacework. By the eighth day the appearance is more compact, although large cavities are still present, and the corpora lutea do not become completely solid until they are about 18 days old. The condition of the uterus indicates that they remain functional for 30 days after heat has ended, but histologically the cells remain in good condition for some time longer. When the decline sets in, it is so gradual a process that one cannot give a time for their effective life on histological grounds alone (10, 22). Alkaline phosphatase may be found in the theca interna but it disappears after ovulation. Only a trace can be found in the granulosa cells (23).

The average size of the ova after they have been shed is $77 \times 90\ \mu$ excluding, and $95 \times 110\ \mu$ including, the zona pellucida.

VAGINA. One feature of proestrum is the marked edematous swelling of the vulva, which increases throughout this period and remains during heat, after which the decline is somewhat gradual. The vestibule is coated with low stratified epithelium, which increases a little in height during proestrum, but never to any great extent. The vagina proper exhibits well-marked changes during the cycle. During anestrum the epithelium is columnar in type, 2 to 3 layers thick. By the beginning of proestrum it has become flat stratified, 6 to 8 layers thick. It continues to grow, so that by first acceptance it is 12 to 20 cells thick, and the superficial layers have become cornified. Cells are being lost from the superficial layer throughout, but marked desquamation does not set in until the third or fourth day of heat, when reduction is rapid. At this time there is also some infiltration of leucocytes. Repair sets in soon after heat is over, and by the tenth day postestrum the epithelium is of the anestrous high-columnar type.

The vaginal smears are clear-cut, with red blood cells and epithelial cells in early proestrum. The latter type gives way entirely to large cornified cells and, after bleeding ceases, these are the only cells present. Toward the end of heat a few leucocytes are found, but they do not become abundant until 2 to 3 days after heat has ceased. The anestrous smear consists of epithelial cells with a few leucocytes (10). Cornified cells do not disappear from the smear until 3 to 5 days after the last day of acceptance, but they are gradually replaced by noncornified cells and leucocytes. Ovulation occurs 24 hours before the appearance of leucocytes in the smear, and the male is regularly accepted at 24 to 48 hours after they first appear (24).

UTERUS. During anestrum the endometrium is low and compact, but the glands are poorly developed. The lumen of the uterus is H-shaped in cross section. In proestrum the endometrium becomes very edematous, the lumen flattens out, and hyperemia is very apparent. Focal bleeding occurs, but, as the epithelium does not rupture, the erythrocytes escape into the lumen by diapedesis. Mitoses are abundant throughout, but the glands remain simple. During heat the edema persists, but bleeding ceases and hyperemia is not so apparent. The glands are increasing in complexity and show some secretory activity. Leucocytes are abundant at this time. During pseudopregnancy the endometrium is in two well-defined zones, a superficial "compacta" and a deep "spongiosa" with numerous branched glands. The zona compacta develops villous processes which obstruct the uterine cavity. Retrogressive changes set in at the twentieth day after heat has ceased, and they become marked by the thirtieth day; but the process of involution is gradual, and the anestrous condition is not reached until 85 days (8, 10, 13, 22). It is said that the end of pseudopregnancy is marked by a further extravasation of blood (13), but this has not always been found (10).

Much dark pigment is present in the superficial layers of the uterus in anestrum and proestrum (12), and during the latter period many pigment-laden macrophages are found (22).

During proestrum the epithelium of the oviduct is tall-columnar and it is secreting. There are variable numbers of cilia. During heat the picture is the same. After heat is over, secretion ceases; the cells are not granulated but are vacuolar. Cilia are much less apparent. In anestrum the cells are low, nonciliated, and they contain some granules (8, 25).

True basophil cells are not present in the anterior pituitary, but two types that seem to be essentially basophilic have been described. A large granulated type increases during proestrum, and these cells become heavily granulated. They degranulate before ovulation and are present in small numbers during the lutein phase of the cycle. Eosinophils are also degranulated during this phase. All granular types increase gradually during anestrum. There are no specialized pregnancy cells (26).

PHYSIOLOGY OF THE FEMALE TRACT

The pH of the fluids secreted into a closed uterine segment of a bitch brought into heat with stilbestrol was 6.09 with a range from 5.19 to 6.26 (27).

The uterine muscle responds to epinephrin by contraction if the bitch is pregnant, i.e., if a corpus luteum is present, and by relaxation at other times (28, 29). This is a peculiarity shared by the cat and the cow.

Estrogens can be detected in the urine during pregnancy after 15 to 18 days, but the level is not high at any time (30). Twenty I.U. daily of estrogen is without effect in the immature bitch, but 200 I.U. produces cornification in the vagina. The same daily dose in the mature spayed female produces the usual changes of heat and of proestrum (31). Brief microscopical bleeding follows ovariectomy in metestrum, after an interval of 7 to 16 days, and the injection of 1,100 R.U. or more of estrogens into spayed females produces macroscopic bleeding from an endometrium resembling that of proestrum (32). This reaction occurs in the absence of the pituitary and is therefore probably a direct one upon the uterus (33).

Since cornification of the vagina has been produced by the injection of estrogens during anestrum without vulval swelling or proestrous bleeding, it probably has a much lower threshold dose than these other changes (34).

Glandular growth in the uterus may be induced by the injection of 3 to 5 Rab.U. daily of progesterone for 5 to 6 days (31). Ovariectomy during pregnancy is followed by reabsorption of the embryos or by abortion. However, none of the operations were performed during the second half of the period (12). Hypophysectomy at 5 to 7 weeks produces the same effect (35).

The injection of 100 M.U. of gonadotrophic hormone during anestrum produces heat and ovulation (36).

THE MALE

The amount of semen ejaculated varies with the breed or, more accurately, with the size of the dog. In the Pekingese it amounts to 3.6 ml., while in large breeds like the boxer, great Dane, and mastiff it may amount to 36 ml. (37). In the beagle it may be from 4 to 7 ml. It is ejaculated in three waves (38), about 10 per cent in the first, which does not contain spermatozoa, 20 per cent in the second which contains most, if not all, the spermatozoa, and the rest in a final fraction which is prostatic in origin and does not contain spermatozoa. The second fraction is more viscid than the others. In view of the fact that the dog possesses no vesiculae seminales the amount of ejaculate is large. The pH of the prostatic fraction is 7.2, compared with 6.75 for the second (39, 40). Another account gives $6.3 \pm .23$ for the sperm fraction and $6.8 \pm .33$ for the prostatic fraction (41). The number of spermatozoa varies from 69 to 1,726 million per ejaculate or from 88 to 588 million per ml. (42). The length of a spermatozoon is 55.3 μ; the head measures 5.6 μ; the neck 1 μ; the midpiece and tail 50 μ (37). The total length of seminiferous tubules has been variously reported as 150 meters (43) and 552 meters (44). Doubtless it varies with the size of the dog.

By the uterine fistula method it has been found that spermatozoa are at the uterine entrance to the oviducts 25 seconds after ejaculation (45). Copulation is a somewhat prolonged process and spermatozoa have been found throughout the oviduct 20 minutes after it began (2). They are able to live in the female tract for 48 ± 12 hours (2).

The ejaculate of the dog does not coagulate, and it has been found to contain a fibrinogenase which destroys the clotting power of blood plasma (46). The total reducing substances, as hexose, in the semen is 20.5 mg. per 100 ml., and the total protein 3,000 mg. (47). The urethral glands produce less than 2 per cent of the amount of fluid secreted by the prostate. After castration the latter gland ceases to secrete in 7 to 23 days (48).

The injection of 50 I.U. of androgen daily in 1-month-old puppies causes growth of the accessory organs, mostly in the prostate and ductus deferens (49).

1. Zuckerman, S. PZS., 122: 827–950, 1952–3.
2. Whitney, L. F. How to Breed Dogs. New York, 1937.
3. Engle, E. T. JM., 27: 79–81, 1946.
4. Asdell, S. A. Unpublished work.
5. Hancock, J. L., and I. W. Rowlands. Vet. Rec., 61: 771–776, 1949.
6. Fuller, J. L. J. Hered., 47: 179–180, 1956.

7. Scott, J. P., J. L. Fuller, and J. A. King. J. Hered., 50: 254–261, 1959.
8. Gerlinger, H. Le Cycle Sexuel chez la Femelle des Mammifères. Recherches sur la Chienne. Strasbourg, 1925.
9. Griffiths, W. F. B., and E. C. Amoroso. Vet. Rec., 51: 1279–1284, 1939.
10. Evans, H. M., and H. H. Cole. Mem. Univ. California, 9: 66–118, 1931.
11. Stricht, O. van der. AB., 33: 229–300, 1923.
12. Marshall, F. H. A., and W. A. Jolly. TRS., 198B: 99–141, 1905.
13. Marshall, F. H. A., and E. T. Halnan. PRS., 89B: 546–559, 1917.
14. Winzenburger, W. Zeit. Zucht., 36B: 227–236, 1936.
15. Druckseis, H. Diss., Munich, 1935.
16. Sierts-Roth, U. Zool. Garten, 22: 204–208, 1958.
17. Rowlands, I. W. Proc. Soc. Study Fertility, 2: 40–55, n.d.
18. Burns, M. The Genetics of the Dog. Farnham Royal, Slough, Bucks., 1952.
19. Kelley, R. B. Sheep Dogs. Sydney, 1947.
20. Briggs, E. C., and N. Kaliss. J. Hered., 33: 222–228, 1942.
21. Schotterer, A. AA., 65: 177–192, 1928.
22. Mulligan, R. M. J. Morphol., 71: 431–438, 1942.
23. Corner, G. W. CE., 575: 1–8, 1948.
24. Newberry, W. E., and H. T. Gier. Vet. Med., 47: 390–392, 1952.
25. Courrier, R., and H. Gerlinger. CRSB., 87: 1363–1364, 1922.
26. Wolfe, J. M., R. Cleveland, and M. Campbell. Zeit. f. Zellforsch. Mikros. Anat., 17: 420–452, 1933.
27. Shih, H. E., J. Kennedy, and C. Huggins. AJP., 130: 287–291, 1940.
28. Sharaf, A., and A. Dabash. Am. J. Vet. Res., 19: 935–939, 1958.
29. Reynolds, S. R. M. Physiology of the Uterus. New York, 1939.
30. Lesbouyries, —, and — Berthelon. Bull. Acad. Vet. de France, 9: 62–64, 1936.
31. Sammartino, R., and N. Arenas. CRSB., 133: 321–323, 1940.
32. Meyer, R. K., and S. Saiki. PSEBM., 29: 301–303, 1931.
33. Robson, J. M., and W. R. Henderson. PRS., 120B: 1–14, 1936.
34. Asdell, S. A., and F. H. A. Marshall. PRS., 101B: 185–192, 1927.
35. Votquenne, M. CRSB., 122: 91–93, 1936.
36. Lesbouyries, —, and — Berthelon. Bull. Acad. Vet. de France, 10: 126–130, 1937.
37. Perez-Garcia, J. Rev. Patron. Biol. Anim. (Madrid), 3: 97–150, 1957.
38. Boucher, J. H., R. H. Foote, and R. W. Kirk. Cornell Vet., 48: 67–86, 1958.
39. Bartlett, D. J. J. Reproduction and Fertility, 3: 173–189, 1962.
40. Harrop, A. E. Vet. Rec., 67: 494–498, 1955.
41. Wales, R. G., and I. G. White. J. Physiol., 141: 273–280, 1958.
42. Nooder, H. J. Tijdschr. Diergeneesk., 75: 81–94, 1950.
43. Bascom, K. F., and H. L. Osterud. AR., 31: 159–169, 1925.
44. Knepp, T. H. Proc. Pennsylvania Acad. Sci., 10: 39–42, 1936.
45. Evans, E. I. AJP., 105: 287–293, 1933.
46. Huggins, C., and V. Vail. AJP., 139: 129–134, 1943.
47. Bartlett, D. J. Nature, 182: 1605–1606, 1958.
48. Huggins, C., M. H. Masina, L. Eichelberger, and J. D. Wharton. J. Exp. Med., 70: 543–556, 1939.
49. Itho, M., and T. Kon. CRSB., 120: 678–681, 1935.

Canis latrans Say

NORTHERN COYOTE

The common coyote of North America experiences one breeding season a year (1), and the age of puberty is usually 2 years. In one case bleeding from the vagina was first noted December 11; it became more rapid and brighter on January 19, but attempts by the male to mount were rejected. On February 6 the vulva began to swell; it reached its maximum development by the eighth. The female allowed coitus for the first time on February 24, and on February 27. This suggests a long proestrum and a heat period lasting about 4 days (2). Normally the female sexual season begins in February and ends in April. The testes begin to show activity in November. In January spermatozoa are present and the epididymis is enlarging. The testes gradually recede from March onward, and by the end of May there are no spermatozoa. The time of minimal activity is reached in November (1).

The average number of embryos is 6.23 (1,330 cases), and the average number of den young is 5.70 (1,582 cases) (1). The period of gestation is 60 to 65 days (1, 2). Another series of embryo counts gave a mean of 6.66 ± .11 (3).

1. Hamlett, G. W. D. United States Dept. Agric., Tech. Bull. 616, 1938.
2. Whiteman, E. E. JM., 21: 435–438, 1940.
3. Bailey, V. NAF., 55, 1936.

Canis lupus L.

WOLF

The wolf mates late in January to the beginning of March, or later if the weather is unusually cold (1). The litters are born in March to June after a gestation of 60 to 63 days (2). The litter size varies from 1 to 13 (3). In the London zoo the number has been from 1 to 7, mean and mode 4 (4). This seems to be the usual number (5). Puberty is reached at 2 years old (6).

1. Seton, H. T. Life Histories of Northern Animals. New York, 1909.
2. Brown, C. E. JM., 17: 10–13, 1936.

3. Heptner, W. G., L. G. Morosowa-Turowa, and W. I. Zalkin. Die Säugetiere in der Schutzwaldzone. Berlin, 1956.
4. Zuckerman, S. PZS., 122: 827–950, 1952–3.
5. Sutton, G. M., and W. J. Hamilton, Jr. Mem. Carnegie Mus., 12(2): 1–111, 1932.
6. Blanford, W. T. Fauna of British India. Mammals. London, 1888–91.

Canis aureus L. This Asiatic jackal breeds in January and February (1). The number of young varies from 2 to 6, mode 5, and mean 4 (2). The gestation period is 60 to 63 days (3).

C. azarae Wied. This South American jackal breeds in winter (4). In the London zoo litters of 1 and 4 young have been born in May and June (2).

C mexicanus L. This wolf breeds at 2 years old. The young are born in March and usually number 6 to 10 (4).

C. (*Thos*) *adustus* Sundevall. The side-striped jackal gives birth to 3 to 7 young in the winter after a gestation of 57 to 60 days (5). Three to 6 is probably the usual number (6).

C. (*T.*) *mesomelas* Schreber. The black-backed jackal has 2 to 6 young, usually 3 or 4 (7), or 6 or 7 (6), born in November or December in South West Africa (7). In the London zoo young have been born in March and April (2).

1. Heptner, W. G., L. G. Morosowa-Turowa, and W. I. Zalkin. Die Säugetiere in der Schutzwaldzone. Berlin, 1956.
2. Zuckerman, S. PZS., 122: 827–950, 1952–3.
3. Phillips, W. W. A. Ceylon J. Sci., 13B: 143–183, 1924.
4. Bailey, V. NAF., 49, 1926.
5. Haagner, A. K. South African Mammals. London, 1920.
6. Roberts, A. The Mammals of South Africa. Cape Town, 1951.
7. Wilhelm, J. H. J. South West Africa Sci. Soc., 6: 51–74, 1933.

Alopex lagopus L.

ARCTIC FOX

The males reach puberty at age 10 months. Spermatogenesis begins in January and mature spermatozoa may be found in the epididymis in March and until after the end of July. Large follicles can be found in the ovaries from February and March. The corpora lutea are well developed until the beginning of August. Mating begins in Siberia at the end of April or early in May (1). On Southampton Island mating begins in March or April (2). Two types of fox have been recognized in Greenland, a coast fox with a small litter and a lemming fox with a large litter (3). For the coast type one report gives a range from 1 to 10, mean 6.4 ± .3, mode 6 to 7 (4). Embryo counts up to 21 have been given for the lemming fox (3). The gestation period has been given as 52 days (4), and as 60 days (5). The minimal weight of functioning testes is 2.6 g. (1). On the Adelaide Peninsula the young are born in June (6).

1. Sokolov, N. N. ZZ., 36: 1076–1083, 1957.
2. Sutton, G. M., and W. J. Hamilton, Jr. Mem. Carnegie Mus., 12(2): 1–111, 1932.
3. Braestrup, F. W. Meddel. om Grönland, 131(4): 1–101, 1941.
4. Barabash-Nikiforov, I. JM., 19: 423–429, 1938.
5. Rand, A. L. Canada Nat. Mus., Bull. 100, 1945.
6. Macpherson, A. H., and J. H. Manning. Canada Nat. Mus., Bull. 161, 1959.

Vulpes fulva Desmarest

RED FOX, SILVER FOX

The specific name, *Vulpes fulva,* is taken to include most domesticated foxes, and it probably should include several wild varieties, such as *V. regalis* Merriam and *V. macroura* Baird, which have sometimes been described as subspecies.

The fox is monestrous, with one season a year, from December to March with most matings in late January and February, though the blue fox is said to be somewhat later than the silver fox, i.e., from February to March (1). Foxes breed in the first year following their birth (2). Three-year-olds breed an average of 3½ days earlier than 2-year-olds and older vixens 6½ days earlier (3). In New York State wild red foxes breed from mid-December to the end of March. Of these, 76 per cent had mated from mid-January to mid-February. Proestrum and estrus lasts 12 to 14 days, and during this time there is some swelling and whiteness of the vulva. Sexual receptivity lasts for 2 to 4 days (5). Fifty-three per cent of matings occur between 5 A.M. and 10 A.M., and 20 per cent from 10 A.M. to 1 P.M. (6). Ovulation is spontaneous, and it occurs most frequently on day 1 or early on day 2 of sexual receptivity. The effects of mating on specific days upon the chance of pregnancy and upon litter size are given (7):

	Percentage of pregnancies	*Average litter size*
Mating on day 1	54	1.77
Mating on day 2	96	3.98
Mating on day 3	86	3.79

It has also been stated that mating on the first day of vaginal cornification produces significantly larger litters than matings either before or after this day (8).

There is no bleeding during proestrum, which is, therefore, not well defined. If pregnancy does not result from mating, or if the female is not mated, she becomes pseudopregnant, denoting a fairly long life for the corpus luteum (2, 9).

The mean litter size is 4.52 ± .03, with a distribution of 1 to 8, mode 4 to 5, and standard deviation, 1.35 (2). On German fox farms the average was 3.94 ± .03 (10). The litter size rises steadily to 5 to 7 years of age, from an initial level of 4.34 to 4.84, and litters born early in the season tend to be larger than those born later (2). The sex proportion taken during the first week or so after birth was 52.93 ± 1.00 per cent males (2). The length of gestation is from 49 to 55 days, with a pronounced mode at 52 days in all the data. Means are 51.99 ± 0.036 (2), 52.74 (8), 52.07 (11), and 52.12 ± .006 (6). There is a slight tendency for small litters to have a longer gestation (2). The mean litter size of wild foxes in New York State, on the basis of embryo counts and placental sites, was 5.4 (4). In Michigan the actual litter size increased from north to south, rising from 4.6 ± .07 to 5.5 ± .09 as one proceeded south (12).

HISTOLOGY AND PHYSIOLOGY OF THE FEMALE TRACT

The fox resembles the dog in that the first polar body is not extruded from the egg until after ovulation. Accordingly, fertilization does not occur until at least one day after ovulation (13). At 2 to 3 days before acceptance of the male the largest follicles are 4 mm. in diameter. The granulosa is thin and is folded in the manner found in the ripe follicle of the dog. At first acceptance the follicle measures 7 mm. in diameter. The ovum measures $104 \times 76\ \mu$. Fertilization takes place in the middle section of the oviduct (14). Distribution between the uterine horns is even (15).

The vaginal smear always contains some epithelial cells, and there is a gradual increase in cornified cells during proestrum, with a maximum at day 1 of heat. On that day leucocytes are absent; then they appear and gradually increase in numbers (16).

The pH of the vaginal secretions in early proestrum is 6.3, and during heat it is 7.9 (5), an unusual reversal.

Vixens may be brought in heat in September and October by the injection of 250 to 340 M.U. of prolan (pregnancy urine extract), judging by the vulval swelling and vaginal smears (17). However, heat is only induced by this means late in anestrum when the follicles are in the antrum stage. Ovulation is rarely induced (18). Other results have not been promising as it is said that single doses of gonadotrophe will not induce heat and that repeated doses are harmful to the ovaries (19).

THE MALE

The testes retrogress and spermatogenesis ceases during the anestrous period. Spermatogenesis begins in November, and the male is fertile 8 to 10 weeks before the breeding season begins. In their first season, in England, spermatogenesis in young foxes is complete by the end of November (20, 21). The left testis is the longer and heavier, and the right is broader and thicker. The sperm head measures $8 \times 5\ \mu$, and the entire length is $60\ \mu$ (21). According to one report, in the United States spermatozoa can be found at all seasons, but outside the breeding season most tubules are inactive (22). During January there is a marked development of the prostate gland, which increases from 0.5 cc. to 5 to 6 cc. in volume. In young males the development of the testes and accessories lags 2 weeks behind that in mature males (23).

In mature males the average ejaculation is 6 cc., with a range from 0.1 to 33 cc. The concentration of spermatozoa is 55×10^6 per cc. (5), and the pH of the semen is 6.2 to 6.4 (23).

Mating usually lasts 15 to 25 minutes, and spermatozoa may be found in the oviduct within 8 minutes after unlocking. Artificial insemination into the vagina is unsuccessful, and it is impossible to deposit semen in the cervix because of the minute curved passage in the spongy papilla (8).

In October the basophil cells in the anterior pituitary increase in numbers, reaching 28 per cent by December; they begin to recede in February until they are lowest, 10 to 14 per cent in April. The chromophobes exhibit the reverse trend. Eosinophils are low, 23 per cent, at the end of November; then they rise through the mating season (24).

1. Schulze, G. Deutsche Pelztierzucht, 13: 97–100, 126–128, 1938.
2. Johanssen, I. Ann. Agric. Coll., Sweden, 5: 179–200, 1938.
3. Pearson, O. P., and C. F. Bassett. Am. Nat., 80: 45–67, 1946.
4. Sheldon, W. G. JM., 30: 236–246, 1949.
5. Starkov, I. D. Usp. Zooteh. Nauk, 3: 385–401, 1937.
6. Schmidt, F. Der Silberfuchs und Seine Zucht. Munich, 1938.
7. Bassett, C. F., F. Wilke, and O. P. Pearson. Black Fox Mag., n.d.
8. Exp. Fox Ranch, Summerside, P. E. I., Progress Rpt., Ottawa, 1941.
9. Stoss, A. O. Landw. Pelztierzucht, 4: 181–189, 1933.
10. Richter, J. Landw. Pelztierzucht, 3: 81–86, 97–101, 1932.
11. Klemola, V. M. Finsk. Vet. Tidskr., 46: 19–25, 41–47, 1940.
12. Schofield, R. D. JWM., 22: 313–315, 1958.
13. Pearson, O. P., and R. K. Enders. Fur Trade J., Canada, 21: 14–15, 30–31, 1943.
14. Pearson, O. P., and R. K. Enders. AR., 85: 69–83, 1943.
15. Hoffman, R. A., and C. M. Kirkpatrick. JM., 35: 504–509, 1954.
16. Bassett, C. F., and J. R. Leekley. North Am. Vet., 23: 454–457, 1942.
17. Kakuskina, E. A. Probl. Zooteh. Eksp. Endokrinol., 2: 128–143, 1935.
18. Kakuskina, E. A. Bull. Biol. Med. Exp., U.R.S.S., 4: 29–31, 1937.
19. Koch, W. Deutsche Pelztierzucht, 14: 2–6, 30–34, 1939.
20. Rowlands, I. W., and A. S. Parkes. PZS., 823–841, 1935.
21. Beck, A. Diss., Leipzig, 1936.
22. Bishop, D. W. AR., 84: 99–115, 1942.
23. Starkov, I. D. 1st All-Union Confr. Artif. Insemination, 122–130, 1935.
24. Hillesund, C. M. Norsk Vet.-Tidskr., 55: 174–394, 1943.

Vulpes vulpes L.

ENGLISH RED FOX

The English red fox, like *Vulpes fulva,* is monestrous, with the mating season at the end of January. It breeds first at about 10 months of age. Ovulation is spontaneous (1). In Rusia the rutting season follows the latitude. In the Ukraine it begins early in January, in middle Russia in February, and in the north not until the end of March or the beginning of April. Some adults may be found to be pregnant from February to July, but most in March, April, and May. A very few may also be found pregnant in November and December (2). There are few or no vulval signs of approaching heat, and no premature follicular growth during anestrum. The ovum measures $110 \times 125\ \mu$. The unmated vixen passes into pseudopregnancy after she has been in heat. The litter size varies from 3 to 7, and corpus luteum counts in 6 vixens averaged 5.7 (1). In proestrum the vaginal epithelium is many-layered, and the stroma of the uterus is dense and vascular. In pseudopregnancy the uterine stroma is not so dense; the glands are coiled, mainly at their bases, and their lumens are practically obliterated, as is also the uterine lumen. At the end of this period the surface epithelium of the uterus breaks down (1). The observed gestation period is 51 to 53 days (3).

1. Rowlands, I. W., and A. S. Parkes. PZS., 823–841, 1935.
2. Bernard, J. Säugetierk. Mitt., 7: 110–113, 1959.
3. Zuckerman, S. PZS., 122: 827–950, 1952–3.

Vulpes bengalensis Shaw. The Indian fox mates from November to January and has 4 young, born from February to April (1).

V. chama A. Smith. The young of this African fox are born in August and September (2), after a gestation of probably 7 to 8 weeks (3).

V. corsac L. The steppe fox mates in February, and from 2 to 6 young are born in March or April after a gestation of 50 to 60 days (4).

V. velox Say. The kit fox has 4 or 5 young (5). In Mexico birth takes place usually in February, with litters ranging from 1 to 7 (6).

1. Blanford, W. T. Fauna of British India. Mammals. London, 1888–91.
2. Shortridge, G. C. The Mammals of South West Africa. London, 1934.
3. Roberts, A. The Mammals of South Africa. Cape Town, 1951.
4. Heptner, W. G., L. G. Morosowa-Turowa, and W. I. Zalkin. Die Säugetiere in der Schutzwaldzone. Berlin, 1956.
5. Seton, W. T. Life Histories of Northern Animals. New York, 1909.
6. Leopold, A. S. Wildlife of Mexico. The Game Birds and Mammals. Berkeley, Calif., 1957.

Fennecus zerda Zimmermann

FENNEC

In the Sahara the young are born in March or early April. In captivity a pair bred in April and a single young was born after a gestation of 51 days (1). Another pair mated in April and gave birth to 2 young after a gestation of 50 days. They mated again 6 days later but had no young (2).

1. Petter, F. Mammalia, 21: 307–309, 1957.
2. Volf, J. Mammalia, 21: 454–455, 1957.

Urocyon cineroargenteus Schreber

GRAY FOX

The males of this fox produce spermatozoa from December until March and breeding begins late in December, with a peak late in January or early in February (1, 2). This was in southern Illinois, Georgia, and Florida. In New York State mating takes place from mid-January to mid-May, and 71 per cent of adult females have mated by the last week of February to mid-March (3). In Alabama mating is most frequent from mid-February to the end of the month, although spermatozoa may be found in the testes from mid-November to the end of April (4). Of the young vixens, 92.3 per cent breed in their first season (2). The mean embryo count was 4.9 ± .1, and a post-partum placental scar count gave 4.3 ± .1 (2). Combined embryo and placental scar counts in New York State gave 3.7 as the litter size (3). The largest litters were produced by the 3- to 4-year old females (2). The gestation period is about 63 days (5).

1. Layne, J. N. JWM., 22: 157–163, 1958.
2. Wood, J. E. JM., 39: 74–86, 1958.
3. Sheldon, W. G. JM., 30: 236–246, 1949.
4. Sullivan, E. G. JM., 37: 346–351, 1956.
5. Grinnell, J., J. S. Dixon, and J. M. Linsdale. Fur Bearing Mammals of California. Berkeley, Calif., 1937.

Nyctereutes procyonides Gray

DOG RACCOON

The mating season of this Asiatic species is from February to March; one litter a year is produced in April or May (1). Estrus lasts 6 to 8 days and is shown externally by swelling of the vulva and nipples (2). The gestation period is usually given as 61 to 63 days, and the litter size varies from 5 to 12, with 7 and 8 as the modal numbers (3). Two observed gestations lasted 59 and 79 days (1).

1. Schneider, K. M. Zool. Anz., Suppl., 14: 373–387, 1950; 15: 271–285, 1951.
2. Belter, M. Deutsche Pelztierzucht, 14: 85–86, 1939.
3. Schmidt, F. Deutsche Pelztierzucht, 12: 235–237, 1937.

Chrysocyon brachyurus Illiger

MANED WOLF

This South American wolf has 2 young to the litter (1).

1. Miller, F. W. JM., 11: 10–22, 1930.

SIMOCYONINAE

Cuon alpinus Pallas

DHOLE

This Indian dog breeds at all seasons, but the young are mostly born in January and February. It usually has 4 or 5 pups (1). In the London zoo young have been born in March, April, and November. Their number averaged 3.5. One observed gestation lasted 70 days (2).

1. Brander, A. A. D. Wild Animals in Central India. London, 1923.
2. Zuckerman, S. PZS., 122: 827–950, 1952–3.

Lycaon pictus Temminck

CAPE HUNTING DOG

This dog has no fixed season, but spring and fall, especially the month of April, are the most likely times for the young to be born. The number varies from 2 to 6 (1). In Northern Rhodesia the litters are born from May to July (2). A female in the Naples zoo came in heat in October and had 5 young after a gestation that lasted from 60 to 63 days. The heat period lasted for 9 days (3).

1. Shortridge, G. C. The Mammals of South West Africa. London, 1934.
2. Ansell, W. F. H. PZS., 134: 251–274, 1960.
3. Cuneo, F. Personal communication, 1954.

OTOCYONINAE

Otocyon megalotis Desmarest

BIG-EARED FOX

The young of this fox are born in November (1) or December and April (2). They are usually 3 to 5 in number, born after a gestation of probably 60 to 70 days (3).

1. Shortridge, G. C. The Mammals of South West Africa. London, 1934.
2. Wilhelm, J. H. J. South West Africa Sci. Soc., 6: 51–74, 1933.
3. Roberts, A. The Mammals of South Africa. Cape Town, 1951.

URSIDAE

Selenarctos thibetanus Cuvier

ASIATIC BLACK BEAR

This bear probably mates in the autumn and the young, usually 2, are born in the spring (1). A January cub has been reported in the Cleveland zoo (2).

1. Lydekker, R. The Game Animals of India, Burma, Malaya and Tibet. London, 1924.
2. Reuther, R. T. JM., 42: 427–428, 1961.

Ursus americanus Pallas

BLACK BEAR

The black bear of North America mates in June or early July. Implantation is probably in November, and the young are born, very immature, late in January and early in February. The time of ovulation is not known; hence, instead of delayed implantation, this may be a case of delayed ovulation (1). Puberty is reached at about 3 years of age, gestation from the time of coitus is about 7 months, and the number of young born is 1 to 4, but usually 2 to 3 (2).

1. Hamlett, G. W. D. Quart. Rev. Biol., 10: 432–447, 1935.
2. Grinnell, J., J. S. Dixon, and J. M. Linsdale. Fur Bearing Mammals of California. Berkeley, Calif., 1937.

Ursus arctos L.

EUROPEAN BROWN BEAR

The common brown bear of the Old World reaches puberty at 6 years and remains fertile until its thirtieth year. Mating is in April–June, and the period of gestation is about 7 months (1). The young are born from December to January, depending on the locality, and it is believed that delayed implantation is the rule (2). In the Himalayas the mating season is said to be from the end of September to November, and the young, usually 2, or 1 with young females, are born in April or May (3). In the London zoo the cubs have been born from December to February, most of them in January (4). In the Augsburg zoo a pair has been observed mating at intervals of 30, 24, and 18 days. On each occasion heat lasted for 2 to 3 days (5). Another report describes a breeding season from April to June and a second minor one that does not produce pregnancies from June to August (6). Seventeen gestations were spread evenly in length from 151 to 177 days. The mean length was 164.6 days and the mean litter size was 2.3, usually 3, and range 1 to 4 (7).

1. Popoff, N. Compt. Rend. Assn. Anat., 29: 471–484, 1934.
2. Hamlett, G. W. D. Quart. Rev. Biol., 10: 432–447, 1935.
3. Blanford, W. T. Fauna of British India. Mammals. London, 1888–91.
4. Zuckerman, S. PZS., 122: 827–950, 1952–3.
5. Steinbacher, G. Säugetierk. Mitt., 6: 27–28, 1958.
6. Prell, H. Biol. Zent., 50: 257–271, 1930.
7. Lindemann, W. Säugetierk. Mitt., 2: 1–8, 1954.

Ursus horribilis Ord. The grizzly bear breeds in June or July and the young, 1 to 4, but usually 2, are born in January after a gestation that has lasted for 6½ to 7 months. Delayed implantation is probable (1). For *U. h. gyas* 8 breeding seasons were in May and June and 3 gestations were

216, 245, and 258 days long (2). The age of puberty has been given as 4 to 5 years old (3).

U. middendorfi Merriam. The testis weight is given as 32.0 g., and the length of the seminiferous tubules as 901 m. The spermatozoon was 30.7 μ long (4).

U. ornatus Cuvier. A set of cubs was born in July after a gestation of 8.5 months (5).

1. Hamlett, G. W. D. Ouart. Rev. Biol., 10: 432–447, 1935.
2. Svihla, A. The Murrelet, 30: 53–54, 1949.
3. Leopold, A. S. Wildlife of Mexico. The Game Birds and Mammals. Berkeley, Calif., 1959.
4. Knepp, T. H. Proc. Pennsylvania Acad. Sci., 13: 58–62, 1939.
5. Saporiti, E. J. Ann. Soc. Cient. Argentina, 147: 3–12, 1949.

Thalarctos maritimus Phipps

POLAR BEAR

The polar bear mates in midsummer and the young are born in midwinter. A summary of 118 births showed that they were all between November 9 and the end of December, with most from November 23 to December 15 (1). In the London zoo 10 births all took place in November and December after a gestation of about 240 days. All but 2 (singles) were twin births (2). On Southampton Island January is the usual month of birth, which is believed to occur every other year (3). In eastern Greenland the young are born in March, and in western Greenland in April (4). The number of young is usually 2, but 3 and 4 have been known (3). Delayed implantation is believed to occur (5). The weight of a testis was 31.7 g., and the length of the seminiferous tubules, 2,635 m. (6).

1. Dittrich, L. Säugetierk. Mitt., 9: 12–14, 1961.
2. Zuckerman, S. JZS., 122: 827–950, 1952–3.
3. Sutton, G. M., and W. J. Hamilton, Jr. Mem. Carnegie Mus., 12(2): 1–111, 1932.
4. Erdbrink, D. P. A Review of Fossil and Recent Bears of the Old World. Deventer, 1953.
5. Hamlett, G. W. D. Quart. Rev. Biol., 10: 432–447, 1935.
6. Knepp, T. H. Zoologica, 24: 329–332, 1939.

Helarctos malayana Raffles

SUN BEAR

In the Cleveland zoo a single cub has been born in September (1).

1. Reuther, R. T. JM., 42: 427–8, 1961.

Melursus ursinus Shaw

SLOTH BEAR

This species has no set breeding season but mates usually in the first half of the year; it has usually 2, but sometimes 1, cubs after a gestation of about 7 months (1).

1. Phillips, W. W. A. Ceylon J. Sci., 13B: 143–183, 1924–6.

PROCYONIDAE

PROCYONINAE

Bassariscus

RING-TAILED CAT

Bassariscus astutus Lichtenstein. This cat breeds in April or earlier (1). The young are born in May or June (2). The number of young is usually 3 or 4 (1).

B. sumichrasti Saussure. The cacomistle breeds in January and has a litter of 2 to 4 young (3). In Mexico parturition is usually in March (4).

1. Taylor, W. P. JM., 35: 55–63, 1954.
2. Grinnell, J., J. S. Dixon, and J. M. Linsdale. Fur Bearing Mammals of California. Berkeley, Calif., 1937.
3. Hall, E. R., and K. R. Kelson. The Mammals of North America. New York, 1959.
4. Gaumer, G. F. Mamiferos de Yucatan. Mexico City, 1917.

Procyon lotor L.

RACCOON

The male raccoon is not fertile until he is about 2 years old (1), at which time he weighs about 15 lb. (2). The females breed at 10 months, and during their first year about half of them conceive (1). Weight is not a good criterion of breeding condition in the female (2). In southern Illinois males were in full breeding condition in November (3). In southern Iowa some sign of sexual activity is apparent in the adult males in November and December, but they are not fully active until January. No juvenile males show any sign of activity in November or December (4). Mating is in early February to early March, though there may be another heat later in the season (1). In New England mating begins in the last half of January and there may be a later season for some females, as some are born in August (5). The female is receptive at 1 to 2 weeks after the onset of vulval swelling and she remains in heat for 3 days. The average gestation period is 63 days (1). The litter size is from 1 to 6, average $3.5 \pm .2$, mode of 3 and 4 (6). In Florida the season is more extended; births occur from April to early October, though half of them take place in May, and the embryo number averages $3.2 \pm .2$ (7). In the southern Iowa data, which extend over several years, placental scars gave an average of 3.7, range 1 to 8, and mode 3 and 4 (4).

At the time of heat the vulva is red and much swollen and, after heat is over, it takes 3 to 4 weeks to regain its resting appearance (1). There is usually, but not always, a trace of blood in the vaginal smear for a few days, then a white or cream-colored discharge. Ovulation is induced by coitus and only one copulation is needed to cause it (8). Large follicles are present in the ovary in February (9). The placental scars do not persist for more than a year. When they are present they are visible from the outside of the uterus (2). The ova are capable of trans-uterine migration (10).

If raccoons are exposed to an increased length of light in the fall they will breed in December and may be induced to produce two litters in the year (6).

1. Stuewer, F. W. JWM., 7: 60–73, 1943.
2. Sanderson, G. C. JWM., 14: 389–402, 1950.
3. Layne, J. N. AMN., 60: 219–254, 1958.
4. Sanderson, G. C. Quart. Biol. Rpts., Iowa, 5(2): 8–21, 1953; 7(1): 18–24, 1955.
5. Whitney, L. F. JM., 12: 29–38, 1931.

6. Bissonnette, T. H., and A. G. Csech. PRS., 122B: 246–254, 1937.
7. McKeever, S. JWM., 22: 211, 1958.
8. Whitney, L. F., and A. B. Underwood. The Raccoon. Orange, Ct.: 1952.
9. Wood, J. E. JWM., 19: 409–410, 1955.
10. Llewellyn, L. M., and R. K. Enders. JM., 35: 439, 1954.

Procyon cancrivorus Cuvier. In the London zoo this raccoon has produced litters of 2 and 3 young in May (1).

1. Zuckerman, S. PZS., 122: 827–950, 1952–3.

Nasua narica L.

COATI

This South American species probably breeds in July. The young have been recorded in February, June (1), and July (2). In the London zoo young, usually 2, have been born in April, May, June, August, and November, so the season may be more extended (3). The gestation period is 77 days (4).

1. Enders, R. K. HCMZ., 78: 385–502, 1935.
2. Allen, J. A. AM., 22: 191–262, 1906.
3. Zuckerman, S. PZS., 122: 827–950, 1952–3.
4. Brown, C. E. JM., 17: 10–13, 1936.

Potos flavus Schreber

KINKAJOU

The young are born from May to September (1). A single embryo has been recorded for April (2), and 2 young were born in the Milwaukee zoo in September (3). Two pregnant females, each with a single embryo, have been recorded in Mexico for the month of December (4).

1. Davis, W. B., and P. W. Lukens. JM., 39: 347–367, 1958.
2. Enders, R. K. HCMZ., 78: 385–502, 1935.
3. Goodwin, G. G. AM., 87: 271–474, 1946.
4. Alvarez del Toro, M. Los Animales Silvestres de Chiapas. Tuxtla Gutierrez, Mexico, 1952.

AILURINAE

Ailurus fulgens Cuvier

PANDA

The panda gives birth to a single young in the spring (1); one record gives July (2). The gestation period lasts 90 days (3). In the London zoo three births, one a single and two sets of twins, have occurred in June (4). Another report gives 2 young, born in the spring (5).

1. Blanford, W. T. Fauna of British India. Mammals. London, 1888–91.
2. Bentham, T. Rec. Indian Mus., 2: 304, 1908–9.
3. Pocock, R. I. Fauna of British India. Mammals. London, 1939–41.
4. Zuckerman, S. PZS., 122: 827–950, 1952–3.
5. Jerdon, T. C. The Mammals of India. Roorkee, India, 1867.

Ailuropoda melanoleuca David

GIANT PANDA

This panda mates early in the spring and the young are born in January (1). Usually 1 young is born, but sometimes 2 (2).

1. Sheldon, W. G. JM., 18: 13–19, 1937.
2. Pocock, R. I. Fauna of British India. Mammals. London, 1939–41.

MUSTELIDAE

MUSTELINAE

Mustela erminea L.

STOAT, ERMINE

The European stoat or ermine comes in heat during the season after its birth, i.e., at age 1 year. The periods begin in May or June, and ovulation is spontaneous. They continue at intervals of not less than a month, but young stoats do not become pregnant until the following year. Pregnant females are found only in March and April. They come in heat and ovulate but do not become pregnant during lactation. Further ovulations and infertile cycles continue later in the year. At these times 8 to 10 follicles rupture, but the corpora lutea of these infertile cycles are much smaller than are those of pregnancy. The vulva swells during heat, but it does not become as conspicuous as that of the ferret, and the vaginal epithelium cornifies. The average number of embryos is 9, with a range of 6 to 13. The male is sexually mature at 1 year. There is a rapid enlargement of the testes to 13 times their former weight from February to March, but retrogression sets in from July onward, and no spermatozoa can be found by October (1).

Birth is in April and an observed pregnancy lasted for 10 months (2). The fertilized ova develop for 2 months and then remain latent until January or February when they implant. Birth takes place from March on (3). The corpus luteum begins to develop in February, but in some stoats that conceive early in the spring its development is not arrested and implantation is immediate (4), with an 8- or 9-week pregnancy (5). In New Zealand introductions have done well and the young are born at practically any time of the year. The litter size is usually 4 to 6, but may be as high as 9 or 10 (6). Probably this species is very sensitive to changes in day length so that its behavior differs in the various environments. The American subspecies, *M. e. cicognanii,* comes in heat in the early summer (7) and the young are born between mid-April and early May. The testes begin their growth in February or March (8). The Southampton Island subspecies, *M. e. arctica,* has its young in June (9).

1. Deanesly, R. TRS., 225B: 459–492, 1935.
2. Grigorjev, N. D. ZZ., 17: 811–814, 1938.
3. Lavrov, N. P. Trud. Cent. Lab. Biol. Ohot. Prom., 6: 124–150, 1944.
4. Watzka, M. Zeit. Anat. Entw. Ges., 114: 366–374, 1949.
5. Watzka, M. Zeit. Mikros. Anat. Forsch., 48: 359–374, 1948.
6. Wodzicki, K. A. New Zealand DSIR, Bull. 98, 1950.
7. Hamilton, Jr., W. J. The Mammals of Eastern United States. Ithaca, N.Y., 1943.
8. Knight, P. L. J. Exp. Zool., 91: 103–110, 1942.
9. Sutton, G. M., and W. J. Hamilton, Jr. Mem. Carnegie Mus., 12(2): 1–111, 1932.

Mustela frenata Lichtenstein

LARGE BROWN WEASEL

The testes enlarge in March and are active by April or May. Most matings occur in July and August; then the testes begin to decline in August and quickly regress in September (1). At the onset of heat the vulva is markedly swollen. If she is bred the female remains receptive for 3 or 4 days; otherwise she remains in heat for several weeks. Coitus-induced ovulation is suggested by this pattern. Delayed implantation is usual. For 19 litters born in captivity gestation averaged 279 days. Lactation continues for about 5 weeks and the females come in heat immediately after their litters have been weaned (2). The litter size is from 4 to 8 (3).

The second polar division of the ova takes place at 74½ hours after the beginning of the first period of copulation. Morulae are found in the extreme lower end of the oviduct at 11 days and early blastocysts in the upper uterus at 15 days. At 251 days, or 27 days before birth, activated but unimplanted blastocysts may be found. The rate of cleavage and of travel through the oviduct is definitely slower than in the ferret and probably slower than in the mink (4).

1. Wright, P. L. JM., 28: 343–352, 1947.
2. Wright, P. L. AMN., 39: 338–344, 1948.
3. Hamilton, Jr., W. J. The Mammals of Eastern United States. Ithaca, N.Y., 1943.
4. Wright, P. L. AR., 100: 593–607, 1948.

Mustela nivalis L.

EUROPEAN WEASEL

In the male weasel there are no vesiculae seminales and no obvious prostate. The testes descend to the scrotum at an early age and remain there throughout life. The male is sexually mature when 4 months old, i.e., in August, as birth is in April. By December the testes are quiescent in young weasels, but adults have no spermatozoa by the end of October, though spermatocytes are present at all times. The interstitial cells increase to twice their former diameter during the breeding season, and the accessory organs change in harmony with them (1). Since the reported periods of gestation are most variable, delayed implantation may be suspected. In Morocco it is believed to be about 100 days, after which 3 to 8, more usually 4 to 6, young are born (2). In Europe the mating time begins in March and the gestation period is probably about 7 weeks (3).

1. Hill, M. PZS., 109B: 481–512, 1939.
2. Panouse, J. B. Trav. de l'Inst. Chérifien, Ser. Zool., 5, 1957.
3. Heptner, W. G., L. G. Morosowa-Turowa, and W. I. Zalkin. Die Säugetiere in der Schutzwaldzone. Berlin, 1956.

Mustela putorius L.

STOAT, POLECAT

This common European stoat has one litter a year, in March or April, followed by mating and ovulation. Implantation is delayed and pregnancy lasts for 9 or 10 months (1). In New Zealand, where the species has been introduced, the females give birth to their young throughout the year, but mostly from October to January. They number 5 to 9, but may be as many as 12 (2). The subspecies, *M. p. eversmanni,* breeds in March, and the gestation period is said to last from 40 to 42 days (3).

1. Deanesly, R. Nature, 151: 365–366, 1943.
2. Wodzicki, K. A. New Zealand DSIR, Bull. 98, 1950.
3. Heptner, W. G., L. G. Morosowa-Turowa, and W. I. Zalkin. Die Säugetiere in der Schutzwaldzone. Berlin, 1956.

Mustela putorius furo L.

FERRET

The ferret has a limited breeding season, from March or April until August, and can have 2 or 3 litters during the year (1). Ferrets transported from England to the Southern Hemisphere bred in October, immediately upon their arrival in Pretoria. In Kenya, almost on the Equator, they experienced heat at the normal time for Great Britain. Afterwards some came in heat twice in the year, others at any time (2). Much work has been done upon the effects of varying the length and intensity of light in this species, and it is summarized later in the section on physiology. In the absence of the male the female may remain in heat for a long time, e.g., from April to August.

Ovulation is induced by coitus, and it occurs about 30 hours afterward (3). Pseudopregnancy lasts for 5½ to 6 weeks. It is followed by a new heat if it terminates early enough in the season. There is no heat during lactation, but, if the young are removed, heat ensues 7 to 10 days later (4, 5). Proestrum is somewhat slow in developing at the beginning of the season. It is characterized by a marked swelling of the vulva, which takes about 2 to 3 weeks to reach its full development. This development is maintained throughout heat, and the vulva at this time is about 50 times the anestrous size. It declines gradually after coitus (4).

The average duration of coitus is about 2 hours (range ¼ to 3 hours). For 36 hours afterwards the vulva remains turgid but it becomes flaccid by about 60 hours. The ovum remains capable of fertilization for not more than 30 hours; from 18 to 30 hours after ovulation only a small proportion are capable of being fertilized, and small litters are produced from matings made at this time (3).

The mean litter size is $8.5 \pm .3$, with a range from 5 to 13, and the duration of gestation is 42 days from mating with practically no variation (4).

HISTOLOGY OF THE FEMALE TRACT

OVARY. During anestrum the ovary contains only small follicles, but during proestrum these grow at a rate corresponding to that of the vulva. The maximum size of the ripe follicle is about 1.4 mm. The fully formed corpus luteum measures about 2 mm., and during pregnancy or pseudopregnancy the ovaries consist almost entirely of luteal tissue. The corpora lutea persist throughout pregnancy and degenerate rapidly at its end. There is no corpus luteum of lactation. The ovary contains much interstitial tissue, but it is not so abundant as it is in the rabbit (4).

VAGINA. The vaginal epithelium just after the end of the breeding season consists of 2 to 3 layers of low-columnar epithelial cells. Just before the beginning of the season the epithelium grows and cornifies to some extent, while the connective tissue is compact. In proestrum several squamous layers of epithelium develop with some cornification. This is at the time that the vulval swelling begins. In heat the epithelium is thicker, and there is much cornification. The connective tissue is very spongy, and it has many elastic fibers embedded in it. In pseudopregnancy the epithelium is stratified high-columnar. Few cornified cells remain, and the surface is rough. The connective tissue remains spongy. At the end of pseudopregnancy the appearance is similar except that many lymphocytes are present (6). When cornification sets in, it begins in the upper part of the vagina and progresses toward the vulva. Growth is most active during early heat, and there is no sudden appearance of leucocytes at its end (7). There is no true cornified stage in the vaginal smear (5).

UTERUS. In anestrum the endometrium is thin, with compact, undeveloped glands, while the epithelium is low-cubical. In proestrum cell divisions are abundant. Congestion is marked, and there may be some extravasation of blood, which passes through lacunae in the epithelial layer into the lumen. There is some suggestion, however, that this occurs only in the heat period soon after parturition. During heat the glands have large lumens and are secreting. The glandular cells are large, and those of the epithelium are tall-columnar. The mucosal layer is somewhat folded. In pseudopregnancy there is a great development of the glands with the result that the endometrium becomes lacelike, resembling the condition in the rabbit. At the height of this condition giant cells tend to form in the crypts of the glands. Degeneration, indicated by the sloughing off of cells and accumulation of debris, begins at about 5½ weeks of pseudopregnancy (1, 4).

PHYSIOLOGY OF THE FEMALE TRACT

If ferrets receive illumination during the evenings beginning October 12, most females come in heat in December, 3 months earlier than usual, and they may become pregnant (8). This response is limited to a band in the spectrum extending from red λ 6,500 to λ 3,650 in the near ultraviolet. Within this range intensity is more important than the wave length. Females subjected to incomplete darkness did not come in heat at the proper time unless proestrum had already begun, in which case the usual changes followed (9). The degree of acceleration is related to the intensity of light (10), and it is not produced if the optic nerves are cut (11). However, intensity or duration of light is not wholly the explanation of emergence from anestrum, since females kept in darkness for 23½ hours each day from the end of January mostly underwent the usual vulval development but took longer than usual (12).

Ovulation may be induced in the unmated estrous ferret by the injection of a gonadotrophic preparation. The dose needed is about the same as that required to induce ovulation in the rabbit; hence, weight for weight, the ferret requires more than the rabbit (13). The considerable length of copulation in the normal course of events raises the question how soon sufficient stimulus has been given to cause the release of an adequate quantity of gonadotrophe from the pituitary. If coitus is interrupted after 15 to 20 minutes ovulation follows in 90 per cent of cases, and it has been found that sexual excitement, i.e., the presence of the male with the female in heat, without coitus, occasionally produces ovulation. If hypophysectomy is performed 1¾ hours after the start of coitus, which lasted 1¼ hours in these cases, ovulation is not usually interrupted, but if it is done within an hour of the start, following an interrupted coitus, ovulation does not occur. Sufficient gonadotrophe has not been released, therefore, within an hour of the beginning of copulation (14, 15). Hypophysectomy during anestrum causes but little further atrophy of the genitalia. The onset of new heat periods at the usual time is prevented by the operation, and increased light is without effect. Estrogens produce the usual effects upon the uterus and the vulva. The latter requires more than the uterus, which is also the case in normal anestrous ferrets (16, 17).

The injection of P.M.S. during anestrum causes the rapid development of quiescent follicles. With moderate doses the response is almost purely follicle-stimulating without ovulation or corpus-luteum development. With larger amounts thecal luteinization follows the injections (18).

Hysterectomy does not cause prolongation of the life of the corpora lutea, and it has no effect upon other reproductive phenomena (19).

The fuchsinophil acidophils in the anterior pituitary are most numerous and widely distributed in early estrus. They are scanty or even lacking from November to January (20).

If females become pregnant or pseudopregnant at the end of the usual breeding season, these conditions are maintained and last the normal length of time, despite the fact that normally the female would be in anestrum (21).

THE MALE

In October the testis and epididymis are at their lowest level. In November and December their weights begin to rise to a peak, which is reached early in March. Their weights fall again rapidly in August. The weight of the penis follows a similar curve, but the individual variability is greater. There is no spermatogenesis in October, when the tubules are small, and only Sertoli cells and spermatogonia are present. Renewed activity begins in November; spermatocytes are found in December, spermatids in January, and spermatozoa at the beginning of February. Interstitial cells are largest in February and smallest in November (22). After hypophysectomy the testes retrogress to the anestrous level within about a month (23).

Immature ferrets kept on a "short-day" light regime reach puberty later than usual; a "long day" accelerates its onset, but this is followed by an early regression (24), even though the increased amount of light be continued. The interstitial tissue of the testis is more sensitive than the spermatogenic; it is earlier to respond and later to regress (25).

1. Marshall, F. H. A. QJMS., 48: 323–345, 1904.
2. Bedford, Duke of, and F. H. A. Marshall. PRS., 130B: 396–399, 1942.
3. Hammond, J., and A. Walton. J. Exp. Biol., 11: 307–319, 1934.
4. Hammond, J., and F. H. A. Marshall. PRS., 105B: 607–629, 1930.
5. Muir, E. Zeit. Zucht., 32B: 269–290, 385–408, 1935.
6. Marshall, F. H. A. Quart. J. Exp. Physiol., 23: 131–135, 1933.
7. Hamilton, W. J., and J. H. Gould. Trans. Roy. Soc. Edinburgh, 60: 87–106, 1940.
8. Allanson, M., I. W. Rowlands, and A. S. Parkes. PRS., 115B: 410–421, 1934.
9. Marshall, F. H. A., and F. P. Bowden. J. Exp. Biol., 11: 409–422, 1934.
10. Marshall, F. H. A. J. Exp. Biol., 17: 139–146, 1940.
11. Le Gros Clark, W. E., T. McKeown, and S. Zuckerman. PRS., 126B: 449–468, 1939.
12. Hill, M., and A. S. Parkes. PRS., 115B: 14–17, 1934.
13. McPhail, M. K. PRS., 114B: 124–128, 1933.
14. Hill, M., and A. S. Parkes. PRS., 112B: 153–158, 1932.
15. McPhail, M. K. PRS., 114B: 124–128, 1933.
16. Hill, M., and A. S. Parkes. PRS., 113B: 530–544, 1933.
17. Parkes, A. S. PRS., 109B: 425–434, 1932.
18. Rowlands, I. W. J. Physiol., 92: proc. 8, 1938.
19. Deanesly, R., and A. S. Parkes. J. Physiol., 78: 80–84, 1933.
20. Holmes, R. L. Nature, 183: 1521, 1959.
21. Hammond, J., and A. Walton. J. Exp. Biol., 11: 320–325, 1934.
22. Allanson, M. J. Physiol., 71: proc. 20, 1931.
23. Hill, M., and A. S. Parkes. PRS., 112B: 146–152, 1932.
24. Bissonnette, T. H. Biol. Bull., 68: 300–313, 1935.
25. Bissonnette, T. H. J. Exp. Zool., 71: 341–373, 1935.

Mustela vison Schreber

MINK

The North American mink is apparently polyestrous in the sense that several waves of follicles ripen. The breeding season begins in March and may continue for a month or more. Ovulation is coitus-induced and implantation is sometimes slightly delayed. The gestation period is therefore somewhat variable; a range of 45 to 70 days has been reported. One litter a year is the rule.

The mink breeds in the season following its birth, at about 10 months old. In the wild the young are born in April or May (1). Among ranch mink the breeding season begins about mid-March for the Quebec strain, and a slightly shorter season begins a week later for the Alaska (Yukon) strain. The duration of mating averages 64 minutes, with a slight mode at 31 to 60 minutes. It becomes more prolonged as the season advances but its interruption does not affect fertility. There is a tendency for females to repeat matings 8 or 9 days after the first. For once-mated Quebec mink the gestation period averaged 51.3 days (S.D., 5.4), and for the Alaska strain 49.0 days (S.D., 4.2). There is a slight tendency for shorter gestations with larger litters, but the difference is not more than a day. Gestations after matings made late in the season are about 2 days shorter (2). The modal pregnancy length was 48 days, with a spread from 40 to 73 days. In this sample the mean was 49.7 ± .2 days and the litter size averaged 4.4. The time from implantation to parturition was 30 ± 1 days (3). The sex proportion at birth was 50.75 ± .47 per cent males. In stillborn mink and those that died shortly after birth it was 56.6 ± 3.5 per cent males (4).

Ovulation takes place 42 to 50 hours after mating. The average number of ova shed was 4.9 per ovary, but it increased as the season progressed (3). A mating lasting 15 to 29 minutes is sufficient to produce ovulation. The spermatozoa take about 15 minutes to traverse the uterus, and 30 the oviduct (5). If coitus is not allowed, 4 or more cycles of overlapping degenerating and developing follicles may be found (3). A sterile copulation is followed by pseudopregnancy (6). Of each 100 ova shed 83.7 implant and 50.2 produce cubs (2).

HISTOLOGY OF THE FEMALE TRACT

OVARY. The diameter of the ripe follicle is 1.1 mm., and of the ovum, 100 μ. The egg is full-sized in follicles of 0.5 mm. and up. It is difficult to ascertain by gross examination of the ovaries whether ovulation has occurred. At 14 to 17 hours after mating capillaries in the theca interna rupture; by 30 hours the granulosa cells are proliferating. The stigma forms about 36½ hours after mating (2). A small papilla, Corinth red in color, pushes out from the avascular area while the rest of the follicle appears to be darker. After the follicle has ruptured the stigma becomes "brightish" red. The follicular fluid is very viscous and the corona cells persist round the ovum unless it has been fertilized (3). Passage of the ova through the first half of the oviduct is rapid, but they remain in it for about 6 days (2).

At first the corpus luteum measures 0.5 to 0.9 mm. in diameter, reaching 1.3 mm. by the eleventh day. In early ovulators growth is slower than it is in later ones. At implantation, about day 23 to 25, the diameter is 1.4 to 1.8 mm. It continues to grow throughout pregnancy and early lactation. Development in pseudopregnancy resembles that observed in pregnancy. Unruptured follicles may become typical corpora lutea except for the enclosed ovum. They are functional, as they are able to maintain pregnancy in the absence of sufficient normal corpora lutea (3). Interstitial tissue is abundant in the ovary (2). Before implantation the corpora lutea enter an inactive phase in which the cells are small, with dense and strongly basophilic nuclei.

VAGINA. In proestrum the vaginal smear shows an increase in squamous cells and cornification begins. Round cells increase up to the day before copulation and are present in the ratio of 10 round cells to 1 cornified cell. At estrus the round cells disappear and the cornified cells increase. After copulation there is a massive desquamation and leucocytes appear in the smear. Mucus is heavy and stringy in late February and early March; the females are reluctant to mate while it is in this condition. Later in March and early in April it becomes abundant and more fluid. The cornified stage coincides with large follicles in the ovary but, as these develop even when corpora lutea have formed, the smear is not clear-cut. At the time of heat the vulva increases in size but to a smaller degree than in the ferret and marten (3).

UTERUS. As the breeding season approaches, the uterus elongates and its walls thicken. All the embryos implant in the horns and trans-uterine migration occurs. The circular muscle is not well developed (3).

OVIDUCT. The ovary is completely surrounded by a bursa except for a small slit, which is blocked by the fimbria so that fluid injected into the bursa does not escape. During or following coitus the bursa fills slowly with fluid. Immediately after a prolonged copulation about 5 ml. of fluid may be present. At ovulation the ovary floats in the fluid. The distention continues while the eggs are within the oviduct. The ova are fertilized within the fimbriated end, and not beyond the first loop of the oviduct. This organ enters the uterus somewhat obliquely at right angles to the cornu. A fold projects into a pocket formed by the uterine wall. Fluid may be forced from the uterus to the oviduct or in the reverse direction at any time in the cycle with very little pressure (3).

PHYSIOLOGY OF THE FEMALE TRACT

If the mink is remated after an interval of less than 5 days no new ovulation occurs, but a mating more than 5 days after the previous one is followed by a new ovulation. Apparently the new mating causes expulsion of the blastocysts that resulted from the first one (2). One corpus luteum is insufficient to maintain pregnancy with a single kit. Ovariectomy causes resorption of the kits or abortion. If one ovary is removed, compensatory hypertrophy of the remaining one follows (3). The females remain fecund until age 7 years or older (3).

THE MALE

The volume of ejaculate is small. The spermatozoa remain fertile for at least 48 hours in the female tract. They reach the utero-tubal junction within 15 minutes of the beginning of copulation (3).

1. Bailey, V. NAF., 55, 1936.
2. Hansson, A. Acta Zool., 28: 1–136, 1947.
3. Enders, R. K. Proc. Am. Philos. Soc., 96: 691–755, 1952.
4. Venge, O. Acta Zool., 34: 293–302, 1953.
5. Johansson, I. Våra Pälsdjur., 3: 42–45, 1954.
6. Anon. Am. Fur Breeder, 14(4): 6–14, 1941.

Mustela altaica Pallas. The alpine weasel probably has a second season, for one was found rearing young in September (1). The mating season is usually February and from 1 to 5 young are born in May (2).

M. flavigula Boddaert. A female suckling 2 young was found in April (3).

M. kathiac Hodgson. The yellow-bellied weasel of the Himalayas has been recorded with 2 young born in midwinter (1).

M. lutreola L. This European mink begins to breed in February. From 2 to 7 young are born after a gestation of 6 to 8 weeks (4).

M. rixosa Bangs. The least weasel has been recorded with young, numbering from 3 to 6, in October, January, and February (5). But there is a record of a female with 12 fetal young in June (6).

M. sibirica Pallas. This weasel mates early in the spring and has 4 to 10 young after a gestation of about 30 days (4). In India it breeds in the summer (1).

1. Pocock, R. I. Fauna of British India. Mammals. London, 1939–41.
2. Blanford, W. T. Fauna of British India. Mammals. London, 1888–91.
3. Banks, E. J. Malay Br., Roy. Asiatic Soc., 9(2): 1–139, 1931.
4. Heptner, W. G., L. G. Morosowa-Turowa, and W. I. Zalkin. Die Säugetiere in der Schutzwaldzone. Berlin, 1956.
5. Hamilton, Jr., W. J. The Mammals of Eastern United States. Ithaca, N.Y., 1943.
6. Allen, J. A. AM., 19: 101–184, 1903.

Vormela peregusna Güldenstaedt

MARBLED POLECAT

This species mates in March, and 4 to 8 young are born after a gestation of about 8 weeks (1).

1. Heptner, W. G., L. G. Morosowa-Turowa, and W. I. Zalkin. Die Säugetiere in der Schutzwaldzone. Berlin, 1956.

Martes americana Turton

PINE MARTEN

The pine marten of North America mates in late July and August (1). A single heat period may last 2 weeks (2). At heat the vulva swells and darkens; the swelling is maintained throughout the breeding season and does not regress after an early copulation (3). The litter size is 3 to 5, and the length of gestation, according to some workers, is 220 to 230 days, the young being born in March and April (4). Others have reported 245 to 265 days (5) so there is apparently considerable variation. The long gestation is caused by delayed implantation of the embryos, and it may be shortened by 3 months if the martens are exposed to increased daily length of light throughout the fall (1).

1. Pearson, O. P., and R. K. Enders. J. Exp. Zool., 95: 21–35, 1944.
2. Cocks, A. H. PZS., 836–837, 1900.
3. Enders, R. K., and J. B. Leekley. AR., 79: 1–5, 1941.
4. Brassard, J. A., and R. Bernard. Canadian Field. Nat., 53: 15–21, 1939.
5. Salversen, S. Deutsche Pelztierzucht, 12: 371, 1937.

Martes foina Erxleben

STONE MARTEN, BEECH MARTEN

In England the stone marten mates in July and August, and the gestation period is 8½ to 9½ months (1). In India mating is said to occur in February, and the gestation period is 9 weeks. The litter size is 4 to 5 (2). If the Indian record is correct, it would suggest that delayed implantation is the rule in the northern part of the range and not in the southern. In view of the fact that increased daylight in the fall shortens the gestation period in *M. americana* this would not be surprising. The females are believed to be in heat in January and February and, again, in July and August. Pregnancy occurs only after the later heat. Spermatogenesis and testicular activity are minimal in January and are maximal in June (3).

1. Fairfoul, D. British Fur Farmer, Jan., 84–85, 1934.

2. Blanford, W. T. Fauna of British India. Mammals. London, 1888–91.
3. Ehrlich, I. RSZ., 56: 621–626, 1949.

Martes martes L.

EUROPEAN PINE MARTEN

The European pine marten is said to be seasonally polyestrous with mating in July. Heat lasts 3 to 4 days and recurs 3 to 4 times at intervals of 3 to 7 days. The duration of pregnancy is 270 to 285 days, and the litter size 1 to 4 (1).

1. Schmidt, F. Zeit. f. Säugetierk., 9: 392–403, 1934.

Martes pennanti Erxleben

FISHER

The fisher, a native of North America, reaches puberty in both male and female at 1 year of age, but conception does not occur until it is 2 years old (1). Mating is in April, and gestation lasts 51 weeks. The females mate again a week after whelping. The litter size is 1 to 5, with an average of 3 (2). In 15 cases the range of gestation was 338 to 358 days, with a mean of 352 days (3). The mean corpus-luteum count is 2.7, mode 3. These are strongly colored, orange-yellow bodies with a mean diameter of 1.25 mm. In December their cells are highly vacuolated; this is at the time the blastocyst is free. Implantation probably occurs in February. Puberty in the female is at 1 year old (4).

1. Douglas, W. O. Am. Fur Breeder, 16: 18, 20, 1943.
2. James, C. S. Am. Fur Breeder, 13(8): 14–15, 1941.
3. Hall, E. R. California Fish and Game, 28: 143–147, 1942.
4. Eadie, W. R., and W. J. Hamilton, Jr. New York Fish and Game J., 5(1): 77–83, 1958.

Martes zibellina L.

SIBERIAN SABLE

The Siberian sable has a mating season from mid-June to the beginning of August (1). Heat lasts 2 days, and, during the season, it recurs 3 to 4 times at 9- to 12-day intervals. At the time of heat the vulva reddens. The testes swell enormously during the rutting season. Mating is lengthy, lasting on the average 50 minutes, with a range from 10 to 150 minutes. The duration of pregnancy is 270 to 285 days, and the litter size is 1 to 4 (2). Delayed implantation is the rule, with gestation varying from 249 to 299 days, usually 270 to 275 days (1). Sexual maturity is reached at 2 years (3), and breeding continues until 15 years, but the litter size decreases from 10 years onward (4).

Reduction of light to 7 hours daily in mated sables from the end of August until October, and then a gradual increase to 15 hours daily, caused earlier birth than usual. Sixteen of 20 sables littered before the end of March, while the controls did not give birth until April and May. Two of the experimental females came in heat in April but were not then mated. They again came in heat in June and July (5).

1. Zitkov, B. M. ZZ., 21: 245–250, 1942.
2. Schmidt, F. Zeit. f. Säugetierk., 9: 392–403, 1934.
3. Ponomarev, A. L. ZZ., 17: 482–504, 1938.
4. Starkov, I. D. Karakulevodstvo i Zverovodstvo, 6: 58–62, 1949.
5. Beljaev, D. K., N. S. Pereljdik, and N. T. Portnova. Zobsc. Biol., 12: 260–265, 1951.

Galera barbara L.

TAYRA

In the London zoo this species has given birth to 2 young in September (1). The usual number is believed to be 3 or 4 (2). In Mexico 2 young are usual, born in February (3), but there is a record of young, probably 3 or 4 days old, in November (4).

1. Zuckerman, S. PZS., 122: 827–950, 1952–3.
2. Cabrera, A., and J. Yepes. Historia Natural Ediar. Mamiferos Sud-Americanos. Buenos Aires, 1940.

3. Gaumer, G. F. Mamiferos de Yucatan. Mexico City, 1917.
4. Allen, J. A. AM., 22: 191–262, 1906.

Grison canaster Nelson

GRISON

One record gives litters of 2 born in March (1), and another records newborn young in August and September (2).

1. Gaumer, G. F. Mamiferos de Yucatan. Mexico City, 1917.
2. Dalquest, W. W., and J. H. Roberts. AMN., 46: 359–366, 1951.

Zorilla striatus Perry

AFRICAN POLECAT

This polecat has 2 to 3 young, born from January to March in the Giza zoo (1, 2).

1. Shortridge, G. C. The Mammals of South West Africa. London, 1934.
2. Flower, S. PZS., 369–450, 1932.

Poecilictis libyca Hemprich and Ehrenberg

A captured female had 3 young in June. The gestation may have been for 37 days or less, or more than 11 weeks (1).

1. Pretter, F. Mammalia, 23: 378–380, 1959.

Poecilogale albinucha Gray

This species usually has 2 young to the litter (1).

1. Hill, J. E. JM., 22: 81–85, 1941.

Gulo gulo L.

WOLVERINE, GLUTTON

The wolverine breeds sometime between April and October, probably about midsummer, but implantation is delayed until January. The young, numbering 2 to 4, are born late in March or early in April. The female does not become pregnant during her first summer. Spermatogenesis occurs from February on, and spermatozoa may be found in the epididymis in April and May. None could be found from November to January (1). In the Archangel region of Russia mating has been observed late in July and on into September. Young have been found from February on. The females reproduce in their third year of life (2).

The ovary is enclosed in a bursa with an ostium. No corpora lutea could be found in immature females in the fall or winter. In mature ones 1 to 3 are present and they measure from 1.65 to 3.6 mm. The corpora lutea of pregnancy measure 3.4 to 3.6 mm. There is no post-partum heat. Ova migrate from one uterine horn to the other (1).

1. Wright, P. L., and R. Rausch. JM., 36: 346–355, 1955.
2. Parovshchikov, V. J. ZZ., 39: 1111, 1960.

MELLIVORINAE

Mellivora capensis Schreber

HONEY BADGER, RATEL

This African species has 2 litters of 2 to 4 young in the year (1). The gestation period is about 6 months (2).

1. Verheyen, R. Inst. Parcs Nat. du Congo Belge, Upemba, 1951.
2. Wilhelm, J. H. J. South West Africa Sci. Soc., 6: 51–74, 1933.

MELINAE

Meles meles L.

EUROPEAN BADGER

The European badger breeds late in July and early in August. Implantation does not occur until January, and in the interval the blastocyst lies free in the uterus. Birth is in March (1, 2). Mature and recently ruptured follicles are present in the ovaries from February to July and the average number of blastocysts is 2.3, range 1 to 3, while the corpora lutea average 4.0, range 3 to 6. In September and October new follicles ripen and rupture. At this time degenerating follicles may be observed in the ovaries. It is believed, also, that some follicles may rupture in June, when the vaginal epithelium is stratified and cornified. The number of embryos in January and February was 3.1 while the number of young born was 2.2 $\pm$ .1, range 1 to 4, with a pronounced mode of 2. All this follicular activity with formation of accessory corpora lutea occurs before implantation. The corpora lutea of pregnancy are twice as large and better vascularized than are those of "delay." Some involution takes place at mid-pregnancy but they persist until it ends (3).

In another account free blastocysts have been recorded for all months except August and September. Birth is in February, and in 80 per cent of females it is followed by a post-partum estrus. The blastocysts remain free until December, when they implant. The corpora lutea were small and the cells appeared to be inactive until that time (4).

The females reach puberty at age 12 to 15 months (5), or at 2 years (4). The males possibly do not reach puberty until 2 years (5). The ovaries show little change until the females are nearly a year old. A few anovular follicles may be found. Ripe follicles measure about 3 mm., and the corpora lutea 3.0 to 3.5 mm. Apparently ovulations continue until June and then recur in September or October with implantation in November, December, or January and births from late January to March. The length of lactation does not appear to affect the time of implantation, nor does ovariectomy at 25 weeks (5). Bilateral ovariectomy did not cause death of the embryos, and injection of ovarian hormones did not assist implantation (4).

During the period of delayed implantation the uterine glands are straight and the lumen is not obvious. There is no evidence of secretion. The endometrial stroma is cellular and uniformly dense, without any edema and with little fibrous tissue. The blood vessels are not conspicuous and are not engorged with blood. The epithelium cells are tall and columnar, measuring 25 to 45 μ. This lack of activity on the part of the uterine glands is regarded as indicating that no progesterone is being secreted. The true gestation period is 6 weeks (5).

At the time of heat the vaginal wall consists of stratified squamous epithelium which sloughs off after ovulation. The blood vessels are enlarged and engorged (5).

The testes are active from late January to July, at least. They decline in late September (5).

1. Fischer, E. Verh. Anat. Ges., Anat. Anz., 71: 22–34, 1931.
2. Fries, S. Zool. Anz., 3: 486–492, 1880.
3. Harrison, R. J., and E. G. Neal. Nature, 177: 977–979, 1956.
4. Canivenc, R. Ann. Endocrinol., 18: 716–736, 1957.
5. Neal, E. G., and R. J. Harrison. TZS., 29: 67–130, 1958.

Taxidea taxus Schreber

AMERICAN BADGER

The American badger mates late in summer, from August to September; implantation does not occur until about February 15, and the young are born about April 1 (1). In the west birth usually occurs in February at sea level, but not until April or May in the mountains. The litter size is 1 to 4 or 5 (2).

1. Hamlett, G. W. D. AR., 53: 283–303, 1932.
2. Grinnell, J., J. S. Dixon, and J. M. Linsdale. Fur Bearing Mammals of California. Berkeley, Calif., 1937.

Helictus personata Geoffroy

FERRET-BADGER

This Asiatic species usually has 3 young, born in June (1).

1. Pocock, R. I. Fauna of British India. Mammals. London, 1939–41.

MEPHITINAE

Mephitis mephitis Schreber

SKUNK

The female skunk is first in heat at age 10 months (1), and breeding is mostly in the first two weeks of March. The young are born about the second week of May (2). In one instance the young of a captive skunk died and the female mated again. She gave birth to 3 kittens 64 days later (1). Proestrum lasts for 4 days, but the vulva is not swollen. Gestation lasts for about 62 days from the first mating. The number of young is from 4 to 7 (3).

1. Shadle, A. R. JWM., 17: 388–389, 1953.
2. Wight, H. M. JM., 12: 42–47, 1931.
3. Hamilton, Jr., W. J. The Mammals of Eastern United States. Ithaca, N.Y., 1943.

Mephitis macroura Lichtenstein. The western skunk has been recorded with 3 embryos in June (1), but other specimens with 5 have been recorded (2).

1. Davis, W. B., and P. W. Lukens. JM., 39: 347–367, 1958.
2. Bailey, V. NAF., 53, 1931.

Spilogale

SPOTTED SKUNK

Spilogale gracilis Merriam. This western American skunk usually has 4 or 5 young at a birth (1).

S. putorius L. This species mates late in the winter and the young, 4 or 5, are born early in the spring (2). One had young in May after a gestation of at least 120 days (3). In captivity a female gave birth to 7 young early in June (4).

1. Bailey, V. NAF., 55, 1936.
2. Hamilton, Jr., W. J. Mammals of Eastern United States. Ithaca, N.Y., 1943.
3. Constantine, D. G. JM., 42: 421–422, 1961.
4. Crabb, W. D. JM., 25: 213–221, 1944.

Conepatus

Conepatus chinga Molina. This skunk usually has 3 or 4 young (1).

C. mapurito Gmelin. This skunk has a gestation of 42 days (2).

C. mesoleucas Lichtenstein. On March 24 a female was found with a single fetus near birth. Two, with litters of 2, were found to be lactating in April (3).

1. Moojen, J. Bol. Mus. Nac. Rio de Janeiro, Zool., N5: 1943.
2. Kenneth, J. H. Gestation Periods. Edinburgh, 1943.
3. Bailey, V. NAF., 25, 1905.

LUTRINAE

Lutra canadensis Schreber

CANADA OTTER

This North American species reaches sexual maturity at age 2 years but the males are not successful in mating until they are 5 or 6 years old. The testes begin to descend in November and the females come in heat from December to early April. The heat period lasts for 42 to 46 days unless mating takes place. The days of maximum receptivity are about 6 days apart. The gestation period varies from 9½ to 12½ months and the cubs, from 1 to 4, usually 4, are born toothless and blind. This immature condition implies that delayed implantation is usual (1). In the wild the young are born in April in the northern part of the range and earlier towards the south (2). One may infer from the data on estrus that in this otter coitus-induced ovulation is the rule.

1. Liers, E. E. JM., 32: 1–9, 1951.
2. Bailey, V. NAF., 55, 1936.

Lutra annectens Major. This otter has 1 to 5 young, usually 2 or 3 in a litter (1).

L. lutra L. The Old World otter breeds once a year in winter, but young have been born at any season. In captivity the female experiences heats at monthly intervals throughout the year (2). In Ceylon the 2 or 3 young are born at any season (3). The gestation period has been observed as 61 days or a few more (4).

L. maculicollis Lichtenstein. This African species has 2 or 3 young after a gestation period of about 2 months (5).

1. Goodman, G. G. AM., 87: 271–474, 1946.
2. Cocks, A. H. PZS., 249–250, 1881.
3. Phillips, W. W. A. Ceylon J. Sci., 13B: 143–183, 1924–6.
4. Brown, C. E. JM., 17: 10–13, 1936.
5. Malbrant, R., and A. Maclathy. Faune de l'Equateur Africain Française. Paris, 1949.

Pteronura braziliensis Zimmermann

GIANT OTTER

In the Cologne zoo this species has been observed in heat in January and April. Heats lasted for 14 days (1).

1. Zeeller, F. Inter Zoo Year Book, 2: 81, 1960.

Amblonyx cinerea Illiger

CLAWLESS OTTER

This Asiatic species has usually 2 young at a birth. One record gave a single young born in April (1).

1. Webb-Peploe, C. G. J. Bombay Nat. Hist. Soc., 46: 629–644, 1946–7.

Aonyx capensis Schinz

CLAWLESS OTTER

This African species has 2 to 5, usually 2 or 3, young (1). The gestation period is about 63 days (2). One record gives a young "baby" seen in April (3).

1. Shortridge, G. C. The Mammals of South West Africa. London, 1934.
2. Wilhelm, J. H. J. South West Africa Sci. Soc., 6: 51–74, 1933.
3. Allen, G. M., and A. Loveridge. HCMZ., 75: 47–140, 1933–4.

Enhydris lutris L.

SEA OTTER

This otter has no fixed season, but mates most frequently in late spring (1). The single young is born after a gestation of 240 to 270 days (2). A specimen well advanced in pregnancy had a single corpus luteum which measured $5\frac{1}{2} \times 3$ mm. The ovaries were both scarred by irregular surface fissures (3). Mating is in June and July while the young are born from March to May after a gestation of 9 to 10 months. They probably reproduce every other year (4).

1. Barabash-Nikiforov, I. JM., 16: 255–261, 1935.
2. Kenneth, J. H. Gestation Periods. Edinburgh, 1943.
3. Pearson, O. P. JM., 33: 387, 1952.
4. Nikolaev, A. M. Trudy Soveshchanii Komiss. Akad. Nauk, S.S.S.R., 12: 214–217, 1961.

VIVERRIDAE

VIVERRINAE

Genetta

GENET

Genetta genetta L. In South Africa the young, 2 or 3, rarely 4, are born in October and May (1). In the Praha zoo most have been born in April or August and September. The litter size has varied from 1 to 3, mean 2.25, and the gestation has been 10 to 11 weeks long (2).

G. tigrina Schreber. This genet has 2 or 3 young in October (3) or May (1). Some young have been recorded as born probably in early January, during the dry season of the Belgian Congo (4). In the London zoo young have been born in July and August, and hybrids in June and October (5).

1. Shortridge, G. C. The Mammals of South West Africa. London, 1934.
2. Volf, J. Mammalia, 23: 168–171, 1959.
3. Wilhelm, J. H. J. South West Africa Sci. Soc., 6: 51–74, 1933.
4. Verschuren, J. EPNG., 9, 1958.
5. Zuckerman, S. PZS., 122: 827–950, 1952–3.

Viverricula

RASSE

Viverricula indica Desmarest. In Ceylon this species usually has 3 to 5 young to the litter. It breeds all the year (1).

V. rasse Horsfield. This species has its young from March to June. This record is for introductions into Madagascar (2).

1. Phillips, W. W. A. Ceylon J. Sci., 13B: 143–183, 1924–6.
2. Kaudern, W. AZ., 9(1): 1–22, 1914.

Viverra

CIVET

Viverra megaspila Blyth. This civet has from 1 to 3 young to the litter (1). *V. zibetha* L. This species has 3 to 4 young, born in May or June (2).

1. Blanford, W. T. Fauna of British India. Mammals. London, 1888–91.
2. Pocock, R. I. Fauna of British India. Mammals. London, 1939–41.

Civettictis civetta Schreber

AFRICAN CIVET

This civet of tropical Central Africa has two breeding seasons a year in Tanganyika, in March and October. The litter size is 2 to 5. More secretion can be obtained from the scent glands of the male at the time of rut than at other times (1).

1. Vandenput, R. Bull. Agric. Congo Belge, 28: 135–146, 1937.

Prionodon pardicolor Hodgson

LINSANG

This Asiatic species breeds from February to August and has 2 litters of 2 young each year (1).

1. Pocock, R. I. Fauna of British India. Mammals. London, 1939–41.

PARADOXURINAE

Nandinia binotata Reinwardt

AFRICAN PALM CIVET

This species has 2 young as a rule (1). In the London zoo a single young was born in July (2).

1. Dekeyser, P. L. Les Mammifères de l'Afrique Noire Française. n.p., 1956.
2. Zuckerman, S. PZS., 122: 827–950, 1952–3.

Arctogalidia trivirgata Gray

PALM CIVET

This Indian species has been recorded with 2 young, born early in the summer (1).

1. Pocock, R. I. Fauna of British India. Mammals. London, 1939–41.

Paradoxurus

PALM CIVET

Paradoxurus hermaphroditus Schreber. This Indian civet breeds all the year, but most of the young are born in the latter part of the year, before the onset of the northeast monsoon. The usual litter is from 3 to 4, sometimes as many as 6 (1). Suckling young may be found from September to March. Possibly 2 litters a year are produced (2).

P. jerdoni Blanford. This civet usually has 2 young (3). Suckling young may be found from September to March and there may be 2 litters a year (2). The weight of one testis was 2.4 g. and the length of seminiferous tubules was 70 meters. The spermatozoon was 42.4 μ long (4).

P. zeylonensis Schreber. The golden palm civet has 2 or 3 young, mostly born in October or November (5). Suckling young may be found from September to March, and the females may bear 2 litters a year (2).

1. Phillips, W. W. A. Ceylon J. Sci., 13B: 143–183, 1924–6.
2. Pocock, R. I. J. Bombay Nat. Hist. Soc., 36: 855–877, 1933.
3. Webb-Peploe, C. G. J. Bombay Nat. Hist. Soc., 46: 629–644, 1946–7.
4. Knepp, T. H. Proc. Pennsylvania Acad. Sci., 13: 58–62, 1939.
5. Pocock, R. I. Fauna of British India. Mammals. London, 1939–41.

Paguma

PALM CIVET

Paguma lanigera Hodgson. Litters of 2 to 4 have been found (1). In the London zoo births have been recorded in April and July (2).

P. larvata Bennett. One record gives 4 young (3), another 2 mature fetuses in October. (4).

1. Pocock, R. I. PZS., 621–622, 1911.
2. Zuckerman, S. PZS., 122: 827–950, 1952–3.
3. Blanford, W. T. Fauna of British India. Mammals. London, 1888–91.
4. Banks, E. J. Malay Br. Royal Asiatic Soc., 9(2): 1–139, 1931.

Arctictis binturong Raffles

BINTURONG

In the London zoo litters have been born in March, July, and November (1).

1. Zuckerman, S. PZS., 122: 827–950, 1952–3.

HEMIGALINAE

Hemigalus derbyianus Gray

A fetus was found in a female taken in February (1).

1. Banks, E. J. Malay Br. Roy. Asiatic Soc., 9(2): 1–139, 1931.

Cynogale bennetti Gray

Young of the otter civet, usually 2, have been found in May (1, 2).

1. Banks, E. J. Malay Br. Roy. Asiatic Soc., 9(2): 1–139, 1931.
2. Lyon, M. W. Proc. United States Nat. Mus., 34: 619–679, 1908.

Eupleres goudotii Doyère

In Madagascar males with spermatozoa were found at the end of June; young were found in November. A single young is the rule (1).

1. Kaudern, W. AZ., 9(1): 1–22, 1914.

GALIDIINAE

Galidia elegans Geoffroy

MADAGASCAR MONGOOSE

The young, 2 or 3, are born in May (1).

1. Kaudern, W. AZ., 9(1): 1–22, 1914.

HERPESTINAE

Suricata suricatta Erxleben

MIERKAT

Pregnant females have been found in November (1) and February (2). The number of young is usually 2 but may be as many as 4.

1. Shortridge, G. C. The Mammals of South West Africa. London, 1934.
2. Haagner, A. K. South African Mammals. London, 1920.

Herpestes auropunctatus Hodgson

MONGOOSE

This mongoose has been introduced into Hawaii and Puerto Rico and all the breeding records of the species are from these islands. The male reaches puberty at age 4 months, when the testes are 13 mm. long or longer. They produce spermatozoa all the year but the testes are largest in February and March (1). The first breeding season in Hawaii is from February to April, with another in May and June. Apparently 2 litters a year are produced in the wild but there is no breeding from October to January (2). In Puerto Rico the season extends from January to October, but most litters are born in March and April and in July and August (3). Yearlings are somewhat late in breeding (1). There is a suggestion of a 4-day estrus period at 20-day intervals, and possibly the females ovulate only on coitus. The estrus of the second season is post-lactation and not post-partum. There is no evidence of reproductive failure in extremely old individuals. The duration of pregnancy is possibly 7 weeks (1); it is certainly over 32 days (3). The average fetus count is 2.1, range 2 to 4 (3). Corpus-luteum counts have given 2.7, range 1 to 4, and ovulation was as frequent from the left ovary as from the right. Ova frequently crossed and implanted in the opposite horn of the uterus (1).

The ovary is almost completely encapsulated. The ripe follicle, which protrudes slightly from the ovarian surface, measures 1.4 mm. or more.

The corpus luteum is yellow, has a small stigma, and is 3 mm. in diameter late in pregnancy (1).

1. Pearson, O. P., and P. H. Baldwin. JM., 34: 436–447, 1953.
2. Baldwin, P., C. W. Schwartz, and E. R. Schwartz. JM., 33: 335–356, 1952.
3. Pimental, D. JM., 36: 62–68, 1955.

Herpestes edwardsi Geoffroy. The gray mongoose has no regular season. The female reaches puberty at age 9 months. The young, usually 2, are born after a gestation of 60 to 65 days (1, 2).

H. fuscus Waterhouse. This mongoose breeds mostly early in the year but is not restricted to this season. From 3 to 4 young are usual (1).

H. ichneumon L. This African species has 2 to 5 young (3).

H. javanicus Geoffroy. A mating in July was followed by a litter 43 days later (1).

H. smithi Gray. This species has 2 to 3 young (1).

H. urva Hodgson. The crab-eating mongoose has been known to give birth to 2 young (1).

H. viticollis Bennett. This mongoose has given birth to 3 young, according to one record (1).

H. (Myonax) cauui A. Smith. This African species has 2 to 3 young at a time (4).

H. (M.) sanguineus Rüppell. Two or 3 young is the rule in this mongoose (5, 6). One pregnant female has been reported in June (7).

1. Pocock, R. I. Fauna of British India. Mammals. London, 1939–41.
2. Frere, A. G. J. Bombay Nat. Hist. Soc., 33: 426–428, 1929.
3. Palouse, J. B. Trav. de l'Inst. Sci. Chérifien, Ser. Zool., 5, 1957.
4. Shortridge, G. C. The Mammals of South West Africa. London, 1934.
5. Cowles, R. B. JM., 17: 121–130, 1936.
6. Hill, J. E. JM., 22: 81–85, 1941.
7. Verschuren, J. EPNG., 9, 1958.

Helogale

DWARF MONGOOSE

Helogale dybowskii Pousargues. In the Belgian Congo this species gave birth to 4 young (1).

H. ivori Thomas. The 4 to 5 young are born in October and possibly in November (2).

H. parvula Sundevall. This mongoose has 2 to 3 young (3) in March (4).

H. vetula Thomas. Puberty is at about 8 months. Litters of 2 to 6 have been born in April, May, June, July, October, and December. The gestation period is believed to be about 5 weeks (5).

1. Verheyen, R. Inst. Parc Nat. du Congo Belge, Upemba, 1951.
2. Loveridge, A. J. East Africa and Uganda Nat. Hist. Soc., 17: 36–69, 1922.
3. Hamilton, J. S. Animal Life in Africa. London, 1912.
4. Shortridge, G. C. The Mammals of South West Africa. London, 1934.
5. Taylor, S., and C. S. Webb. Zoolife, 10: 70–72, 1955.

Atilax paludinosus Cuvier

WATER MONGOOSE

This African species usually has 2 young to the litter. In one instance these were born in August (1). A June pregnancy has also been recorded (2). In the London zoo a litter has been born in June (3).

1. FitzSimons, F. W. The Natural History of South Africa. London, 1919.
2. Frechkop, S. EPNU., 1944.
3. Zuckerman, S. PZS., 122: 827–950, 1952–3.

Mungos mungo Gmelin

STRIPED MONGOOSE

In South West Africa pregnant females have been found in November and December (1). In the Belgian Congo a heat period was recorded in December in a female aged 1 year. Newborn young were found in December and January (2). They have also been found in May, early June (3), and August (4). From 2 to 6 young have been recorded.

1. Shortridge, G. C. The Mammals of South West Africa. London, 1934.
2. Verschuren, J. EPNG., 9, 1958.
3. Allen, G. M., and A. Loveridge. HCMZ., 75: 47–140, 1933–4.
4. Loveridge, A. J. East Africa and Uganda Nat. Hist. Soc., 17: 39–69, 1925.

Ichneumia albicauda Cuvier

This African species has from 2 to 4 young at a time (1).

1. Jeannin, A. Les Mammiféres Sauvages du Cameroun. Paris, 1936.

Cynictis

MIERKAT

Cynictis penicillata Cuvier. This species has 1 to 4 young (1). In the London zoo they have been born from July to September (2).

C. selousi De Winton. This species has been recorded with 2 young (3).

1. Shortridge, G. C. The Mammals of South West Africa. London, 1934.
2. Zuckerman, S. PZS., 122: 827–950, 1952–3.
3. Hill, J. E. JM., 22: 81–85, 1941.

CRYPTOPROCTINAE

Cryptoprocta ferox Bennett

FOSSA

This Madagascar species usually has 2 young, and a record of a litter about 2 weeks old was made in February (1).

1. Kaudern, W. AZ., 9(1): 1–22, 1914.

HYAENIDAE

PROTELINAE

Proteles cristatus Sparrman

AARD WOLF

The African aard wolf bears its young in the southern part of its range during the months of November and December. The litter size is usually 2 to 4, but as many as 6 may be born at one time (1).

1. Shortridge, G. C. The Mammals of South West Africa. London, 1934.

HYAENINAE

Crocuta crocuta Erxleben

SPOTTED HYENA

The hyena of Africa is polyestrous all the year, with a cycle of about 14 days. There is no seasonal sexual activity in the male. The female reaches puberty at a later age than the male. She has a peniform clitoris, scrotal pouches, and perineal and anal glands, resembling those of the male. The resemblance is so great that the ancients regarded the hyena as hermaphroditic. The corpus luteum is functional throughout pregnancy and persists during lactation. The number of young is 1, rarely 2, but a larger number of eggs are shed, apparently from only one ovary (1). The duration of pregnancy is given as 110 days (1) but London zoo records give 90 days as the more usual period (2). In South Africa the young are born in winter (3).

1. Matthews, L. H. TRS., 230B: 1–78, 1939.
2. Zuckerman, S. PZS., 122: 827–950, 1952–3.
3. Shortridge, G. C. The Mammals of South West Africa. London, 1934.

Hyaena hyaena L.

HYAENA

This hyaena has from 2 to 4 young. In India they are born in the hot weather after a gestation period given as 7 months (1). An African record gives the probable duration as 3 months (2). The birth of 2 cubs to a young primipara at the end of May has been reported (3).

1. Pocock, R. I. Fauna of British India. Mammals. London, 1939–41.
2. Palouse, J. B. Trav. de l'Inst. Chérifien, Ser. Zool., 5, 1957.
3. Shulov, A. Mammalia, 22: 595–597, 1958.

Hyaena brunnea Thunberg. The brown hyaena has 2 to 4 young dropped after a gestation of 3 months (1).

1. Hamilton, S. Animal Life in Africa. London, 1912.

FELIDAE

FELINAE

Felis catus L.

DOMESTIC CAT

The cat reaches puberty usually at from 7 to 12 months old. She is, in a sense, seasonally polyestrous, since in the absence of the male several heat periods are experienced, in spring and in early fall. Ovulation is induced by coitus and rarely occurs in its absence. Sterile copulation produces a condition of pseudopregnancy which lasts for 36 days. The modal litter size is 4, and the period of gestation is about 63 days from coitus.

THE ESTROUS CYCLE

Occasionally the female reaches puberty at 5 months, although 7 to 12 months is the more usual age. Puberty is always a little later in the male.

The weights at this epoch are, for the female, 2½ kg., and for the male, 3½ kg. (1).

In northern Europe heat periods are experienced twice a year, in spring and early fall (2), but in Algiers the seasons are from December 15 to the end of January, and from July 15 to the end of August. Some females, however, may be found in heat at any time from January to July (3). In captivity the first series of matings for the year are from January to March and the second from May to June. The principal littering seasons are from mid-March to mid-May and July to August (1). In the northern United States anestrum lasts from September to January (4). Heat, in the presence of the male, lasts about 4 days, with the time of greatest acceptance on the third day. In the absence of the male it lasts 9 to 10 days (5), or more (2), and it recurs 15 days to 3 weeks later (5). Ovulation is induced by coitus, and in 100 cases without coitus it did not occur (5). It follows coitus after 24 to 30 hours (6), 27 hours (5), or 40 to 54 hours (2). Sterile copulation is followed by pseudopregnancy, which lasts for 36 days, but the corpus luteum regresses from 28 days onward (1). Another report gives about 30 days as the usual duration, but it may last 44 days (4). Mating may occur during lactation (1).

The ovum takes 95 to 103 hours from the time of shedding to pass through the oviduct into the uterus. The number of eggs shed is 4.12, with a mode at 4 and 5 (5), and the mean litter size is 3.88. The latter is correlated with the weight of the mother (7). The duration of gestation is 63 days from coitus (5), or 65 days in highly bred cats with little exercise and less in those which are free to roam (8). One record gives a spread of 58 to 71 days, mean 65 days, and a mean litter size of 4.03 young (1).

HISTOLOGY OF THE FEMALE TRACT

OVARY. During the period between coitus and ovulation there are no obvious changes in the graafian follicle in the first six hours; then the cumulus oöphoron enlarges because of intercellular accumulation of secretion. Later, the whole granulosa layer shares in this enlargement and loosening. The secretion of this secondary liquor folliculi ceases by 24 hours, and then the granulosa cells begin to appear active; they lose their pycnotic appearance and become vesicular. The first spindle in the ovum is formed at 18 hours, and the polar body is extruded at 22½ hours (6). At ovulation there is no hemorrhage or leucocytic invasion into the cavity, but the walls are deeply folded. The theca cells form the interstitial tissue of the corpus luteum; mitoses are numerous in these cells by 18 hours after ovulation, but few can be found after 7 days. The lutein cells are formed exclusively from the granulosa layers. The maximum size of the corpus luteum is reached at 10 to 16 days after coitus. During pregnancy gradual regression sets in after 20 days (9). This is said to occur after 28 days in the corpus luteum of pseudopregnancy. Follicles which do not rupture retrogress by theca atresia (1), and the interstitial tissue of the ovary originates in the thecal cells (10). There is no evidence for neogenesis of ova during adult life, of for linking the degree of atresia to events in the estrous cycle (11). Lactation causes an increase in the vascularity of the corpus luteum and disappearance of the small vacuoles in the lutein cells. This condition lasts for 28 days (12).

VAGINA. The anestrous vaginal epithelium consists of a few layers of epithelial cells. In proestrum there is rapid growth and the superficial cells become flattened. At heat the epithelium grows until it is 24 cells thick, with several layers of cornified cells. This continues for 3 days after coitus. Toward the end of the period gradual sloughing and leucocytic invasion occurs (4, 5). The anestrous smear contains nucleated epithelial cells, which become very numerous in proestrum. The smear of heat contains mostly large, nonnucleated, cornified cells with a few nucleated cells. This is said to be the smear if heat occurs in the spring. Nucleated cells are said to predominate in the smear of a fall heat. During metestrum large numbers of leucocytes appear (13, 14). The vaginal smears show that estrus may apparently occur during pregnancy and some females do undergo full mating in this period. There is no immediate post-partum heat, but cycles recommence at the fourth to the sixth week of lactation (15).

UTERUS. In the anestrous uterus the glands are straight and do not extend deeply into the mucosa. Their epithelium is 6 to 12 μ high, and their diameter, 20 to 30 μ. The surface cells are 6 to 12 μ in height. Mitoses are absent. During proestrum the surface epithelium increases to 20 to 24 μ. At heat this layer does not grow further, but the glands almost double in diameter and remain uncoiled. In the lutein phase the epithelial cells revert to their height in anestrum. The glands increase to 120 μ in diameter and the cells to 40 μ. This growth is apparent within 48 hours or less from ovulation and is maintained to from 12 to 15 days, the normal time of implantation. At this time secretion begins in the necks of the glands and proceeds toward the bases, which are actively secreting by the twenty-second day (4). The degree of coiling, also, is very marked. The great glandular growth causes the uterus to resemble that of the rabbit during its lutein phase, but the changes are slower than in that species. If ovulation has not occurred, the uterus retrogresses and there is a marked invasion of leucocytes. The secretory changes in the uterus occur about 3 days earlier in the nonpregnant cat, and, in either case, the collapse of the glands as secretion sets in causes the inner surface of the uterus to become fringed (5).

OVIDUCT. The tubal epithelium during anestrum is 12 to 15 μ high, and cilia are not prominent. In proestrum the cells measure 24 to 40 μ, and cilia are much more prominent. During heat mitoses become appreciable. In the lutein phase the height of the cells decreases, but the extrusion of protoplasm is not so prominent as it is in most other species (4). The cells lose their cilia except in the crypts (5).

The oviduct enters the uterus at its tip on the mesenteric side, then it turns and ends on a low papilla without villi or thickening of the muscle. It is impossible to force fluid from the uterus into the tube (16).

PHYSIOLOGY OF THE FEMALE TRACT

The cat differs from other species with induced ovulation as in this species it can be induced by stimulating the distal portion of the reproductive tract with a glass rod. The interval between stimulation and ovulation is 25 hours or less (3).

If the amount of illumination is increased by exposing cats to longer periods of light beginning in October, they come in heat at the end of November, much earlier than the usual time. It requires about 50 days of treatment to produce an effect (17, 18).

The placenta does not appear to be so good a barrier to hormones as it is in most species since the uterine glands of the new-born kitten are active. This activity ceases soon after birth (19). If the ovaries are removed up to 46 days after coitus, abortion results. If the operation is performed at 49 days or later, the young are carried to term (5).

The removal of one ovary causes a compensatory hypertrophy of the remaining ovary (5).

The myometrium does not make spontaneous contractions during anestrum and estrus. When a corpus luteum is present, slow and regular contractions occur at the rate of 10 to 18 every 10 minutes (20). The uterine muscle has a very low spontaneous motility during anestrum, and the response to drugs at this time is fainter or negative. The muscle always responds to pituitrin. Epinephrin is inhibitory during heat and motor in the lutein phase (4). In the cat ovariectomized for 15 days, the response given is that of relaxation (21). In the pseudopregnant cat the epinephrin reversal response is given from 48 hours after ovulation to 44 days (4).

The anterior pituitary of the mature, anestrous female contains 65 B.U. of lactogen per gram of tissue. During the breeding season this is increased to 224 B.U. (22).

The injection of 100 R.U. of estrogen daily into immature cats, 5 to 13 weeks old (340 to 810 g.), caused hypertrophy of the vaginal mucosa by 5 days and a little increase in the endometrium. By 9 to 10 days the proestrous type of endometrium had been produced, and by 14 days, the estrous type. These changes could not be hastened by increasing the daily dose to as high as 1,000 R.U. (23). Before nidation pregnancy can very easily be terminated by the injection of estrogens, afterwards a dose of 1,000 R.U. or more, given over 6 days, is needed to produce this result (5).

Heat and ovulation may be induced most effectively during anestrum by injecting F.S.H. together with a very small amount of L.H. The resulting pseudopregnancy lasts 40 to 44 days, judging by the epinephrin reversal test with uterine muscle. This is also the length of normal pseudopregnancy, according to these workers (4). Anestrous cats injected in July were brought in heat by from 22 to 84 R.U. of F.S.H., together with a trace of L.H., given over 6 days. The mean dose for 5 cats was 54 R.U. of F.S.H. Sometimes ovulation and sometimes luteinization of follicles resulted (24). In immature cats the injection of menopause urine at the rate of 20 to 25 R.U. daily causes follicle stimulation. The right ovary tends to respond more than the left, and at this dosage the endometrium is brought into the estrous state (25). In the anestrous cat F.S.H. is more effective in inducing ovulation than is either P.U. or P.M.S., but it is difficult to adjust the dose so that ovulation and pregnancy may occur (26). Gonadotrophic hormones in doses of 50 to 100 M.U. are said to be effective (27).

THE MALE

At birth the testes are fully descended. Spermatogonia begin to divide at 20 weeks of age and the testis weight rises from 40 to 500 mg. Spermatids appear at 700 mg. Spermatozoa appear at 30 to 36 weeks of age, when the testis weighs 1 g. or more. The body weight is then 2½ to 3 kg. (28). Two reports give the length of seminiferous tubules in a single testis as 25 and 27 meters respectively (29, 30).

Early castration inhibits the growth of the penis, and no penial spines are formed (31). The normal male excretes more ketones in the urine than does the female. Castration abolishes this difference, and the injection of testosterone restores it (32).

The lactogen content of the pituitary of the male during the nonbreeding season is 37 B.U. per g., about two thirds that of the anestrous female (22).

The injection daily of 100 R.U. of estrogen into immature males causes hypertrophy of the glandular epithelium of the prostate (33).

1. Scott, P. P., A. C. da Silva, and M. A. Lloyd-Jacob. In A. N. Worden and W. Lane-Petter, eds., The UFAW Handbook on the Care and Management of Laboratory Animals. London, 1957.
2. Liche, H. Nature, 143: 900, 1939.
3. Greulich, W. W. AR., 58: 217–224, 1934.
4. Foster, M. A., and F. L. Hisaw. AR., 62: 75–93, 1935.
5. Gros, G. Thesis, Algiers, 1936.
6. Dawson, A. B., and H. B. Friedgood. AR., 76: 411–430, 1940.
7. Hall, V. E., and G. N. Pierce, Jr. AR., 60: 111–124, 1934.
8. Soame, E. B. H. Fur and Feather, 96(2403): 10, 1936.
9. Dawson, A. B. AR., 79: 155–173, 1941.
10. Kingsbury, B. F. AJA., 65: 309–331, 1939.
11. Kingsbury, B. F. J. Morphol., 63: 397–419, 1938.
12. Dawson, A. B. AR., 95: 29–51, 1946.
13. Liche, H., and K. Wodzicki. Nature, 144: 245–246, 1939.
14. Scott, P. P., and M. A. Lloyd-Jacob. Studies in Fertility, 7: 123–129, 1953.
15. Scott, P. P. J. Physiol., 130: 47P–48P, 1955.
16. Andersen, D. H. AJA., 42: 255–305, 1928.
17. Dawson, A. B. E., 28: 907–910, 1941.
18. Scott, P. P., and M. A. Lloyd-Jacob. Nature, 184: 2022, 1959.
19. Courrier, R., and G. Gros. CRSB., 110: 1021–1022, 1932.
20. Handowsky, H. Acta Brevia Neerl. Physiol., 15: 1–3, 1947.
21. Robson, J. M., and H. O. Schild. J. Physiol., 92: 1–8, 1938.
22. Reece, R. P., and C. W. Turner. Missouri Agric. Exp. Sta., Res. Bull. 266, 1936.
23. Starkey, W. F., and J. H. Leathem. Proc. Pennsylvania Acad. Sci., 14: 87–90, 1940.
24. Friedgood, H. B. AJP., 126: 229–233, 1939.
25. Starkey, W. F., and J. H. Leathem. PSEBM., 41: 503–507, 1939.
26. Windle, W. F. E., 25: 365–371, 1939.
27. Lesbouyries, —, and — Berthelon. Bull. Acad. Vet. Française, 10: 126–130, 1937.
28. Scott, M. G., and P. P. Scott. J. Physiol., 136: 40P–41P, 1957.
29. Bascom, K. F., and H. L. Osterud. AR., 31: 159–169, 1925.
30. Knepp, T. H. Proc. Pennsylvania Acad. Sci., 10: 39–42, 1936.
31. Reisinger, L. Tierarztl. Rundschau, 43: 25, 1937.
32. Chamberlain, P. E., W. H. Furgason, and V. E. Hall. J. Biol. Chem., 121: 599–606, 1937.
33. Starkey, W. F., and J. H. Leathem. AR., 75: 85–89, 1939.

Felis silvestris Schreber

EUROPEAN WILD CAT

The wild cat in Scotland is apparently seasonally polyestrous, with one heat in the first half of March, a second at the end of May or the beginning of June, and another, rarely, in late fall. Anestrum then sets in until the end of February. The female reaches puberty at less than 12 months (1).

The ovaries are in open capsules. From 2 to 10 follicles rupture at a time, with 5 as the modal number. The corpora lutea are slow to retrogress, and large ones from the first heat persist when mature follicles are present at the second heat. The fertilized ova may transfer from one horn to the other across the body of the uterus (1).

The uterus and vagina are ridged longitudinally. During heat the endometrium has large glands, and at the same time there is extensive desquamation of cornified cells in the vagina (1). Gestation lasts 68 days (2).

1. Matthews, L. H. PZS., 111B: 59–77, 1941.
2. Tetley, H. PZS., 111B: 13–23, 1941.

Felis (Lynx) rufus Schreber

BOBCAT

The main breeding season is from February to April. The females are probably polyestrous, and those failing to become pregnant in early spring come in heat later in the spring or early summer. Some of the early breeders may have a second litter later in the year. Mature males are probably fecund all the year round (1). The young are mostly born in May or June (2) and they number from 1 to 6, with a mean of $2.8 \pm .3$ and modes of 2 and 4. The gestation period is about 50 days (3). An extensive series from Utah gave an observed heat towards the end of March and pregnancies from January to September, but mostly in March and April. Of 356 pregnant females the embryo count ranged from 1 to 8, mode 3, average $3.2 \pm .07$. In 47 newborn litters the range was 1 to 6, mode 4, and average 3.5 (4).

The ovaries are large for the size of the female and they are often markedly asymmetric. There are two types of corpus-luteum cells, one of large cells associated with patches of small epithelial cells, and one of large cells containing acidophil pigment and seldom associated with patches of smaller cells. The corpus luteum appears to persist for years (5).

1. Duke, K. L. AR., 120: 816–817, 1954.
2. Bailey, V. NAF., 55, 1936.
3. Hamilon, Jr., W. J. The Mammals of Eastern United States. Ithaca, N.Y., 1934.
4. Gashwiler, J. S., W. L. Robinette, and O. W. Morris. JM., 42: 76–84, 1961.
5. Duke, K. L. AR., 103: 111–131, 1949.

Felis (Puma) concolor True

PUMA, MOUNTAIN LION

In Utah and Nevada the mountain lion may have young at any time of the year, but 60 per cent are born in June through September with the peak in July. At the time of capture 41 per cent of mature females were pregnant. Most have young at 2-year intervals, a few at 12 to 15 months. Some reach puberty at 80 lb. weight (1), which may be in the second or third year (2). In November in the Chicago zoo a female had her first heat at the age of 2 years and 5 months. Heat, judging by the vocalization, lasted 8 to 11 days and there was an interval of 14.4 days between the end of one and the beginning of another, which gave a mean cycle length of 22.8 days. A male was introduced on one occasion after she had been in heat for 2 days. She remained in heat for 8 more days, and had 2 young after a gestation of between 82 and 90 days. Her young died and she came in heat again 17 days later for 8 days and was mated. This gestation was between 88 and 92 days, so 90 days may be taken as the gestation period. There was no heat for 2½ months while the cubs were with her (3). In the London zoo births have been from April to December and the litter size has varied from 1 to 3, usually 1 and averaging 1.6 (4). In the Utah-Nevada data the mean embryo count was 3.4, range 1 to 6 in 66 litters. For 258 litters born the average was 2.9, range 1 to 5 (1).

1. Robinette, W. L., J. S. Gashwiler, and O. W. Morris. JM., 42: 204–217, 1961.
2. Leopold, A. S. Wildlife of Mexico. The Game Birds and Mammals. Berkeley, Calif., 1959.
3. Rabb, G. B. JM., 40: 616–617, 1959.
4. Zuckerman, S. PZS., 122: 827–950, 1952–3.

Felis chaus Güldenstaedt. In India the jungle cat has 2 litters a year, usually 3 to 4 young (1). They breed first in the early spring and gestation lasts about 2 months (2). Young are born in May and November in south India (3).

F. libyca Forster. A female experienced estrous periods at intervals of about 6 weeks from September to May. No observations were made in the intervening period. Heat lasted for 2 or 3 days (4). Mating is usually at the end of February and the young are 2 to 5 in number (2). They are born after a gestation of about 56 days (5).

F. margarita Loche. There are records of 2 and 4 young born early in April (6, 7).

F. (*Herpailurus*) *geoffroyi* D'Orbigny. This South American species of jaguar has 2 or 3 young once a year (8).

F. (*H.*) *yaguarondi* Desmarest. This jaguar breeds towards the end of the year and has 2 to 3 young after a gestation of 9 weeks (8).

F. (*Ictailurus*) *planiceps* Vigors and Horsfield. A single young was born to a female in January (9).

F. (*Leopardus*) *pardalis* L. The ocelot mates from the end of the year into January. Two or 3 young is the usual number (8). In Texas breeding probably takes place all year (10). In the London zoo young have been born in April, June, and December (11).

F. (*Leptailurus*) *serval* Schreber. The serval has 2 to 4 young (12). New-born cubs have been recorded for September and November (4), and a pregnant female in May (1).

F. (*L.*) *caracal* Schreber. The caracal has litters of 2 to 4 young all year (13). In South West Africa most are born in July and August (14). In the London zoo litters have been born in January, March, and June (11). A record for northern Africa gives 3 to 4 young born early in April (15).

F. (*Lynx*) *canadensis* Kerr. This bobcat mates in late winter and has 1 to 4 young after a gestation of about 2 months (16).

F. (*L.*) *lynx* L. This lynx mates from the end of January to April and has 2 to 3 young, occasionally more, after a gestation of 9 to 10 weeks (2).

F. (*Microfelis*) *nigripes* Burchell. This African cat usually has 2 or 3 young to the litter (17).

F. (*Noctifelis*) *tigrina* Erxleben. The margay usually has 2 young (17).

F. (*Prionailurus*) *bengalensis* Kerr. The leopard cat of India mates in May and has 3 to 4 young (1) after a gestation of 56 days (18).

F. (*P.*) *rubiginosa* Geoffroy. This Asiatic cat has 2 or 3 young to the litter (19).

1. Blanford, W. T. Fauna of British India. Mammals. London, 1888–91.
2. Heptner, W. G., L. G. Morosowa-Turowa, and W. I. Zalkin. Die Säugetiere in der Schutzwaldzone. Berlin, 1956.
3. Webb-Peploe, C. G. J. Bombay Nat. Hist. Soc., 46: 629–644, 1946–7.
4. Verschuren, J. EPNG., 1958.
5. Wilhelm, J. H. J. South West Africa Sci. Soc., 6: 51–74, 1933.
6. Dementier, G. P. Mammalia, 20: 217–222, 1956.
7. Sapozhenkov, Y. F. ZZ., 40: 1086–1089, 1961.
8. Cabrera, A., and J. Yepes. Historia Natural Ediar. Mamiferos Sud-Americanos. Buenos Aires, 1940.
9. Banks, E. J. Malay Br. Roy. Asiatic Soc., 9(2): 1–139, 1931.
10. Petrides, G. A., B. O. Thomas, and R. B. Davis. JM., 32: 116, 1951.
11. Zuckerman, S. PZS., 122: 827–950, 1952–3.
12. Dekeyser, P. L. Les Mammifères de l'Afrique Noire Française. n.p., 1956.
13. Panouse, J. B. Trav. de L'Inst. Chérifien, Ser. Zool., 5, 1957.
14. Shortridge, G. C. The Mammals of South West Africa. London, 1934.
15. Sapozhenkov, Y. F. ZZ., 41: 1111–1112, 1962.
16. Hamilton, Jr., W. J. The Mammals of Eastern United States. Ithaca, N.Y., 1943.
17. Elliott, D. G. A Monograph of the Felidae or Family of the Cats. London, 1883.
18. Kenneth, J. H. Gestation Periods. Edinburgh, 1943.
19. Phillips, W. W. A. Ceylon J. Sci., 13B: 143–183, 1924–6.

Panthera pardus L.

LEOPARD

The leopard has 2 to 4 young, and no fixed season in Africa (1). In India there is no fixed season but most births occur in April (2). In the London zoo births have been distributed throughout the year and the number of young has been from 1 to 3, usually 1, mean 1.75 (3). In Northern Rhodesia births are probably all year and the usual number is 2 (4). The gestation period lasts about 3 months (5).

1. Shortridge, G. C. The Mammals of South West Africa. London, 1934.
2. Brander, A. A. D. Wild Animals in Central India. London, 1923.
3. Zuckerman, S. PZS., 122: 827–950, 1952–3.
4. Ansell, W. F. H. PZS., 134: 251–274, 1960.
5. Blanford, W. T. Fauna of British India. Mammals. London, 1888–91.

Panthera (Leo) leo L.

LION

The lion has no fixed breeding season and is polyestrous. Heat may last for a week, and it recurs at intervals of 3 weeks. Two litters may be born in a year. The gestation period is about 108 days (1). Observed gestations have varied from 105 to 113 days. Up to 6 young may be born at a time, but 2 to 3 is the usual number (2). A compilation of 323 births in the Pretoria and other zoos gave a mode of 3 and a mean of 3.04 ± .08 young per litter. In general, males breed at 5 to 6 years old, and females at 3 or 4. The fifteenth year is the usual breeding limit (2).

The anestrous vaginal smear consists of nucleated epithelial cells of various types. During heat it shows mostly large, flat, nonnucleated, cornified cells with a smaller number of cells whose nuclei are pycnotic (3).

1. Marshall, F. H. A. The Physiology of Reproduction. London, 1922.
2. Steyn, T. J. Fauna and Flora (Pretoria), No. 2, 1951.
3. Liche, H., and K. Wodzicki. Nature, 144: 245–246, 1939.

Panthera (Jaguarius) onca True

JAGUAR

In Yucatan this species mates in August and September and the young are born from November to December (1), but in Sinaloa the young are born in July and August (2). In the London zoo they have been born at all times of the year. Usually one is the number, but twins occurred twice in 6 births (3). The gestation period is about 100 days (4).

1. Gaumer, G. F. Mamiferos de Yucatan. Mexico City, 1917.
2. Allen, J. A. AM., 22: 191–262, 1906.
3. Zuckerman, S. PZS., 122: 827–950, 1952–3.
4. Brown, C. E. JM., 17: 10–13, 1936.

Panthera (Tigris) tigris L. The tiger breeds all the year and has 2 to 5 young to the litter (1). In the London zoo, while young are born at any time, most are born from June to August. The modal litter size is 3, and the mean 2.3 (2). The gestation period is about 113 days (3).

P. (Uncia) uncia Schreber. The snow leopard or ounce has a gestation of 93 days (4).

1. Blanford, W. T. Fauna of British India. Mammals. London, 1888–91.
2. Zuckerman, S. PZS., 122: 827–950, 1952–3.
3. Davis, M. JM., 27: 393, 1946.
4. Kenneth, J. H. Gestation Periods. Edinburgh, 1943.

Acinonyx jubatus Schreber

AFRICAN CHEETAH

In the Transvaal this species gives birth to from 2 to 4 cubs in the last half of the year (1). Litters have also been found in March and November (2), while in the Krefeld zoo a litter of 4 cubs was born in April (3). The gestation period is about 95 days (4).

1. Hamilton, J. S. Animal Life in Africa. London, 1912.
2. Ansell, W. F. H. PZS., 134: 251–274, 1960.
3. Encke, W. Inter Zoo Year Book, 2: 85–86, 1960.
4. Kenneth, J. H. Gestation Periods. Edinburgh, 1943.

OTARIIDAE

Arctocephalus pusillus Schreber

CAPE FUR SEAL

This seal breeds at age 1 year. There is a heat period from August to November for the virgin cows, who then have a longer gestation than usual, of 14 to 15 months duration. The period of birth is from November into early December and the females mate within 6 days after calving. This gives a gestation of almost 12 months. Ovulation takes place spontaneously once a year. The right ovary ovulates 61 per cent of the time and the left, 39 per cent. Only 3 of 62 pregnancies were twins. The ovum takes 6 days to traverse the oviduct, and no trans-uterine migration has been observed. Delayed implantation is the rule, but there is slow and continuous growth of the blastocyst. In the pregnant female large follicles are present in the ovaries at 3 months (1).

The maximum size of the corpus luteum cells is 70 μ. This body retrogresses quickly, and the corpus albicans has disappeared by 25 months. At the time of heat the outer layer of the vaginal wall is sloughed off; an invasion of leucocytes follows. At implantation there is another influx of leucocytes and some apparent sloughing. At estrus the uterine glands are active (1). Suckling lasts for 10 months (2).

1. Rand, R. W. PZS., 124: 717–740, 1954–5.
2. Rand, R. W. Union of South Africa Dept. Commerce, Indust. Div., Fish Invest. Rpts., 34: 1–74, 1959.

Arctocephalus australis Zimmermann. On the Atlantic coast the pups, usually 1, are born from December to January. On the coast of Chile the time of birth is November and December. The gestation period is about 11 months (1).

A. forsteri Lesson. The New Zealand fur seal breeds at age 3 years and the young, usually 1, are born from November 10 to December 19 (2).

1. Cabrera, A. and J. Yepes. Historia Natural Ediar. Mamiferos Sud-Americanos. Buenos Aires, 1940.
2. Lucas, A. H. S., and W. H. Le Souef. The Animals of Australia. Melbourne, 1909.

Callorhinus ursinus L.

NORTHERN FUR SEAL

Females of this seal of Arctic waters begin to mate successfully at age 3 years but some not until 6 years (1). The earliest breeding age is 3 years. At 4 years 11 per cent are pregnant. The peak, reached at 7 years, is 78 per cent and then the rate decreases. The pregnancy rate of 4-year-olds and upwards is 60 to 70 per cent (2). The males breed at 4 years (3). The peak of the pupping season is in the second week of July, but the whole season spreads over 6 weeks or more. Heat follows parturition by 4 to 7 days; it lasts no more than a day and there is only one heat period. At this time the vulva is swollen and protruding while the vestibular mucosa and vaginal orifice are bright pink (4). The number of young is usually 1, and females approaching their first ovulation have large follicles in one ovary only. In the adults ovulation and implantation occur on the side that was inactive the previous year (5). Seven per cent of females are pregnant at the end of their third year; after the fourth year 57 per cent are pregnant, and at above 5 years old 79 per cent carry pups. The sex proportion of 1,172 fetuses examined in April and May was 52.7 per cent males (6). Another figure for fetuses from January to July gave 50.9 per cent males (7). The loss between ovulation and parturition is about 24 per cent. Delayed implantation is the rule, and the blastocyst is free in the uterus for 3.5 to 4.5 months. The corpus luteum persists for 21 months to 2 years (1).

The males produce spermatozoa at 3 to 4 years and may be regarded as sexually mature at 5 or 6 years, but they do not assemble a harem until later in life (6). The breeding males go without food and water for 2 months or more. At the height of the season a male copulates at the rate of over 3 an hour and maintains a rate of over 1 an hour for three weeks (4).

1. Pearson, A. K., and R. K. Enders. AR., 111: 695–711, 1951.
2. Abegglen, C. E., and A. Y. Roppel. JWM., 23: 75–81, 1959.
3. Enders, R. K. Trans. 10th North Am. Wildlife Conf., 92–94, 1945.
4. Bartholomew, G. A., and P. G. Hall. JM., 34: 417–436, 1953.
5. Enders, R. K., O. P. Pearson, and A. K. Pearson. AR., 91(4): proc. 9, 1945.
6. Kenyon, K. W., V. B. Scheffer, and D. G. Chapman. United States Fish and Wildlife Service, Sp. Sci. Rpt., Wildlife, 12: 1–77, 1954.
7. Niggol, K. JWM., 24: 428–429, 1960.

Zalophus californianus Lesson

CALIFORNIA FUR SEAL

The breeding season is from June 15 to July 15; the single pup is born in June (1). A report from the Galapagos Islands suggests a more extended season in that region, as pups of all ages were found on the rookeries during an observation period of 6 months (2). In the London zoo 12 births, all single, have occurred late in May or early in June (3).

1. Rowley, J. JM., 10: 1–36, 1929.
2. Heller, E. Proc. California Acad. Sci., 3: 233–250, 1904.
3. Zuckerman, S. PZS., 122: 827–950, 1952–3.

Eumetopias jubata Schreber

NORTHERN SEA LION

The single pups are born from late May to late June and the females breed again soon afterwards. Over 70 per cent of adult females become pregnant (1). Another account gives the period of mating as from May 31 to July 10, and parturition from May 24 to June 27 (2).

1. Pike, G. C., and B. E. Maxwell. J. Fisheries Res. Bd., Canada, 15: 5–17, 1958.
2. Mathisen, O. A., R. T. Blade, and R. J. Lopp. JM., 43: 469–477, 1962.

Otaria byronia de Blainville

SOUTHERN SEA LION

The males attain puberty at 5 years old and the females at the end of their fourth year. Parturition begins in December and continues to early January. Mating takes place soon after the pup has been born. Birth is during the night or early morning (1). Ovulation is probably spontaneous (2).

The ripe follicle does not project appreciably from the surface of the ovary, but its presence is indicated by a darkening of the wall at the point where it will break through, and the thinning tissues in this region enable one to see the follicular fluid. The corpus luteum, too, does not project, and it disappears fairly rapidly after parturition. In its formation the granulosa cells and many leucocytes advance from different points on the periphery. There is no central cavity. The roughly wedge-shaped masses converge on a more or less central point and are supported by fibrous tissue derived from the theca externa. The mature corpus luteum measures 1.5 to 3.2 cm., while the follicles range from 0.07 to 2.3 cm. in diameter. The ovary is enclosed by a capsule with a small aperture to the body cavity, but immediately after ovulation the capsule fills with liquor folliculi so that at least during this period the aperture is almost entirely nonfunctional (2). From our knowledge of conditions in certain other species it is probable that much of this fluid is secreted by the cells of the oviduct.

The resting uterine epithelium is low-columnar to cubical, about 8.7 μ high. At the time of estrus this layer may reach a height of 52 μ. The layer is marked by fairly widespread degeneration and cytoplasmic vacuolization but recovery is rapid during estrus, when mitoses are frequent. The stroma is edematous during the heat period but glands are few. Mitotic activity is considerable but it is a little later than in the epithelium. After ovulation the glands increase greatly but their lumens are smaller due to the larger cell size. Secretion is plentiful (2).

The cycle of vaginal activity is one of growth, stratification, and destruction. Cornification and mucification do not occur. The resting epithelium consists of 2 layers of almost cubical epithelium divided by a layer that stains less deeply. As the middle layers grow, the outer ones are sloughed off and leucocytes invade the epithelial coat (2).

No indication of seasonal variation in fertility or testis size was found in the males, although, from June to August, spermatozoa were found in the ducts as well as in the testes. In winter spermatogenesis is greatly reduced but it is never entirely suspended (2).

1. Hamilton, J. E. *Discovery* Rpts., 8: 269–318, 1934.
2. Hamilton, J. E. *Discovery* Rpts., 19: 121–164, 1938–40.

Neophoca hookeri Gray

NEW ZEALAND SEA LION

In the Auckland Islands the height of the pupping season is in the last week of December (1).

1. Turbott, E. G. In F. A. Simpson, ed., The Antarctic Today: A Mid-century Survey. Wellington, 195–215, 1952.

ODOBENIDAE

Odobenus rosmarus L.

WALRUS

In Alaska waters the walrus breeds early in June and the young are born after a gestation of 11 months. The males attain sexual maturity at 5 or 6 years and the females at 4 or 5 years (1). In the Kara Sea the mating season is from mid-April to the end of May (2). Other records give the gestation period as 250 to 255 days (3), and it is believed that delayed implantation is not usual in this species (4).

1. Belopoljskii, L. O. ZZ., 18: 762–778, 1939.
2. Capskii, K. K. Trud. Arkt. Inst., 67, 1936.
3. Kenneth, J. H. Gestation Periods. Edinburgh, 1943.
4. Scheffer, V. B. Seals, Sea Lions and Walruses. Stanford, Calif., 1958.

PHOCIDAE

PHOCINAE

Phoca groenlandica Erxleben

HARP SEAL

The males reach puberty at 3 years old and the females at about 4 years. In the White Sea parturition is spread over a 2½-month period from late January to early April, but most are born from February 20 to March 5 (1). In the Canadian breeding regions the pups are born in the first week of March. Mating takes place 2 weeks after the young are born. Usually the left ovary ovulates first and then the two alternate. One pup is generally produced at a breeding period, and about 80 per cent of the adults become pregnant each year (2). The sex proportion of the pups is 53 per cent males (1). Delayed implantation is usual. The corpus luteum persists for amost 10 years (2).

1. Sivertsen, E. Hvalrådets Skrifter, 26: 1–166, 1941.
2. Fisher, H. D. Canadian J. Comp. Med. Vet. Sci., 17: 305–313, 1953.

Phoca hispida Schreber

RINGED SEAL

Puberty in the male is at 7 years and in the female in her seventh year. The first ovulation may be outside the usual breeding season (1), but the females do not always become pregnant at this time (2). In Canada mating is from March to mid-May (1) while in European waters it is said to be from June to August (3). On Southampton Island the young are born mostly in March, with a few in April (4). The female ovulates during lactation, shortly after parturition, and the blastocyst implants 3½ months later, in Canadian waters in early August. The corpus luteum changes its structure and increases in size at this time (1). In the southwest of Baffin Island about 10 per cent of the females produce pups at 6 years old, 50 per cent at 7 years, and 75 per cent at 8 years. Almost 100 per cent of older females are fecund but there is a slow decrease with age. The average adult pregnancy rate is 90 to 95 per cent and they probably cease to reproduce in their late twenties or early thirties (2). Another account states that the females reproduce only every other year and that the gestation period lasts for 8 months (3).

1. McLaren, I. A. Bull. Fisheries Res. Bd. Canada, 118: 1–97, 1958.
2. McLaren, I. A. In E. D. Le Cren and M. W. Holdgate, eds., The Exploitation of Natural Animal Populations. Oxford, 1962.
3. Selptsov, M. M. ZZ., 22: 109–128, 1943.
4. Sutton, G. M., and W. J. Hamilton, Jr. Mem. Carnegie Mus., 12(2): 1–111, 1932.

Phoca vitulina L.

HARBOR SEAL

In the Wash region of England the majority of the pups are born in the period from June 19 to 23, and mating is in late July and August, about 2 weeks after lactation ceases. Lactation lasts for 3 weeks only. Implantation is probably in late October or November. Puberty is reached probably at 5 or 6 years old, at least in the bulls (1). In Nova Scotia the whelping period is in May and June, with its peak early in the latter month. The mating period is from late June to early July (2). In British Columbia the mating period is September and October, while the young are born late in May or in June (3). On Southampton Island the whelping season is late June or July (4). In Nova Scotia implantation takes place in September (2).

Mature follicles measure 2.5 to 3.0 cm. and are present in late July. Near term the corpus luteum measures 1.8 to 1.0 cm.; it is degenerate at this time. It shows some signs of post-partum regeneration, as it does in the cat, but in a week it definitely retrogresses. Ovulation can be spontaneous in the captive female and it may occur in the wild as early as May (1). In the last half of their second year of life the follicles of the young animal have begun to grow. The mature follicle measures 1.0 to 1.3 cm. and the ovum 110 to 135 μ. The presence of a corpus luteum in one ovary suppresses follicular development in that ovary, so that they function alternately. The ovaries are enclosed in a bursa, and consequently the fimbria do not project. The larger follicles have a marked theca interna enlargement (5).

Evidence for the production of placental gonadotrophes during pregnancy is the existence of two periods of follicle stimulation. The first of these is in the week after implantation and until the embryo is 5 cm. long. The second begins before the fetus is 48 cm. long and it continues through the remainder of gestation and parturition (5). In addition the gonads of the newborn pups are enlarged. In the male pup the testes contain much interstitial cell tissue. The cells are 15 μ in diameter, nonvacuolated, granular, and eosinophil. In the first two weeks they lose two-thirds of their weight. The epididymis and prostate also are enlarged and soon diminish. The infantile ovaries are also hypertrophied and weigh 4.0 g. but they soon diminish to 1.2 g. The hypertrophy is entirely in interstitial cells; there are no large follicles. The uterine mucosa is active but it rapidly subsides and the glands become small, lose their lumens, and assume an inactive condition. At birth the vaginal epithelium is stratified but it sloughs off by the third day. The mammary glands are unaffected by these stimuli (1).

The adult seminiferous tubules measure 200 to 250 μ in diameter. They show seasonal activity and are at their maximum in June (1).

1. Harrison, R. J. Mammalia, 24: 372–385, 1960.
2. Fisher, H. D. Nature, 173: 879–880, 1954.
3. Fisher, H. D. Bull. Fisheries Res. Bd. Canada, 93: 1–58, 1952.
4. Sutton, G. M., and W. J. Hamilton, Jr. Mem. Carnegie Mus., 12(2): 1–111, 1932.
5. Harrison, R. J., L. H. Matthews, and J. M. Roberts. TZS., 27: 437–540, 1952.

Phoca caspica Gmelin. The Caspian seal has a mass pupping season in January followed by a mating season from the end of February to the middle of March (1).

P. fasciata Zimmermann. The ribbon seal mates at the end of July and in August. The young are born after a gestation of 9 months (2).

1. Bobrinskii, N. A. Mammals of U.S.S.R. Moscow, 1944.
2. Selptsov, M. M. ZZ., 22: 109–128, 1943.

Halichoerus grypus Fabricius

GRAY SEAL

In the Scilly Isles this seal mates from mid-September into the first week of October and the young are born in September (1). On the Pembroke coast by far the greatest number of calvings were between September 13 and October 16; they were evenly distributed during the period (2). But occasional spring puppings, in late March, have also been observed in this region (3). On North Rona the peak of births is on October 9 and 10, with the commencement in September and the finish in mid-November (4). The age at puberty is 2 years (2) or 4 to 5 (5). Gonadotrophic stimulation of a placental type is evident, as the neonatal testes and ovaries are enlarged by hypertrophy of the interstitial tissue (6). The uteri of the newborn young, also, have thickened mucosa and several layers of epithelial cells. The glands are numerous and enlarged and they extend almost to the base of the endometrium. The vagina is filled with mucus and its epithelium is multilayered (7).

The ovaries are enclosed in a large bursa with a small opening. In consequence the fimbria do not protrude (6). As the pupping season procedes the proportion of male pups decreases. From October 14 to 27 it was 58.2 per cent males, for the next two weeks 52.6 per cent, from November 11 to 24 it was 50.1 per cent, and after that date 42.5 per cent. For the entire season the proportion in the 1,433 pups was 51.0 per cent male (8).

1. Davies, J. L. PZS., 127: 161–166, 1956.
2. Davies, J. L. PZS., 119: 673–692, 1949–50.
3. Backhouse, K. M., and H. R. Hewer. PZS., 128: 593–594, 1957.
4. Boyd, J. M., and J. D. Locke. PZS., 138: 257–277, 1962.
5. Locke, J. D. In E. D. Le Cren and M. W. Holdgate, eds., The Exploitation of Natural Animal Populations. Oxford, 1962.
6. Amoroso, E. C., R. J. Harrison, L. H. Matthews, and I. W. Rowlands. Nature, 168: 771, 1951.
7. Harrison, R. J., L. H. Matthews, and J. M. Roberts. TZS., 17: 437–540, 1952.
8. Coulson, J. C., and G. Hickling. Nature, 190: 281, 1961.

Erignathus barbatus Erxleben

BEARDED SEAL

The males reach puberty at age 7 years and the females at 6. The pups are born about May 1 and the mating season is later in May. The embryos implant during a 1½-month period around early August after a delay of about 2½ months (1). Another account places the mating season in June and July, after the end of lactation, and gives a gestation period of 9 months with whelping every other year (2). Those which have given birth apparently fail to ovulate until the rut of the male is over, so that a 2-year cycle of pup production is established (3). One young is the prevalent number, and on Southampton Island the pups are born in April (4).

1. McLaren, I. A. J. Fisheries Res. Bd., Canada, 15: 219–227, 1958.
2. Selptsov, M. M. ZZ., 22: 109–128, 1943.
3. McLaren, I. A. In E. D. Le Cren and M. W. Holdgate, eds., The Exploitation of Animal Populations. Oxford, 1962.
4. Sutton, G. M., and W. J. Hamilton, Jr. Mem. Carnegie Mus., 12(2): 1–111, 1932.

LOBODONTINAE

Lobodon carcinophagus Jacquinot and Pucheran

CRAB-EATING SEAL

Some females of this seal may become pregnant at the age of a few months more than a year. They reach puberty at a length of about 90 inches (1), or 81 inches (2). Spermatozoa are present in the testes in October and November. In Graham Land the mating season is mid-December but a good many females mate before this time. The pups are born in September, after a gestation of 9 months. Singles are the rule; there is no record of twins. About 80 per cent of the adults become pregnant each year. The corpus luteum is large in the first half of pregnancy; there are no data for the second half but it persists for long after parturition, probably for life (1). Delayed implantation is probable (3).

The ovaries are enclosed by a bursa except for a minute opening near each cornu; this opening is plugged by the fimbria. There is no scar on the surface of the corpus luteum, which reaches 3 cm. in diameter. It is lobulated but the septa are short and vascularization seems to be slight until implantation, though the granulosa lutein cells are well developed. Theca lutein cells are confined to the periphery. In color the corpus luteum is pale yellow and a central cavity seems to be common. The lutein cells measure about 30 μ, and their nuclei 8 μ. Late in pregnancy there is a heavy invasion of connective tissue and shrinkage (3). The neonatal testis is enlarged and interstitial cells are prominent (3).

1. Bertram, G. L. C. British Graham Land Exped., Sci. Rpt. 1: 1–139, 1940.
2. Laws, R. M. PZS., 130: 275–288, 1958.
3. Harrison, R. J., L. H. Matthews, and J. M. Roberts. TZS., 27: 437–540, 1952.

Hydrurga leptonyx de Blainville

LEOPARD SEAL

In this sub-Antarctic seal the male reaches puberty in its third year and is mature by 4 years (1) or a year later (2). There is no indication of seasonal spermatogenesis. The female probably begins ovulating in her second year, but this is not always so. She is usually pregnant in her third year. The mating season is in January, February, and perhaps March, with several ovulations at short intervals. The gestation period is about 8 months and the pups are born over an extended season. Lactation is short (1).

1. Harrison, J. E. *Discovery* Rpts., 18: 239–264, 1938–9.
2. Laws, R. M. Säugetierk. Mitt., 5: 49–55, 1957.

Leptonychotes weddellii Allen

WEDDELL'S SEAL

The females of this seal reach puberty at age 26 months, when they are about 85 inches long (1). Spermatozoa are present in the testes in September, October, and November. During the rest of the year the tubules have regressed (2). No spermatozoa can be found in January. In the Ross Sea the pupping season is at its height by the third week of October (1). In the Bay of Whales the range is from October 5 to November 10, with October 23 as the median date (3). In Graham Land it is a month earlier and it lasts for about a month (1). In the South Orkneys newborn pups may be found from the end of August on, but the season ends by September 22 (4). The season is later as one approaches the South Pole. The earliest ovulations occur in the second half of November, and mostly early in December. Gestation lasts for about 10 months. Twins are rare; one case of binovular twins was observed in more than 100 births (1). Delayed implantation is probable (2). About 16 per cent of females fail to become pregnant each season (1). Lactation persists for 50 days (4).

The ovarian bursa has a minute opening through which parts of the fimbria project. This opening enlarges during pregnancy (2). The female has an average of 6 pups during her lifetime. Ovulation alternates between the ovaries and the blastocyst usually implants on the side on which the ovum is shed. The atretic follicles are larger in the ovary with the large follicle than in the one with the old corpus luteum. The mature follicle measures 14 to 15 mm. in diameter. In the later stages of follicular growth the granulosa develops lobes that project into the antrum. The earliest signs of follicular atresia may be detected in the granulosa, particularly in the discus proligerus. The diameter of the corpus luteum is 25 mm. and this length is maintained throughout pregnancy. Healthy lutein cells measure 25 μ, and the nuclei 7 to 10 μ. They become vacuolated and the diameter rises to 45 to 50 μ at the time of parturition and just after, yet the corpus luteum as a whole is shrinking at this time. The corpora albicantia do not persist throughout life (4). The fully developed corpus luteum is ellipsoidal in shape, with its greatest diameter about 25 mm. It remains full-sized throughout pregnancy and then rapidly declines after parturition, but it may be recognized in ovarian slices throughout the life of the animal. There is no new ovulation until lactation is completed, which takes about 7 weeks; then the discharge is usually from the ovary opposite to the previous one and the fetus lies in the corresponding horn (1).

The uterine and vaginal pictures are complicated by the close association between parturition and breeding, but the height of the epithelium and glandular activity in the uterus is reduced in the intervening period (4).

The neonatal testis and ovary are enlarged to the size of the maternal gonad, owing to hypertrophy of the interstitial tissue (2). Milk is present in the mammae of the new-born (1).

1. Bertram, G. C. L. British Graham Land Exped., Sci. Rpt., 1: 1–139, 1940.
2. Harrison, R. J., L. H. Matthews, and J. M. Roberts. TZS., 27: 437–540, 1952.
3. Lindsey, A. A. JM., 18: 127–144, 1937.
4. Mansfield, A. W. Falkland Islands Dependency Survey, Sci. Rpts. 18, 1958.

MONACHINAE

Monachus monachus Hermann

MONK SEAL

Breeding commences at 4 years old. The pups are born from September to October and lactation lasts for 6 to 7 weeks. Mating takes place 7 or 8 weeks after the pups are born but the females have a pup only every other year. The gestation period is 11 months. The evidence does not suggest that any spring births occur (1).

1. Troitzky, A. Bull. Inst. Oceanogr. Monaco, 1032, 1953.

Monachus schauinslandi Matschie. The Hawaiian monk seal mates from early March to early July and has 1 young (1). The pupping rate is 22 per cent or lower. They are born from late December to the end of June, mostly from mid-March to the end of May. The pups nurse for about 35 days (2).

M. tropicalis Gray. The young are born about the beginning of December; 1 is usual (3).

1. Kenyon, K. W., and D. W. Rice. Pacific Science, 13: 215–252, 1959.
2. Rice, D. W. JM., 41: 376–385, 1960.
3. King, J. E. Bull. British Mus. Nat. Hist., 3: 201–256, 1955–6.

CYSTOPHORINAE

Cystophora cristata Erxleben

HOODED SEAL

The age of puberty is 4 years (1). In the Newfoundland area the pups are born on the ice floes about the end of February (2).

1. Colman, J. S. J. Anim. Ecol., 6: 145–159, 1937.
2. Lillie, H. R. Papers and Proc. 5th Tech. Meeting Internat. Union Protection of Nature, Copenhagen, 62–63, 1954.

Mirounga leonina L.

SOUTHERN SEA ELEPHANT

The males of this species probably reach puberty at about 47 months but the testes continue to increase in size to 9 years, at least. The females usually begin to breed at 2 years, as the 3-year-olds are mostly pregnant (1). At this time they are about 2.6 m. long (2). There is a good deal of variation in the age at puberty; males in the Macquarie Islands are not fecund until 6 years old while those on South Georgia, in the same latitude, are fecund at 4 years. At the former site one-third of the females are pregnant in their fourth year but a quarter defer breeding until 7 years. On south Georgia all 3-year-olds are pregnant (3). The testes and epididymides increase in August and reach a peak of growth in October. They are again reduced by the end of December. The period of mating begins about October 10 (2), about 13 to 24 days after parturition. The females are probably polyestrous and, if pregnancy does not ensue at the first mating, the heat periods are repeated (1, 4). Coitus-induced ovulation is suspected (5). Six calves in a lifetime is the maximum reproductive life (2). The right ovary usually ovulates first (it did in 14 of 19 instances), and after the first the ovaries alternate in producing the single ovum. No case of transuterine migration was observed. Delayed implantation is usual (1, 4), and it takes place 4 months after fertilization. The pregnancy lasts for nearly 12 months but in any one year 14 per cent of cows do not bear young. The newborn are 54.9 per cent males. The corpus luteum persists for over a year following parturition (1). The young are nursed for 3 weeks only (6).

The females usually give birth to the single pup (twins are most infrequent) 5 days after they haul ashore, and the majority of them mate again during the last week of suckling. The onset of estrus is dramatically sudden and copulation lasts for about 5 minutes (5).

The ovaries are enclosed in a bursa with a fairly large opening into the peritoneal cavity (1). Involution of the corpus luteum after parturition is rapid and several follicles develop in the opposite ovary, but only one reaches maximum development. The site of rupture becomes a slightly bulging area marked with small blood vessels. The follicle is supported by several lines of thickened tissue forming buttresses. In follicles up to 12 mm. the liquor folliculi is jellylike; then it changes to watery with a greenish tinge. This change is regarded as a sign of impending ovulation. The mature follicle is 15 to 16 mm. in diameter, and the corpus luteum 16 to 18 mm. The presence of surface scars on the ovaries indicates previous ovulations (5). In another account the measurement of the ripe follicle at estrus is given as 2.0 cm., and the corpus luteum 2.3 cm. The lutein cells measure 30 to 50 μ. They contain a crescentic region around the nucleus that stains darkly with hematoxylin. At the time of implantation a number of small granules staining red with Heidenhain's Azan appear. The corpus luteum has begun to regress at parturition. Subsurface crypts have been observed, but they are absent in specimens with ova in the free blastocyst stage. The germinal epithelium is not a source of new ova (1).

The uterine epithelium is tall and pseudostratified at the time of estrus. Some ciliated cells are present. The glands measure up to 1 mm. long and 50 to 75 μ in external diameter. The stroma is slightly edematous at the time of heat. In nonpregnant mature females the uterus has small cornua, very smooth cuboidal epithelium, and straight, narrow glands with little or no secretory activity (1). At the time of implantation, in February, an implantation cavity is formed in the uterine horn. The placental scar may still be recognized 3 or 4 months after parturition (5).

At estrus the vaginal epithelium is stratified with tall, columnar mucified cells at the surface. Many leucocytes and lymphocytes are present and the stroma is slightly edematous. As ovulation time approaches, cornification sets in. At implantation there is desquamation and an extreme leucocytic invasion (1).

The female fetal gonad hypertrophies and is mostly composed of interstitial cells containing much phosphatase. In the male fetus there is less hypertrophy and it is mainly confined to the seminiferous tubules (7).

Virgin females are not found at the mating grounds and it is not known where they mate. It has been suggested that they mate at sea with younger males at the same time as the parous females mate (8).

In anestrum giant cells are found in the seminiferous tubules of the males. They are first formed in December and can be seen until April or May. Interstitial tissue is scanty but the cell size increases during the rutting season. The proboscis is a secondary sexual character and it is not completely grown until much later than 4 years old (1). The average harem consists of 40 to 50 females. Motile spermatozoa were present throughout the uterus within an hour of copulation (5).

1. Laws, R. M. Falkland Islands Dependency Survey, Sci. Rpts., 15: 1–66, 1956.
2. Angot, M. Mammalia, 18: 1–111, 1954.
3. Carrick, R., S. E. Csordas, and S. E. Ingham. CSIRO Australia Wildlife Res., 7: 161–197, 1962.
4. Gibbney, L. Nature, 172: 590–591, 1953.
5. Gibbney, L. F. Australian National Antarctic Res. Exped., Ser. B., 1: 1–26, 1957.
6. Carrick, R., and S. E. Ingham. Mammalia, 24: 325–342, 1960.
7. Bonner, W. N. Nature, 176: 982–983, 1955.
8. Laws, R. M. Falkland Islands Dependency Survey, Sci. Rpts., 8: 1–62, 1953.

Mirounga angustirostris Gill. The northern sea elephant breeds from January to early March and the young are born in mid-December after a gestation of about 11 months. The heat period is relatively prolonged and the vulva is noticeably enlarged and swollen, especially ventrally (1).

1. Bartholomew, G. A. Univ. California Publ. Zool., 47: 369–471, 1940–52.

Tubulidentata

ORYCTEROPODIDAE

Orycteropus afer Pallas

AARD VARK

IN the Belgian Congo this ant bear bred in April and May, at the end of the rainy season. The young, usually 1 but very rarely 2, are born in October or November after a gestation of about 7 months (1). Further south they are born in the winter, from May to July (2). In a specimen with one fetus two corpora lutea were present in the right ovary and three in the left. All appeared histologically to belong to the same gestation. These corpora lutea were pear-shaped, about 12 mm. long by 7 mm. wide. The free end was the larger (3).

1. Urbain, A. Ann. Mus. Congo Belge, Tervueren, N.S. in 4°, Zool., 1: 101–105, 1954.
2. FitzSimons, F. W. The Natural History of South Africa. London, 1919.

Proboscidea

ELEPHANTIDAE

Loxodonta africanus Blumenbach

AFRICAN ELEPHANT

THE male African elephant reaches puberty at age 8 to 12 years and is capable of breeding at all times of the year. The age of the female at puberty is about the same (1). The mating season is mainly in January or February (2). The single young are born after gestations of about 22 months. A lactation anestrum ensues and then a number of brief cycles until the female conceives. Lactation continues into pregnancy. The interval between parturition and conception is usually about the same length as the gestation period so that the calving interval is rather less than 4 years. Judging from vaginal smears the estrous cycle lasts about 6 weeks. During pregnancy multiple ovulations occur. The corpus luteum of conception is replaced in mid-pregnancy by a second set produced by luteinization of all the follicles possessing antra, but some of the larger ones ovulate. Follicular growth is suppressed in late pregnancy (1).

The ovarian sac is composed of two compartments. The ovary is in the inner one while the oviduct opens on the inner, fimbrial, surface of the outer wall of the outer compartment. The ovarian periphery has numerous subsurface crypts and papillose projections (1). The corpora lutea of pregnancy vary from 40 to 2 mm. in diameter. Usually 6 of them are above 20 mm. and there are from 20 to 30 smaller ones. All are histologically alike. The mid-pregnancy replacement comes near the normal mating season (3).

The uterine horns fuse for some distance before their lumens join and the embryo implants in the region of fusion. In the nonpregnant female the endometrium is only slightly vascular. No fluid could be forced from the oviduct into the uterus. The utero-tubal junction contains a thick muscular collar (1).

The long urogenital canal is like that of the hyaena except that it does not penetrate the clitoris. At the time of estrus the urogenital smear contains more mucus than usual. Sometimes considerable amounts of blood were found (1).

The testes weigh 2 to 3 kg. There is no clear connection between the activity of the musth glands and breeding condition. The tusks of the males continue to grow after puberty, but at a slower rate than before. The tusks of the female cease to grow soon after puberty (4).

The late fetus shows a large development of interstitial tissue in the testis or ovary (1).

1. Perry, J. S. TRS., 237B: 93–149, 1954.
2. Shortridge, G. C. The Mammals of South West Africa. London, 1934.
3. Perry, J. S. JE., 7: liii–lv, 1951.
4. Perry, J. S. PZS., 124: 97–104, 1954–5.

Elephas maximus L.

INDIAN ELEPHANT

The Indian elephant is said to be polyestrous, and in captivity estrus lasts for 3 to 4 days (1). The young are born, one usually and rarely twins, from September to November, with a few at other times (2). The gestation period varied from 607 to 641 days in 6 observed cases, with a mean of 623.5 days. An average of 25 cases gave 21 months, 3 days, with a range from 17 to 24 months (3).

The age at puberty is variously stated. One record shows that a male and a female were 9 and 8 years old respectively (4), another that a male rutted at 15 years and a female calved at 13 years (5). More extensive records give 14 to 15 years for the male, and a calving age of 15 to 16 years for the female (6). The zoo at Portland, Oregon, has had several successes in breeding elephants. Two females bred first at 5 and 7 years old, respectively. A female was born after a gestation of 634 days and a male after 635 days (7). The average elephant, in India, produces 4 calves in her lifetime (8).

In the wild the preorbital glands of the male swell chiefly from November to February (9). A male elephant which was dissected had testes, which are inguinal, weighing 2 kg. each. The seminal vesicles contained 1.5 liters of an opalescent gray fluid. The penis had no prepuce. The head length of the spermatozoa averaged 8.3 μ, and the tails about 42 μ (10).

1. Marshall, F. H. A. The Physiology of Reproduction. London, 1922.
2. Blanford, W. T. The Fauna of British India. Mammals. London, 1888–91.
3. Burne, E. C. PZS., 113A: 27, 1943.
4. Foot, A. E. J. Bombay Nat. Hist. Soc., 38: 392, 1935.
5. Robinson, G. C. J. Bombay Nat. Hist. Soc., 37: 950, 1935.
6. Flower, S. S. PZS., 113A: 21–26, 1943.
7. Marks, Jack L. Personal communication, 1963.
8. Williams, J. H. Elephant Bill. London, 1950.
9. Sanderson, G. P. Thirteen Years among the Wild Beasts of India. London, 1893.
10. Schulte, T. L. AJA., 61: 131–157, 1937.

Hyracoidea

PROCAVIIDAE

Dendrohyrax

TREE HYRAX

DENDROHYRAX arboreus A. Smith. Birth of a single young is recorded and the gestation period given as 225 days (1). There seems to be no particular breeding season and the number of young is 1 or 2, occasionally 3 (2).

D. dorsalis Fraser. The number of young is 1 or 2 and the gestation period probably less than 2 months (3). Pregnant females have been captured from November to January (4).

1. Webb, C. S. The Odyssey of an Animal Collector. London, 1954.
2. Shortridge, G. C. The Mammals of South West Africa. London, 1934.
3. Malbraut, R., and A. Maclathy. Faune de l'Equateur Africain Français. Paris, 1949.
4. Hatt, R. T. AM., 72: 117–141, 1936–7.

Heterohyrax

ROCK HYRAX

Heterohyrax chapini Hatt. A female with 2 fetuses was captured in December (1).

H. syriacus Schreber. In the Giza zoo the young have always been born in March or early April (2). The number has nearly always been 2. Fetuses well developed or near term have been reported in June (3, 4) and August (5).

1. Hatt, R. T. AM., 72: 117–141, 1936–7.
2. Flower, S. PZS., 369–450, 1932.

3. Allen, G. M., and A. Loveridge. HCMZ., 75: 47–140, 1933–4.
4. True, F. W. Proc. United States Nat. Mus., 15: 445–480, 1892.
5. Lawrence, B., and A. Loveridge. HCMZ., 110: 1–80, 1953.

Procavia capensis Pallas

ROCK DASSIE

This South African dassie has a pedunculated corpus luteum (1). Pooled data on the number of embryos (2, 3) gave a mean of 2.44 ± .054, with a mode of 2, and a range from 1 to 6. The young are born in November and December (3). The gestation period (observed) is 7½ months (4). In the London zoo births have been spread from June to November (5).

1. Wislocki, G. B. JM., 9: 117–125, 1927.
2. Wislocki, G. B., and O. P. van der Westhuysen. CE., 518: 65–88, 1940.
3. van der Horst, C. J. Science, 93: 430, 1941.
4. Murray, G. N. J. South Africa Vet. Med. Assn., 13: 27–28, 1942.
5. Zuckerman, S. PZS., 122: 827–950, 1952–3.

Procavia habessinica Hemprich and Ehrenberg. Specimens with 2 or 3 large fetuses have been taken in January (1).

P. johnstoni Thomas. The young are born from the end of December to the beginning of January, but observations do not cover the entire year (2). In Kenya the single young is born in the period August to November. Spermatozoa are present in the testis during December (3).

P. ruficeps Hemprich and Ehrenberg. This species has at least 2 young to the litter (4).

1. Allen, G. M., B. Lawrence, and A. Loveridge. HCMZ., 79: 31–126, 1935–7.
2. Hatt, R. T. AM., 72: 117–141, 1936–7.
3. Coe, M. J. PZS., 138: 639–644, 1962.
4. Rode, P. Mammifères Ongulés de l'Afrique Noire. Paris, 1943.

Sirenia

DUGONGIDAE

Dugong dugon Müller

DUGONG

IN Egyptian waters mating is in winter and the single young is born in the following winter (1). In Ceylon waters it breeds in the northeast monsoon, in the first few months of the year (2).

1. Anderson, J. The Mammals of Egypt. London, 1902.
2. Phillips, W. W. A. Ceylon J. Sci., 14B: 1–5, 1927–8.

TRICHECHIDAE

Trichechus manatus L.

MANATEE

The manatee probably breeds all year (1). The single young, rarely 2, is born after a gestation period of at least 152 days (2).

1. Moore, J. C. Am. Mus. Novitates, 1811: 1–24, 1956.
2. Moore, J. C. JM., 32: 22–36, 1951.

Perissodactyla

TOO few species of Perissodactyla have been studied to afford material for generalization. The Equidae all appear to be seasonally polyestrous. In both the horse and the ass the corpus luteum declines early in pregnancy, and gonadotrophins appear in the blood serum. Other changes during pregnancy are noteworthy and have been described in the section on the horse.

EQUIDAE

Equus caballus L.

HORSE

The horse is seasonally polyestrous, with heat periods commencing early in the spring and continuing, if unbred, well into the summer. The periods are very variable in length, averaging about 22 days, with a heat lasting about 7 days. Ovulation is spontaneous, and it occurs toward the end of heat. The mare has a very high F.S.H. content of the anterior pituitary, higher than that in any other domestic animal. One young is usually born at a time, and twinning very frequently results in abortion. The period of gestation is about 330 days, but seasonal or feed factors cause great variation, more so than in any other species. Histological changes in the reproductive tract are not very marked during the cycle. Ovulation occurs from a restricted portion of the surface of the ovary known as the ovulation fossa. The time of service within the heat period has a great influence upon the chance of conception; the later it occurs, the greater is the chance.

There is very little recorded information on the age at puberty. The testes of Anglo-Norman horses in Japan weigh about 10 g. at birth. They do not increase in weight until 10 months; then they increase slowly to 16 months, when a rapid increase sets in. However, there are considerable individual differences. Fifty per cent of males with single testis weight of 30 g. are fertile and 100 per cent at 90 g., when the stallion is about 24 months old (1). In 3 Korean mares the first estrous cycles appeared at 16 to 17 months old and they were soon fertile. The usual age of puberty in the female may be taken as 20 months (2). A record from Senegal for Barb-Arabs and local mares gives the age at first heat as 8 months (3).

THE ESTROUS CYCLE

The seasonal cycles begin about March and usually continue, in the unbred mare, into August, but many will breed in the fall and winter in England. It is said that in the tropics 2 well-defined breeding seasons tend to occur (4).

At Onderstepoort, South Africa, light farm mares maintained in stables and with restricted exercise had estrous cycles all the year, but there was a decline in April and May. From June to March they were regular but tended to be much less so at other times (5). The length of the cycle is very variable, and the means given (see Table 6) vary from 19 to 22¾ days. From the data it is impossible to draw the conclusion, as some have done, that heavier mares have a longer cycle than light ones; more extensive data gathered under comparable conditions are needed to settle the point. The duration of the heat period is also very variable, and the averages range from 4½ to 9 days; but a great deal depends upon the method of testing for heat and on the statistical analysis of the data, since the mare is apt to go out of heat for a short period and to come in again during what is evidently one full heat period (6). The length of life of the corpus luteum appears to be about 15 to 17 days, and one set of data in which the coefficients of variation are given shows that the length of the heat period is about twice as variable as that of the interval between heats (7). A similar calculation of another set of data (6) gives a C.V. of 49 for the heat period, and of 26.4 for the interval between heats. Thus, the corpus luteum has a relatively constant length of life in the mare. It is said that heat is 1 day shorter if the mare is suckling, and that it tends to be 1½ days longer in March, at the beginning of the season, than in July. There is no constant difference with age or breed (8).

Whatever the length of heat, ovulation bears a closer relation to its end than to its beginning. The symptoms of heat begin to decrease 6 to 12 hours after rupture of the follicle (9), and this, i.e., rupture, occurs in Korean mares about 1.6 days before heat totally ceases (10). It has been found also that 76 per cent of all corpora lutea are first detected from the day before to the day after heat ceases. These figures apply to Belgian and grade Thorobred mares (6). In an extensive series, mainly of ponies, 64.5 per cent of mares

TABLE 6. Length of the Estrous Cycle in the Mare.

	LENGTH OF CYCLE, DAYS			LENGTH OF ESTRUS, DAYS		
	Mean	Standard Deviation	Range	Mean	Standard Deviation	Range
Heavy mares (11)	19.1			7.0 ± .16	2.22	2–16
Belgian draft (12)	20.1		Mode 20, 60 per cent 18–22	5.2		Mode 5, 1–14
Thorobreds (13)	20.3 ± .5	7.5	78 per cent 13–25	4.96 ± .13	2.47	Most 4–7
Percherons (7)	20.94				5.00	
Percherons (14)	21.4				8.9	
Light (12)	22.0		Mode 19, 60 per cent 18–21	5.5		Mode 3, 1–37
Mixed breeds (15)	22.52		56 per cent 19–24	4.52 ± .04		73 per cent 1–5
Semiwild Korean (8)	22.86 ± .33	2.78		8.98 ± .25	2.68	Mode 10
Thorobreds, etc. (8)	22.79 ± .12	4.44		7.53	2.53	Mode 6–8
Ponies, etc. (1)	22.55		14–32	6.18		Mode 6

ovulated one day before the end of estrus and 25.3 per cent on the second day before. On the last day 8.4 per cent ovulated. The right ovary ovulated in 55.6 per cent of cases (1). The first signs of estrus occur when the largest follicle measures 1 to 2 cm. At the time of ovulation it is about 4 cm. (1, 16). If the graafian follicle is ruptured artificially on the second day of heat, no corpus luteum is formed, and a new heat begins in a few days. On the other hand, artificial rupture on the fifth day is followed by corpus luteum formation, and the next heat period occurs at the usual time (17). This may demonstrate the importance of the maturation process in preparing the granulosa cells to luteinize, or it may mean that the pituitary is not able to secrete luteinizing hormone early in estrus.

Various anomalies in the heat period, such as the split estrus already mentioned, physiological estrus in which the usual anatomical changes occur but not the psychological state of receptivity, and ovulation after the cessation of heat, have been described (18).

Breeding has no effect upon the length of heat nor upon the time of ovulation (15, 19). The left ovary functions more frequently than the right. In a series of 185 ovulations 61 per cent were from the left (6).

The "foal" heat follows soon after parturition, and the mean interval has been given as $9.13 \pm .05$ (15), $9.7 \pm .3$ (11), 11.3 (7), and 11.4 (12) days. It has begun in 75 per cent of cases by the fourteenth day after foaling (12). In another series foal heat followed parturition by 8.6 days; it lasted 5.6 days, and the mean ovulation time was 12.9 days after parturition (1).

The notorious infertility of mares has raised the question of the relationship of the time of service to ovulation. In a long heat period can the spermatozoa deposited early survive until ovulation has occurred? There is a tendency for services early in the heat period to be less successful than those made later, though mid-heat appears to be the best time. One record gives 29 per cent fertility on the first day of heat against 45 per cent on the second to fourth days (20). Another record gives practically no difference on the first 3 days of heat and after that a falling off (21). Neither of these trials takes into account the lengths of the individual heat periods. In a study in which this was considered there was no fertility from services earlier than 7 days before the end of heat, 75 per cent success at 4 days before, and later than that a reduction (22). Insemination prior to ovulation was 86 per cent successful; at the time of ovulation, 74 per cent; and 2 to 10 hours after ovulation, 30 per cent (23). Successes have even been reported if insemination was delayed to 12 to 14 hours after ovulation (24). Another report gives no success 2 to 4 hours after ovulation (25), but it seems that in some cases the egg can be fertilized for at least 20 hours after it has been shed. Yet another gives 12 to 15 hours as the limit for the ovum, because insemination at 18 hours after ovulation was without success. At from 2 to 10 hours after, 39 per cent conceived. From 1 to 2 hours after, the rate was 55.2 per cent, and up to 48 hours before ovulation it was 71.2 per cent (26).

The time required for the passage of the egg through the oviduct was found in the case of one mare artificially ovulated by means of pregnancy urine to be 96 hours. This ovum was found in the last part of the oviduct (27).

Double ovulations, i.e., the rupture of more than one follicle in a heat period, have been observed in 3.8 per cent of cases (6), and twin pregnancies in 3.2 per cent (28) and 1.6 per cent (29). In the first instance only 0.5 per cent twins were born, the remainder were aborted or one of the twins died; and in the second, two thirds terminated in abortion or miscarriage. A report of English Thorobreds in Russia gives 5 per cent of twin conceptions and 1 per cent births, and also records that two-thirds of the sets of twins born are of opposite sexes (30). This report also states that the average incidence of twin births in general is 0.57 per cent. A very extensive series gives 1.5 per cent of twin births (31); another, 1.1 per cent (32); and another, 1.23 per cent (33). Evidently the mare is unable, as a rule, to carry twins through gestation. One account states that of 443 twin pregnancies 281 aborted and only 297 viable foals were born. Of these 130 were twin pairs and 37 singles. Seventeen pairs may have been monozygotic (34).

The sex proportion of the horse at birth has been extensively studied, and the literature has been reviewed in a paper dealing with the sex proportion of hybrids. This work points out that most of the earlier data, obtained from stud books, give a proportion of less than equality, $49.69 \pm .002$ per cent males, in a series of more than a million records. A summary of records obtained from experiment stations gave $52.52 \pm .95$ per cent males, so the proportion is probably a little above equality and similar to that found in other species (32).

The duration of gestation is extremely variable, with breed averages ranging from 329 to 345 days; the standard deviation is usually about 9.5 days. There is a tendency for the lighter breeds to have the longer gestation; hence the saddle horses tend to carry their foals longer than the draft horses. There is general agreement that a gestation terminating in the winter is about 20 days shorter than one ending in spring, a surprising difference. Gestations ending in fall and summer are a little shorter than those ending in spring. Males tend to be carried 1.6 to 2.0 days longer than females, and twins 10 days less than singles. Age has little effect, but the gestation period tends to be slightly shorter between the ages of 11 and 13 years (33). A mare carrying a mule foal tends to have a period 10 days longer than she would have had if she had been served by a stallion of her own breed (35, 36). This is a little less than midway between the periods of the horse and ass. As the reduction in an ass carrying a hinney is about the same, it suggests a maternal influence on the length of gestation, since if the difference were due to the inheritance of the hybrid, each gestation should be midway between those of the parents.

HISTOLOGY OF THE FEMALE TRACT

OVARY. The ovary of the mare presents several interesting features, one of which is the ovulation fossa, or groove, from which the eggs are liberated. Another is its hypertrophy in the fetus during gestation, and another, the limited life of the corpus luteum during pregnancy. At birth the ovary weighs about 20 g., and at this time a great deal of the interstitial tissue which develops in the fetal ovary has undergone lipid degeneration (37). Graafian follicles are fully formed and the filly usually comes in heat for the first time at about 11 months of age (6). Eight to nine large follicles are produced at each cycle, as a rule, but only one normally ruptures; the rest undergo degeneration (19). The weight of the mature resting ovary is about 40 to 70 g.; it is largest at 3 to 4 years of age, and then gradually diminishes (38).

The ovary of the young mare is much like that of other mammals, with the follicles distributed over the whole ovary. The poles, however, grow inward toward each other, and the ovary becomes covered by a serous coat, leaving only the groove between the poles in which ovulation can occur. This is the ovulation fossa. Small follicles and ova are distributed throughout the stroma, but they migrate to the fossa as they mature (39).

The maturing follicles are from 1 to 3 cm. in diameter at the beginning of heat, and the one which will rupture rapidly grows to about 5 to 6 cm. Palpation of follicles is fairly easy as they are large in relation to the rest of the ovary, but they cannot usually be detected until 1 to 2 days before the beginning of heat (6). During heat the ovary becomes intensely congested, a condition which may be felt as a softening of the stroma (19), but the follicles themselves are tense, even on the first day (6).

Ovulation is spontaneous and is accompanied by a breakdown of blood vessels and considerable bleeding into the cavity. The corpus luteum rapidly forms and is buff-colored when it is new. It persists for 2 to 3 cycles but can be distinguished from the latest one by its darker color. When it is fresh, secondary liquor folliculi is found at the center (38). It is pear- or toadstool-shaped with the stem toward the fossa, and its diameter is one-half to three-fourths that of the follicle (19). At the time of follicular maturity the granulosa cells are increasing in size. Theca interna cells show a few mitoses and they lose their glandular appearance as they invade the granulosa. By 6 days the cavity is filled and theca externa cells begin to pass into the corpus luteum. The organ is at its maximum at 14 days; then it gradually diminishes (40).

VULVA AND VAGINA. The mucous membranes of the labia, vagina, and cervix become more intensely pink during heat, and congestion is more marked, but the best diagnostic sign is the moist and glistening appearance of the vaginal mucosa at the time of heat. This passes off rapidly after ovulation (41).

The growth of the vaginal epithelium is not so cyclical in nature in the horse as it is in many species. Mitoses may be found at any time, but they are most frequent during heat. Cornified cells tend to increase in number at this time, but as a layer they are never very prominent. The polyhedral cells of the mid-epithelial zone are largest during heat and smallest at 5 to 10 days of diestrum. Leucocytes are present at all times, but they are most numerous during heat and are least so at 5 to 10 days of diestrum (6).

The vaginal smear is indefinite, as would be expected from the epithelial changes. It is most abundant during heat, and at that time the cornified cells tend to increase, but none of the cellular changes are constant enough to be of diagnostic value (6, 42, 43, 44). The amount of mucus begins to increase 1 to 2 days before heat, the increase continues to ovulation, and at the same time its viscosity is lowered. The volume is lowest and the viscosity is greatest at 5 to 10 days of diestrum (6).

UTERUS. There is no well-marked cervical canal, but the os uteri is well defined, with large folds on the ventral side and a powerful sphincter muscle. There are no tubular glands in the cervix, but epithelial mucous cells are abundant. The os uteri is partly open during the second half of the heat period, and at this time the mucous cells are swollen. Two days after ovulation these cells have become cubical, but they still contain a fair amount of mucus. At 8 days postestrum there is little secretion, and the subepithelial layer of blood vessels has become reduced. At the beginning of heat the cells lengthen, more secretion is present, the connective tissue becomes looser, the blood vessels increase in number, and many leucocytes invade the cervix (38).

The uterine mucosa is brownish yellow to pinkish yellow. The muscle layers merge into each other, and the long, tubular glands are more numerous in the horns than in the body of the uterus. The connective tissue tends to be loose, with large lymph spaces. The uterine cells are not ciliated. Changes during the cycle are minor only. The uterus is turgid at the beginning of heat, and flaccid toward the end (38, 45). The epithelium is highest during the later stages of heat and at 5 to 8 days of postestrum. On the first day of heat the cells are columnar with some pseudostratification, a condition which increases throughout heat. Minor changes only occur in the glands; they are most active late in heat and at 5 to 8 days postestrum, and least so between these times. The cells are highest and the diameter of the glands is greatest, at 5 days postestrum. Leucocytes are present at all times (6).

OVIDUCT. The oviducts are 30 to 70 mm. long, and are 4 to 8 mm. thick at the ampulla and 2 to 3 mm. thick at the uterine end. They have a dense muscular coat and very many mucosal folds. The subepithelial capillaries are small 2 days before heat and enlarge throughout that period. In diestrum the epithelial cells are cubical, but during heat they become columnar and are filled with secretion. Leucocytes invade the stroma in the heat period (38).

PHYSIOLOGY OF THE FEMALE TRACT

As was mentioned above, the vaginal mucus is most abundant during heat. The pH of this mucus is 8.15 on the first day and 7.9 on the last day of estrus. Corresponding figures for the cervical mucus are 7.75 and 7.85 (6). In diestrum the vaginal pH was from 7 to 7.5; in proestrum it was 8; and in estrus, 9. It began to fall rapidly at the moment of ovulation. During estrogen-induced heat the pH rapidly increased to 9 (46).

The amount of estrogen in the liquor folliculi has been studied in follicles ranging from 1 cm. to 5.5 cm. in diameter. The fluid is aspirated fairly easily from the follicles by a hypodermic needle inserted through the rectum. The amount of estrogen in a follicle varies with its volume, but the quantity per 100 cc. remains about the same regardless of the volume. The results found vary from 2,030 to 2,625 R.U. per 100 cc. The pH of the liquor folliculi varies from 7.6 to 8.1, and it tends to rise with the size of the follicle (47). The estrogens in the follicle are estrone and estradiol-17β. The level of the former is 2.3 μg/100 ml., and of the latter, 33.8 μg/100 ml. (48). The excretion of estrogens in the urine rises during and just following heat. There is evidence of two peaks, one during heat and one at 10 to 15 days postestrum (4), the latter of which may be related to the decline of the corpus luteum. It is noteworthy that a similar peak has been reported by some workers as occurring in man a short time before menstruation. The decline of the sexual skin in other primates is further evidence that there is a fall in the circulation of estrogens at this time. At the time of estrus 17-ketosteroid excretion reaches the level of 160 mg. per 24 hours, range 51 to 314 (49).

The excretion of estrogens during pregnancy can first be detected at from 30 to 40 days, when it is about 250 R.U. per liter. It rises to 17,000 R.U. per liter, or more, as the 260th day, after which the quantity rapidly falls as pregnancy nears its end (50). Estrone and equilin are the principal components, with equilenin a minor one. The equilin level rose from 4 to 5 months until it equaled or surpassed that of estrone in the later months of pregnancy (51).

It is a disputed point whether progesterone or pregnanediol is present to an appreciable degree in the blood or urine of the pregnant mare. One report on serum progesterone is negative (52), and one gives the level of pregnanediol in the urine as rising rapidly after 100 days of gestation (53), while another worker was unable to detect it (54).

The mare is remarkable in that her blood serum contains at a certain stage of pregnancy large quantities of a gonadotrophic hormone. This first appears at about the thirtieth day, is at a maximum at 45 days, remains high to 115 days, then rapidly declines and cannot be detected after 160 days (50, 55–58). At its peak 50,000 R.U. per liter, or more, are present in the serum. It appears to be formed in the tissue comprising the endometrial cups (59). It is found in the highest concentration in the smaller breeds of mares and may be four times as high in Welsh ponies as it is in draft mares (60). Its level has been studied in many breeds. It was found to be highest in Thorobreds and lowest in Jutland mares (61). The blood serum of mares that have been covered by asses and are pregnant with mules does not contain this gonadotrophe (62, 63). The active substance closely resembles F.S.H. in its biological activity.

Other peculiar phenomena occurring during pregnancy may be related to the appearance of this substance. At about the thirty-fifth day the corpora lutea of pregnancy degenerate. They are followed by a large crop of corpora lutea which, in their turn, degenerate by 150 days. These are formed partly by direct luteinization and partly by the rupture of follicles (64), though some of them appear to be formed by the luteinization of theca interna cells with degeneration of the ovum and granulosa. They are a rich source of progesterone. After 150 days the ovaries become atrophic and fibrotic (65). These organs may be removed at 200 days without interruption of the pregnancy. After the operation estrogen excretion falls but it soon rises again to its normal level, and it is maintained within the normal range for the remainder of the pregnancy (66).

The ovaries of female fetuses from 5 to 9 months after conception weigh 120 to 150 g., compared with 20 g. at birth. During this time they are almost

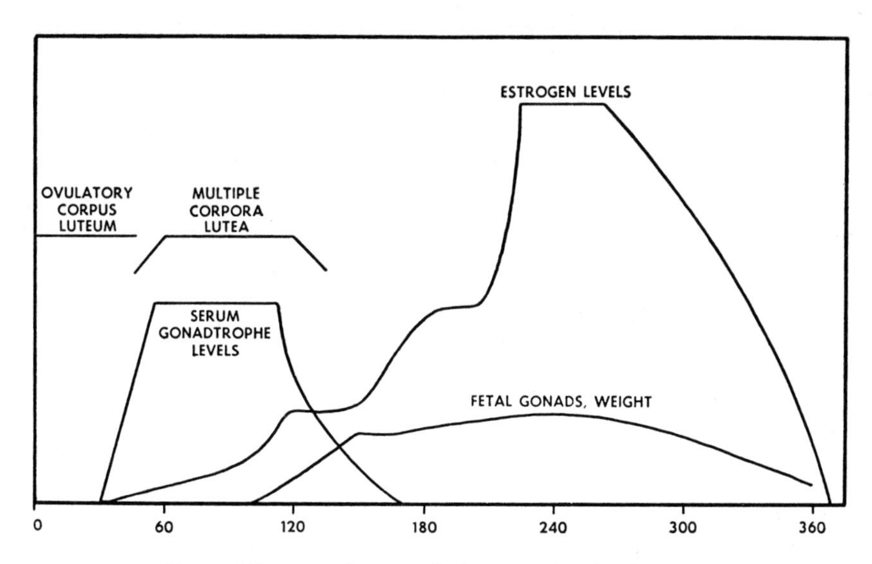

FIG. 1. Diagram of events during gestation in the mare.

a solid mass of interstitial cells (37). Fetal testes follow a similar curve of growth and decline (67) and contain large islands of interstitial tissue. These increases occur at the time when the maternal ovaries are devoid of lutein tissue. These fetal gonads do not contain detectable amounts of progesterone, but estrogens are present in a fairly high concentration (37).

The general relationship is, therefore, that the corpus luteum of pregnancy degenerates as serum gonadotrophin appears. The latter apparently causes new corpora lutea to form. When the serum gonadotrophin disappears, these corpora lutea degenerate and the maternal ovaries become inactive. The fetal gonads now increase enormously in size and remain large until near the end of pregnancy, but there is no evidence that they have taken over the functions of the maternal ovaries. The growth and decline of the fetal gonads correspond with changes in estrogen excretion in the mother's urine. It would be interesting to know whether their growth and decline are associated with changes in the kind and quantity of gonadotrophic hormones in the fetal pituitaries, especially as the reaction resembles a response to the luteinizing hormone.

The anterior pituitary of the adult horse is exceptionally high in F.S.H., but is somewhat low in L.H. content (68). Pituitaries of old mares and old geldings have a strong F.S.H. effect when they are implanted into immature rats, while stallions' pituitaries give a less strong F.S.H. effect. Young nonpregnant mares and young geldings give an L.H. effect, but that given by the pituitaries of colts and fetuses is mixed. Pituitaries from mares early in pregnancy are low in gonadotrophins. As pregnancy advances their potency rises, followed by a slight decline at about mid-pregnancy (69).

By exposing mares to increasing amounts of light each day during winter they may be brought into estrus by as much as two months earlier than usual (1, 70).

Estrus may be induced in the ovariectomized mare by a single injection of between 30 and 50 mg. of estrone benzoate or of 0.25 to 2.5 mg. of stilbestrol. For the latter substance 67 per cent of mares came in estrus with the lower dose but 100 per cent success was not achieved until the higher one was given (1).

The optimum dose of estrogen to produce heat is 5,000 M.U.; higher doses are less effective. The optimum dose of "prolan" (P.U.) to induce ovulation is 1,000 to 2,000 M.U. (71). At the beginning of heat two injections of 500 M.U. of prolan given at a 5- to 6-hour interval induced ovulation at an average of 44 hours after the first injection. The mares remained in heat an average of 46 hours after ovulation, and the total length of heat was about half the normal time. The injections accelerated the ripening of the follicle and ovulation, and the treated mares were slightly more fertile than untreated ones (72). Pony mares may be ovulated at any time, if moderately large follicles are present, by the injection of 1,000 M.U. of prolan. If this is done during heat, the ovulation occurs 22 to 30 hours after injection and the mare goes out of heat afterwards. If the injections are made outside estrus, ovulation occurs without heat 36 to 60 hours later (27). P.M.S. in doses of 500 to 1,000 M.U. given on 3 to 4 successive days produced heat and ovulation if the injections were made during early or mid-diestrum. It was without effect during anestrum, and so were injections with horse pituitary extract at this time (25).

The size of the living tubal egg is 133 μ, and the zona pellucida is 13.2 μ thick (73).

THE MALE

The ejaculate of the stallion varies from 50 to 200 cc., and the number of spermatozoa varies from 4 to 13 billions. The average ejaculate is about 125 cc., and the average number of spermatozoa is about 8 billion (74). The pH of the semen varies from 6.94 to 7.51, with an average of 7.31 (75). Another observer gives an average pH of 7.23 (76). Three fractions may be observed; the first is a grayish fluid, about 19 ml., which contains no spermatozoa. The second, the spermatozoon fraction, is from 30 to 75 ml., and the third, a viscous fluid mixed with gelatinous material and containing few spermatozoa, represents from one-third to two-thirds of the entire ejaculate (77). The spermatozoa are able to travel at the rate of 3 to 4 mm. per minute. They live for less than 6 hours in the vagina and for 12 hours in the uterus, but the data are insufficient for us to be sure that these times are even approximate (75).

A single observation has shown that spermatozoa may remain alive in the male tract for at least 3 weeks, as they were found in a stallion gelded for that length of time (78). The length of the epididymis is 72 to 86 meters, and the number of ejaculatory ducts is 12 to 18 (79). The dimensions of the spermatozoon are as follows: head, 7.0 μ (6.1–8.0) $\times$ 3.9 μ (3.3–4.6); midpiece, 9.8 μ (9.2–11.5); and total length, 60.6 μ (57.4–63.0). The spermatozoon has a characteristically curved tip and lacks a head cap (80).

The urine both of the stallion and of the mare contains a large variety of sterols which appear to be metabolic products of the gonadic hormones. Stallions' urine is an exceptionally rich source of estrogens, since that of immature males contains 1,000 to 10,000 M.U. of estrogen per liter, and that of adults, 20,000 to 50,000 M.U. The titer is highest in spring and summer and is lowest during the winter (81). The free estradiol content per liter is 5 mg. (82). This is about equally divided between estradiol-17β and estradiol-17α, while the content of estrone is 20.7 mg. per liter. There are only traces of all three in the urine of geldings (83).

The semen contains citric acid to the extent of 15.2 mg. per 100 ml. and ergothionine, 6.7 mg. per 100 ml. A little of each of these compounds may be found in the oviducts by 50 minutes after mating (84). Inositol is present in the semen plasma to the extent of 38 mg. per 100 ml. In the ampullar secretion there are 93 mg. per 100 ml. (85).

1. Nishikawa, Y. Studies on Reproduction in Horses. Tokyo, 1959.
2. Sugie, T., and Y. Nishikawa. Bull. Nat. Inst. Agric. Sci., Chiba, G8: 151–159, 1954.
3. Redon, P., and L. Fayolle. Rev. Elev. Méd. Vet. Pays Tropicales, N.S., 10: 257–262, 1957.
4. Hammond, J. Yorkshire Agric. Soc. J., 95: 11–25, 1938.
5. Quinlan, J., S. W. J. van Rensburg, and H. P. Steyn. Onderstepoort J. Vet. Sci., 25: 105–119, 1951.
6. Andrews, F. N., and F. F. McKenzie. Missouri Agric. Exp. Sta., Res. Bull. 329, 1941.
7. Cummings, J. N. J. Anim. Sci., 1: 308–313, 1942.

8. Sato, S., and S. Hoshi. J. Japanese Soc. Vet. Sci., 13: 237–260, 1934.
9. Zivotkov, H. L., K. S. Goncarenko, and A. G. Krivoscekov. Probl. Zivotn., 3: 71–76, 1936.
10. Sato, S., and S. Hoshi, J. Japanese Soc. Vet. Sci., 12: 200–223, 1933.
11. Vladescu, D. Rev. Vet. Milit., 8: 157–167, 1937.
12. Andrews, F. N., and F. F. McKenzie. Proc. Am. Soc. Anim. Prod., 365–369, 1939.
13. Krat, A. T. Konevodstvo, No. 1: 55–59, 1933.
14. Trowbridge, E. A., and H. C. Moffett. Missouri Agric. Exp. Sta., Bull. 387: 30–31, 1937.
15. Constantinescu, G. K., and A. Mauch. Ann. Inst. Nat. Zootech. Roumanie, 5: 9–82, 1936.
16. Varadin, M., and S. Kamhi. Veterinaria (Sarajevo), 2: 83–92, 1953.
17. Hammond, J. Proc. Physiol. Congr., Moscow. In Sechenov J. Physiol. of the U.S.S.R. 21: 193–194, 1938.
18. Andrews, F. N., and F. F. McKenzie. Proc. Am. Soc. Anim. Prod., 228–232, 1938.
19. Aitken, W. A. J. Am. Vet. Med. Assn., 70: 481–491, 1927.
20. Schtschjekin, V. Zeit. Tierzucht u. Zuchtungsbiol., 18: 447–454, 1930.
21. Caslick, E. A. Cornell Vet., 27: 187–206, 1937.
22. Hammond, J. Proc. XVI Internat. Congr. Agric., Budapest, June 1934.
23. Zwotkov, H. I. Sovetsk. Zooteh., No. 1: 108–109, 1940.
24. Saltzman, A. A. Sovetsk. Zooteh., No. 4: 77–80, 1939.
25. Day, F. T. J. Agric. Sci., 30: 244–261, 1940.
26. Skatkin, P. N. Dokl. Akad. Selskohoz. Nauk, Leningrad, 9: 7–10, 1947.
27. Day, F. T. J. Agric. Sci., 29: 459–469, 1939.
28. Blakeslee, L. H., and R. S. Hudson. J. Animal Sci., 1: 155–158, 1942.
29. Wagnaer, H. Deutsche Tierarztl. Woch., 42: 610–613, 1934.
30. Koroljkov, V. Konevodstvo, No. 5: 31–34, 1939.
31. Uppenborn, W. Zeit. f. Zucht., 28B: 1–27, 1933.
32. Craft, W. A. Quart. Rev. Biol., 13: 19–40, 1938.
33. Mauch, A. Zeit. f. Zucht., 39B: 31–42, 1937.
34. Bergmeister, E. Diss., Hannover, 1951.
35. Bilek, F. Neue Forsch. in Tierz. u. Abstammungslehre, 17–23, 1936.
36. Braccini, P. Clin. Vet., 60: 335–344, 1937.
37. Cole, H. H., G. H. Hart, W. R. Lyons, and H. R. Catchpole. AR., 56: 275–293, 1933.
38. Hammond, J., and K. Wodzicki. PRS., 130B: 1–23, 1941.
39. Kupfer, M. Union of South Africa, Dept. of Agric., Dept. of Vet. Education and Res., 13th and 14th Rpt., 1211–1270, 1928.
40. Harrison, R. J. J. Anat., 80: 160–166, 1946.
41. Day, F. T., and W. C. Miller. Vet. Rec., 52: 711–716, 1940.
42. Schatalov, P. Berliner Tierarztl. Woch., 49: 81–84, 1933.
43. Chernov, V. I. Probl. Zivotn., No. 5: 102–104, 1933.
44. Mirskaja, L. M., and A. A. Salzmann. Usp. Zooteh. Nauk, 1: 157–168, 1935.
45. Barone, R., and J. Poirier. Rev. de Méd Vét. Lyon et Toulouse, 106: 441–455, 1955.
46. Berthelon, M., and J. Juteau. CRSB., 139: 459–460, 1945.
47. Mayer, D. T., F. N. Andrews, and F. F. McKenzie, E., 27: 867–872, 1940.

48. Knudsen, O., and W. Velle. J. Reprod. and Fertility, 2: 130–137, 1961.
49. Santesteban-Garcia, F. Arch. Zootec., 9: 303–347, 1960.
50. Cole, H. H., and G. H. Hart. J. Am. Vet. Med. Assn., 101: 124–128, 1942.
51. Savard, K. E., 68: 411–416, 1961.
52. Suzuki, Y. Japanese J. Vet. Sci., 9: 149–162, 1947.
53. Hosi, T. Japanese J. Vet. Sci., 7: 29–40, 1945.
54. Stevenson, W. G. Canadian J. Comp. Med., 11: 25, 1947.
55. Cole, H. H., and G. H. Hart. AJP., 93: 57–68, 1930.
56. Catchpole, H. R., and W. R. Lyons. AJA., 55: 167–227, 1934.
57. Day, F. T., and I. W. Rowlands. JE., 2: 255–261, 1940.
58. Stamler, S. M. Probl. Zooteh. Eksp. Endokrinol., 2: 186–220, 1935.
59. Cole, H. H., and H. Goss. Essays in Biology, in Honor of Herbert M. Evans. Berkeley, Calif., 1943. Pp. 105–119.
60. Cole, H. H. PSEBM., 38: 193–194, 1938.
61. Bielanski, W., Z. Ewy, and H. Pigoniowa. Roczn. Nauk Rol., 65: 245–251, 1952.
62. Bielanski, W., Z. Ewy, and H. Pigoniowa. Bull. Acad. Polon. Sci., Cl. II, 3(2): 37–39, 1955.
63. Calisti, V., and O. Oliva. Atti Soc. Ital. Sci. Vet., No. 10: 249–251, 1956.
64. Cole, H. H., C. E. Howell, and G. H. Hart. AR., 49: 199–209, 1931.
65. Kimura, J., and W. R. Lyons. PSEBM., 37: 423–427, 1937.
66. Hart, G. H., and H. H. Cole. AJP., 109: 320–323, 1934.
67. Brill, J. CRSB., 113: 951–952, 1933.
68. Witschi, E. E., 27: 437–446, 1940.
69. Hellbaum, A. A. AR., 63: 147–157, 1935.
70. Burkhardt, J. Vet. Rec., 243–248, 1948.
71. Vorontzovskaya, M., and V. Chernov. Probl. Zivotn., No. 2: 49–54, 1933.
72. Mirskaja, L. M., V. K. Kedrov, and A. N. Lihacev. Probl. Zivotn., No. 3: 66–76, 1938.
73. Amoroso, E. C., W. B. Griffiths, and W. J. Hamilton. Vet. Rec., 51: 168, 1939.
74. Day, F. T. Vet. Rec., 52: 597–602, 1940.
75. Anderson, W. S. Kentucky Agric. Exp. Sta., Bull. 239, 1922.
76. Sergin, N. P. Probl. Zivotn., No. 12: 100–122, 1935.
77. Berliner, V. In E. T. Engle, ed., The Problem of Fertility. Princeton, N.J., 1946.
78. Steffens, M. Berliner Tierarztl. Woch., 53: 320–322, 1937.
79. Ghetie, V. AA., 87: 369–374, 1939.
80. Nishikawa, Y., Y. Waide, and H. Onuma. Bull. Nat. Inst. Agric. Sci. Chiba, G1: 29–36, 1951.
81. Barulin, K. I. Racion. Korml. Isoljz Provizvod. S.-H. Zivotn., 95–100, 1939.
82. Levin, L. J. Biol. Chem., 158: 725–726, 1945.
83. Pigou, H., T. Lunaas, and W. Velle. Acta Endocrinol., 36: 131–140, 1961.
84. Mann, T., C. Polge, and L. E. A. Rowson. JE., 13: 133–140, 1955–6.
85. Hartree, E. F. Bioch. J., 66: 131–137, 1957.

Equus asinus L.

ASS, DONKEY

The female ass reaches puberty at one year of age. The breeding season is limited as it is in the horse; some jennets experience cycles from March to August, and some for a longer period. In South Africa the breeding season is from the second half of October to the first half of April (1). In Japan the ass breeds from April to September. During December and January there is absolute rest; the other months are transitional (2). The estrous cycle lasts 21 to 28 days, and heat 2 to 7 days (3). In Japan the mean duration of estrus was 6.0 ± .6 days, range 3 to 14. The duration of the whole cycle was 22.8 ± .1 days, range 13 to 31 (2). Another account gives a mean cycle length of 22 days, mean duration of heat 4.7 days, with ovulation usually in the last third (4). In a limited series ovulation occurred about 6.6 days after the beginning of estrus, or 1.4 days before its end. Most females ovulate once but a few shed 2 or 3 ova. At the beginning of heat the largest follicles measured 1.0 to 3.0 cm., and on the day of ovulation the largest were 3.0 to 4.5 cm. in diameter. Ovulation is spontaneous (5). The foal heat occurs modally 17 and 18 days after parturition, with a spread from 6 to 69 days (4). In Mississippi foal heat sets in 2 to 8 days after foaling and lasts 2 to 6 days; ovulation is approximately 48 hours before the end of heat (3). One young is usually born at a time. The period of gestation is about 365 days, but it is 8 to 12 days less if the jennet is carrying a hinney (6).

The ovaries are relatively small. Follicles are arranged in rows, develop at the cranial pole, and move caudally. There is no ovulation fossa at first; later the polar ends of the ovaries grow disproportionately, curve toward each other, and the ovulation fossa comes into being. Ovulation does not occur at each heat period, especially in younger females. Blood is extravasated in the ruptured follicle, as it is in the mare. The corpus luteum is very inconspicuous at parturition (1), so that one may suspect that the peculiar series of reactions observed in the mare during pregnancy occur also in the jennet.

The labia swell during heat, and at this time blood-stained mucus is produced (1). The vagina during heat contains abundant mucus, which liquefies and flows as heat progresses. The cervix becomes relaxed, flabby, and dark red, and it frequently shows hemorrhagic spots during this period. It is slightly narrower and longer than it is in the mare. It protrudes far into the vagina and is situated close to the floor. In open jennets it is rather tortuous, even during heat, but during the foal heat it is wide open (3). The vaginal smear is similar to that of the mare and it is not a reliable criterion of the reproductive state of the animal (7).

Gonadotrophins may be detected in the blood from 40 to 200 days of pregnancy (8). On the fiftieth day, 5,000 to 10,000 M.U. per liter have been found, but the level falls too low for accurate pregnancy diagnosis after the 150th to 160th day of gestation (9). Another report gives a positive reaction for serum gonadotrophins from 47 to 117 days. It lacks consistency at other times (10).

Estrogens can be found in the urine from 50 days of pregnancy onward. At this time about 1,000 M.U. per liter are present. The amount increases to 70,000 M.U. per liter at 101 days; it then decreases but does not fall much below 1,000 M.U. per liter (9). By the Cuboni test estrogens may be detected between 132 and 301 days of pregnancy. Before and after these times the results were either doubtful or negative (11). It would be interesting to know whether the ass has the same variety of estrogens as the horse.

The semen is more milky and more turbid than that of the stallion. In color it is brownish yellow and it contains almost no gelatinous material. The average volume of an ejaculate was 49 ml., range 10 to 80; and the total number of spermatozoa was 23.75×10^9, range 6.1 to 44. The pH of the semen ranged from 7.2 to 7.7. The volume and sperm count are lowest from October to December (12). In other reports the volume has been from 70 to 115 ml. (13), or 40 to 100 ml. (3), or a mean of 54 ml. The concentration of spermatozoa was from 95 to 264 million per ml. in a group of jacks in which the mean volume was apparently on the low side (14). In species in which a large proportion of the ejaculate comes from the accessory glands the mean number of spermatozoa per unit volume is likely to be very variable. The sperm concentration of jack's semen is higher than that of stallions (3), and the total output at an ejaculation seems to be greater, as it varies from 8 to 21 billions (13).

The dimensions of the spermatozoon are as follows: head length, 6.9 μ (6.5–7.6); head width, 3.96 μ (3.3–4.6); midpiece, 9.9 μ (9.4–10.4); total length, 64.1 μ (61.8–66.4) (15).

1. Kupfer, M. Union of South Africa, Dept. of Agric., Dept. of Vet. Education and Res., 13th and 14th Rpt., 1211–1270, 1928.
2. Nishikawa, Y., and Y. Yamasaki. Japanese J. Zootech. Sci., 19: 119–123, 1949.

3. Berliner, V. R., E. W. Sheets, R. H. Means, and F. E. Cowart. Proc. Am. Soc. Animal Prod., 295–298, 1938.
4. Svecin, K. Konevodstvo, No. 4: 42–45, 1939.
5. Nishikawa, Y., and Y. Yamasaki. Japanese J. Zootech. Sci., 20: 28–32, 1949.
6. Koch, W. Deutsche Landw. Tierzucht, 40: 56–57, 1936.
7. Nishikawa, Y. Studies on Reproduction in Horses. Tokyo, 1959.
8. Samodelkin, P. A. Voprosy Plodov. Rabotosp. Losad., Moscow, Seljhozgiz, 100–104, 1939.
9. Svecin, K. Sovetsk. Vet., No. 3: 107–108, 1939.
10. Ajello, P. Clin. Vet., Milano, 73: 144–148, 1950.
11. Ajello, P., and E. Ioppolo. Clin. Vet., Milano, 72: 364–367, 1949.
12. Nishikawa, Y., and Y. Waide. Bull. Nat. Inst. Agric. Sci., Chiba, G1: 37-45, 1951.
13. Mirianshvili, D. Sborn. Trud. Zoo-Vet. Inst. Gruzii, 2: 41–46, 1941.
14. Salzman, A. A. Dokl. Akad. Seljskohoz Nauk, No. 7: 37–38, 1940.
15. Nishikawa, Y., Y. Waide, and H. Onuma. Bull. Nat. Inst. Agric. Sci., Chiba, G1: 37–45, 1951.

Equus burchelli Gray

BURCHELL'S ZEBRA

In South Africa births take place from July to September. The period is apparently more prolonged in the Okovango district and rather later in the Transvaal (1). In the London zoo births have been distributed throughout the year except for November to March (2). The gestation period is between 11 and 12 months (3).

1. Ansell, W. F. H. PZS., 134: 251–274, 1960.
2. Zuckerman, S. PZS., 122: 827–950, 1952–3.
3. Brown, C. E. JM., 17: 10–13, 1936.

Equus hemionus Pallas

ONAGER

In the region west of the Indus River the onager usually foals in the months from June to August (1). In the London zoo the single young has always been born in June or July, mostly in the latter month (2). In eastern Tibet the breeding season is from August to September and foaling about

July (3). The gestation period is about 11 months (4). Puberty is reached in the females at age 2 to 3 years, and in the males a year later (5).

1. Blanford, W. T. The Fauna of British India. Mammals. London, 1888–91.
2. Zuckerman, S. PZS., 122: 827–950, 1952–3.
3. Schäfer, E. Zool. Garten, 9: 122–139, 1937.
4. Sallim, A. J. Bombay Nat. Hist. Soc., 46: 472–477, 1946–7.
5. Bannikov, A. G. Zeit. Säugetierk., 23: 157–168, 1958.

Equus grevyi Oustalet. Grevy's zebra has given birth to young in the London zoo at all times of the year, with a peak in July and August (1). The male has no special rutting season and the gestation period is 390 days (2).

E. quagga Gmelin. The quagga has an extended foaling season but most young are born in October or November (3). In the male the age at puberty is 3 years and in the female, 4 years (4).

E. zebra L. The mountain zebra has one young after a gestation of 12 months (5). In the London zoo they have been born at all seasons except from September to January (1).

1. Zuckerman, S. PZS., 122: 827–950, 1952–3.
2. Riley, E. H. United States Dept. Agric., Bur. Animal Ind., 26th Ann. Rpt., 229, 1909.
3. Hamilton, J. S. Animal Life in Africa. London, 1912.
4. Trumler, E. Säugetierk. Mitt., 5 (suppl. 1): 1957.
5. FitzSimons, F. W. The Natural History of South Africa. London, 1919.

TAPIRIDAE

Tapirus

TAPIR

Tapirus bairdii Gill. This Central American species usually has 1 young; twins are rare (1). They breed in March and the period of gestation is about 4 months (2).

T. indicus Desmarest. The Asiatic tapir has 1 young, born in November or December, probably every other year (3). The gestation period is about 13 months long (4).

T. terrestris L. The Brazilian tapir breeds before the beginning of the rains. The single young (5) is born after a gestation of 397 ± 2 days (6).

In the London zoo young have been born from February to May and in August and October (7).

1. Hall, E. R., and K. R. Kelson. The Mammals of North America. New York, 1959.
2. Gaumer, G. F. Mamiferos de Yucatan. Mexico City, 1917.
3. Sanborn, C. C., and A. R. Watkins. JM., 31: 430–433, 1950.
4. Brown, C. E. JM., 17: 10–13, 1936.
5. Cabrera, A., and J. Yepes. Historia Natural Ediar. Mamiferos Sud-Americanos. Buenos Aires, 1940.
6. Baker, A. B. JM., 1: 143–144, 1920.
7. Zuckerman, S. PZS., 122: 827–950, 1952–3.

RHINOCEROTIDAE

Dicerorhinus sumatrensis Fischer

ASIATIC TWO-HORNED RHINOCEROS

This species is said to reach puberty at age 20 years. It mates from July to August and has a single young after a gestation of 7 months (1).

1. Thom, W. S. J. Bombay Nat. Hist. Soc., 44: 257–274, 1943.

Ceratotherium

WHITE RHINOCEROS

Ceratotherium simum Burchell. This African species breeds at any time of year and the single young is born after a gestation of 17 to 18 months. Puberty is reached at 4 to 5 years old (1).

C. unicornis L. This Asiatic species reaches puberty at 4½ years old; it has heat periods at intervals of 40 to 50 days and a gestation of about 488 days. These observations were made at the Whipsnade zoo (2). In India matings have been observed from February to the end of April (3).

1. Lang, H. New York Zool. Soc. Bull., 23: 66–92, 1920.
2. Tong, E. PZS., 130: 296–299, 1958.
3. Gee, E. P. J. Bombay Nat. Hist. Soc., 51: 341–348, 1952–3.

Diceros bicornis L.

BLACK RHINOCEROS

This African species has no regular season (1), but breeds mostly from November to December (2). The gestation period is about 530 to 550 days (2).

1. Roosevelt, T., and E. Heller. Life Histories of African Game Animals. New York, 1914.
2. Wilhelm, J. H. J. South West Africa Sci. Soc., 6: 51–74, 1933.

Artiodactyla

MUCH more work is needed before one can make many generalizations regarding reproduction in Artiodactyla. In general, it may be said that species living in temperate regions tend to be seasonally monestrous or seasonally polyestrous, but in the tropics reproduction may take place at any time of the year. There are, however, many exceptions.

In Cervidae, the deer, species from temperate regions are seasonally monestrous, breeding in late fall or winter in most cases, but tropical species tend to breed at any time of the year, even when they are brought into temperate regions. It is not clear whether they are monestrous or polyestrous in the tropics, and more information is needed on this point. Antler shedding is due to high testosterone levels; antler growth is independent of sex steroid hormones except that the growing buds need an initial stimulus that is given by testosterone. In the tropical species the antlers are shed at any time of the year, but this is said not to occur annually in all species. Possibly each individual has is own rhythm of testicular activity.

In the Bovinae all species investigated tend to be polyestrous throughout the year, but in their natural habitats they probably all breed at a fairly definite time. The Caprini, sheep and goats, are seasonally polyestrous, mainly in the fall, but some breeds of sheep have a much more extended season than others. In general, antelopes breed at any time of the year, but this may not be true of the Neotragini, the pygmy antelopes, and of the Alcelaphini, the hartebeests and gnus.

The Bovinae and, to a lesser extent, the Caprini are remarkable for the low level of sex-hormone production and excretion which seems to prevail, though more information is desirable on these facts in all mammals. At present they are certainly at the bottom of the list.

SUIDAE

Potamochoerus porcus L.

RIVER HOG

In French Guinea from 4 to 6 young were born in February or March (1). They usually farrow in the rains, from October to March, and have 2 to 6 young, usually 3 or 4 (2). An August pregnancy was reported from the Belgian Congo (3). In the London zoo births of from 1 to 4 young have been recorded in January and from June to October (4). The gestation period is 4 months (5).

1. Prunier, —. Mammalia, 10: 146–148, 1946.
2. Ansell, W. F. H. PZS., 134: 251–274, 1960.
3. Frechkop, S. EPNU., 1944.
4. Zuckerman, S. PZS., 122: 827–950, 1952–3.
5. Rode, P. Mammifères Ongulés de l'Afrique Noire. Paris, 1943.

Potamochoerus larvatus Cuvier. This hog has 5 to 8 young, born usually in December or January (1). In Madagascar pregnancies have been observed from April to June and young from September on (2). The gestation is about 130 days (3).

1. FitzSimons, F. W. The Natural History of South Africa. London, 1919.
2. Kaudern, W. AZ., 9(1): 1–22, 1914.
3. Kerbert, C. Bijd. Dierkunde, 22: 185–191, 1922.

Sus scrofa L.

PIG

In Morocco the wild pig has from 5 to 8 young, usually born in winter or spring (1). In Russia puberty is during the second year; most breeding is from November to January and 3 to 5 is the usual litter size, though it may be as high as 10. Gestation lasts 4 months (2). In Iraq breeding is from April to May and the litter is 4 to 5 (3). Domestic pigs were released by Captain Cook in New Zealand and doubtless their descendants have been joined by others. They have flourished in a wild state and breed all the year round, with an average of 6 to 10 to the litter. The males breed at 1 year old and the females at 10 months (4).

THE DOMESTIC PIG

The pig has a cycle of about 21 days, which recurs all the year round. The duration of heat is about 2 to 3 days. Ovulation is spontaneous. Heat does not occur during lactation. The changes in the reproductive tract are not great, but, owing to the ease of obtaining material, excellent accounts, some of which are classical, have been written on the cyclical changes. The sow occupies an intermediate position among the domestic animals in regard to gonadotrophic hormone content of the pituitary. The duration of gestation is about 112 to 115 days. Many hermaphrodites are found in this species.

THE ESTROUS CYCLE

Age and weight at puberty vary considerably with breed and management. For Large Black and Large Whites in South Africa the age for sows was 219 ± 1.7 and 217 ± 2.1 days, respectively. For both the mode was 210 days and the range 161 to 346 days. The mean weight was 177 ± 3 lb. for both breeds and the mode 170 lb. with a range from 100 to 310 lb. (5). For Large Whites in Britain the average age was from 6½ to 7½ months, but gilts born in spring and early summer were in breeding condition earlier than those born later in the year (6). In Germany an average age of 6.37 months has been reported (7). An average age for Chester Whites and Poland Chinas was 204 days (8), and the weight was 212 lb. for the former and 224 lb. for the latter (9). For crossbred Landrace × Poland-Duroc 186 ± 16 days has been given (10). For Duroc-Jerseys the age was 195 ±16.5 days and the weight 196 ± 27 lb. (11). In the Philippines Berkshires have averaged 5 months, 16 days old at puberty (12). In small type pigs the average was 208 days and the weight 189.5 lb. For large type pigs the corresponding figures were 199 days and 199.3 lb. (13).

The length of the estrous cycle and its component parts have been studied in some detail; the results are summarized in Table 7. The modal length is 20 to 22 days (18). In South Africa, as in Europe and America, the cycles extend throughout the year; they tend to be a little shorter in gilts than in mature sows. Ovulation can occur on either day of heat but it is more frequent on the second day (19). A slaughter experiment showed that in 7 sows killed before 30 hours from the beginning of heat no follicles had ruptured, but when killed at from 30 to 36 hours, 6 of 7 sows had ruptured follicles, though the process of ovulation was not complete in all cases (20). The South African data for Large Blacks gave the time of ovulation as from 40 to 54 hours; the Large White data varied from 18 to 36 hours (5). In Japan the time was found to be 31 ± 5.5 hours after the onset of heat (14). Another

TABLE 7. Length of the Estrous Cycle in the Sow

	CYCLE LENGTH, DAYS	DURATION OF PHASES:		
		PROESTRUM, HOURS	ESTRUS, HOURS	METESTRUM, HOURS
Large Blacks in South Africa (5)	21.7 ± .14	13.5 ± .75	Gilts 62 ± .6 Sows 68.2 ± .6	10.0 ± .32
Large Whites in South Africa (5)	21.0 ± .22	14.1 ± .66	Gilts 47.6 ± .5 Sows 49.0 ± .9	9.6 ± .3
Pigs in Japan (14)	20.6		Gilts 54.7 Sows 70	
Large Blacks in France (15)	21.2 ± .2			
Large Whites in France (15)	20.4 ± .6			
German pigs (16)	20.7 ± .2			
Poland China (17)	21–22		Gilts 40–46 Sows 65	

account gives the beginning as 32 hours after the onset of heat and the duration as 1 to 3 hours (21).

The time of ovulation has been studied, not only by the method of slaughter and observation, but also by the use, at definite intervals, of double matings involving two inherited colors. By this means it was found that most eggs are shed at about 36 hours from the onset of heat and that all have been shed by the forty-eighth hour (22). Another similar study gave much the same results (23).

The left ovary appears to be slightly more active than the right. Observations on 14 sets of corpora lutea gave a mean ovulation rate of 15, of which 51.9 per cent were shed from the left ovary (8). A more extensive series which included 469 sows gave 55.3 per cent of ovulations from the left (9).

There is frequently a post-partum heat, usually not accompanied by ovulation. It occurs at about 44 hours after parturition but the time is variable (5). Apart from this, lactation heats are rare; cycles are resumed soon after the young are weaned. The mean length of the interval is 7 days, and it tends to be 1 day longer if the lactation was short (26).

Pigs inseminated on the day before estrus began gave 9.8 per cent of fertilized ova; on the first day 68.8 per cent were fertilized; on the second, 98.2 per cent; and on the third day, only 15.2 per cent (27). Matings made a few hours after the onset of heat were 63.0 per cent fertile; those made on the second day were 65.8 per cent fertile (28). Matings made the day after the sow had gone out of heat were 20 per cent fertile, and later matings were completely infertile (29). Mating on the second day as well as the first increased conceptions from 66 per cent to 89 per cent, but the increase in litter size was small and uncertain. Fertilization appeared to be an all-or-none process (29).

The rate of travel of the ovum through the oviduct has been worked out on material in which the oviducts were divided into five parts. The number of eggs in each part was obtained and expressed as a percentage of the whole. As the average time for the complete passage is 3 days (30), this percentage gave an approximate idea of the rate of travel in the different portions of the tube. Passage is rapid through the first two-fifths, very slow through the next two-fifths, and rapid through the last fifth (31). This method probably gives too short a duration for the first part of the travel owing to a time lag in obtaining the specimens after slaughter, but the error is not great enough to invalidate the general conclusion. Another account gives the time of passage through the utero-tubal junction as at 24 to 48 hours after ovulation instead of 72 hours and suggests that the large number of corpora lutea in the ovaries hastens the passage (32).

The fecundity of the sow varies much with the breed. The litter size of the wild pig is said to be usually 4 (33), and from this low point the average extend to 12.0 for the Large White Yorkshire. Climate or some other environmental factors influence the fecundity, since in the Philippine Islands several breeds average from 1 to 2 pigs less than they do in the United States (34). Young pigs have smaller litters than older ones; the maximum fecundity is reached at the fourth to sixth litters. The first litter averages about two less than the maximal size (35, 36). This is partly due to the age of the sow, since those having their first litters at the usual age average one less than those littering for the first time a year later. It has been found in the rat that reproduction is, in itself, a stimulus to fecundity, and this seems to be true also for the pig, as those which have been continuously reproducing are more fecund than those of the same age which are producing their first litter. An interval of 9½ to 10½ months between the first and second litters increases the litter size by 1.6 pigs over sows with an interval of 4½ to 5½ months between litters, but the difference is not statistically significant. Litters born in the early spring tend to be larger than those born at other times of the year, and the smallest litters are those produced toward the end of the year (37).

One of the major problems in pig breeding is the wastage caused by loss of eggs and by fetal atrophy. This may be as high as 33 per cent or more. There is some reason to believe that it may be largely due to the action of genetic factors (38). Another is the number of stillborn pigs. The loss from this cause averages about 6 per cent. It is greater in litters with prolonged farrowings, and most of the dead pigs are born in the second half of the litter to be farrowed. Many of them have attempted to breathe but have smothered while they were still in the uterus. There is some evidence that the incidence of stillbirths may be reduced by an injection of pituitrin when farrowing begins (39).

The duration of gestation shows a difference with the breed, but the average for the lowest recorded, Middle White, 112.1 ± .6 days is not far from the highest, Poland China, 116.0 ± .08 days (40, 41). All the data are in agreement on this point, and it is hardly worth while further to summarize them. However, the wild pig is said to have a mean gestation period between 127 and 128 days (42). The mode for domestic pigs is close to the mean, the standard deviation is usually about 3 days, and the differences for age of sow, litter sequence, and litter size are not significant (43). Seventy-three per cent of farrowings occur between 2 P.M. and 4 A.M. (44).

The sex ratio of pigs at birth is 49.56 per cent males for a large series of records. The sex ratio has attracted some attention because it has been found that stillborn pigs have a much higher ratio of males, while that for fetuses increases at the younger ages. Extrapolation of the curve of fetal sex ratios suggests a ratio at conception of 60 per cent males, sufficiently remote from the equality demanded by current genetic theory to be interesting (45). The phenomenon is, as yet, unexplained. Another unexplained peculiarity of the sex ratio is that more litters with nearly equal proportions of the sexes are born than would be expected if the results were due to chance (46).

HISTOLOGY OF THE FEMALE TRACT

OVARY. Multilayered graafian follicles appear in the ovaries of the gilt at about 7 weeks of age. At 15 weeks vesicular follicles appear. The blood supply to the medullary region increases markedly when multilayered follicles appear, and to the theca externa of individual follicles when the antra are formed. Stimulation with gonadotrophes is without effect before vesicular follicles are present (47). In the breeding sow the follicles about to rupture measure 7 to 10 mm. in diameter. The granulosa layer is 6 to 9 cells deep and the theca interna is 0.09 to 0.1 mm. thick. The ripe follicles are clear and semitransparent and stand out from the ovary like a bunch of grapes. When the follicle ruptures, a considerable amount of liquor folliculi escapes and the follicle collapses. There is a slight eversion of the granulosa and theca interna and a slight oozing of blood. The membrana propria breaks down, with the result that the sharp line of demarcation between granulosa and theca is lost. The latter grows inwards, starting at the apices of the folds into which it has been thrown by the contraction of muscular fibers in the theca externa. The thecal cells divide by mitosis, and the granulosa cells enlarge. By the seventh day of the cycle the corpora lutea measure 8 to 9 mm., and in the second week their maximum size of 10 to 11 mm. is reached. Degeneration of the corpus luteum is rapid at the end of heat and at parturition (48). At first the corpus luteum is pinkish in color, but at the end of diestrum a rapid change takes place with the degeneration of the capillaries, and the color becomes a primrose yellow, changing to yellow-brown as resorption advances. The cellular changes in the corpus luteum of pregnancy have been described in great detail (49). At the time of ovulation the theca interna cells hypertrophy and the capillaries are dilated while this layer folds into the cavity (50). During development of the corpus luteum fine argyrophil fibers penetrate at the vascularization stage. These move from the theca to the granulosa layer and then encircle each lutein cell. In the regression stage they disappear and the amount of collagen fibers increases (51). Alkaline phosphatase is present in both theca and granulosa cells (52). Between the fifteenth and eighteenth days the change from purplish to yellow or cream yellow is rapid. The rupture point is visible until day 12, and vascularization collapses at day 18 (5). Prior to rupture the follicle is not marked by a papilla but there is a change in color from light red to dark red (5).

There is a double wreath of blood vessels in the theca externa and theca interna. That of the theca interna grows directly into the granulosa at ovulation, but that of the theca externa merely enters the developing corpus luteum in the folds of tissue. No lymphatic vessels are present until 2 days or more after ovulation; these structures are the first to degenerate and are not found later than 16 to 17 days after ovulation. The capillaries and venules are the next to degenerate (53).

VULVA. During diestrum the labia are contracted, the muscles are tight, and the walls are pink and moist. The latter swell during early heat and become red internally. Late in heat they become flabby, the muscles are fully relaxed, the mucosa is still red, and some mucus is present. During metestrum the labia revert to the diestrous stage, but a good deal of mucus is present (24).

VAGINA. During diestrum the basement membrane of the epithelial layer is indistinct; there are 3 to 6 layers of cells with but few leucocytes between them. In early heat the basement membrane is distinct, the epithelial layer is actively growing and is now 10 to 15 cells thick, with low-columnar cells at the surface. Blood vessels are distended and the stroma is edematous. In late heat the basement membrane has again become indistinct, the number of layers of epithelial cells has increased to 14 to 26, cornification is evident, and leucocytes have wandered between the cells. During metestrum desquamation is in evidence, and leucocytes are abundant, but cornified cells are not present in great numbers. Many of the cells show vacuolar degeneration (24, 54).

The vaginal smear is not a good indication of the reproductive state. There is more mucus during heat, especially late in the period, and the proportion of leucocytes tends to increase. In early diestrum many of the epithelial cells are vacuolated, more so than at other times (54, 55).

UTERUS. The uterine stroma becomes edematous in late diestrum. This increases during proestrum, becomes less intense during heat, and rapidly disappears during metestrum. Leucocytes are to be found at all stages but are more frequent from late diestrum to 4 days postestrum. The epithelial cells grow and become pseudostratified, owing to the crowding of the nuclei, during proestrum and continue in this state to about 1 week postestrum. They show vacuolar degeneration during proestrum but have become normal in appearance by late heat. Cilia are present only in the crypts and glands (24, 30). They do not appear to fluctuate in numbers during the cycle (56).

The amount of cervical mucus is greatest in proestrum and early estrus, while reducing substances are greatest in amount near estrus. Large, vacuolated epithelial cells are much in evidence during metestrum (57). The periodic-Schiff positive material is absent from the uterine wall in the 8 days immediately following estrus. Then, until the next estrus, it may be found in the glandular epithelium and the myometrium. In the surface epithelium it is present at 12 days after estrus; then it increases for a time but has disappeared by estrus. It is entirely absent from the uterus of the ovariectomized sow (58). The course of endometrial development in the immature female has been described (59).

The oviduct enters the horn of the uterus as a continuous passage, i.e., without any angle. There is no sphincter, but the mucosa projects in a number of fingerlike folds. The whole structure is like a rosette with the oviduct opening through the center. These projections are edematous just after heat, and at this time the pressure needed to force fluid from the uterus into the oviduct increases from 50 to 155 mm. of mecury. The pressure needed to force fluid from oviduct to uterus varies only slightly. It is about 25 mm. of mercury (60).

The ciliated epithelium of the oviduct is 25 μ high during and just after heat; it declines to 10 μ in the second week of diestrum and then gradually grows to the maximum during the third week. The nonciliated cells are smooth in the first week of the interval, and in the second numerous cytoplasmic processes, like those described for other species, including the cow, are extruded (61).

PITUITARY. During the heat period the basophilic cells of the anterior pituitary lose their granules, which were abundant during proestrum. They also contain many granules during the early lutein phase, but not in the middle and late lutein phases. Degranulation of the eosinophils also occurs during heat (62). The gonad-stimulating hormone can first be detected in the anterior pituitary of the fetus at a crown-rump length of 17 to 18 cm.

(63). Its appearance coincides with a large increase in the number of basophilic cells (64).

PHYSIOLOGY OF THE FEMALE TRACT

The uterine muscle of the sow has been intensively studied. During heat there is great spontaneous activity. Two kinds of contractions are found, large ones at intervals of 1.5 to 2.5 minutes, with superimposed smaller contractions of short duration. During metestrum (tubal ova) they are more irregular, and the minor waves are more pronounced than the major ones. In mid-diestrum the irregularity has increased, and the short waves are even more in evidence (65, 66). The muscle uses most oxygen at 3 days before ovulation, and least at 4 to 18 days after ovulation (67). The muscle contracts in response to pituitrin and relaxes to adrenalin at all stages of the cycle (68).

The muscle of the oviduct is most active during, and just after, heat (69). At this time the contractions occur about 13 to 15 per minute as measured during insufflation. In diestrum they occur 5 to 9 per minute, but their amplitude is greater (70).

Intra-uterine migration of the ova is a usual phenomenon in the sow. If one ovary is removed, 42 per cent of the embryos which develop are found in the horn of that side. There is also a tendency in the normal sow for the number of embryos on each side to be equalized (25).

The estrogen content of large graafian follicles is about 900 R.U. per kg. of liquor folliculi, while corpora lutea contain from 25 to 78 R.U. per kg. (71). The occurrence of estrogenic substances in the urine during pregnancy is peculiar. From 19 to 30 days 1,000 to 3,000 M.U. per liter are found, their first appearance being rather sudden. Then they disappear almost entirely, to reappear about the seventieth day. After this they rise gradually in amount and just before parturition 5,000 to 10,000 M.U. per liter are found. After parturition they disappear rapidly. These facts have been amply confirmed (72, 74). The content of chorionic tissue exactly parallels that of the urine, though at a lower level (75), and the gonadotrophic hormone content of the pituitary is in inverse relationship. This is highest during proestrum, falls abruptly at ovulation, and is very low during diestrum. During pregnancy it reaches a maximum from 30 to 75 days (76). These relationships raise some interesting questions which deserve further study.

The estrogen content of urine in the nonpregnant sow is, for estriol, 0.4

to 1.1 μ/l.; for estrone, 0.5 to 1.0 μ/l.; and for estradiol, 0 to 0.5 μ/l. urine. In pregnancy the level rises to a maximum of 17.4 μ/l. for estriol; 131.0 μ/l. for estrone; and 3.2 μ/l. for estradiol. The estrone is almost entirely in conjugated form (77). Another report states that there is no significant quantity of estradiol or of estriol in nonpregnant urine (78). In ovariectomized sows at least 0.5 mg. or, better, 1.0 to 1.5 mg. of stilbestrol is needed to induce estrus (79). If a single dose of 30 to 50 mg. of estrogen is injected in the period from days 5 to 18 of the cycle a condition of anestrum lasting for several months ensues (80).

Pregnanediol excretion of the nonpregnant sow is 5.0 mg./l.; about two-thirds of this is in conjugated form. In pregnancy the level rises to 34 mg./l., and the amount is related to the number of fetuses (81). The corpus luteum contains appreciable quantities of progesterone during the first 3 days after ovulation, before the lutein cells are fully differentiated. The content rises until 15 days, after which it falls abruptly and soon cannot be estimated. During pregnancy it rises rapidly to 20 days, then continues to rise slightly to 105 days, after which the fall is rapid (82). The amount of progesterone secreted by a single corpus luteum is about 4 mg. daily. At least 5 are needed to support pregnancy whatever the number of fetuses. After hysterectomy the corpora lutea persist for 87 days (83, 84). The follicular fluid of a sow near ovulation contains 8 μg of progesterone per ml. (85). If progesterone is injected from day 15 of the cycle onward, 25 mg. delays estrus for 3 to 5 days, while 50 mg. suppresses it entirely for some time (86).

The anterior pituitary contains less F.S.H. than that of the horse, but more than that of the sheep. Its L.H. content is variable, but it is always below that of the sheep (87, 88). The F.S.H. level is lower at 2 days before heat than at heat or early proestrum or in pregnancy. The level is high at the eighty-fifth day of pregnancy (89). The gonadotrophe level is high in day-old females; then it decreases to 225 days, after which it is constant. The lowest titer is at the time of puberty (90).

There is no relaxin in the ovary at the follicular phase, but a little is present in the lutein phase. The amount increases rapidly in early pregnancy to a level of 10,000 G.P.U. per gram of tissue. It may also be detected in the blood at this time (91).

Maturation of the ovum begins 5 hours after the onset of estrus and the second metaphase is reached just before ovulation (92). The cumulus is shed by 8 hours after mating (93). The ova remain fertilizable for 12 to 18 hours after they have been shed, but if they are fertilized as late as this the rate of embryonic mortality rises (5). By the use of sex-chromatin techniques it has been shown for 7 intersexes that all were genetic females (94). Tables have been constructed giving the prenatal growth of pigs (95).

THE MALE

Primary spermatocytes first appear in the testes at about 84 days and secondary spermatocytes at 105 days, and by 147 days spermatozoa are present (96). Another account gives the appearance of spermatids at 90 days and of spermatozoa in the epididymis at 121 days (97). The greatest development of the testis occurs from 4 to 8 months of age, while the accessory organs grow at 7 to 8 months. The first ejaculation takes place at 6 to 7 months, when the boar weighs from 50 to 80 kg. (98). Another study gave the time of first appearance of spermatozoa at 180 days (99). The two testes are about 1/250 of the body weight at maturity, the left usually being the larger; and the epididymides are about one third the weight of the testes. The vasa efferentia number from 1 to 5, and the vas deferens is about 25 to 30 cm. long (100). The length of the epididymis of a Mangalita boar was 62 to 64 meters, and the number of efferent ducts was 8 to 12 (101). In one testis the length of seminiferous tubules was between 3,000 and 6,000 meters (102). The epididymides contain from 175,000 to 200,000 million spermatozoa. Of these, 68.1 per cent were in the tail and 1.4 per cent in the anterior lobe of the head (103). The accessory glands are very large: the seminal vesicles weigh from 151 to 844 g. when full, and the contents weigh 39 to 507 g. Cowper's glands weigh 146 to 209 g. when empty, and the contents, 20 to 178 g. The prostate weighs 15 to 26 g. The boar also possesses a disseminate gland not at all like the prostate in structure, but in close proximity to it (100).

Coitus takes 5 to 8 minutes, and the average volume of ejaculate is 255 cc. (104). The semen penetrates directly into the cervix during mating (105).

The average ejaculate contains about 85 billion spermatozoa. Ejaculation occurs in stages; it increases to the third minute and is maintained to 5 minutes, then there is a fall in the rate and a final rise. Most spermatozoa are found in the first ejaculate. Ejaculation at 24-hour intervals does not materially reduce the volume, but greater frequency does. The pH of normal semen is 7.3 to 7.8. If it is allowed to stand, a large quantity of gelatinous and waxy materials settles out. This gelatinous material is secreted partly by the vesiculae seminales. Removal of these glands has little effect upon the number of spermatozoa, although the glands provide about 25 per cent of the ejaculate. Cowper's glands secrete the waxy material, and their secretion reacts with that of the seminal vesicles to produce the gelatinous material. They provide about 19 per cent of the total ejaculate. Without either seminal vesicles or Cowper's glands, the boar produces a watery, thin semen containing no gelatinous or syrupy material. The prostate and urethral glands produce about 56 per cent of the total secretion. This is clear and almost as thin as water. The testes produce about 2 per cent of the total volume (100).

Removal of Cowper's glands, or of the seminal vesicles, or both, is without effect upon fertility (100), in contrast to the condition found in the rat.

A few points of interest in the composition of these secretions follow:

Vesiculae seminales, pH 6.4 to 6.8, relatively high in K and glucose, low in Cl.

Cowper's glands, pH 7.2 to 7.3, relatively high in Na and Ca, very high in Mg.

Epididymal secretion, pH 6.7 to 6.9, high in P.

Prostate and urethral glands, pH 7.5 to 8.5 (100).

The mean length of a spermatozoon is 54.6 μ (49.2–62.4); the head is 8.5 μ (7.2–9.2) $\times$ 4.2 μ (3.6–4.8) (106). The midpiece is 10 μ long while the tail is 30 μ. The acrosome has two components—an outer larger and an inner smaller one (107). The cryptorchid testis contains 1 B.U. of androgen to 87 g. of tissue, and the normal testis contains I B.U. to 39 g. (108). The following levels of gonadic hormones have been found in the urine: androgens, 90 to 134 M.U. per liter; estrogens, 1,170 to 1,640 M.U. per liter (109). Estrone and estradiol—17β have been found in the urine, but not estriol (110). The semen contains 2.4 to 12.6 mg./100 ml. of fructose; 5.9 to 23.1 mg./100 ml. of ergothioneine, and 32 to 156 mg./100 ml. of citric acid (111). Inositol is secreted by the seminal vesicles and 1.8 g. is present in 100 ml. of secretion (112). It is also present in seminal plasma to the extent of 600 to 700 mg./100 ml. and in the epididymal secretion at 52 mg./100 ml. (113).

Castration at 200 days of age, after the accessory organs have grown somewhat under the influence of androgens, does not produce organs as small as those which result if the operation is performed at 50 or 100 days (114).

The spermatozoa are in the oviduct by 30 minutes after mating and are viable in the female tract for 42.5 hours, but they retain their fertilizing capacity for 25 to 30 hours only. The ova may be fertilized for from 10 to 21 hours after they were shed (14). The minimum volume of fluid needed to ensure complete fertility is from 30 to 50 ml. (115).

1. Panouse, J. B. Trav. de l'Inst. Sci. Chérifien, Ser. Zool., 5: 1957.
2. Heptner, W. G., L. G. Morosowa-Turowa, and W. I. Zalkin. Die Säugetiere in der Schutzwaldzone. Berlin, 1956.
3. Hatt, R. T. Univ. Michigan Mus. Zool., Misc. Publ. 106, 1959.
4. Wodzicki, K. A. New Zealand DSIR, Bull. 98, 1950.
5. Burger, J. F. Ondestepoort J. Vet. Res., Suppl. 2: 1–218, 1952.
6. Pomeroy, R. W. J. Agric. Sci., 54: 57–66, 1960.
7. Schmidt, K., and W. Bretschneider. Tierzucht, 8: 119–125, 1954.
8. Zimmerman, D. R., H. G. Spies, E. M. Rigor, H. L. Self, and L. E. Casida. J. Animal Sci., 19: 687–694, 1960.
9. Robertson, G. L., R. H. Grimmer, L. E. Casida, and A. B. Chapman. J. Animal Sci., 10: 647–656, 1951.
10. Reddy, V. B., J. F. Lasley, and D. T. Mayer. Missouri Agric. Exp. Sta., Res. Bull. 666, 1958.
11. Haines, C. E., A. C. Warnick, and H. D. Wallace. J. Animal. Sci., 18: 347–354, 1959.
12. Guinto, M. G. Philippine Agr., 41: 19–26, 1957.
13. Phillips, R. W., and J. H. Zeller. AR., 85: 387–400, 1944.
14. Ito, S., A. Kudo, and T. Niwa. Ann. Inst. Nat. Agron. Japan, Ser. D, Ann. Zootech., 8: 105–107, 1959.
15. Joubert, D. M. South African J. Sci., 54: 183–184, 1958.
16. Struve, J. Fühlings, Landw. Zeit., 60: 833–838, 1911.
17. McKenzie, F. F., and J. C. Miller. Missouri Agric. Exp. Sta., Res. Bull., 285: 43, 1930.
18. Berge, S. Meld. Norg. Landbr. Högsk., 20: 82–119, 1940.
19. Kupfer, M. Union of South Africa, Dept. Agric., Dept. Vet. Education and Res., 13th and 14th Rpt., 1211–1270, 1928.
20. Lewis, L. L. Oklahoma Agric Exp. Sta., Bull. 96, 1911.
21. Pitkjanen, J. G. Vet. Med. Praha, 7: 487–494, 1962.
22. Haring, F. Zuchtungskunde, 12: 1–19, 1937.
23. McKenzie, F. F. Missouri Agric. Exp. Sta., Bull. 310: 15–16, 1932.
24. McKenzie, F. F. Missouri Agric. Exp. Sta., Res. Bull. 86, 1926.
25. Warwick, B. L. AR., 33: 29–33, 1926.
26. Krallinger, H. Arch. Tiernährung u. Tierzucht, 8: 436–452, 1933.
27. Hancock, J. L., and G. J. R. Hovell. Anim. Prod., 4: 91–96, 1962.
28. Krallinger, H. F., and H. Schott. Arch. Tiernährung u. Tierzucht, 9: 41–49, 1933.
29. Squiers, C. D., G. E. Dickerson, and D. T. Mayer. Missouri Agric. Exp. Sta., Res. Bull. 494, 1952.
30. Corner, G. W. CE., 276: 119–146, 1921.
31. Andersen, D. AJP., 82: 557–569, 1927.
32. Pomeroy, R. W. J. Agric. Sci., 45: 327–330, 1955.
33. Lush, J. L. J. Hered., 12: 57–71, 1921.
34. Minano, G. Philippine J. Animal Ind., 4: 225–232, 1937.
35. Carmichael, W. J., and J. B. Rice. Illinois Agric. Exp. Sta., Bull. 226, 1920.
36. Olbrycht, T. M. J. Agric. Sci., 33: 28–43, 1943.
37. Johansson, I. Zeit. Tierzucht u. Zuchtungsbiol., 15: 49–86, 1929.

38. Hammond, J. J. Agric. Sci., 6: 263–277, 1914.
39. Asdell, S. A., and J. P. Willman. J. Agric. Res., 63: 345–353, 1941.
40. Schmidt, J., E. Lamprecht, and H. Staubesand. Zeit. Tierzucht u. Zuchtungsbiol., 36: 55–100, 1936.
41. Johnson. L. E. Proc. South Dakota Acad. Sci., 24: 27–31, 1944.
42. Dechambre, P., and J. Malterre. Rev. Zootech., No. 12: 443–444, 1934.
43. Krizenecky, J. Sbornik Ceskoslovenske Akad. Zemed., 10: 351–358, 1935.
44. Deakin, A., and E. B. Fraser. Scientific Agric., 15: 458–462, 1935.
45. Parkes, A. S. J. Agric. Sci., 15: 285–299, 1925.
46. Parkes, A. S. Biol. Rev., 2: 1–51, 1926–7.
47. Casida, L. E. AR., 61: 389–396, 1935.
48. Corner, G. W. AJA., 26: 117–183, 1919.
49. Corner, G. W. CE., 222: 69–94, 1915.
50. Yamashita, T. Japanese J. Vet. Res., 7: 177–201, 1959.
51. Yamashita, T. Japanese J. Vet. Res., 8: 107–125, 1960.
52. Corner, G. W. CE., 592: 1–8, 1948.
53. Andersen, D. H. CE., 362: 107–123, 1926.
54. Wilson, K. M. AJA., 37: 417–429, 1926.
55. Altmann, M. J. Comp. Psychol., 31: 481–498, 1941.
56. Snyder, F. F., and G. W. Corner. Am. J. Obst. Gyn., 3: 358–366, 1922.
57. Betteridge, K. J. J. Reprod. and Fertility, 3: 410–421, 1962.
58. Anstod, R., and O. Garm. Nature, 184: 999–1000, 1959.
59. Hodek, R., and R. Getty. Am. J. Vet. Res., 20: 573–577, 1959.
60. Andersen, D. H. AJA., 42: 255–305, 1928.
61. Snyder, F. F. Bull. Johns Hopkins Hospital, 34: 121–125, 1923.
62. Cleveland, R., and J. M. Wolfe. AJA., 53: 191–220, 1933.
63. Smith, P. E., and C. Dortzbach. AR., 43: 277–297, 1933.
64. Nelson, W. O. AJA., 52: 307–332, 1933.
65. Keye, J. D. Bull. Johns Hopkins Hospital, 34: 60–63, 1923.
66. King, J. L. AJP., 81: 725–737, 1927.
67. King, J. L. AJP., 99: 631–637, 1932.
68. Adams, E. E., 26: 891–894, 1940.
69. Wislocki, G. B., and A. F. Guttmacher. Bull. Johns Hopkins Hospital, 35: 246–252, 1924.
70. Whitelaw, M. J. Am. J. Obst. Gyn., 25: 475–484, 1933.
71. Allen, E., and E. A. Doisy. Physiol. Rev., 7: 600–650, 1927.
72. Roth, S. Y., D. T. Mayer, and R. Bogart. Am. J. Vet. Res., 2: 436–438, 1941.
73. Kust, D., and M. Struck. Deutsche Tierarzt. Woch., 42: 54–56, 1934.
74. Faiermark, S. E., and V. Bigos. Probl. Zivotn., No. 1: 73–75, 1933.
75. Faiermark, S. E. Probl. Zooteh. Eksp. Endokrinol., 2: 71–77, 1935.
76. Faiermark, S. E., and L. S. Singerman. Bull. Biol. Med. Exp. U.R.S.S., 6: 89–92 1938.
77. Velle, W. Am. J. Vet. Res., 19: 405–408, 1958.
78. Velle, W. Vet. Rec., 72: 116–118, 1960.
79. Nishikawa, Y., Y. Waide, and A. Soejima. Bull. Nat. Inst. Agric. Sci., Chiba, G10: 221–240, 1953.

80. Spörri, H. Experientia, 7: 267–268, 1951.
81. Mayer, D. T., B. R. Glasgow, and A. M. Gawienozski. J. Animal Sci., 20: 66–70, 1961.
82. Kimoura, G., and W. S. Cornwell. AJP., 123: 471–476, 1938.
83. du Buisson, F. du M., and L. Dauzier. Ann. de Zootech., Suppl., 147–159, 1959.
84. Anderson, L. L., R. L. Butcher, and R. M. Melampy. E., 69: 571–580, 1961.
85. Edgar, D. G. Nature, 170: 543–544, 1952.
86. Ulberg, L. C., R. H. Grummer, and L. E. Casida. J. Animal Sci., 10: 665–671, 1951.
87. Witschi, E. E., 27: 437–446, 1940.
88. West, E., and H. L. Fevold. PSEBM., 44: 446–449, 1940.
89. Day, B. N., L. L. Anderson, L. N. Hazel, and R. M. Melampy. J. Animal Sci., 18: 675–682, 1959.
90. Hollandbeck, R., B. Baker, Jr., H. W. Norton, and A. V. Nalbandov. J. Animal Sci., 15: 418–427, 1956.
91. Hisaw, F. L., and M. X. Zarrow. PSEBM., 69: 395–398, 1948.
92. Spalding, J. F., R. O. Berry, and J. G. Moffitt. J. Animal Sci., 14: 609–620, 1955.
93. Hancock, J. L. J. Reprod. and Fertility, 2: 307–331, 1961.
94. Cantwell, G. E., E. F. Johnston, and J. H. Zeller. J. Hered., 49: 199–202, 1958.
95. Warwick, B. L. J. Morphol. and Physiol., 46: 59–84, 1928.
96. Phillips, R. W., and F. N. Andrews. Massachusetts Agric. Exp. Sta., Bull. 331, 1936.
97. Niwa, T. Bull. Nat. Inst. Agric. Sci., Chiba, G8: 17–29, 1954.
98. Niwa, T., and A. Mizuho. Bull. Nat. Inst. Agric. Sci., Chiba, G9: 141–159, 161–177, 1954.
99. Hauser, E. R., G. E. Dickerson, and D. T. Mayer. Missouri Agric. Exp. Sta., Res. Bull. 503, 1952.
100. McKenzie, F. F., J. C. Miller, and L. C. Bauguess. Missouri Agric. Exp. Sta., Res. Bull. 279, 1938.
101. Ghetie, V. AA., 87: 369–374, 1939.
102. Bascom, K. F., and H. L. Osterud. AR., 31: 159–169, 1925.
103. Novoseljeev, D. V. Sovetsk. Zooteh., 6(7): 76–83, 1951.
104. Rodolfo, A., and T. Timojeeva. Probl. Zivotn., No. 9–10: 101–102, 1932.
105. Milovanof, V. K. Probl. Zivotn., No. 4: 31–34, 1932.
106. Niwa, T., and A. Mizuho. Bull. Nat. Inst. Agric. Sci., Chiba, G8: 31–42, 1954.
107. Hancock, J. L. J. Roy. Micros. Soc., 76: 84–97, 1957.
108. Hanes, F. M., and C. W. Hooker. PSEBM., 35: 549–550, 1937.
109. Marsman, W. S. Acta Neerl. Morph. Norm. et Path., 1: 115–128, 1937.
110. Velle, W. Acta Endocrinol., 28: 255–261, 1958.
111. Glover, T., and T. Mann. J. Agric. Sci., 44: 355–360, 1954.
112. Mann, T. Nature, 168: 1043–1044, 1951.
113. Hartree, E. F. Biochem. J., 66: 131–137, 1957.
114. Baker, J. R. Brit. J. Exp. Biol., 5: 187–195, 1927–8.
115. Ito, S., T. Niwa, and A. Kudo. Res. Bull. Zootec. Exp. Sta., Japan, 55: 1948.

Sus barbatus Müller. In Borneo this hog has no fixed season for breeding.

The number of young is usually 2, sometimes 3 (1). However, litters of from 4 to 11 have been reported (2).

S. salvanius Hodgson. In the London zoo the pygmy hog has had litters of from 1 to 4 in May and June (3). In the wild 3 to 4 to the litter is usual (4).

S. vittatus Müller and Schlegel. In the London zoo this Sumatran pig has had a litter of 7 pigs in May (3). From 4 to 7 has been given as the usual number (5).

1. Pfeiffer, P. Mammalia, 23: 277–303, 1959.
2. Lyon, M. W. Proc. Nat. Mus., United States, 40: 53–146, 1911.
3. Zuckerman, S. PZS., 122: 827–950, 1952–3.
4. Jerdon, T. C. The Mammals of India. Roorkee, India, 1867.
5. Kerbert, C. Bijd. Dierk., 22: 185–101, 1922.

Phacochoerus aethiopicus Pallas

WART HOG

In Rhodesia this hog mates in late June and early July. The young, usually 4, are born from late October to early November, or later (1). In South West Africa the usual season of birth is October and November (2), while in French Guinea it was February and March (3). In the Belgian Congo pregnancies were found from March to May (4); further north the young are born at all seasons, but most during the rains (5). The observed gestation periods have been from 171 to 175 days (6).

1. Hubbard, W. D. JM., 10: 294–297, 1929.
2. Shortridge, G. C. The Mammals of South West Africa. London, 1934.
3. Prunier, —. Mammalia, 10: 146–148, 1946.
4. Frechkop, S. EPNU., 1944.
5. Rode, P. Mammifères Ongulés de l'Afrique Noire. Paris, 1943.
6. Brown, C. E. JM., 17: 10–13, 1936.

Hylochoerus meinertzhageni Thomas

In French Guinea this hog had from 6 to 8 young, as a rule. They were born in February or March (1). The gestation period may be 125 days (2).

1. Prunier, —. Mammalia, 10: 146–148, 1946.
2. Kenneth, J. H. Gestation Periods. Edinburgh, 1943.

Babirussa babyrussa L.

BABIRUSA

In the London zoo this hog of the Celebes has given birth to a single young in each of the months January, March, and April. The gestation period has been from 125 to 150 days (1).

1. Zuckerman, S. PZS., 122: 827–950, 1952–3.

TAYASSUIDAE

Tayassu angulatus Cope

PECCARY

The peccary probably breeds in June in Panama, and the number of young is usually 2 (1). The uterus is bicornuate with relatively short and coiled cornua. The oviducts are short and coiled, and the fimbria form well-defined leaflets. The ovaries are not encapsulated. The old corpora lutea seem to persist much longer than is usual, since two sets of two each were found during pregnancy. The larger ones, 10 mm. in diameter, contained two cell types, large cells with dense cytoplasm, and much smaller cells with less dense cytoplasm. The smaller corpora lutea, 7 mm. in diameter, consisted mainly of the second type of cells, but they stained more deeply than they did in the larger corpora lutea (2).

A single testis weighed 11.3 g., and the seminiferous tubules were 677 meters long. The length of a spermatozoon was 23.4 μ (3).

1. Enders, R. K. HMCZ., 78: 383–502, 1935.
2. Wislocki, G. B. JM., 12: 143–149, 1931.
3. Knepp, T. H. Proc. Pennsylvania Acad. Sci., 13: 58–62, 1939.

Tayassu tajacu L.

COLLARED PECCARY

This peccary usually has 2 young but may have as many as 6. Births have been recorded in April, August, and November after gestation of from 112 to 116 days (1). In one account, 34 of 44 births were in July and August (2). The species is also said to breed all year. Six gestations were between 142 and 148 days, 3 of them 144 days (3). In the London zoo births have occurred throughout the year, mostly twins but with a few singles (4).

1. Hall, E. R., and K. R. Kelson. Mammals of North America. New York, 1959.
2. Neal, B. J. AMN., 61: 177–190, 1959.
3. Sowls, L. K. JM., 42: 425–426, 1961.
4. Zuckerman, S. PZS., 122: 827–950, 1952–3.

Tayassu pecari Fisher. This species usually has 2 young to the litter (1), and the females breed at all times of the year (2).

1. Goodwin, G. G. AM., 87: 271–474, 1946.
2. Leopold, A. S. Wildlife of Mexico. Game Birds and Mammals. Berkeley, Calif., 1951.

HIPPOPOTAMIDAE

Hippopotamus amphibius L.

HIPPOPOTAMUS

The hippopotamus has no fixed season for breeding (1). In the London zoo births have occurred throughout the year (2). The heat period lasts from 4 to 7 days (3), and it occurs at monthly intervals. Heat periods are resumed about 40 days after calving (3). Mating is usually early at night and is repeated at least two nights consecutively (4). Usually one young is born but 2 sets of twins have been recorded (4). Combined data on the duration of gestation give a mean value of 237 $\pm$ 1.2 days with a spread from 225 to 257 days. Puberty in the male is reached at 5 years and in the female at 4 (5), but one report gives 36 months as the usual age (6). In the Cleveland zoo it has occurred at 3 years in both sexes (7).

1. Verschuren, J. EPNG., 1958.
2. Zuckerman, S. PZS., 122: 827–950, 1952–3.
3. Sailer, O. AA., 145: 835–839, 1950.
4. Verheyen, R. Monographie Ethologique de l'Hippopotame (*Hippopotamus amphibius* Linné). Brussels, 1954.
5. Rode, P. Mammifères Ongulés de l'Afrique Noire. Paris, 1943.
6. Kerbert, C. Bijd. Dierk., 22: 185–191, 1922.
7. Anon. Inter Zoo Year Book, 2: 90, 1960.

Choeropsis liberiensis Morton

PYGMY HIPPOPOTAMUS

This species breeds the year-round and estrus recurs at intervals of about a month. Heat lasts for 3 days, the act of mating about an hour. The duration of gestation is from 201 to 210 days, and the female comes in heat about 12 to 16 days after the young have been weaned (1). In the London zoo births of single young have occurred throughout the year (2). The weight of a single testis from an immature male was 4.3 g., and the length of the seminiferous tubules 918 meters. A spermatozoon was 31.5 μ long (3).

1. Steinmetz, H. Zool. Garten, 9: 255–263, 1937.
2. Zuckerman, S. PZS., 122: 827–950, 1952–3.
3. Knepp, T. H. Proc. Pennsylvania Acad. Sci., 13: 58–62, 1939.

CAMELIDAE

Lama

Lama glama L. The llama breeds from spring to early fall and the single young, or occasional twins (1), are born after an 11-month gestation (2). In the London zoo births have been distributed throughout the year (3).

L. vicugna Molina. The vicuña breeds all year, but mostly from April to June (1). Births are mostly in February or early in March (4). A female with a single fetus, taken in September, was also lactating (5).

1. Cabrera, A., and J. Yepes. Historia Natural Ediar. Mamiferos Sud-Americanos. Buenos Aires, 1940.
2. Brown, C. E. JM., 17: 10–13, 1936.
3. Zuckerman, S. PZS., 122: 827–950, 1952–3.
4. Koford, C. B. Ecol. Monogr., 27: 153–219, 1957.
5. Pearson, O. P. HCMZ., 106: 117–174, 1951–2.

Camelus bactrianus L.

BACTRIAN CAMEL, TWO-HUMPED CAMEL

The bactrian camel is polyestrous, having cycles all the year round. These are variable in length; the usual interval between heats is from 10 to 20 days; and the duration of heat 1 to 7 days, with a mode of 3 to 4 days (1). The mean cycle length is about 14 days, and the foal heat is experienced the day after calving, more rarely after 2 to 3 days. The suckling female comes in heat regularly but the duration is shorter than it is in the dry camel (2).

At the beginning of heat the graafian follicle measures from 1 to 1.5 cm. in diameter, and at ovulation it has grown to 2.5 to 3 cm. The optimum time of mating is from 3 to 5 days before the end of heat. There is little obvious change in the appearance of the vagina and cervix during heat, but the vaginal smears show a decrease in leucocytes and an increase in epithelial cells (2).

The mean duration of pregnancy in 850 cases was 406 days with a spread from 370 to 440 days, calculated from the last mating. From single matings the mean was 410 to 411 days. One young is usually born at a time, and males are carried for 2 days longer than the females, but this difference is not statistically significant (3).

Both the males and females have well-developed glands at the summit of the neck, just below the occiput. In London these are active in March in the males, but the glands of the females are not noticeably active during the breeding season (4). Rut in London and in Mongolia is in spring (5).

1. Bošaev, J. Konevodstvo, No. 4: 44–50, 1938.
2. Barmincev, J. B. Konevodstvo, No. 8–9: 26–32, 1938.
3. Barmincev, J. Konevodstvo, No. 1: 42, 1939.
4. Pocock, R. I. PZS., 840–986, 1910.
5. Heape, W. QJMS., 44: 1–70, 1901.

Camelus dromedarius L.

DROMEDARY

In one account puberty is said to be reached in both sexes at age 3 years (1); in another the females mate at age 4 years and will breed until they are 30 years old, the males are used for breeding at 6 years and continue for 15 to 20 years (2). The breeding season is from December to March. Estrus may be recognized by restlessness and swelling and discharge from the vulva. They calve every other year and the gestation period lasts 12 to 13 months (2). Thirty-three pregnancies averaged 389.9 ± 2.1 days, and females were carried 1.3 days longer than males (3).

The ovary has no ovulation fossa and there are many more double ovulations than twin births. The left ovary ovulates more frequently than does the right. The corpus luteum persists throughout pregnancy; its maximum size is about 2 cm., and weight about 4 g. Externally it appears to be gray or bluish-gray in color, but the lutein tissue is at first reddish-orange, changing to orange-brown as it gets older. It retrogresses quickly after the calf is born (4). The glands in the neck below the occiput are active in both sexes. In London this activity is noticeable in June (5).

1. Iwema, S. Veeteelten Zuivelber., 3: 390–394, 1960.
2. Yasin, S. A., and A. Wahid. Agric. Pakistan, 8: 289–297, 1957.
3. Mehta, V. S., A. H. A. Prakash, and M. Singh. Indian Vet. J., 39: 387–398, 1962.
4. Tayes, M. A. F. Vet. J., 104: 179–186, 1948.
5. Pocock, R. I. PZS., 840–986, 1910.

TRAGULIDAE

Hyemoschus aquaticus Ogilby

WATER CHEVROTAIN

In the London zoo a female has given birth to a single calf in January (1).

1. Zuckerman, S. PZS., 122: 827–950, 1952–3.

Tragulus

CHEVROTAIN

Tragulus javanicus Osbeck. In the London zoo the single young has been born at all seasons except winter, but mostly in October and November (1). In Borneo birth is usually in December after a gestation of at least 4 months (2). A pregnant female has been found in March (3).

T. meminna Erxleben. In India, rut is in June and July and the 2 young are born at the close of the rainy season (4). The gestation period is 120 days (5). In Ceylon there is no regular season, but in the southwest young are usually running with the females from October to December (6).

T. pretullius Miller. All females were pregnant or had recently given birth in July (7).

T. russeus Miller. Most of the females taken from January 22 to February 5 were pregnant (8).

T. stanleyanus Gray. In the London zoo single calves have been born in January, March, August, and October (1).

1. Zuckerman, S. PZS., 122: 827–950, 1952–3.
2. Banks, E. J. Malay Br. Roy. Asiatic Soc., 9(2): 1–139, 1931.
3. Sanborn, C. C. Fieldiana Zool., 33: 89–158, 1952.
4. Blanford, W. T. Fauna of British India. Mammals. London, 1888–91.
5. Kenneth, J. H. Gestation Periods. Edinburgh, 1943.
6. Phillips, W. W. A. Ceylon J. Sci., 14B: 1–50, 1927–8.
7. Miller, G. S. Proc. Nat. Mus. United States, 31: 247–286, 1907.
8. Miller, G. S. Proc. Nat. Mus. United States, 26: 437–483, 1903.

CERVIDAE

MOSCHINAE

Moschus moschiferus L.

MUSK DEER

This deer attains puberty before it is 1 year old. Rut takes place in January and the young are born in June after a gestation of 160 days (1). One or 2 is the usual number at a birth (2). In Siberia mating is in December and January. The young, usually 1, occasionally 2, are born late in April or in June (3).

1. Blanford, W. T. Fauna of British India. Mammals. London, 1888–91.
2. Jerdon, T. C. The Mammals of India. Roorkee, India, 1867.
3. Flerov, K. K. Fauna of U.S.S.R. Mammals, I(2): 1952. Washington, 1960.

MUNTIACINAE

Muntiacus muntjac Zimmermann

MUNTJAC

In the north of India this deer mates mostly in January and February. The 1 or 2 young are born in July and August. In other districts it has no regular breeding season (1). The males shed their antlers in May (1, 2) and renew them in August but it is doubtful whether this occurs annually (3). In the London zoo births have occurred throughout the year (4).

1. Lydekker, R. The Game Animals of India, Burma, Malaya and Tibet. London, 1924.
2. Lydekker, R. The Deer of All Lands. London, 1898.
3. Banks, E. J. Malay Br. Roy. Asiatic Soc., 9(2): 1–139, 1931.
4. Zuckerman, S. PZS., 122: 827–950, 1952–3.

Muntiacus reevesi Ogilby. At Woburn this deer has bred at all times of the year (1). In the London zoo a single fawn has been born at a time during the months from April to December (2).

1. Bedford, Duke of, and F. H. A. Marshall. PRS., 130B: 396–399, 1942.
2. Zuckerman, S. PZS., 122: 827–950, 1952–3.

CERVINAE

Dama dama L.

FALLOW DEER

In Britain the females breed at age 2 years. The period of rut is October and the young, usually 1, but rarely twins, are born in June or early July (1), after a gestation of 8 months (2). In Russia, rut is from the end of September into October. The older does drop their fawns in June and the younger ones in July (3). They are monestrous (4). The velvet is shed from the antlers in August and the antlers are shed at the end of April (1). The necks of the males swell during the rutting season (5).

1. Whitehead, G. K. Deer and Their Management in the Deer Parks of Great Britain and Ireland. London, 1950.
2. Brown, C. E. JM., 17: 10–13, 1936.
3. Flerov, K. K. Fauna of U.S.S.R. Mammals, I(2): 1952. Washington, 1960.
4. Lydekker, R. The Deer of All Lands. London, 1898.
5. Millais, S. G. The Mammals of Great Britain and Ireland. London, 1906.

Dama mesopotamica Brooke. This deer has produced its fawns from June to September, according to the London zoo records (1).

1. Zuckerman, S. PZS., 122: 827–950, 1952–3.

Axis axis Erxleben

SPOTTED DEER, CHITAL

In the wild this deer breeds at any time of year, but there is a distinct rut at the end of April or the beginning of May. Corresponding to the extended breeding season is the fact that the antlers are shed at any time but mostly in August (1). In Ceylon young are born throughout the year but the majority at its beginning. The velvet is shed at any time and so are the antlers, though in southern Ceylon 75 per cent are shed in April and May (2). In captivity a series of diestrous cycles, each lasting 3 weeks, is experienced throughout the year (3), and at Woburn chitals will breed at any time of year (4), though most fawns are born between Christmas and Easter (5). In the London zoo births have been evenly distributed through the year, and the number born is nearly always 1; twins have been born only once in 80 births (6). The period of gestation is 7 to 7½ months.

A single testis weighed 14.9 g., and the seminiferous tubules were 1,071 meters long. The spermatozoon measured 40.6 μ (7). A male was castrated at 8 years old. The antlers grew in irregular and they remained in the velvet (8).

1. Brander, A. A. D. Wild Animals in Central India. London, 1923.
2. Phillips, W. W. A. Ceylon J. Sci., 14B: 1–50, 1927–8.
3. Heape, W. QJMS., 44: 1–70, 1901.
4. Bedford, Duke of, and F. H. A. Marshall. PRS., 130B: 396–399, 1942.
5. Whitehead, G. K. Deer and Their Management in the Deer Parks of Great Britain and Ireland. London, 1950.
6. Zuckerman, S. PZS., 122: 827–950, 1952–3.
7. Knepp, T. H. Proc. Pennsylvania Acad. Sci., 13: 58–62, 1939.
8. Bullier, P. Mammalia, 22: 271–274, 1958.

Axis (Hyelaphus) porcinus Zimmermann

HOG DEER

In India this species ruts in September and October (1). In Ceylon the probable time is April and May (2). The London zoo has recorded births all year, and usually a single fawn is born, though twins have been recorded twice in 55 births (3). Gestation lasts about 8 months (1). In India the antlers are shed in April (1), but in the Philadelphia zoo there is a record of a February shedding (4), and in the Washington zoo one in July (5).

1. Blanford, W. T. Fauna of British India. Mammals. London, 1888–91.
2. Phillips, W. W. A. Ceylon J. Sci., 14B: 1–50, 1927–8.
3. Zuckerman, S. PZS., 122: 827–950, 1952–3.
4. Brown, C. E. JM., 2: 39, 1921.
5. Hollister, N. JM., 1: 244–245, 1920.

Cervus canadensis Erxleben

WAPITI, ELK

In the Yellowstone region the does breed at about 31 months old, though a few have bred earlier. In any year from 74 to 94 per cent of those aged 2 to 3 years and older are pregnant (1). The yearling bucks produce spermatozoa (2). In the wild, rut is from late September well into October, and the young are born in June. In captivity the female experiences a continuous series of diestrous cycles lasting 3 weeks (3). The cycles are 21.2 days long and heat may be up to 17 hours (4). Pregnancy at any time of year is prevented only by the fact that the males do not rut during the time of casting and growth of antlers (3). These are shed in mid-March and they begin to grow again immediately. The velvet is lost in August (5) but young bucks often retain it until the antlers are shed (6). One young is usual. In the London zoo, of 67 births all in the period from May to October, 1 pair of twins was recorded (7); in the Yellowstone 2 pairs were reported in 896 pregnancies (1). The fetal sex proportion ($n = 1{,}629$) was 103 males to 100 females (1). The gestation period varies from 249 to 262 days (8).

It is believed that only one follicle ruptures at the time of estrus and a single corpus luteum is formed. Later, another follicle matures and ovulates, as a postconceptional corpus luteum appears at about the time the embryo becomes macroscopic. The two corpora lutea persist throughout gestation (9). One such recently ruptured follicle was found on December 10. There was a bloody exudate from the cavity, and large blood vessels had penetrated the follicle from several points in the theca. The mature corpus luteum of pregnancy is about 14 mm. in diameter, range 12.9 to 16.5 mm., while that of the cycle is about 5 mm. It often retains a central cavity that is later replaced by a plug of connective tissue. The corpus luteum becomes orange-colored as it degenerates. In pregnancy it increases in size up to 70 days (10).

A 2-year-old elk was castrated; the antlers were shed and new ones grew. These were small and crooked; they never lost their velvet or shed (11).

1. Kittams, W. H. JWM., 17: 177–184, 1953.
2. Conaway, C. JWM., 16: 313–315, 1952.
3. Heape, W. QJMS., 44: 1–70, 1901.
4. Morrison, J. A. Behaviour, 16: 84–92, 1960.
5. Seton, E. T. Life Histories of Northern Animals. New York, 1909.
6. Schwartz II, J. E., and G. E. Mitchell. JWM., 9: 295–319, 1945.
7. Zuckerman, S. PZS., 122: 827–950, 1952–3.
8. Rust, H. J. JM., 27: 308–327, 1946.
9. Halazon, G. C., and H. K. Buechner. Trans. North Am. Wildlife Conf., 21: 545–554, 1956.
10. Morrison, J. A. JWM., 24: 297–307, 1960.
11. Skinner, M. P. JM., 4: 252, 1923.

Cervus cashmiriensis Adams

HANGLU

This deer mates in October and the young are born in April after a gestation of 6 months (1). In the London zoo births have all been in June or July after a gestation period of 183 days (2). In Kashmir the antlers are shed in March (3). There are records of April shedding in the Washington and Philadelphia zoos (4, 5).

1. Blanford, W. T. Fauna of British India, Mammals. London, 1888–91.
2. Zuckerman, S. PZS., 122: 827–950, 1952–3.
3. Lydekker, R. The Deer of All Lands. London, 1898.
4. Hollister, N. JM., 1: 244–245, 1920.
5. Brown, C. E. JM., 2: 39, 1921.

Cervus elaphus L.

RED DEER

The females usually breed in their third year (1). In the males puberty is reached towards their fourth year but they try to bugle and to cover the does from the end of their second year (2). In captivity the females have a very extensive series of diestrous cycles (3). In the London zoo births have spread from May to September (4). The young is almost always single; twins are rare and the sex proportion of 881 newborn fawns was 59.3 per cent males (1). The gestation period varies from 225 to 245 days (4). In the Crimea, rut begins at the end of August and continues to October; in the Caucasus it begins a little later, while in the Altai region it is also later and is confined to September (2). In the females there is a heavy layer of apocrine glands at the side of the vagina. These glands show a most striking increase at the time of estrus (5).

The antlers are shed in the spring, from mid-March onward; older bucks shed a month earlier than the young ones (1). In the Altai shedding begins at the end of March but in the Crimea it begins at the end of February (2).

1. Whitehead, G. K. Deer and Their Management in the Deer Parks of Great Britain and Ireland. London, 1950.
2. Flerov, K. K. Fauna of U.S.S.R. Mammals, I(2). Washington, 1960.
3. Heape, W. QJMS., 44: 1–70, 1901.
4. Zuckerman, S. PZS., 122: 827–950, 1952–3.
5. Frankenberger, Z. Ceskoslovenska Morfol., 5: 255–265, 1957.

Cervus (Rucervus) duvauceli Cuvier

BARASINGHA

This deer is monestrous (1) but the rut is ill defined and mating is mostly from mid-December to mid-January (2). In Kashmir mating may be earlier, from late September to late November (3). In the London zoo the births of single young have been from June to August, and the gestation period was 340 to 350 days (4). Antlers are shed in March in the Kashmir district (2), and earlier in India (5). In the Washington and Philadelphia zoos there have been December, January, and April sheddings (6, 7).

1. Lydekker, R. The Deer of All Lands. London, 1898.
2. Brander, A. A. D. Wild Animals in Central India. London, 1923.
3. Heape, W. QJMS., 44: 1–70, 1901.
4. Zuckerman, S. PZS., 122: 827–950, 1952–3.
5. Blanford, W. T. Fauna of British India. Mammals. London, 1888–91.
6. Hollister, N. JM., 1: 244–245, 1920.
7. Brown, C. E. JM., 2: 39, 1921.

Cervus (Rucervus) eldi Guthrie

THAMENG

This deer mates from March to May in Burma, and the age at puberty is 18 months (1). In England rut is in August (2). In Manipur the antlers are shed in June, but in Lower Burma this occurs in September (1). In Great Britain they are shed in March (2). In Thailand the single young is born in October and the antlers are shed in mid-July (3).

1. Blanford, W. T. Fauna of British India. Mammals. London, 1888–91.
2. Whitehead, G. K. Deer and Their Management in the Deer Parks of Great Britain and Ireland. London, 1950.
3. Sanborn, C. C., and A. R. Watkins. JM., 31: 430–433, 1950.

Cervus (Rusa) timoriensis Blainville

RUSA

The females reach puberty at 27 months old. Rut extends from June to September and the single fawn is born after a gestation that has varied from 249 to 284 days. The antlers are shed from December to January and they are in velvet from April to May. In zoos the antlers are usually shed from May to July (1). In Great Britain the period of rut is at the end of winter (2), but in the London zoo single fawns have been born at all periods of the year (3).

1. Lydekker, R. The Game Animals of India, Burma, Malaya and Tibet. London, 1894.
2. Whitehead, G. K. Deer and Their Management in the Deer Parks of Great Britain and Ireland. London, 1950.
3. Zuckerman, S. PZS., 122: 827–950, 1952–3.

Cervus (Rusa) unicolor Bechstein

SAMBAR

This deer has a wide distribution in southern Asia and its reproductive behavior varies with the region. In the Peninsula of India the period of rut is from October to November (1). In the Ceylon hills rut is also at this time of year, but in the lowlands there is no definite season (2). In Sumatra a very extended season seems to be the rule (3). In the London zoo births have been spread throughout the year, with a peak at the end of May (4). Fawns are also dropped at any time of year at Woburn (5). One to 2 fawns are usually born at a time after a gestation of about 8 months (6). In the native habitat of the sambar birth is usually in May or early June (7).

The antlers are not shed regularly every season but only every 2 or 3 years. In the Himalayas they are shed in April, while March is the more usual month in the plains. At Woburn they are shed at any time of year (8). In European zoos they have been shed from April to September, but mostly in May and June (3). In the Washington zoo there is a record of a June shedding (9).

1. Blanford, W. T. Fauna of British India. Mammals. London, 1888–91.
2. Phillips, W. W. A. Ceylon J. Sci., 14B: 1–50, 1927–8.
3. van Bemmel, A. C. V. Treubia, 20: 191–262, 1949–50.
4. Zuckerman, S. PZS., 122: 827–950, 1952–3.
5. Lydekker, R. The Deer of All Lands. London, 1898.
6. Brander, A. A. D. Wild Animals in Central India. London, 1923.
7. Cahalane, V. H. Nat. Geographic Mag., 76: 463–510, 1939.
8. Lydekker, R. The Game Animals of India, Burma, Malaya and Tibet. London, 1924.
9. Hollister, N. JM., 1: 244–245, 1920.

Cervus (Sika) nippon Temminck

JAPANESE DEER

In this species puberty is reached in the second year, i.e., at an age between 16 and 18 months (1). It ruts in September to mid-November and the fawns are born at the end of May, after a gestation of 8 months (2). In Great Britain, rut is in autumn and the young are born from June to September (3). Estrus lasts for several days (1). In the London zoo births have been observed from April to October, with a peak in June. Five sets of twins have been born in 108 births (4). In Great Britain antlers are cast in spring (3), and likewise in the United States (5, 6).

1. Flerov, K. K. Fauna of U.S.S.R. Mammals, I(2). Washington, 1960.
2. Heptner, W. G., L. G. Morosowa-Turowa, and W. I. Zalkin. Die Säugetiere in der Schutzwaldzone. Berlin, 1956.
3. Whitehead, G. K. Deer and Their Management in the Deer Parks of Great Britain and Ireland. London, 1950.
4. Zuckerman, S. PZS., 122: 827–950, 1952–3.
5. Hollister, N. JM., 1: 244–245, 1920.
6. Brown, C. E. JM., 2: 39, 1921.

Cervus (Rusa) alfredi Sclater. In the London zoo this Philippine deer has had single fawns in December, March, and April (1).

1. Zuckerman, S. PZS., 122: 827–950, 1952–3.

Elaphurus davidianus Milne-Edwards

PÈRE DAVID'S DEER

At Woburn this domesticated species used to breed at the end of May and in June. The first fawn was born at the end of March. After the lapse of 50 years the period of rut now begins late in June and the fawns are born mostly from the third week of April into May. Only 1 fawn is born; there is no record of twins (1). The antlers are shed in October, but there is a great year-to-year variation (1). With special winter feeding antlers were shed in October and November. New ones grew, but these were shed at the end of January and a new set had grown by June and July (2). The gestation period is 250 days (3). This does not fit the Woburn data.

1. Wood-Jones, F., and Duke of Bedford. PZS., 121: 327–331, 1951–2.
2. Whitehead, G. K. Deer and Their Management in the Deer Parks of Great Britain and Ireland. London, 1950.
3. Kenneth, J. H. Gestation Periods. Edinburgh, 1943.

ODOCOILEINAE

Odocoileus hemionus Rafinesque

MULE DEER

This deer ruts from November to mid-December and the fawns are born in May and June. The antlers are shed in March (1). In northern Idaho the season is from mid-October through November, a little earlier than the usual one (2). In the London zoo the season seems to be a little later as most fawns are born early in August, with a scattering into September (3). In California the antlers have completed their growth by mid-July. The velvet is shed from July 20 to August 10. The necks of the males are swollen early in September; rut is late in October and early in November. The antlers are shed suddenly one month after the rut. Yearling males are not active in this area (4). In Utah the testes are swelling by the last week of October (5).

Some yearling does breed and carry a single fawn, but it is more usual for breeding to begin with 2-year-olds (6). The peak of mating is from November 15 to December 6, when most of the does are impregnated. There is a subsequent period from December 18 to January 3 when the remainder, about 12 per cent of all those impregnated, are successfully mated (7). In one California herd, in two consecutive years the number of pregnant does, including yearlings, has changed from 78 to 92 per cent. The number of embryos to each pregnant doe varied during these years from 1.34 to 1.50 (8). In one large sample examined in all months of pregnancy, the number of corpora lutea was 1,281 and of live fetuses 1,156, a loss of about 10 per cent. The heaviest prenatal mortality was in the first 2 or 3 months. The sex proportion of 1,408 fetuses was 108 males to 100 females. There was no indication of a differential mortality under normal circumstances, but in a period of malnutrition a greater proportion of males apparently perished. It was also noted that the smaller the fetus count, the higher was the proportion of males (9). There is some evidence that identical twins may occur: in two instances, twins have been associated with single corpora lutea. In another, triplets were found of which two had a common chorion (10). An average of 5 gestations gave 203 days, range 199 to 207 days (11).

In Montana an average of 1.66 fetuses was found for the pregnant does. Forty had 1 fetus; 56 had 2; 4 were with 3; and 1 had 4 fetuses (12). Eighty-seven per cent of does over 1 year old were found to be pregnant. Probably there is a second estrus if no conception occurs during the first. The embryo implants at 27 to 29 days after the doe was bred (13). In Idaho 66 per cent of yearlings were pregnant; of 2-year-olds, 93.5 per cent; while for older ones the figure was 97.6 per cent. The average corpus-luteum count was 0.9 for yearlings, and 1.86 for 3-year-olds and up. Two was the modal number (14).

The gonadotrophic-hormone potency of the anterior pituitary of female fawns and yearlings is high during their first summer. In winter it is low, but it gradually increases in spring. In adult does the peak is reached by late November and early December. Then the potency declines sharply to its lowest point in February and March. Afterwards it rises slowly for the remainder of pregnancy (12).

Castrated bucks either do not renew their antlers or retain them permanently in the velvet with remarkable distortions. They are known locally as "cactus bucks" (1). The tarsal and metatarsal glands of the males are markedly more active during rut than at other times. An antlered "doe" sheds her antlers and grows them at the usual time for the male. Her neck swells during the rutting season; she attacks bucks in rut and attempts to mount a doe (16).

1. Mearns, E. A. United States Nat. Mus., Bull. 56, 1907.
2. Rust, H. J. JM., 27: 308–327, 1946.
3. Zuckerman, S. PZS., 122: 827–950, 1952–3.
4. Dasmann, R. T., and R. D. Taber. JM., 37: 143–164, 1956.
5. Robinette, W. L., and J. S. Gashweiler. JWM., 14: 257–269, 1950.
6. Taber, R. D. California Fish and Game, 39: 177–186, 1953.
7. Chattin, J. E. California Fish and Game, 34: 25–31, 1948.
8. Lassen, R. W., C. M. Ferrel, and H. Leach. California Fish and Game, 38: 211–224, 1952.
9. Robinette, W. L., J. S. Gashweiler, J. B. Low, and D. A. Jones. JWM., 21: 1–16, 1957.
10. Robinette, W. L., J. S. Gashweiler, D. A. Jones, and H. S. Crane. JWM., 19: 115–136, 1955.
11. Golley, F. B. JM., 38: 116–120, 1957.
12. Hudson, P. JWM., 23: 234–235, 1959.
13. Hudson, P., and L. G. Browman. JWM., 23: 295–304, 1959.
14. McConnell, B. R., and P. D. Dalke. JWM., 24: 265–271, 1960.
15. Grieser, K. C., and L. G. Browman. E., 58: 206–211, 1956.
16. Cowan, I. McT. California Fish and Game, 22: 155–246, 1936.

Odocoileus virginianus Boddaert

WHITE-TAILED DEER, VIRGINIA DEER

In New York State the bucks reach puberty in their second year. In the southern part of the state many does breed in their first year, but about 4 weeks later than the adults. Further north it is unusual for yearling does to breed (1). The height of the season is from November 10 to 23, but to the north of the Adirondack Mountains it is from November 10 to 16, and, to the south, November 17 to 23. If a doe is not impregnated in her first heat a second estrus follows about 4 weeks later, to be succeeded by a third if she again fails to become pregnant (1, 2). In Florida mating is earlier and the fawns are born from January to March, in contrast to May and June further north (3). In Texas the testis volume increases from August to December and then decreases from January to March so that it is lowest for the year in April and May. Breeding is rather late, as the height of the season is from December 15 to January 5 (4). It is later still in Alabama, stretching from December 26 to February 21, with most matings in the last 8 days of January. This was among captive deer (5). In the northeastern part of

Alabama breeding was from late November through early January with the peak in early December (6).

The number of young is usually 2, but it varies greatly according to conditions of nourishment and the like. In the Adirondacks 78 per cent of adult does became pregnant in one season and, of these, 81 per cent carried single fawns, 18 per cent twins and 1 per cent triplets. Further south 92.3 per cent of adults were pregnant, 33 per cent with singles, 60 per cent with twins, and 7 per cent with triplets (7). The gestation period averages 204 days with a range from 197 to 222 days (8).

The males shed their antlers from late December to February in the north, earlier in the south (3). They begin to grow again after an interval of 2 to 6 weeks, and the velvet is lost in August and September. Rut, lasting about 2 months, is accompanied by swelling of the neck (9).

The weight of the testes fluctuates seasonally, being greatest at the time of rut and least from late winter to early summer. Spermatogenesis begins in July, is maximal in October, and diminishes in December and January. The lowest point is reached in June when the seminiferous tubules consist of a single layer of cells. Interstitial cells are larger and have clearer nuclei in October than in May and June. The prostate complex shows no seasonal variation, but the seminal vesicles are least in size in June and July, enlarge in August, and are maximal in October. The height of the secretory cells follows the same rhythm. Secretion is abundant in October, whereas the lumen of the gland contains black dried masses in June and July (10).

The antlers begin their annual growth when the testes and accessory organs are inactive, harden and lose their velvet when these glands are enlarging, and are shed when they begin to decline (10). Castration following loss of the velvet results in shedding within 30 days. New growth, which occurs at the normal time, is abnormal in shape, and the velvet is not lost again. Growth ceases at the usual time and part of the growth, being somewhat fragile, may be lost by accident. Renewed growth activity follows in the spring. Eventually an exaggerated burr is produced (11). These events have been interpreted as indications that antler growth is under the influence of a nontesticular hormone, possibly from the anterior pituitary, though direct evidence for the latter is lacking. Hardening and consequent loss of the antlers is believed to be due to the action of a testicular hormone (10). Some bucks were submitted to a regime of 16 hours daily of light. They came into rut, shed the velvet and antlers and renewed them about 2 weeks earlier than the normal (12).

In the fall the interstitial cells of the testes contain much steroid material.

In June the reaction for its detection is much less intense. In the fall the seminiferous tubule cells are high in acid and alkaline phosphatase, but in June their content is low (13). The weight of a single testis was 36.2 g., and the length of the seminiferous tubules was 2,121 meters (14).

The Mexican subspecies, *O. v. acapulcensis,* ruts in January and the fawns are born from July to August (15).

1. Cheatum, E. L., and G. H. Morton. JWM., 10: 249–263, 1946.
2. Gardiner, E. J. JM., 29: 184–185, 1948.
3. Hamilton, Jr., W. J. The Mammals of Eastern United States. Ithaca, N.Y., 1943.
4. Illige, D. JM., 32: 411–421, 1951.
5. Haugen, A. O. JM., 40: 108–113, 1959.
6. Adams, W. H. Ecol., 41: 706–715, 1960.
7. Cheatum, E. L. New York State Conservationist, 1(5): 18–32, 1947.
8. Haugen, A. O., and L. A. Davenport. JWM., 14: 290–298, 1950.
9. Seton, E. T. Life Histories of Northern Animals. New York, 1909.
10. Wislocki, G. B. Essays in Biology, in Honor of Herbert M. Evans. Berkeley, Calif., 1943. Pp. 631–653.
11. Caton, J. D. The Antelope and Deer of North America. New York, 1884.
12. French, C. E., L. C. McEwen, N. D. Magruder, T. Rader, T. A. Long, and R. W. Swift. JM., 41: 23–29, 1960.
13. Wislocki, G. B. E., 44: 167–189, 1949.
14. Knepp, T. H. Zoologica, 24: 329–332, 1939.
15. Davis, W. B., and P. W. Lukens, Jr. JM., 39: 347–367, 1958.

Mazama

BROCKET

Mazama americana Erxleben. The red brocket of South America has no known fixed breeding season, but the single fawns are usually born from December to April. A July birth has been recorded for the Matto Grosso (1), and a fetus near term, also in July (2). The antlers are shed at all times of the year (1).

M. simplicicornis Illiger. The wood brocket has been recorded as usually bearing two fawns at a time (3).

M. tema Rafinesque. This species usually has twins (4).

1. Cabrera, A., and J. Yepes. Historia Natural Ediar. Mamiferos Sud-Americanos. Buenos Aires, 1940.

2. Miller, F. W. JM., 11: 10–22, 1930.
3. Lydekker, R. The Deer of All Lands. London, 1898.
4. Goodwin, G. G. AM., 87: 271–474, 1946.

Hippocamelus antisensis d'Orbigny

HUEMUL

This deer was found pregnant with well-advanced fetuses late in December (1).

1. Koford, C. B. Ecol. Monogr., 27: 153–219, 1957.

Blastocerus

PAMPAS DEER

Blastocerus besoarticus L. This deer ruts at the end of summer and the single fawn is born in September and October in the Matto Grosso, and in April in Argentina (1). In Brazil the antlers are shed in May; they grow again immediately and the velvet is shed by September (2). In the London zoo single young have been born in May and August (3).

B. dichotoma Illiger. This deer breeds in October or November and has a single fawn (1).

1. Cabrera, A., and J. Yepes. Historia Natural Ediar. Mamiferos Sud-Americanos. Buenos Aires, 1940.
2. Miller, F. W. JM., 11: 10–22, 1930.
3. Zuckerman, S. PZS., 122: 827–950, 1952–3.

Pudu pudu Molina

PUDU

In the London zoo this South American deer has had its fawns in May (1).

1. Zuckerman, S. PZS., 122: 827–950, 1952–3.

Alces alces L.

ELK, MOOSE

Puberty is at age 16 to 28 months and females give birth first at 2 years while most males rut first at 2½ years (1). In Newfoundland 60 per cent of yearlings are in breeding condition (2). Rut is from mid-September to mid-October and the young are born late in May (3). They are probably conceived over 4 estrous periods with intervals of 30 days between them. Eighty-five per cent of does conceive at the first of them (4). The incidence of twin pregnancies is about 22 per cent, but in those with high summer range more are produced (4). In Newfoundland the twinning rate is 12 per cent, but twice as many does have two corpora lutea as produce twins in the uterus. The pregnancy rate among adults is 81 per cent (2). The antlers are shed in January or February, begin to grow in April, and the velvet is shed in July or August (5). In Russia the old bulls shed at the end of November, or, more frequently, in December; stags of age 3 to 4 years shed in January and February; younger ones shed still later (1). The gestation period is from 240 to 250 days (6).

1. Flerov, K. K. Fauna of U.S.S.R. Mammals, I(2). Washington, 1960.
2. Pimlott, D. H. JWM., 23: 381–401, 1959.
3. Hamilton, Jr., W. J. The Mammals of Eastern United States. Ithaca, N.Y., 1943.
4. Edwards, R. Y., and R. W. Ritcey. JWM., 22: 261–268, 1958.
5. Seton, E. T. Life Histories of Northern Animals. New York, 1909.
6. Kenneth, J. H. Gestation Periods. Edinburgh, 1943.

Rangifer tarandus L.

REINDEER, CARIBOU

There are many races of the reindeer, some of which have been given specific rank. In this account Lydekker's classification has been followed. This groups them all into one species, but, where possible, the subspecies has been indicated. Both sexes are antlered and shed annually.

The breeding season in Siberia is in September and October, and the calving season is from May to June. The males breed first at 1½ years of age (1). In the middle of Kamchatka, rut is towards the end of October, with the peak in the middle of November. On South Georgia, in the Antarctic, estrus has shifted towards March and calving towards October and November (2). Females breed until age 20 years (2). Normal bucks shed their antlers from November to January, and castrated ones from March to April (1). The number of young is usually 1, rarely 2, and the period of gestation is from 7 to 8 months (3). These remarks presumably refer to *R. t. sibiricus* Schreber.

In *R. t. caribou* Gmelin, the woodland caribou of North America, rut is in October. The males shed their antlers in early spring and the does in summer (4).

In *R. t. arcticus* Richardson, the barren-ground caribou of North America, rut is in October and the calving season is from May 15 to June 15. The bucks usually shed their antlers in November, but old and very young ones may carry them until late April. The does usually shed in May and June (5). The velvet is rubbed off beginning in August, and the antlers are clean by the end of September. Antler growth begins in May, even though they may have been shed in December (6). In Russia old stags shed shortly after rutting, and younger ones later. Growth begins in April. Females lose them 3 to 7 days after calving. Barren females shed before the pregnant ones, in March or April (2). Authorities are agreed that castrated or ovariectomized reindeer shed and regrow their antlers each year, contrary to the habit in other deer which have been investigated (7).

The testis weight increased from 15 g. in winter to 50 g. in September. During the winter a few spermatogonia and occasional spermatocytes may be found. As the first antler buds appear in March or April spermatogenesis begins. Retrogression begins as soon as the antlers are shed. Zimmermann-positive steroids are at a minimum in the urine during January and February, when the level is 12 to 15 mg. per liter. From March to July it reaches 20 to 23 mg./l. From August to December the level has risen to 30 to 45 mg./l. This report also states that the species is monestrous (8).

1. Sokolov, I. I. Trud. Arkt. Inst., 24: 67–128, 1935.
2. Flerov, K. K. Fauna of U.S.S.R. Mammals, I(2). Washington, 1960.
3. Mertz, P. A. Trud. Biol. Inst. Tomsk Univ., 6: 175–208, 1939.
4. Seton, E. T. Life Histories of Northern Animals. New York, 1909.
5. Murie, O. J. NAF., 54, 1935.
6. Strong, W. D. JM., 11: 1–10, 1930.

7. Wislocki, G. B. Essays in Biology, in Honor of Herbert M. Evans. Berkeley, Calif., 1943. Pp. 631–653.
8. Meschaks, P., and M. Nordkvist. Acta Vet. Scand., 3: 151–162, 1962.

Hydropotes inermis Swinhoe

CHINESE WATER DEER

In England, rut is in December and the fawns are born during the midsummer months, but some are dropped in May. Twins and triplets are common; of 8 pregnant does, 5 had 2 fetuses, 2 had 3, and 1 had 4 (1).

1. Whitehead, G. K. Deer and Their Management in the Deer Parks of Great Britain and Ireland. London, 1950.

Capreolus capreolus L.

ROE DEER

The roe deer is monestrous, mating in July and August, and the fawns, usually twins, are dropped in May (1). Implantation is delayed, the only case so far known in Cervidae. The period of gestation is about 40 weeks, with implantation of the embryo in December, but exceptional cases are known of immediate implantation with a reduced gestation period of 20 weeks (2). The time of breeding is exceptional among deer, but implantation and birth are at about the usual time. In Russia, rut and estrus begin in August and extend into September and occasionally into October. Rut begins in older bucks earlier than in younger ones, and it is earlier in the plains than in the mountains (3).

The antlers begin to grow in January and are shed in December (4). In Central Europe shedding may be earlier, in late August or September, while the velvet is shed in May (5). Castration before the velvet has been shed is not followed by casting of the antlers, but their further growth is abnormal; castration after the velvet has been shed leads to immediate casting and the growth of abnormal forms. Ovariectomy of the female does not result in any antler growth (6).

Bucks that are 2 years old or more experience testis development each February to May. Semen is then abundant until the beginning of August. Later in that month spermatogenesis ceases and the testes are quiescent until the following February. The interstitial cells develop from April to July; they are scanty from September to March. The epididymis contains spermatozoa from May to December, and sometimes into January so that the buck is able to fecundate females at that late date (5).

The embryo develops slowly from July to December, but from that time to the end of gestation, in May, development is rapid. Apparently some does conceive in November and December and, in these, there is no delayed development (5).

A castrated stag had incipient peruke formation. After two injections of 50,000 M.U. of progynon the growth began to shrink and it was shed. It began to grow again in August and September, although it was not the normal season for growth (7). Injections of 50,000 M.U. of progynon at 7- to 10-day intervals in October and November caused an immediate onset of growth; the velvet dried and the antlers were shed in December (8).

1. Heape, W. QJMS., 44: 1–70, 1901.
2. Prell, H. Zuchtungskunde, 13: 325–345, 1938.
3. Flerov, K. K. Fauna of U.S.S.R. Mammals, I(2). Washington, 1960.
4. Lydekker, R. The Deer of All Lands. London, 1898.
5. Stieve, H. Zeit. Mikros. Anat. Forsch., 55: 427–530, 1950.
6. Tandler, J., and S. Grosz. Die Biologischen Grundlagen der Sekundaren Geschlechtscharaktere. Berlin, 1913.
7. Blauel, G. Endokrinol., 15: 321–329, 1935.
8. Blauel, G. Endokrinol., 17: 369–372, 1936.

GIRAFFIDAE

PALAEOTRAGINAE

Okapia johnstoni Sclater

OKAPI

A pair mated at about 20-day intervals for periods of 3 to 7 days each (1) In the Basel zoo a male was born in March after a gestation of 446 days (2). Another gestation lasted 426 days (3).

1. Nouvel, J. Mammalia, 22: 107–111, 1958.
2. Lang, E. M. Inter Zoo Year Book, 2: 94, 1960.
3. Kenneth, J. H. Gestation Periods. Edinburgh, 1943.

GIRAFFINAE

Giraffa camelopardalis L.

GIRAFFE

In the Cleveland zoo both sexes have reached puberty at age 3 years (1). In the wild the species is said to mate from July to September and to drop the young from October to February (2), but accounts differ. One report gives young from May to September (3); another states that there is no fixed season and that the young are dropped at any time (4). In the London zoo births have been from February to October, with most from March to May (5). A series of estrous cycles of 12 to 15 days have been observed, stretching from June into October (6). One young, rarely twins, is born at a time, and the period of gestation has been observed to be between 14 and 15 months, or a week more (1).

Gonadotrophic hormones are present in the urine of the pregnant female (7). The sex proportion at birth for 117 giraffes has been 61.5 per cent males (8).

1. Anon. Inter Zoo Year Book, 2: 90, 1960.
2. Shortridge, G. C. The Mammals of South West Africa. London, 1934.
3. Dasmann, R. F., and A. S. Mossman. JM., 43: 533–537, 1962.
4. Innes, A. C. PZS., 131: 245–278, 1958.
5. Zuckerman, S. PZS., 122: 827–950, 1952–3.
6. Lang, E. M. Säugetierk. Mitt., 3: 1–5, 1955.
7. Wilkinson, J. F., and P. de Fremery. Nature, 146: 491, 1940.
8. Boulière, F. Mammalia, 25: 467–471, 1961.

ANTILOCAPRIDAE

Antilocapra americana Ord

PRONGHORN

The pronghorn has a limited breeding season in September and October, and the horns are shed immediately afterwards, from October to December (1). The new ones have already begun to sprout (2). The young are usually born from late May to early June, though in Texas they are born over a wide range of time, but mainly in June. Gestation lasts for 230 to 240 days (4). Records of young give 2 kids or fetuses on 29 occasions and singles on 2. The sex distribution of fetuses has been unusual. One record of 6 pairs of twins states that each pair were of like sex (5). Another of 17 pairs gave only 3 of like sex and 14 of unlike sex (6). The female breeds at 15 or 16 months old (4). In view of the relationship of this species to both the antelopes and the deer it would be interesting to learn the connection of the testis cycle to horn shedding. A report on one animal states that in a castrated buck horn development was abnormal and that the seasonal separation of the horns was incomplete (7).

1. Skinner, M. P. JM., 3: 33–39, 1922.
2. Seton, E. T. Life Histories of Northern Animals. New York, 1909.
3. Mearns, E. A. United States Nat. Mus., Bull. 56, 1907.
4. Einarsen, A. S. The Pronghorn Antelope and Its Management. Washington, 1948.
5. Tryon, C. A., and P. D. Buck. JM., 31: 192–193, 1950.
6. Chattin, J. E., and R. Lassen. California Fish and Game, 36: 328–329, 1950.
7. Pocock, R. I. PZS., 191–197, 1905.

BOVIDAE

BOVINAE

Strepsiceros strepsiceros Pallas

GREATER KUDU

The first calf is born when the female is 2 years old (1). There is apparently no fixed season, as authorities disagree about the season of birth, but October to March, or January to February (2), appears to be the most common time for the single calf to be born (3). The gestation period is about 214 days (4).

1. Dasmann, R. F., and A. S. Mossman. JM., 43: 533–537, 1962.
2. Ansell, W. F. H. PZS., 134: 251–274, 1960.
3. Selous, F. C. In R. Lydekker, The Great and Small Game of Africa. London, 1899.
4. Dekeyser, P. L. Les Mammifères de l'Afrique Noire Française. n.p., 1956.

Strepsiceros (Tragelaphus) scriptus Pallas

HARNESSED BUSHBOK

On the African coast this species breeds all year, but in the interior October to February is the usual season (1). In the Belgian Congo breeding extended throughout the year (2). In the London zoo births have been spread over the year and gestation has lasted from 214 to 225 days (3).

1. Sclater, W. L. The Mammals of South Africa. London, 1900.
2. Verschuren, J. EPNG., 1958.
3. Zuckerman, S. PZS., 122: 827–950, 1952–3.

Strepsiceros imberbis Blyth. The lesser kudu has 1 calf at a time (1).

S. (Limnotragus) spekei Sclater. In the London zoo this kudu has had its single calves from January to May (2). In the wild it probably breeds all year (3).

S. (Tragelaphus) angasi Gray. The nyala mates usually in April. One calf is born, usually in September or October, but sometimes as early as August or as late as March (1).

1. Shortridge, G. C. The Mammals of South West Africa. London, 1934.
2. Zuckerman, S. PZS., 122: 827–950, 1952–3.
3. Ansell, W. F. H. PZS., 134: 251–274, 1960.

Taurotragus oryx Pallas

ELAND

This species has 3-week cycles from May to July (1), but in the wild has apparently no fixed season, though most calves are born from March to November. The gestation of the single calf is about 8½ to 9 months (2). Puberty in the female is reached at age 3 years and in the male at 4 years (3). In the London zoo calves have been born throughout the year with a slight peak from March to June. Twins were born twice in 94 births (4).

1. Heape, W. QJMS., 44: 1–70, 1901.
2. Brown, C. E. JM., 17: 10–13, 1936.
3. Hamilton, J. S. Animal Life in Africa. London, 1912.
4. Zuckerman, S. PZS., 122: 827–950, 1952–3.

Taurotragus derbianus Gray. This species has a gestation of 260 days for its single calf (1).

1. Dekeyser, P. L. Les Mammifères de l'Afrique Noire Française. n.p., 1956.

Boselaphus tragocamelus Pallas

NILGAI

This Indian antelope is said to have no regular breeding season in the wild. It breeds immediately after dropping its calves (1). The period of gestation has been observed to be 8 months and 7 days (2). In captivity it experiences a series of diestrous cycles from March to May, each lasting 3 weeks (3). In England, at Woburn, it breeds from March to May (4), and in the London zoo births have been evenly distributed throughout the year. At about every other birth twins have been produced (5).

1. Brander, A. A. D. Wild Animals in Central India. London, 1923.
2. Brown, C. E. JM., 17: 10–13, 1936.
3. Heape, W. QJMS., 44: 1–70, 1901.
4. Bedford, Duke of, and F. H. A. Marshall. PRS., 130B: 396–399, 1942.
5. Zuckerman, S. PZS., 122: 827–950, 1952–3.

Tetracerus quadricornis Blainville

FOUR-HORNED ANTELOPE

In India this antelope breeds in the rains and gives birth to 1 or 2 calves in January after a gestation of 6 months (1). In the London zoo three births have taken place in February and others in May and June. Twins were produced in 3 of the 5 births (2).

1. Blanford, W. T. The Fauna of British India. Mammals. London, 1888–91.
2. Zuckerman, S. PZS., 122: 827–950, 1952–3.

Bubalus bubalis L.

EURASIAN BUFFALO

The buffalo reaches puberty usually at age 2 to 3 years but it may do so at 1.5 years with good feeding (1). In Egypt the average age of the cows at first estrus was 406 days and the weight 198 kg. The first fertile mating was at 647 days and the first calving at 963 days. At the time of fertile mating the average weight was 319 kg. (2). They breed all the year, but in Egypt most conceptions under free mating conditions occur from October to December and fewest from July to September (3). In India the Murrah breed has 77.6 per cent of all calvings between June and December (4), which makes the conception period February to August. The swamp buffalo of Malaya has estrous periods throughout the year but more regularly during the hot weather, from mid-April to mid-July (5). The cycle length closely resembles that of the cow, about 21 days, and a spread from 13 to 28 days (6). One report gives it as 22.9 days (4), and another, 21.6 ± .23 days, and the duration of estrus 11.9 hours. The cows ovulate 18 to 42 hours after the beginning of estrus so that, as in *Bos taurus,* it takes place after the end of estrus (7). In Pakistan the average duration of estrus was found to be 19.6 hours, range 3 to 69 hours, and it lasted longer in older cows than in young ones. Heats in which mating took place lasted 16.6 hours; those in which it did not lasted 21.7 hours (8). Another record from Egypt gives a cycle length of 21.1 ± .7 days, mode 21 and range 11 to 30 (9). Proestrum lasted 21.2 hours; estrus 28.5 hours; and metestrum 19.2 hours (10). This is a longer time for heat than is given by other reports, but it is substantiated by one from India which gives 36.0 hours, range 24 to 58. In this work ovulation was found to occur 12 to 24 hours after the expiration of heat (11).

In Malaya the buffalo has been divided into two races. The swamp race has a neckless scrotum, mates nocturnally, and has the first heat after parturition at about 65 days. The other, the river race, has a pendulous scrotum, mates during the day, and has first heat at about 42 days (12). In Bulgaria the average calf heat occurs 118 days after calving (13), and in the Philippines the carabao has an interval of 34 days, range 24 to 50. The Indian buffalo in the same region has an interval of 49 days, range 45 to 53 (14). In Egypt the average interval is 44 days (10). There are many reports on the duration of gestation. The averages vary between 307.7 (15) and 318.7 days (16). A typical report gives 316.15 with 315.7 days for female calves and 316.5 for males (17). However the swamp buffalo differs in one more way from the river type because its gestation lasts for 330 to 340 days (12, 18). Twins have been reported from Italy as occurring in 0.3 per cent of calvings (19), in Egypt, 0.2 per cent (20) or 0.6 per cent (21). Of 247 pregnancies 136 were in the right horn of the uterus and 111 in the left (22). The right ovary ovulates 52.9 per cent of the time (23).

At the time of estrus there is no swelling or hyperemia of the vaginal mucosa and no flow of mucus. The vaginal mucus is relatively pale and anemic (24). The pH of the vaginal secretion at diestrum is 8.1; at estrus, 7.3; and in pregnancy, 8.3 (25). Neither body temperature nor vaginal smear is a good indication of the reproductive state (10).

Estrone may be detected first in the urine of the pregnant cow at 3 months, when the mean amount is 37.5 μg./l. The amount rises to the eighth month when it reaches 340.6 μg./l. Then it falls gradually and cannot be detected during the last 15 days (26). There is no gonadotrophic substance in the urine (27).

The semen is milky-white with a very light tinge of blue. The volume is 3.45 ml., range 1.2 to 6.0 ml. and the sperm content 17.3 per cent by volume. The concentration of spermatozoa is 210 million to 2 billion per ml. (28). The pH of the semen is 6.27 ± .02 (29). Another report gave an average volume of 2.0 to 2.3 ml. with a concentration of 1,075 million per ml. (30). The total reducing substance in the semen is 875 mg/100 ml., of which 615 mg. is fructose and 3.7 mg. is ascorbic acid (31). The spermatozoa take 3 minutes, 20 seconds, or more, to reach the oviduct and are viable in the female tract for 36 to 48 hours (11).

The head of the spermatozoon measures $7.44 \pm .44\ \mu \times 4.26 \pm .52\ \mu$; the midpiece is $11.65 \pm .94\ \mu$; and the tail length is $42.82 \pm 3.04\ \mu$ (28).

1. Gorbelik, V. I. Trud. Azerbaidzan Stanc. Zivotn., 4: 5–26, 1935.
2. Hafez, E. S. E. J. Agric. Sci., 46: 137–142, 1955.
3. Badreldin, A. L. Experientia, 8: 391–392, 1952.
4. Rao, C. K., and T. Murari. Indian Vet. J., 33: 54–57, 1956.
5. Vendargon, X. A. J. Malay Vet. Med. Assn., 1: 13–19, 1955.
6. Kaleff, B. Zeit. Zucht., 24B: 391–408, 1932.
7. Shalash, M. R. Internat. J. Fertility, 3: 425–432, 1958.
8. Ishaq, S. M. Agric. Pakistan, 7: 361–365, 1956.
9. Hafez, E. S. E. Empire J. Exp. Agric., 21: 15–21, 1953.
10. Hafez, E. S. E. J. Agric. Sci., 44: 165–172, 1954.
11. Rao, A. S. P., S. N. Luktuke, and P. Bhattacharya. Indian J. Vet. Sci., 30: 178–190, 1960.
12. MacGregor, R. Vet. Rec., 53: 443–450, 1941.
13. Kaleff, B. Zeit. Tierzucht u. Zuchtungsbiol., 51: 131–178, 1942.
14. Ocampo, A. R. Philippine Agric., 28: 286–307, 1939–40.
15. Ishaq, S. M. Agric. Pakistan, 5: 225–228, 1954.
16. Ahmed, I. A., and A. O. Tantawy. Empire J. Exp. Agric., 24: 213–221, 1956.
17. Shalash, M. R., and F. M. El Mikkawi. British Vet. J. 112: 466–469, 1956.
18. Hua, L. C. J. Malay Vet. Med. Assn., 1: 141–143, 1957.
19. Ferrara, B. Riv. Zootec., 33: 98–102, 1960.
20. Asker, A. A., and A. A. El Itriby. Empire J. Exp. Agric., 25: 151–155, 1957.
21. Tantawy, A. O., and I. A. Ahmed. Empire J. Exp. Agric., 25: 24–28, 1957.
22. Reddy, D. B. Indian Vet. J., 37: 270–272, 1960.
23. Damodaran, S. Indian Vet. J., 32: 227–230, 1955.
24. Goswami, S. K. Personal communication, 1960.
25. Shalash, M. R. J. Agric. Sci., 51: 70–71, 1958.
26. Hafez, E. S. E., and T. Attar. Acta Physiol. Latin-Am., Buenos Aires, 6: 27–32, 1957.
27. Gobba, A. H. J. Am. Vet. Med. Assn., 123: 299–300, 1953.
28. Mahmoud, I. N. Bull. Fac. Agric. Fouad I Univ., Cairo, 15, 1952.
29. Shukla, D. D., and P. Bhattacharya. Indian J. Vet. Sci. and Animal Husbandry, 19: 161–170, 1949.
30. Prabhu, S., and P. Bhattacharya. Indian J. Vet. Sci. and Animal Husbandry, 21: 257–262, 1951.
31. Roy, A., Y. R. Karnik, S. N. Luktuki, S. Bhattacharya, and P. Bhattacharya. Indian J. Dairy Sci., 3: 42–45, 1950.

Anoa depressicornis H. Smith

ANOA

A single calf is usual and, in the London zoo, births have been distributed throughout the year (1). Gestation has been reported as lasting 276 or 315 days (2).

1. Zuckerman, S. PZS., 122: 827–950, 1952–3.
2. Kenneth, J. H. Gestation Periods. Edinburgh, 1943.

Bos taurus L.

DOMESTIC CATTLE

Domestic cattle are polyestrous, breeding all year. Ovulation is spontaneous, and it occurs at intervals of about 21 days. The length of proestrum has not been accurately determined, but probably it lasts for 2 to 3 days. Estrus lasts for less than 1 day, and the cow goes out of heat before she ovulates. Frequently, more so in younger than in older cows, overt bleeding occurs from the uterus about 18 hours after ovulation. The corpus luteum is functional for about 19 days; its decline and proestrum are synchronous. Uterine changes during the cycle are not marked. The vaginal smear is not a very reliable indication of the reproductive state. The pituitary of the cow has the lowest recorded F.S.H. content; L.H. is also low, but prolactin is very high. The period of gestation is about 280 days; 1 calf is the rule, but more are frequently born. When a female is born cotwin with a bull, she is usually sterile, as her gonads have been modified early in development. The bull produces spermatozoa all the year round.

THE ESTROUS CYCLE

There have been many investigations of the length of the estrous cycle in cattle (1–6). A summary of all the available data, which include a considerable series of records from the New York State Dairy Herd Improvement and Artificial Insemination Associations, gives a modal cycle length for unbred heifers of 20 days, and for cows, 21 days (6). The cycle length is less variable in heifers, in which 85 per cent fall between 18 and 22 days, but in cows 84 per cent fall between 18 and 24 days. The mean for heifers is 20.23 ± .05 days, and the standard deviation is 2.33 days. For cows the mean is 21.28 ± .06 days, and the standard deviation is 3.68 days. Although there is a definite difference in the cycle lengths of virgin and parous cattle, it cannot be said that the cycle increases in length with age, as analyses of different sets of data have yielded conflicting results. The season of the year has no effect (5, 6). There are indications that the individuality of the cow has some effect on the cycle length, but more evidence is needed on this point. All these figures refer to dairy or to dual-purpose cattle. For range (beef) cattle, Herefords, the length of the cycle is essentially similar (7). The mean cycle length for all ages of beef cows is 19.6 ± .12 days, the mode is 20 days, and 79 per cent fall between 17 and 23 days. In this group cows came in heat equally at all times of the day.

The figures for the duration of estrus must vary somewhat with the method of testing and the interval between tests. Most place the average as between 12 and 22 hours. One report showed that heifers remained in heat for 15.3 hours and cows 17.8 hours. The range was considerable, from 2½ to 28 hours, but 82.6 per cent of heifers were between 10 and 21 hours and 92.6 per cent of cows were between 13 and 27 hours. In this group there was no significant seasonal difference. More cows tended to go out of heat between 6 P.M. and midnight than at other times (8). It has also been reported that 60 per cent of heats begin in the twelve hours before midday, compared with 40 per cent in the period after (9). One account gives the average duration of estrus as 18.3 hours, with a 5-hour difference between spring, the shorter, and fall, but the difference was not statistically significant (10).

Ovulation is spontaneous and usually occurs about 10 hours after the cow goes out of heat, but there is considerable spread. One account gives 10.7 hours for cows and 10.2 for heifers. In all, 81.8 per cent were between 7 and 14 hours (8). Another report gives 11.5 hours as the mean interval (10). In 25 heifers served by a vasectomized bull ovulation took place 7.73 hours after heat had ended; in 25 that were not served at all the interval was 9.91 hours, a significant difference (11). The right ovary tends to produce more ovulations than the left. A corpus-luteum count showed 60.2 per cent in the right ovary (12), and an embryo count gave 60.5 per cent in the right horn of the uterus (13). This is an unusually large difference in ovulation rates. Removal of the corpus luteum brings the cow into heat 3 to 4 days afterwards in the majority of cases. The mean interval is 4.2 days (14).

The return of heat periods after parturition occurs after a very variable interval. For dairy cows a modal interval of 41 to 60 days has been reported (5), and a mean of 69 days, with a standard deviation of 39 days (15). For beef cattle the mode is 60 to 70 days, with 80 per cent between 20 and 160 days, and the mean, 80.2 ± 1.3 days (7). In this study it was found that the chance of conception at a heat period earlier than 40 days is only 49 per cent, but it rises until average fertility is reached at 70 days. Cows that are nursing their calves first return to estrus in 104 days, while those that are milked return in 74 days. There is an even greater difference in the interval before conception, 152 days compared with 94 days (16).

A little more than a day after ovulation bleeding occurs from the vulva in many cows. The source of this blood is discussed in the section on histology. Bleeding is more frequent in heifers (about 75 per cent) than in cows (48 per cent) (17), but it does not occur at every heat period in any one individual; great variation is shown in this respect (6). There is much division of opinion among cattle breeders concerning the significance of the bleeding; some hold that it is a sign that the cow has not become pregnant, others that she has, provided she had been served. Recent work (18) showed that, of 100 heifers bred at heat, 81 bled; of those that conceived, 85 per cent bled; but of those that did not conceive, 74 per cent bled. In a similar group of 100 cows 61 bled, consisting of 69 per cent of those that conceived and 39 per cent of those that did not.

FERTILIZATION

As the egg is shed some time after the end of heat, it becomes important to know whether the cow can be bred successfully after the end of heat by artificial insemination and whether spermatozoa deposited in the female tract early in the heat period can survive until ovulation. The conception percentage for artificial inseminations at different times is given in Tables 8 and 9, the latter of which relates the degree of success to the actual ovulation time (8, 19). The figures show that the chance of conception with early services is smaller than with later ones, and that after 6 hours have elapsed from the end of estrus it again becomes reduced. As the spermatozoa reach the oviduct in as few as 3 minutes after service or insemination (20), it is

Table 8. The Effect on Fertility of Inseminating Cows at Different Times (19)

TIME OF INSEMINATION	PER CENT FERTILE
Beginning of heat	44.0
Mid-heat	82.5
End of heat	75.0
6 hours after end	62.5
12 hours after end	32.0
18 hours after end	28.0
24 hours after end	12.0
36 hours after end	8.0
48 hours after end	0

Table 9. The Effect of Ovulation Time on Fertility (8)

TIME OF INSEMINATION	PER CENT FERTILE
Over 24 hours before ovulation	53.3
19 to 24 hours before ovulation	73.3
13 to 18 hours before ovulation	85.7
7 to 12 hours before ovulation	78.5
6 hours or less before ovulation	57.1
2 hours or less after ovulation	30.0
6 hours after ovulation	40.0
12 hours after ovulation	25.0

rather puzzling why the drop in fertility begins so soon before the average ovulation time. Perhaps this indicates that the spermatozoa need a stay of 4 or 5 hours in the female tract before capacitation is fully achieved. The figures suggest a short life for the ovum, probably 4 to 6 hours, and of 24 hours or a little more for the spermatozoon.

The ovum remains in the oviduct for 96 hours; in the first 6 hours it travels one-third of the distance, during which time it is fertilized. After this rapid passage the remainder of the journey is made at the rate of about 1 mm. per hour (21).

Twins are more frequent in dairy breeds than in beef breeds, the incidence being 1.88 per cent in the former and 0.44 per cent in the latter. Monozygotic twins account for 6.0 ± 1.9 per cent of the total. The chance of producing twins increases with the age of the dam to 8 to 9 years (22). A compilation of information in the Wurttemburg herdbooks gives a higher incidence of twins, ranging from 3.2 per cent for the spotted breed to 1.7

per cent for the brown breed. Triplets occurred in about 0.03 per cent of conceptions, and quadruplets in about 0.0004 per cent. Multiple births were most frequent from 5 to 7 years of age (23). There is some seasonal variation in the incidence of twins. In dairy cattle it is highest in July and April, and lowest in January and June (22); in beef cattle in the United States it is highest in August and lowest in March (24). Fifty-seven per cent of twins are male and female (23), and of the female twins in this sex combination only 6 per cent are fertile (25, 26).

The sex proportion at birth has been the subject of an extensive investigation (22), and it was found to be 51.52 ± .14 per cent males. The number of the pregnancy had no significant effect except for the first calf, in which the ratio was 52.76 ± .49 per cent males. The proportion in abortions and stillbirths was 58.04 ± 1.58 per cent males, which suggests, in conjunction with the proportion at birth, that many more males are conceived than females. The sex proportion of fetal calves is somewhat variable, but the trend is toward a greater number of males in the younger fetuses (27). Multiple births tend to have a lower proportion of males (22).

The mean duration of gestation varies with the breed, and the range of the breed means is from 277 to 290 days. The Holstein-Friesian is one with a low mean and the Brown Swiss one with a high mean (28). There is general agreement that males are carried about 1 day longer than females and that the gestation of twins lasts 5 to 6 days less than that of singles. A slight increase with the age of the dam up to 6 years of age, perhaps 2.5 days in all (29), has been generally noted, but the month of calving has little or no effect. There is little indication that birth is much more frequent at one time of day than another (30), though there is some suggestion that fewer are born during the early morning and evening (31).

HISTOLOGY OF THE FEMALE

OVARY. At birth the ovaries are in an advanced stage of development; vesicular follicles can be found in young cattle at all ages (32). The number of ova in the ovary of the calf is about 75,000; in the 1½- to 3-year-old cow the number is reduced to 21,000, while at 12 to 14 years only 2,500 survive (33). The age at first heat has received but little attention. One investigation on the Holstein-Friesian calf when exceptionally well fed gave a mean age of 37.4 ± 7.1 weeks and a weight of 262.2 kg. For medium-fed heifers the age at first heat was 49.1 ± 6.3 weeks and the weight was 270.4 kg. (34).

In Japan, Holstein-Friesian heifers reached puberty at 9.6 months and a live weight of 239.4 kg. (35). Large follicles, which have reached the usual maximum, 12 mm. in diameter, before the ovulatory increase sets in, are found in the ovaries of 6-month-old calves well before puberty (1). During the cycle follicles of this size, or a little larger, are found at 8 to 9 days postestrum and also during pregnancy (36). The maturation growth is made mainly during heat, and the size of the follicle at ovulation is 16 to 19 mm. (37). Among the smaller follicles two growth waves have been observed. The first is at 3 to 4 days of the cycle and the second at 12 days. Most of these follicles become atretic (38).

In the mature follicle during heat the blood vessels of the theca interna, which is prominent in the cow, are distended, and there is a slight leakage of erythrocytes into the granulosa and into the outer part of the liquor folliculi. A few granulosa cells are already enlarged. Many leucocytes can be found in the follicle. At ovulation much liquor folliculi and the cumulus are emitted, but most of the granulosa remains. The theca externa contracts markedly, throwing the theca interna and the granulosa into deep folds. There is some hemorrhage at the point of rupture, but any slight hemorrhage in the follicle is confined between the theca interna and the granulosa, where there is a very thin connective-tissue layer. At this time the diameter has shrunk to 5 to 8 mm. The follicle wall protrudes slightly through the rupture point, and occasionally lymph is extruded into the cavity and walled off, producing a small cyst (37). A short time before the follicle ruptures the intrafollicular pressure diminishes and the theca wall becomes slightly infolded. At the same time the follicle feels softer to the touch (39).

The growth of the corpus luteum begins immediately. Theca interna lobes grow into the folds and form connective-tissue strands, the blood vessels are engorged, and there is an invasion of leucocytes, mostly in the theca interna. The lipid content of the granulosa cells rapidly increases, and, to a lesser extent, that of the theca interna cells in which the granules are larger. At 3 days after heat, i.e., in the 2-day-old corpus luteum, the granulosa cells have increased from 10 μ to 15 μ in size. Mitoses are fairly common in the theca cells; eosinophils are common, but the spindle cell invasion (thecal) described by Corner in the developing corpus luteum of the pig has not been found. By the fifth day the granulosa cells increase to 20 μ, and by the sixth arterioles have formed and mitosis has ceased in the theca cells. At 7 days the granulosa cells reach their maximum size, 25 to 30 μ, and on the next day connective tissue elements are becoming more pronounced. At 14 days the lipid increases in both granulosa and theca cells, and the granules begin to coalesce in the former, which, in a few cases, begin to show signs of degeneration. At 17 days both cell types are involved in the degenerative changes. By the twentieth day retrogression has well set in; large fat droplets are seen in the cells, and blood vessels and connective tissue are much more prominent. When the cow is in heat, the granulosa and theca cells of the old corpus luteum cannot be distinguished one from the other, and the capillaries undergo resorption.

The mature corpus luteum is either globular or oblong in shape. Its color is at first light brown to brownish yellow. Gradually it becomes less brown and by the seventh day it is old gold. At 14 days it is bright golden yellow, deepening to orange by the twentieth day. During involution it becomes brick red, and it remains as a bright streak of connective tissue for a year or more. In old cows retrogression is slower than in young ones (37).

The young corpus luteum measures about 6 to 8 mm. By 8 days it has increased to 18 to 20 mm. and when it is mature it measures 20 to 25 mm. There is no marked reduction in size until it is 20 to 30 days old (36). During pregnancy its size and weight remain almost constant, but the lipid globules increase in size and number after 5 months. Its removal at any time during pregnancy results in abortion, even when this operation is performed late in the period (40). The average weight of the corpus luteum of the cycle is 5.3 g. (6); during pregnancy there is little growth after 35 days, when the mean weight is 5.25 g. (41). The follicle content and distribution of several enzymes and other substances have been followed in detail. During the last stages of development phosphatase disappears from the cumulus and zona pellucida. It also is lost from the theca interna just before ovulation, but after that event its distribution in the developing corpus luteum is equalized (42).

VULVA. During proestrum the vulva swells slightly, and this swelling becomes more pronounced during heat. Within the lips the mucosa becomes congested and is bright cherry red in color. The swelling subsides rapidly after heat has passed, and the vulva becomes wrinkled, but the congestion remains until a short while after ovulation (43). In pregnancy the development of transverse folds on the surface has been described. These are most marked at 7 to 8 weeks and disappear at 4½ to 5 months (44).

VAGINA. The vaginal epithelium consists of two distinct regions. In the vestibule, and for a short distance above it, stratified squamous epithelium with frequent patches of lymphoid tissue is found; above that region stratified epithelium with a superficial layer of mucoid cells extends to the cervix uteri.

In the vestibular region during heat the superficial cells react more readily to acid stains, and the number of leucocytes increases. There is also congestion and edema, both of which are also pronounced during proestrum. By 2 days postestrum these changes have subsided, but the epithelial cells of the middle layer have increased in size and number, and this increase is very marked by 8 to 11 days (36). Cornification increases from the tenth to the eighteenth day, after which desquamation sets in (45). Extravasation of erythrocytes is found in the more caudal portion of the region during and just after heat (36).

In the mucoid portion of the vagina the changes with the cycle are more clear-cut in the region 1 to 2 cm. from the cervix. The superficial layer consists of tall-columnar mucus-secreting cells with many goblet cells interspersed between them. The latter are less frequent posteriorly. Below this layer are 2 to 6 layers of polyhedral cells. During proestrum the goblets are clear and free from granules; the cells do not stain with hematoxylin at this time but do with mucicarmine. The stroma is edematous. At heat the congestion and edema are more pronounced and the cells secrete mucin. The initiation of secretion proceeds down the vagina in a wave so that during the postestrous period the posterior portions of the mucoid region are active. The nuclei are compressed into the basal ends of the cells. Edema and congestion disappear rapidly after the end of heat; the epithelium decreases in height as the cells are emptied of their secretion, and the nuclei become oval. By 2 days postestrum the superficial layer takes on a serrated appearance and begins to stain with hematoxylin more intensely than it did before; the polyhedral cells increase in number, and, though there are few leucocytes, many lymphocytes invade the stratified layer. During the mid-diestrum the surface epithelium is low columnar and vacuolated, but this is succeeded by gradual growth to the next proestrum (36).

Smears from the vestibule show increased numbers of cornified cells during proestrum and estrus, but leucocytes are never absent; individual variation is so great that little reliance can be placed on the smear as a method of diagnosis. Deeper smears contain more epithelial cells and lymphocytes. There is a sudden, marked increase in leucocyte content 3 hours after mating (46). This would appear to be due not to friction but to the presence of semen in the tract, since the condition also follows artificial insemination (47). Besides the metestrous bleeding, erythrocytes may be detected in the vaginal smear at the tenth day of the cycle and the discharge is also positive to the benzidine test at this time. There is reason to believe that this is caused by a diminution in circulating estrogen (48). In the cervical region cornified cells are most abundant from 8 to 12 days (49). Phosphatase activity is greatest in the vaginal mucus two days before estrus, and it is again high 8 to 12 days after heat (50).

UTERUS. The cervix uteri is covered with a single layer of mucoid epithelium which is much folded. The changes during the cycle are similar to those described for the upper vagina: congestion and edema during, and just after, heat; cubical cells in the diestrum, becoming columnar and full of mucin during heat, discharging and becoming ragged and cubical about 72 hours after heat (1). The epithelial cells reach their greatest height at 17 to 18 days; secretion begins at the tips of the folds, and exhaustion of the cells is not complete in the crypts until about the fourteenth day (45). The cervix is relaxed at the time of heat and contracted in diestrum (9).

The uterus contains numerous caruncles or cotyledons which aid in the attachment of the fetal membranes. The stroma is somewhat more dense in the cotyledons; there are more blood vessels and no gland openings. The intercotyledonary area is richly supplied with glands, which are scanty before puberty but rapidly develop as sexual life begins (1).

At the end of diestrum the uterine epithelium is tall columnar and pseudostratified, and the stroma becomes congested and edematous. These changes increase during, and for a day after, heat. At the second day the congestion and edema have decreased while the cells, which have begun to secrete, are flatter. At this time, also, many of the congested blood vessels break down, and extravasated blood is present, especially in the cotyledons. At no time does the epithelium break; the erythrocytes find their way in large numbers into the lumen by diapedesis. This is the source of the postestrous bleeding. There is a similar, but much less severe, extravasation of blood at about the tenth day which does not appear as a vaginal discharge (36). This occurs at about the same time as the interheat vaginal congestion. The glands are fairly quiescent and the lumens straight during proestrum and heat. At 2 days postestrum they are more coiled, and the lumens are filled with secretion; they begin to hypertrophy as the corpus luteum grows, and they are at their greatest growth at about 12 days postestrum. At the fifteenth day retrogression begins (9).

The muscle fibers of the uterus are longest at the time of heat and for 2 days afterwards. This growth begins at the tubal end of the uterine horns and spreads toward the cervix. Afterwards the fibers diminish in size and are shortest at about the seventeenth day. They have most spontaneous activity when they are longest (51). Alkaline phosphatase is present in the intercaruncular stroma and in the epithelial cells. In the latter it is highest in mid-cycle, and there is little or none at the beginning and end (52). The glycogen content follows the reverse pattern (52, 53).

OVIDUCT. The tubo-uterine junction of the cow is at the tip of the cornu. It is straight, without a villus, and with only a slight sphincter, which does not prevent the passage of fluid from the uterus into the tube (54).

The tubal epithelium has a distinct cycle. More mucus and leucocytes are found in the lumen at 2 days postestrum than at other times; the epithelium becomes lower at 8 days, and at this time numerous globules of cytoplasm, the centers deeply staining with hematoxylin, are found. These have been interpreted as extruded nuclei, but, to the writer, they appear to be secreted protein which is imbibing fluid at the periphery, thus staining less deeply in that region than at the center. During proestrum the cilia are much more prominent than they are earlier, and the cells enlarge. They are tallest during heat and lowest at from 8 to 12 days, when the protein has been extruded. In the follicular phase of the cycle alkaline phosphatase is high and there is a concentrated cytoplasmic basophilia. In early diestrum more goblet cells are in evidence and more extrusions, interpreted as nuclei. In the luteal phase glycogen is in evidence and small amounts of lipid accumulation (55).

ANTERIOR PITUITARY. A small basophil cell, designated the delta cell, is present in the anterior pituitary. This type of cell is closely associated with the capillaries and it is believed to secrete L.H. Degranulation of these cells occurs during the first hours of estrus, but does not occur in those cows that come into heat but do not ovulate. Acidophil cells discharge their granules at about the third day after estrus and reach a peak of degranulation at about the tenth day. They appear to be very active from the fourth to the eighth month of pregnancy (56). This suggests that they are associated with prolactin secretion.

PHYSIOLOGY OF THE FEMALE TRACT

One of the main external characteristics of the heat period is the secretion of large quantities of stringy mucus, which is derived partly from the mucoid epithelium of the vagina and partly from the seal which plugs the cervix when the cow is not in heat. This liquefaction is brought about by a change in pH which occurs at heat. The literature on this subject has been summarized (57). *In vitro* specimens of the vaginal secretions are rarely acid. During the diestrum they average pH 8.1, dropping to 7.2 at heat, while the cervical mucus is about 1.5 units lower. The injection of estrogens into sterile cows brings about the shift in pH observed during heat (58). The *in vivo* changes are not so marked; the secretions are less alkaline, and they may even be acid (57, 59).

The flow properties of the mucus have been investigated, and they may be used as an indication of physiological heat when psychological signs are slight. The viscosity is least, and the flow elasticity is greatest, at heat. The decrease in viscosity begins 24 hours before heat, and it has risen again by 60 hours postestrum (60). In the ovariectomized cow a flow of mucus can be produced by the injection of small amounts of estrogen, 350 to 1,000 R.U. daily; with larger doses, 10,000 R.U. and upwards, it is suppressed because of the cornifying action of these large doses upon the vaginal epithelium (61). During pregnancy the cervical seal is exceptionally tough in spite of the estrogens present at that time.

It has been suggested that the pattern assumed by the mucus when it is fixed on a slide is an indication of the reproductive state of the female. One such report uses silver nitrate as fixative and follows it with Giemsa stain. In the follicular stage large crystal forms are seen, while in the lutein stage the small crystal forms are noted. At estrus there were always large fernlike crystals (62). During pregnancy a frizzy hairlike pattern was observed. It appeared by the thirty-fifth day (63). Another report gives greatest arborization at estrus and least in mid-cycle (64).

The motility of the uterus during the cycle has been studied by the balloon technique (65). Activity is very marked during heat and for 2 to 3 days after ovulation, but the uterus becomes relatively refractory to pituitrin immediately after ovulation. Spontaneous activity decreases until the eleventh to the sixteenth day after ovulation, but from that time onward spontaneous activity and response to pituitrin increase, to become greatest during heat. Ovariectomy abolishes spontaneous motility. Uterine strips have been studied in detail (55). Spontaneous motility is greatest in these during proestrum and estrus, when the muscle fibers are longest; it becomes irregular during metestrum and dies down in diestrum. Two types of waves have been observed, strong contractions of great amplitude at intervals of 1.5 to 2 minutes, and small contractions at intervals of 20 to 30 seconds. The latter increase in importance during metestrum and are a large factor in the production of the irregularity. In early diestrum long rhythmic changes of tone also occur. Sympathicomimetic drugs are active at all times in the cycle, but parasympathicomimetic drugs are irregular in their action or even inert. Epinephrin is inhibitory during the estrogenic phase and motor in the lutein phase of the cycle. The cow is similar to the dog and cat in this respect. During proestrum there is a tendency for a diphasic response to appear, first a slight contraction, then a marked relaxation. After ovariectomy spontaneous activity is not always abolished, but epinephrin is without effect, a difference from the result found in the cat (61). If the ovariectomized cow is treated with estrogen, the characteristic spontaneous, and also the epinephrin, responses are evoked. If she is treated with 250 R.U. of estrogen and 18 Rab.U. of progesterone together each day for 6 days, a marked diphasic response to epinephrin is given. If the progesterone is increased to 35 Rab.U. daily, the estrogen relaxation with epinephrin no longer occurs and contractions result. With these doses of estrogen the low progesterone cows showed an approximation to the estrogen type of spontaneous activity; with increasing doses of progesterone the activity became less in amount and more irregular.

The level of estrogen in the follicular fluid of cows is about 321 M.U. per kg. of fluid (66) or, in ovaries at the time of heat, 4.6 R.U. per follicle (67). The minimum dose of estrogens that will bring the spayed cow into heat is 350 R.U. of estradiol benzoate for 2 days, and the average dose is 600 R.U. per day for 3 days. The less vigorous the heat period in normal life, the harder it is to bring the cow in heat after ovariectomy. Even though the injections are continued, the cow remains in heat for less than a day. The central nervous system becomes refractory, since the usual changes in the uterus and vagina continue. As this threshold is very low, it is logical to assume that in the normal cow it is reached early in the development of the follicle. As refractoriness sets in quickly, the cow goes out of heat although the follicle is still growing. Hence the fact that the cow, alone of all known animals except the bat, in which the cause may be different, is out of heat before ovulation. Estrogens cannot be detected in the urine at any time of the cycle (61), but in whole jugular blood the amount is greatest at estrus, and again at 9 to 19 days. The levels are much higher in primiparous than in multiparous cows (68).

Near ovulation time the follicle contains 3 μg. of progesterone per ml. of fluid (69). The amount of progesterone in the corpus luteum of the cycle is about 29.2 μ/g. tissue at day 8; and it rises to about 45 μ/g. tissue at day 15 or 16. Then it falls rapidly (70, 71). The daily output is probably between 18 and 35 mg. (61). In the jugular blood there is a trace of progesterone at estrus and the level rises to about 6 μ/ml. when the corpus luteum is active (68).

Even during pregnancy the output of estrogens is low. Estrone is present in the urine to the extent of 0.3 mg./l., and estradiol-17α 0.1 mg./l. There is no estradiol-17β (72). The feces also contain estrogens, especially in the last few weeks of pregnancy (73). The level of progesterone in the corpus luteum is highest during mid-pregnancy; the level in the placenta is very low (74). Relaxin may be detected in the blood at the end of the first month of pregnancy. The level increases to 6 months and then remains steady. There is a sharp fall within 24 hours following parturition (75).

The anterior pituitary of cattle contains very little F.S.H., the lowest yet known (76, 77). The content of L.H. is about the same as that of the hog but is considerably lower than that of the sheep. The cow seems to have a low level of sex-hormone activity throughout. The freemartin, female cotwin to a bull, is usually sterile because the gonads have been modified toward the male. Probably this is due to the low threshold of the hormones involved. The cat also has the anastomosis of blood vessels regarded as necessary for the production of this anomaly, but feline freemartins have not been found. The cat has a higher level of sex-hormone activity throughout, so the threshold which would produce a change may be higher and not attained during pregnancy. The level of gonadotrophic hormones in the anterior lobe is highest at day 11 of the cycle. It remains high until day 18 or 19, when it decreases through estrus. It is lowest at ovulation but begins to increase by 65 hours after the onset of estrus (78). The level of prolactin in the anterior pituitary is high, about 400 to 500 B.U. per gram of fresh tissue (79).

The gonadotrophic hormones are said to be concentrated in the medulla, and the lactogenic hormone in the cortex, of the pituitary (80). The content of the former falls steadily during gestation (81).

It has been reported that 1,500 I.U. of pregnant mare's serum does not produce ovulation in the presence of the corpus luteum, but that 5,000 I.U. is effective (14); another report records that 700 M.U. is the most effective dose. The follicles rupture 48 to 72 hours after injection, and the normal periodicity of the cycle is not disturbed. Injections were made 4 to 8 days after normal heat (82). Daily injections of 25 mg., or more, of progesterone suppress ovulations so that estrus and ovulation do not occur until 4 to 7 days after injections are stopped (83). An injection of 5 to 10 mg., however, during the first 2 hours of estrus shortens its duration and hastens ovulation (84). Ovulation may be delayed by keeping the cow under the influence of atropine during and after heat (85). Daily injection of oxytocin at the rate of 7 U.S.P. units from day 3 to 6 prevents the formation of a normal corpus luteum; a new ovulation follows cessation of the treatment in 2 or 4 days (86). A similar result follows dilation of the uterus in the first third of the cycle (87). The oxytocin method of suppressing the corpus luteum has been used to estimate the level of progesterone needed to maintain pregnancy at 15 days. If the corpus luteum contained 100 μg the embryo survived (88). When corpora lutea were removed by surgery early in pregnancy abortion resulted. Daily injections of 100 mg. of progesterone maintained the pregnancy; 75 mg. were less successful. It was possible to cease the injections as early as 162 days; birth then took place at 267.5 days but in most the placenta was retained (89). Removal of the uterus causes the corpus luteum to be retained for at least 154 days (90).

New ova are produced during adult life from the neogenic stratum below the ovarian tunic and not from the stratum germinativum (91). The cow's ovum has been described and measured; it is 120 μ in diameter without the zona pellucida (92). The pH of the uterus from which this specimen was obtained was 7.1, which agrees well with the pH recorded by others (47). Tables showing the rate of prenatal growth of the embryo and fetus have been prepared (93, 94). The vaginal temperature of the cow is low, 101.1° F., just before estrus begins. On the day of estrus it is high, 101.5° F., but at ovulation it falls to 101.2° F. It remains high, 101.4° F., in the lutein stage of the cycle and in pregnancy. Progesterone is responsible for the high level (95). According to pedometer readings the cow is restless during heat. She begins walking between 5 P.M. and 4 A.M. and the activity increases from 5 A.M. to 4 P.M. (96). In Jersey herds the incidence of postconception heats was from 2.1 to 7.75 per cent (97).

THE MALE

At 1 month of age the testis consists mainly of mesenchyme. Cells of Leydig are formed by 3½ months; from 2 to 5 years of age they are highly vacuolated, then the vacuoles diminish. The androgen content of the testis increases at a uniform rate during the first 2 years. From 2 to 5 years it increases sharply, and thenceforward it decreases. There are no striking changes in the Leydig cells or in the androgen content of the testis at puberty (98).

At 63 days a few primary spermatocytes may be found in the tubules, at 181 days they are abundant in all tubules, and at 224 days spermatozoa are fully formed (99). The testis increases tenfold in size from birth to 3 months and threefold from the fourth to the seventh month. Spermatozoa are formed at 6 months, and at 7 months they are free in the lumen of the tubules (100). In an investigation to compare the rate of development of Holstein-Friesian bulls at different feed levels, those on a high plane of feeding came into semen production at 37 weeks and at 644 lb. weight. Those on a medium plane came into semen production at 43 weeks and 519 lb. weight (101).

Spermatozoa remained potentially motile in the epididymis isolated from the testis for over 2 months (102). It is interesting to note in this connection that Aristotle was led to deny any direct reproductive function for the testis because a bull remained fertile for a while after it had been castrated (103). In the ampulla of the vas deferens spermatozoa remain motile for less than 72 hours (104). It takes them 48 to 50 days to traverse the epididymis and the rate is not related to the frequency of ejaculation (105). The average number of spermatozoa to be found in the ampulla was 418 million, in the ductus deferens 340 million, and in the cauda epididymides 8,806 million. Another count gave 45.4 per cent of the available spermatozoa in the cauda, 18.3 per cent in the corpus, and 36.2 per cent in the caput of the epididymis (106). In 35 mature bulls the average number of extragonadal spermatozoa was 72.6 billion and the gonadal reserve 58.4 million per gram of tissue (107).

The normal ejaculate is an opaque, white to yellowish-white, milky or creamy fluid, and the higher the sperm content, the whiter it is. For dairy bulls the mean volume is 4.6 cc., with a modal ejaculate of 4 and 5 cc. (108). Other records give $3.7 \pm .2$ cc. (109), and 4.2 cc. (110). The mean sperm content is 1,010,800 per cu. mm., with a wide range from 120,000 to 2,144,000, and a modal frequency of 750,000 to 1,250,000 (108). These figures are very close to those given by most other workers, and they may be regarded as typical. The mean pH is $6.9 \pm .07$ (109), also typical of most results. In this case the range was 6.39 to 7.81. The same paper gives the Δ as 0.62° C. Semen is well buffered in the region of pH 4.0 to 5.5 and from pH 9.0 to 10.0, but buffering is higher on the acid than on the alkaline side of neutrality. As the curve for fluid from the seminal vesicles is very similar to that for whole semen, probably the secretion of these glands determines the shape (111).

There is no seasonal variation in pH; the semen volume is least in July to September; and during these months initial motility and power of survival are least, and the proportion of abnormal spermatozoa is greatest. The best semen is produced in the spring (112). There is no difference in effectiveness if artificial insemination is performed with a volume of semen over a range of 0.4 to 6.0 cc. (108).

The pH of the seminal vesicle secretion, taken post mortem, varies from 5.5 to 6.5 (113). The K, Na, and Ca content of some of the secretions of the male tract are given in Table 10 (114).

TABLE 10. Average Composition of Secretions from the Accessory Sex Organs of the Bull

ORGAN	CONTENT, mg. per 100 cc.		
	K	Na	Ca
Epididymis	278.4	115.0	81.7
Ampulla	320.8	165.4	26.7
Vesiculae seminales	326.1	138.5	32.6
Sperm serum	227.8	277.8	33.9

For range (beef) cattle the semen volume was $4.84 \pm .07$ cc., with a modal distribution from 3.5 to 5.5 cc., and 80 per cent between 2.5 and 7.5 cc. The mean concentration of spermatozoa was 1,160,000 per cu. mm., with a mode of 1.1 to 1.3 millions, and 80 per cent between 0.7 and 1.7 millions (7). These figures are a little higher than similar means for dairy bulls.

The amount of inositol in semen plasma is 63 mg. per 100 ml., and in the secretion of the seminal vesicles it is between 26 and 55 mg. per 100 ml. (115). The fructose level is about 300 mg. per 100 ml. in whole semen, and that of citric acid may be as high as 1 per cent (116).

The length of seminiferous tubules in one testis is about 5,350 meters (117). The temperature of the testis is about 4° C. below the body temperature taken in the rectum (118). Androstenedione is present in the testis at all stages of development. Testosterone is also present and the combined content is about the same in young calves and fully mature bulls, but in young calves the predominant substance is androstenedione, while in mature bulls it is testosterone. The change in relative proportions takes place at puberty (119). Bull's urine contains 200 M.U. of androgen per liter (120).

1. Hammond, J. The Physiology of Reproduction in the Cow. Cambridge, 1927.
2. Quinlan, J., and L. L. Roux. Onderstepoort J. Vet. Sci. and Animal Ind., 6: 719–772, 1936.
3. Quinlan, J., L. L. Roux, and W. G. van Aswegen. Onderstepoort J. Vet. Sci. and Animal Ind., 12: 233–249, 1939.
4. Cupps, P. T. Thesis, Cornell University, 1943.
5. Chapman, A. B., and L. E. Casida. J. Agric. Res., 54: 417–435, 1937.
6. Asdell, S. A., J. de Alba, and S. J. Roberts. Cornell Vet., 39: 389–402, 1949.
7. Lasley, J. F., and R. Bogart. Missouri Agric. Exp. Sta., Res. Bull. 376, 1943.
8. Trimberger, G. W. Nebraska Agric. Exp. Sta., Res. Bull. 153, 1948.
9. Roark, D. B., and H. A. Herman. Missouri Agric. Exp. Sta., Res. Bull. 455, 1950.
10. Glod, W. Med. Wet., 17: 353–361, 1961.
11. Marion, G. B., V. R. Smith, T. E. Wiley, and G. R. Barrett. J. Dairy Sci., 33: 855–859, 1950.

12. Reece, R. P., and C. W. Turner. J. Dairy Sci., 21: 37–39, 1938.
13. Schramm, W. Deutsche Tierarztl. Woch., 45: 387–389, 1937.
14. Hammond, Jr., J., and P. Bhattacharya. J. Agric. Sci., 34: 1–15, 1944.
15. Chapman, A. B., and L. E. Casida. Proc. Am. Soc. Animal Prod., p. 57, 1934.
16. Wiltbank, J. N., and A. C. Cook. J. Animal Sci., 17: 640–648, 1958.
17. Krupski, S. Schweiz. Arch. f. Tierheilk., 59, 1917.
18. Trimberger, G. W. J. Dairy Sci., 24: 819–823, 1941.
19. Trimberger, G. W., and H. P. Davis. Nebraska Agric. Exp. Sta., Res. Bull. 129, 1943.
20. Van Demark, N. L., and A. Moeller. AJP., 165: 674–679, 1951.
21. Gerasimova, A. A., N. G. Potapova, M. J. Solovei, and B. P. Hvatov. Dokl. Akad. Seljskohoz Nauk, No. 13: 26–30, 1940.
22. Johansson, I. Zeit. Zucht., 24B: 183–268, 1932.
23. Ruthardt, E. Diss., Univ. Leipzig, 1935.
24. Cole, L. J., and A. Rodolfo. Proc. Am. Soc. Animal Prod., 116–118, 1924.
25. Hewitt, A. C. T. J. Dairy Res., 5: 101–107, 1934.
26. Boenig, —. Deutsche Tierarztl. Woch., 46: 705–707, 1938.
27. Chapman, A. B., L. E. Casida, and A. Cole. Proc. Am. Soc. Animal Prod., 303–304, 1938.
28. Brakel, W. J., D. C. Rife, and S. M. Salisbury. J. Dairy Sci., 35: 175–194, 1952.
29. Knoop, C. E., and C. C. Hayden. Ohio Agric. Exp. Sta., Bimonthly Bull. 166: 8–14, 1934.
30. Richter, J. Berliner Tierarztl. Woch., 49: 517–521, 1933.
31. Rademacher, A. Vet. Med. Diss., Tierarztl. Hochschule, Hannover, 1936.
32. Casida, L. E., A. B. Chapman, and I. W. Rupel. J. Agric. Res., 50: 953–960, 1935.
33. Hediger, H. Acta Anat., Suppl. 5, Vol. 3: 1–196, 1947.
34. Sorensen, A. M., W. Hansel, W. H. Hough, D. T. Armstrong, K. McEntee, and R. W. Bratton. Cornell Univ. Agric. Exp. Sta., Bull. 936, 1959.
35. Masuda, S., N. Onishi, and A. Kudo. Res. Bull. Zootech. Exp. Sta., Japan, No. 56, 1950.
36. Cole, H. H. AJA., 46: 261–302, 1930.
37. McNutt, G. W. J. Am. Vet. Med. Assn., 65: 556–597, 1924.
38. Rajakowski, E. Acta Endocrinol., Suppl. 52: 1–68, 1960.
39. Asdell, S. A. Cornell Vet., 50: 3–9, 1960.
40. Sartoris, P. Nuovo Ercol., 43: 133–154, 1938.
41. Foley, R. C., and R. P. Reece. Massachusetts Agric. Exp. Sta., Bull. 468, 1953.
42. Moss, S., T. R. Wrenn, and J. F. Sykes. AR., 120: 409–433, 1954.
43. Bemis, H. E. J. Am. Vet. Med. Assn., 62: 536–538, 1923.
44. Baumgärtner, G. Tierarztl. Umsch., 7: 415–416, 1952.
45. Murphey, H. S. J. Am. Vet. Med. Assn., 65: 598–621, 1924.
46. Schneerson, S. S. Sborn. Trud. Zooteh. Kaf. s.h. Skol. Kirov., 1: 126–141, 1936.
47. Sergin, N. P., M. P. Kuznecov, V. M. Koslova, and T. N. Nesmejanova. Dokl. Akad. Seljskohoz Nauk, No. 15: 24–28, 1940.
48. Hansel, W., and S. A. Asdell. J. Animal Sci., 11: 346–354, 1952.
49. Hansel, W., S. A. Asdell, and S. J. Roberts. Am. J. Vet. Res., 10: 221–228, 1949.
50. van Klenkenberg, G. A. Nature, 172: 397, 1953.

51. Cupps, P. T., and S. A. Asdell. J. Animal Sci., 3: 351–359, 1944.
52. Moss, S., T. R. Wrenn, and J. F. Sykes. E., 55: 261–273, 1954.
53. Skjerven, O. Acta Endocrinol., 22 (suppl. 26): 1–101, 1956.
54. Andersen, D. H. AJA., 42: 255–305, 1928.
55. Weeth, H. J., and H. A. Herman. Missouri Agric. Exp. Sta., Res. Bull. 501, 1952.
56. Jubb, K. V., and K. McEntee. Cornell Vet., 45: 593–641, 1955.
57. Brown, P. C. Am. J. Vet. Res., 5: 99–112, 1944.
58. Smith, S. E., and S. A. Asdell. Am. J. Vet. Res., 2: 167–174, 1941.
59. Dougherty, R. W. North Am. Vet., 22: 216–219, 1941.
60. Blair, G. W. S., S. J. Folley, F. H. Malpress, and F. M. U. Coppen. Bioch. J. 35: 1939–1049, 1941.
61. Asdell, S. A., J. de Alba, and S. J. Roberts. J. Animal Sci., 4: 277–284, 1945.
62. Higaki, S., and Y. Awai. Bull. Nat. Inst. Agric. Sci., Japan, 7G: 51–59, 1953.
63. Higaki, S., and Y. Awai. Bull. Nat. Inst. Agric. Sci., Japan, 7G: 67–76, 1953.
64. Bane, A., and E. Rajakoski. Cornell Vet., 51: 77–95, 1961.
65. Evans, E. I., and F. W. Miller. AJP., 116: 44–45, 1936.
66. Parkes, A. S., and C. W. Bellerby. J. Physiol., 61: 562–575, 1926.
67. Allen, E. PSEBM., 23: 383–387, 1926.
68. Higaki, S., T. Suga, and T. Fujisaki. Bull. Nat. Inst. Agric. Sci., Japan, 18G: 115–140, 1959.
69. Edgar, D. G. Nature, 170: 543–544, 1952.
70. Bowerman, A. M., and R. M. Melampy. PSEBM., 109: 45–48, 1962.
71. Mares, S. E., R. G. Zimbelman, and L. E. Casida. J. Animal Sci., 21: 266–271, 1962.
72. Klyne, W., and A. A. Wright. JE., 18: 32–45, 1959.
73. Levin, L. J. Biol. Chem., 157: 407–411, 1945.
74. Melampy, R. M., W. R. Hearn, and J. M. Rakes. J. Animal Sci., 18: 307–313, 1959.
75. Wada, H., and M. Yuhara. Japanese J. Zootech. Sci., 26: 215–220, 1955.
76. Witschi, E. E., 27: 437–446, 1940.
77. West, E., and H. L. Fevold. PSEBM., 44: 446–449, 1940.
78. Smirnova, E. I. Bjul. Eksp. Biol. Med., 19(6): 67–69, 1945.
79. Reece, R. P., and C. W. Turner. Missouri Agric. Exp. Sta., Res. Bull. 266, 1937.
80. Friedman, M. H., and S. R. Hall. E., 29: 179–186, 1941.
81. Nalbandov, A., and L. E. Casida. E., 27: 559–566, 1940.
82. Zavadovskii, M. M., I. A. Eskin, and G. F. Ovsjannikov. Trud. Dinam. Razvit., 9: 75–96, 1935.
83. Hansel, W. In H. H. Cole and P. T. Cupps, eds., Reproduction in Domestic Animals. New York, 1959.
84. Hansel, W., and G. W. Trimberger. J. Dairy Sci., 35: 65–70, 1952.
85. Hansel, W., and G. W. Trimberger. J. Animal Sci., 10: 719–725, 1951.
86. Armstrong, D. T., and W. Hansel. J. Dairy Sci., 42: 533–542, 1959.
87. Hansel, W., and W. C. Wagner. J. Dairy Sci., 43: 796–805, 1960.
88. Staples, R. E., and W. Hansel. J. Dairy Sci., 44: 2040–2048, 1961.
89. McDonald, L., R. E. Nichols, and S. H. McNutt. Am. J. Vet. Res., 13: 446–451, 1952.
90. Wiltbank, J. N., and L. E. Casida. J. Animal Sci., 15: 134–140, 1956.

91. Cole, H. H. PSEBM., 31: 241–243, 1933.
92. Hartman, C. G., W. H. Lewis, F. W. Miller, and W. W. Swett. AR., 48: 267–275, 1931.
93. Winters, L. M., W. W. Green, and R. E. Comstock. Minnesota Agric. Exp. Sta., Tech. Bull. 151, 1942.
94. Hamilton, W. J., and J. A. Laing. J. Anat., 80: 194–204, 1946.
95. Wrenn, J. R., J. Bitman, and J. F. Sykes. J. Dairy Sci., 41: 1071–1076, 1958.
96. Farris, E. J. J. Am. Vet. Med. Assn., 125: 117–120, 1954.
97. Rahlmann, D. F., and S. W. Mead. Proc. Western Div. Am. Dairy Sci. Assn., pp. 67–71, 1958.
98. Hooker, C. W. AJA., 74: 1–32, 1944.
99. Phillips, R. W., and F. N. Andrews. Massachusetts Agric. Exp. Sta., Bull. 331, 1936.
100. Michatsch, G. Diss., Berlin, 1933.
101. Bratton, R. W., S. D. Musgrave, H. O. Dunn, and R. H. Foote. Cornell Univ. Agric. Exp. Sta., Bull. 940, 1959.
102. Kirillov, V. S., and V. A. Morozov. Usp. Zooteh. Nauk, 2: 19–22, 1936.
103. Aristotle. De Generatione Animalium. Ed. A. Platt. Oxford, 1910.
104. Kirillov, V. S. Probl. Zivotn., No. 2: 189–198, 1938.
105. Koefoed-Johnsen, H. H. Nature, 185: 49–50, 1960.
106. Bialy, G., and V. R. Smith. J. Dairy Sci., 41: 1781–1786, 1958.
107. Amann, R. P., and J. O. Almquist. J. Dairy Sci., 44: 1537–1543, 1961.
108. Herman, H. A., and E. W. Swanson. Missouri Agric. Exp. Sta., Res. Bull. 326, 1941.
109. Hatzcolos, B. Zeit. Zucht., 38B: 199–254, 1937.
110. Davis, H. P., and N. K. Williams. Proc. Am. Soc. Animal Prod., 232–242, 1939.
111. Smith, S. E., and S. A. Asdell. Cornell Vet., 30: 499–506, 1940.
112. Erb, R. E., F. N. Andrews, and J. H. Hilton. J. Dairy Sci., 25: 815–826, 1942.
113. Dougherty, R. W., and H. P. Ewalt. Am. J. Vet. Res., 2: 419–426, 1941.
114. Nesmejanova, T. N. Usp. Zooteh. Nauk, 5: 65–72, 1938.
115. Hartree, E. F. Bioch. J., 66: 131–137, 1957.
116. Mann, T., D. V. Davis, and G. F. Humphrey. JE., 6: 75–85, 1949.
117. Knepp, T. H. Proc. Pennsylvania Acad. Sci., 10: 39–42, 1936.
118. Bonadonna, T., A. Sferco, and T. Zuliani. Zent. Vet. Med., 4: 697–710, 1957.
119. Lindner, H. R. Nature, 183: 1605–1606, 1959.
120. Marsman, W. S. Acta Neerl. Morphol. Norm. et Pathol., 1: 115–128, 1937.

Bos (*Poephagus*) *grunniens* L.

YAK

The yak in Russian Asia reaches puberty at the age of 24 to 30 months.

The breeding season is well defined lasting from June to November, during which time the female is polyestrous. The gestation period is 258 days (1). In the London zoo births have occurred year-round, but mostly from April to August. A single calf has always been born (2).

1. Denison, V. F. Izv. Akad. Nauk, S.S.S.R., 863–878, 1938.
2. Zuckerman, S. PZS., 122: 827–950, 1952–3.

Bibos gaurus H. Smith

GAUR

In Malaya the gaur has no special breeding season, and it calves at any time of the year, though it does so infrequently from October to December, the wettest months (1). As the period of gestation is 9 months, mating must be relatively infrequent during the period of January to March. In India it breeds at different seasons in different districts, but the most frequent breeding time is in December and January, and births are mostly in August and September. One or two are born (2), though twins are said to be rare (1).

The evidence suggests a polyestrous animal, and this is supported by the fact that *B. frontalis* Lambert, the gayal, which is the domesticated form of the gaur, has a 3-week cycle all the year round (3).

1. Hubback, T. H. JM., 18: 267–279, 1937.
2. Brander, A. A. D. Wild Animals in Central India. London, 1923.
3. Heape, W. QJMS., 44: 1–70, 1901.

Bibos indicus L.

ZEBU, HUMPED CATTLE

In India the first estrus in the Hariana breed was at age 909 ± 43 days. The duration of estrus was 19.0 ± .6 hours for heifers, and for cows, 23.9 ± .9 hours (1). In Peru the cycle length has been 21.5 days and the duration of estrus 16.6 hours. Ovulation was 9.65 hours after its expiration. The duration of estrus was reduced to 11.5 hours after natural service (2). A report from Uganda gives a modal heat length of 24 hours with a spread from 12 to 48 hours (3). In the Philippine Islands the Red Sindi breed has heats at intervals of 21.3 ± 1.2 days and they have lasted 13.4 ± .8 hours (4). In Nyasaland conceptions are most frequent in October or November under conditions of hot weather and low humidity (5). Otherwise the species is described as polyestrous and breeding all year. If the cows are inseminated at the beginning of estrus the conception rate is 16 per cent. At mid-heat the level has risen to 66 per cent, while at the end it is 44 per cent (6). In Tharparkar cows the pH of the vaginal secretions have varied from 6.0 to 8.25, with most about 6.5 (7).

The estrogen excretion during pregnancy is as follows (8):

At 93 days,	33 M.U. per liter
101–150 days,	443 M.U. per liter
151–200 days,	555 M.U. per liter
201–250 days,	833 M.U. per liter
251–term	1,100 M.U. per liter

The sex proportion of 8,770 calves was 50.92 ± .36 per cent males (9). The duration of gestation is about 285 days. Mean durations have varied from 282 to 292 days. A typical report for the Red Sindi gives 286.5 ± .5 days for male calves and 284.5 ± .5 days for females (10). In East Africa, where the shorter gestation was reported, November calvings followed a gestation that is 3.0 days longer than those ending in October, and these are 0.8 days longer than those ending in September. The difference is ascribed to the smaller amount of available nutrition at the end of the dry season (11).

Semen volume has been recorded as 3.16 ml. and 3.8 ml. Sperm concentration was 1,455 and 1,476 million per ml.; the standard error of these figures was 82 and 128 million. The pH of the semen was 6.18 ± .025 and 6.25 ± .025 (12).

1. Ahuja, L. D., S. N. Luktuke, and Bhattacharya. Indian J. Vet. Sci., 31 (suppl. to No. 4): 13–14, 1961.
2. Villacorta Vasquez de Velazko, E. Vet. y Zootec. Lima. 12(32): 4–6, 1960.
3. Rollinson, D. H. L. Nature, 176: 352–353, 1955.
4. Nazareno, L. E. Philippine Agric., 37: 339–352, 1954.
5. Wilson, S. G. J. Agric. Sci., 36: 246–257, 1946.
6. Costa Aroeira, J. A. D. Bol. Insemin. Artif., Rio de Janeiro, 7: 47–53, 1955.
7. Chaudhuri, A. C., and R. B. Prasad. Indian J. Vet. Sci., 24: 81–87, 1954.
8. Anderson, J. Vet. J., 90: 295–298, 1934.

9. Craft, W. A. Quart. Rev. Biol., 13: 19–40, 1938.
10. Singh, R. S., and S. N. Ray. Indian J. Dairy Sci., 14: 1–7, 1961.
11. Hutchinson, H. G., and J. S. Macfarlane. East Africa Agric. J., 24: 148–152, 1958.
12. Shukla, D. D., and P. Bhattacharya. Indian J. Vet. Sci. and Animal Husbandry, 19: 161–170, 1949.

Bibos banteng Raffles. The banteng mates from September to October, and the calves are born in April or May (1).

1. Hermanns, M. J. Bombay Br. Roy. Asiatic Soc., 27: 134–173, 1952.

Syncerus caffer Sparrman

AFRICAN BUFFALO

The evidence regarding the breeding season of the African buffalo is somewhat conflicting, suggesting a fairly prolonged season, the length of which depends upon climatic conditions. In South Africa the breeding season appears to be from September to March, the spring and summer, and as the gestation period is about 11 months, the calving season is about the same (1). In the London zoo births have been distributed throughout the year (2).

1. Shortridge, G. C. The Mammals of South West Africa. London, 1934.
2. Zuckerman, S. PZS., 122: 827–950, 1952–3.

Syncerus nanus Boddaert. This buffalo has been recorded with a 103-cm. fetus in May (1).

1. Frechkop, S. EPNA., 1943.

Bison bison L.

BISON, AMERICAN BUFFALO

The bison in captivity has estrous cycles of about 3 weeks' duration all through the year (1), but records of the numerous herds when they were at large on the North American prairies indicate that in a state of nature their main mating season was from July to September. The period of gestation is about 9 months (2). In Canada the breeding season is restricted, beginning toward the end of July and lasting for 6 weeks. The duration of heat is believed to be fairly long, at least 2 days (3). The males rut from mid-June to the middle or end of September, but the distribution of births indicates that they breed at other times as well. The main calving season is in April and May. Puberty is reached at age 2 years. After the age of 12 years is reached the incidence of pregnancy declines gradually. The newborn are 50.4 per cent males (4). In Jackson Hole the age at puberty is 3 years (5).

1. Heape, W. QJMS., 44: 1–70, 1901.
2. Brown, C. E. JM., 17: 10–13, 1936.
3. Deakin, A. Personal communication, 1944.
4. McHugh, T. Zoologica, 43: 1–40, 1958.
5. Negus, N. C. JM., 31: 463, 1950.

Bison bonasus L.

WISENT

This buffalo reaches puberty in both sexes at about the fourth year of life. The females mate all year but mostly in August and September, so that the majority of calves are born in May or June (1). The gestation period averages 265 days, with about 55 per cent between 260 and 269 days. Single calves are almost invariable, and the newborn are 49.2 per cent males (2). The interval between heat periods is irregular, but is probably mostly between 21 and 26 days (2). On reserves the cows calve twice in 3 years, but in the wild only once every 2 years (3).

1. Glover, R. JM., 28: 333–342, 1947.
2. Jaczewski, Z. Acta Theriol. 1: 333–376, 1958.
3. Scibor, J. Med. Wet., 15: 712–713, 1959.

CEPHALOPHINAE

Cephalophus

DUIKER

Cephalophus dorsalis Gray. The gestation of this species lasts about 4 months (1). In the London zoo the preorbital glands of the male were markedly swollen during the summer months (2).

C. harveyi Thomas. A female with a single fetus is reported for March (3).

C. maxwelli H. Smith. In the London zoo births of single calves have been distributed throughout the year (4). The gestation period lasts about 4 months (5).

C. monticola Thunberg. The blue duiker drops its calves all the year but mostly in September and October (6). The single newborn calf has been found in February (7).

C. natalensis A. Smith. The red duiker usually drops its calves from October to November. One, or rarely 2, is the rule (5).

C. sylvicultrix Afzelius. The calves are born after a gestation of about 4 months (1). A single calf is produced at yearly intervals (7).

1. Jeannin, A. Les Mammifères Sauvages du Cameroun. Paris, 1936.
2. Pocock, R. I. PZS., 840–986, 1910.
3. Lönnberg, E. Kungl. Svenska Vetansk. Akak. Handlingar, 48(7), 1912.
4. Zuckerman, S. PZS., 122: 827–950, 1952–3.
5. Rode, P. Mammifères Ongulés de l'Afrique Noire. Paris, 1943.
6. FitzSimons, F. W. The Natural History of South Africa. London, 1919.
7. Allen, G. M., B. Lawrence, and A. Loveridge. HCMZ., 79: 31–126, 1935–7.

Sylvicapra grimmia L.

DUIKER

In the wild this species breeds all year but most calves are born in spring and early summer. Usually 1 calf is produced at a time (1), and records of twins are very rare. Gestation lasts about 4 months (1). In the London zoo calves have been born throughout the year (2). The preorbital glands of males discharge in July but have become inactive by August (3). In February a pregnant lactating female was taken in Northern Rhodesia (4).

1. FitzSimons, F. W. The Natural History of South Africa. London, 1912.
2. Zuckerman, S. PZS., 122: 827–950, 1952–3.
3. Pocock, R. I. PZS., 840–986, 1910.
4. Allen, W. F. H. PZS., 134: 251–274, 1960.

HIPPOTRAGINAE

Kobus defassa Rüppell

WATERBUCK

This species breeds all the year but most calves are born in the full dry season, or December to May (1). There is probably no fixed season but, in the Kafue National Park the peak of births was from February to March (2). The gestation of the single calf is 243 days (3). A female with 3 embryos has been recorded (4).

1. Verschuren, J. EPNG., 1958.
2. Ansell, W. F. H. PZS., 134: 251–274, 1960.
3. Dekeyser, P. L. Les Mammifères de l'Afrique Noire Française. n.p., 1956.
4. Frechkop, S. EPNU., 1944.

Kobus ellipsiprymnus Ogilby

WATERBUCK

In captivity the females of this species have a series of 3-week cycles from May to July (1). In the wild the single calf is born at any time during the summer after a gestation of about 8 months (2). Matings have been seen from June to August but there is probably no fixed season (3). Another account gives the calving season as from late February to April (4). This agrees with the mating time.

1. Heape, W. QJMS., 44: 1–70, 1901.
2. Hamilton, J. S. Animal Life in Africa. London, 1912.
3. Ansell, W. F. H. PZS., 134: 251–274, 1960.
4. Dasmann, R. F., and A. S. Mossman. JM., 43: 533–537, 1962.

Adenota kob Erxleben

KOB

This species breeds most of the year but births are especially numerous in the first months of the year, i.e., in the dry season (1). Gestation lasts for 5 to 6 months (2). Ovulation takes place at random between the ovaries but implantations are invariably in the right horn of the uterus. There is a record of twins, both in the right horn (3).

1. Verschuren, J. EPNG., 1958.
2. Frechkop, S. EPNA., 1943.
3. Buechner, H. K. Nature, 190: 738–739, 1961.

Adenota vardoni Livingstone. The puku calves in November and December (1). The main lambing season is probably from May to September, with a peak in June to August, though a near-term fetus has been found in November (2).

1. FitzSimons, F. W. The Natural History of South Africa. London, 1912.
2. Ansell, W. F. H. PZS., 134: 251–274, 1960.

Onotragus leche Gray

LECHWE

The season of birth is in October and November (1) and the calves are born after a 7-month gestation (2). A prolonged lambing season, mostly in mid-July to mid-August, has also been reported and a gestation probably nearer 8 months than 7 (3). In the London zoo the single calf has been born at all times of year (4).

1. FitzSimons, F. W. The Natural History of South Africa. London, 1919.
2. Wilhelm, J. H. J. South West Africa Sci. Soc., 6: 51–74, 1933.
3. Ansell, W. F. H. PZS., 134: 251–274, 1960.
4. Zuckerman, S. PZS., 122: 827–950, 1952–3.

Redunca arundinum Boddaert

REEDBUCK

This species has no regular season but in the low country most calves are born in August and September. On the mountain plateaus December to March are the usual months (1). In the London zoo births have been grouped in the months from October to May and gestation has lasted 233 days (2). Usually a single lamb is born, occasionally twins (3).

1. FitzSimons, F. W. The Natural History of South Africa. London, 1919.
2. Zuckerman, S. PZS., 122: 827–950, 1952–3.
3. Ansell, W. F. H. PZS., 134: 251–274, 1960.

Redunca fulvorufula Afzelius. This species has its young from October to December (1).

R. redunca Pallas. This reedbuck breeds at any season of the year (2). A record of a female taken in March gives a single fetus (3).

1. FitzSimons, F. W. The Natural History of South Africa. London, 1919.
2. Schuster, L. Zool. Garten, 2: 114, 1929.
3. Frechkop, S. EPNA., 1943.

Pelea capreolus Bechstein

VAAL RHEBOK

The single or twin calves are born in November and December (1).

1. FitzSimons, F. W. The Natural History of South Africa. London, 1919.

Hippotragus

ANTELOPE

Hippotragus equinus Desmarest. The roan antelope breeds at any time of the year but most calves are born from August to February, or in January to February in the southern part of Africa (1). The gestation period is about 10 months (2).

H. niger Harris. The sable antelope has no fixed season. The calves are born from August to February, but in the south they are dropped mostly in January or February (1). Observed gestations have been 272 and 281 days (3).

1. FitzSimons, F. W. The Natural History of South Africa. London, 1919.
2. Rode, P. Mammifères Ongulés de l'Afrique Noire. Paris, 1943.
3. Hamilton, J. S. South African Mammals. London, 1912.

Oryx

Oryx algazel Oken. The white oryx breeds all year and the gestation period lasts for 9 to 10 months (1). In the London zoo single calves have been born at all periods of the year (2).

O. beisa Rüppell. The beisa oryx breeds in September and has a single calf (3) after a gestation of 260 to 300 days (4).

O. gazella L. The gemsbok breeds in October and November in South West Africa (5). In Angola the calves are born in January (6). At the Oklahoma City zoo a female gave birth to a single calf in September (7).

1. Dekeyser, P. L. Les Mammifères de l'Afrique Noire Française. n.p., 1956.
2. Zuckerman, S. PZS., 122: 827–950, 1952–3.
3. Roosevelt, T., and E. Heller. Life Histories of African Game Animals. New York, 1914.
4. Kenneth, J. H. Gestation Periods. Edinburgh, 1943.
5. Wilhelm, J. H. J. South West Africa Sci. Soc., 6: 51–74, 1933.
6. Blaine, G. PZS., 317–339, 1922.
7. Thomas, W. D. JM., 43: 98–101, 1962.

Addax nasomaculatus Blainville

ADDAX

The single young is born after a gestation of 10 to 11 months (1).

1. Malbrant, R. Faune du Centre Africain Français. Paris, 1952.

Damaliscus

Damaliscus albifrons Burchell. In the London zoo the blesbok has given birth to single calves at all times of the year. The gestation period has lasted from 7 to 8 months (1).

D. korrigum Ogilby. The single calf is born after a gestation of about 214 days (2).

D. lunatus Burchell. The sassaby has its calves in September and October after a gestation of 7½ to 8 months (3). Towards the northern limit of its range they are born in July (4).

D. pygargus Pallas. The bontebok has its young in September and October after a gestation of 9 to 10 months (5), or of 230 days, according to another report (6).

1. Zuckerman, S. PZS., 122: 827–950, 1952–3.
2. Dekeyser, P. L. Les Mammifères de l'Afrique Noire Française. n.p., 1956.
3. Shortridge, G. C. The Mammals of South West Africa. London, 1934.
4. Ansell, W. F. H. PZS., 134: 251–274, 1960.
5. FitzSimons, F. W. The Natural History of South Africa. London, 1919.
6. Kenneth, J. H. Gestation Periods. Edinburgh, 1943.

Alcelaphus lichtensteini Peters

HARTEBEEST

This species breeds in the early summer (1). It calves from September to November (2) after a gestation of about 8 months (1). There seems to be considerable variation with the locality as a July calving peak, with all the calves born within a short period, has also been reported. A few were born in June and September. In Luangwa calving is in October and November. The gestation is about 237 days (3).

1. Hamilton, J. S. Animal Life in Africa. London, 1912.
2. Shortridge, G. C. The Mammals of South West Africa. London, 1934.
3. Ansell, W. F. H. PZS., 134: 251–274, 1960.

Alcelaphus caama Cuvier. This hartebeest has its calves from October to November (1) after a gestation of 242 days (2).

A. cokei Günther. The kongoni probably has an extended season (3, 4), and usually one calf.

A. lelwel Heuglin. The lelwel usually has its calves early in the dry season, in January and February (5).

A. major Blyth. This species has a gestation period of 242 days (6).

1. Wilhelm, J. H. J. South West Africa Sci. Soc., 6: 51–74, 1933.
2. Kenneth, J. H. Gestation Periods. Edinburgh, 1943.
3. Roosevelt, T., and E. Heller. Life Histories of African Game Animals. New York, 1914.
4. Lönnberg, E. Kungl. Svenska Vetansk. Akad. Handlingar, 48(5), 1912.
5. Verschuren, J. EPNG., 1958.
6. Rode, P. Mammifères Ongulés de l'Afrique Noire. Paris, 1943.

Connochaetes gnu Zimmermann

GNU, WILDEBEEST

The black wildebeest ruts in March (1) and the single calves are born from November to January (2) after a gestation of 8 to 8½ months. The female breeds before she is 3 years old (3). In the London zoo births have been spread from January to July (4).

1. Millais, J. G. A Breath from the Veldt. London, 1899.
2. Haagner, A. K. South African Mammals. London, 1920.
3. Sclater, W. L. The Mammals of South Africa. London, 1900.
4. Zuckerman, S. PZS., 122: 827–950, 1952–3.

Connochaetes taurinus Burchell

BLUE WILDEBEEST

This species has not more than 2 cycles per season when in captivity (1). In South West Africa the calves are born from October to January (2). Further north the main season of birth is from September to October, with early and late calves in August and November. The incidence of new grass makes a difference in the date (3). In the London zoo births of single calves have occurred from May to November after a gestation of 255 days (4). In the Augsburg zoo gestations have been 249, 251, and 255 days long (5).

1. Heape, W. QJMS., 44: 1–70, 1901.
2. Shortridge, G. C. The Mammals of South West Africa. London, 1934.
3. Ansell, W. F. H. PZS., 134: 251–274, 1960.
4. Zuckerman, S. PZS., 122: 829–950, 1952–3.
5. Steinbacher, G. Säugetierk. Mitt., 6: 173, 1958; 7: 75, 1959.

ANTILOPINAE

Oreotragus oreotragus Zimmermann

KLIPSPRINGER

This antelope probably breeds in the warm and wet season (1) and calves from July to October (2) after a gestation of 214 days (3).

1. Hamilton, J. S. Animal Life in Africa. London, 1912.
2. FitzSimons, F. W. The Natural History of South Africa. London, 1919.
3. Kenneth, J. H. Gestation Periods. Edinburgh, 1943.

Ourebia ourebi Zimmermann

ORIBI

The season of birth is from September to December, and mostly in the latter month (1). The gestation period is about 7 months (2). In the Belgian Congo a fetus near term was found in March (3) and another in April (4).

1. FitzSimons, F. W. The Natural History of South Africa. London, 1919.
2. Wilhelm, J. H. J. South West Africa Sci. Soc., 6: 51–74, 1933.
3. Verschuren, J. EPNG., 1958.
4. Frechkop, S. EPNU., 1944.

Raphiceros campestris Thunberg

STEINBOK

This species calves in summer and autumn. The number of calves is usually 1, but sometimes twins are born (1). The gestation period is 7 months (2). Breeding may take place at any time of the year, but a peak of lambing seems to occur in the early part of the dry season (3).

1. FitzSimons, F. W. The Natural History of South Africa. London, 1919.
2. Wilhelm, J. H. J. South West Africa Sci. Soc., 6: 51–74, 1933.
3. Dasmann, R. F., and A. S. Mossman. JM., 43: 533–537, 1962.

Raphiceros sharpei Thomas. The calves are usually born at the beginning of the rainy season (1). In the Zambesi district they are dropped at the beginning of January (2).

1. Kirby, F. V. In R. Lydekker, The Great and Small Game of Africa. London, 1899.
2. Lawrence, B., and A. Loveridge. HCMZ., 110: 1–80, 1953.

Nesotragus

SUNI

Nesotragus livingstonianus Kirk. This species calves from mid-November to mid-December (1).

N. moschatus von Düben. A female with a single fetus has been reported at the beginning of February (2).

1. FitzSimons, F. W. The Natural History of South Africa. London, 1919.
2. Lönnberg, E. Kungl. Svenska Vetansk. Akad. Handlingar, 48(7), 1912.

Rhynchotragus kirki Günther

DIK-DIK

In Tanganyika this dik-dik has two breeding seasons each year, at about 6-month intervals. The peaks are from June to July and from November to December, but some females may be found in heat in other months; thus the species may be regarded as polyestrous all the year. The females reach puberty at about 6 months. Males with testes below 2.4 g. weight are sexually inactive; those with testes above this weight are usually active, a condition attained at about 4 kg. body weight. For active males the peaks of testis weight are May and November, with minima in January to February and in July and August (1).

Usually only one follicle ruptures at a heat period. The largest follices measure 4.6 mm., and follicles of this size may be found at all stages of pregnancy. The corpus luteum frequently occupies most of the ovary and it often protrudes from the surface. Its diameter is from 5 to 5.5 mm., but a few may be found as large as 7 mm. Implantation is almost invariably in the right horn of the uterus and, of 109 pregnancies, the fetus was in the left only on 3 occasions. Almost half the fetuses represent ova that have migrated to the opposite horn. The septum of the uterus extends almost to the opening of the cervix. The gestation period lasts from 170 to 174 days (1).

1. Kellas, L. M. PZS., 124: 751–784, 1954–5.

Rhynchotragus guentheri Thomas. A female with a single fetus has been reported in February (1).

1. Lönnberg, E. Kungl. Svenska Vetansk. Akad. Handlingar, 49(7), 1912.

Antilope cervicapra L.

BLACKBUCK

This Indian antelope breeds at all seasons but the chief rutting period is in March. From 1 to 2 young are born at a time (1). In England they breed twice a year (2). In the London zoo births of single calves have been evenly distributed throughout the year (3). One testis weighed 13.0 g., and the length of the seminiferous tubules was 704 meters. The spermatozoon length was 48.4 μ (4). The gestation period has been reported as about 6 months (5), or 5 months (2). In a castrated male the preorbital glands retrogressed to the degree of an immature male (6).

1. Brander, A. A. D. Wild Animals in Central India. London, 1923.
2. Loder, E. G. PZS., 476, 1894.
3. Zuckerman, S. PZS., 122: 827–950, 1952–3.
4. Knepp, T. H. Proc. Pennsylvania Acad. Sci., 13: 58–62, 1939.
5. Brown, C. E. JM., 17: 10–13, 1936.
6. Bennett, E. T. PZS., 34–36, 1836.

Aepyceros melampus Lichtenstein

IMPALA

The impala ruts from April to mid-May and the calves are born from October to December. Twins are rare and the gestation period is about 6½ or 7 months long (1). In Southern Rhodesia the females bear their first young at 2 years old. Rut is early in the dry season, in June, and the young are born early in the wet season with the peak from December 15 to January 1; thus the gestation period would be 180 to 210 days. No twins are born (2). In the Cologne zoo a gestation period of 171 days was observed (3). In northwest Rhodesia there is a sharply defined lambing season, as most young are born from late September into October. In Luangwa the peak is in November and births continue into December (4). Implantations are always in the right horn of the uterus but the ovaries produce the ova in equal numbers (5).

1. Hamilton, J. S. Animal Life in Africa. London, 1912.
2. Dasmann, R. F., and A. S. Mossman. JM., 43: 375–395, 1962.
3. Roberts, A. The Mammals of South Africa. Cape Town, 1951.
4. Ansell, W. F. H. PZS., 134: 251–274, 1960.
5. Mossman, A. S., and H. W. Mossman. Science, 137: 869, 1962.

Gazella thomsonii Günther

GAZELLE

The female gazelle may breed during her first year and the males reach puberty at from 20 months to 2 years. There is no definite rutting season and the males may breed at any time. In Tanganyika calves may be dropped at any time, but on the Serengeti Plain there are two peak calving seasons, one immediately after the short rains, in January to early March, and another following the long rains, from mid-June to late July. On the Sanya Plains fawns were dropped in May, during the rains. The females may breed twice yearly, for pregnant does with fawns have been found. It is suggested that under adverse grazing conditions birth may be delayed (1).

In Kenya most fawns are dropped at the beginning of the rains, in April and November, but births may occur at any time. As a rule a single fawn is dropped; twins are rare (2). In 85 pregnancies not one was with twins (1).

1. Brooks, A. C. British Colonial Office, Colonial Res. Publ., No. 25, 1961.
2. Percival, A. B. A Game Ranger on Safari. London, 1928.

Gazella arabica Lichtenstein. A London zoo record gives the birth of a single fawn in April (1).

G. bennetti Sykes. The Indian gazelle has no regular season and 1 to 2 fawns are born at a time (2). In the London zoo single fawns have been born in July and November (1).

G. cuvieri Ogilby. In the London zoo a single fawn has been born in April (1).

G. dama Pallas. This species has a gestation lasting 5 months and has 2 pregnancies a year (3).

G. dorcas L. This gazelle has 2 to 3 diestrous cycles in the season (4). The number of fawns is 1, rarely 2, born after a gestation of 3 months (3). In the London zoo the single fawn has been born in the months from March to October (1).

G. granti Brooke. This gazelle has no fixed breeding season. One fawn is usual (5).

G. marica Thomas. In the London zoo single fawns have been born in April, May, and June (1).

G. muscatensis Brooke. In the London zoo this species has produced a single fawn in March (1).

G. rufifrons Gray. This gazelle has preorbital glands that are markedly swollen in summer (6). This observation was made in the London zoo.

G. subgutturosa Güldenstaedt. The Persian gazelle has bred freely in the London zoo and the fawns have been born from April to July with a peak at the end of May and early June. A single birth has also been recorded at the end of September. Twins were recorded in one-third of the births (1).

1. Zuckerman, S. PZS., 122: 827–950, 1952–3.
2. Brander, A. A. D. Wild Animals in Central India. London, 1923.
3. Rode, P. Mammifères Ongulés de l'Afrique Noire. Paris, 1943.
4. Heape, W. QJMS., 44: 1–70, 1901.
5. Roosevelt, T., and E. Heller. Life Histories of African Game Animals. New York, 1914.
6. Pocock, R. I. PZS., 840–986, 1910.

Antidorcas marsupialis Zimmermann

SPRINGBOK

This species probably has an extended season but the calves are born from August to January, mostly in November (1, 2). A single calf is usual, with occasional twins. The gestation period is about 171 days (2). A young castrated male developed the slender horns characteristic of the adult female (2). In the London zoo birth of a single calf has been recorded in May (3).

1. Shortridge, G. C. The Mammals of South West Africa. London, 1934.
2. FitzSimons, F. W. The Natural History of South Africa. London, 1919.
3. Zuckerman, S. PZS., 122: 829–950, 1952–3.

Procapra

Procapra gutturosa Pallas. The seren of China usually has 2 calves, born in June. The neck swells during the mating season (1).

P. picticaudata Hodgson. This Mongolian species has its calves in May (1)

1. Wallace, H. F. The Big Game of Central and Western China. London, 1913.

CAPRINAE

Pantholops hodgsoni Abel

CHIRU

This Tibetan antelope ruts in the winter and a single calf is born after a gestation of 6 months (1).

1. Blanford, W. T. The Fauna of British India. Mammals. London, 1888–91.

Saiga tatarica L.

SAIGA

The females reach puberty at 7 to 8 months old. The breeding season is from December 15 to 25 and the young are born in mid-May, or 15 to 30 days later further east (1). In the Volga region they mate from the end of November to December 20, but the season is later further to the east. In zoos gestation has lasted 145 days. Twins are usual; 70 per cent have twins and 30 per cent singles. Occasional triplets are born. The newborn are 50.1 per cent males (2). On the right bank of the Volga birth is from the end of April to mid-May, with a mass lambing early in May (3).

1. Bannikov, A. G. Terre et Vie, 108: 77–85, 1961.
2. Bannikov, A. G. Mammalia, 22: 208–225, 1958.
3. Fandeyev, A. A. ZZ., 39: 906–910, 1960.

Naemorhedus goral Hardwicke

GORAL

The single kid is born in May or June after a gestation of 6 months (1).

1. Blanford, W .T. The Fauna of British India. Mammals. London, 1888–91.

Capricornis sumatrensis Bechstein

SEROW

This species has been variously stated to have its single kid in September or October and in May or June after an 8-month gestation (1). In the northwest of India May or June is the usual season, but September or October is usual in most of its range (2).

1. Blanford, W. T. The Fauna of British India. Mammals. London, 1888–91.
2. Jerdon, T. C. The Mammals of India. Roorkee, India, 1867.

Oreamnus americanus Blainville

ROCKY MOUNTAIN GOAT

In Canada this goat ruts in November and the young, 1, or occasionally 2, are born late in April or May (1). In Montana the earliest kids are born at the end of May (2). The testes develop at age 2 years and the does probably breed at the same age (2). The gestation period is about 147 days (3).

1. Rand, A. L. Canada Nat. Mus., Bull. 100, 1945.
2. Lentfer, J. W. JWM., 19: 417–429, 1955.
3. Kenneth, J. H. Gestation Periods. Edinburgh, 1943.

Rupicapra rupicapra L.

CHAMOIS

The breeding season of the chamois is said to be in September, and at this time the throat of the male swells (1). In the London zoo, however, the postcornual scent glands only begin to enlarge during the latter half of September. They reach their maximum size in November and the first half of December and are fully reduced by the end of the first week of January. They do not enlarge in the female (2). In the Alps the rutting season reaches its climax about November 20 and extends into December (3). In the London zoo kids have been born from May to August after a gestation lasting 150 to 160 days (4). The first heat is at age 1½ years (5).

1. Boner, C. Chamois Hunting in the Mountains of Bavaria. London, 1853.
2. Pocock, R. I. PZS., 840–986, 1910.
3. Niedereder, W. Oestereichs Waidwerk, 177–180, 1953.
4. Zuckerman, S. PZS., 122: 827–950, 1952–3.
5. Couturier, A. L. Terre et Vie, 108: 54–73, 1961.

Budorcas bedfordi Thomas

TAKIN

This species ruts from the end of July to the beginning of August and the single kid is born at the end of March or early in April (1).

1. Wallace, H. F. The Big Game of Central and Western China. London, 1913.

Ovibos moschatus Zimmermann

MUSK OX

The females are in heat from July to September and calves are born from April to June (1) after a gestation of 8 months (2). They breed at age 3 years (1).

1. Rand, A. L. Canada Nat. Mus., Bull. 100, 1940.
2. Glover, R. Oryx, 2: 76–86, 1953.

Hemitragus jemlahicus H. Smith

TAHR

This goat ruts in winter and the kids, usually 1, are born in June or July (1) after a gestation of 6 months (2). It is probably monestrous (2). In the London zoo births have been distributed from May to September, with a peak in June. One birth in every 12 has been of twins (3). The testes weigh 24.4 g., and the seminiferous tubules measured 1,420 meters. The spermatozoon is 46.3 μ long (4).

1. Blanford, W. T. The Fauna of British India. Mammals. London, 1888–91.
2. Lydekker, R. The Game Animals of India, Burma, Malaya and Tibet. London, 1924.
3. Zuckerman, S. PZS., 122: 827–950, 1952–3.
4. Knepp, T. H. Proc. Pennsylvania Acad. Sci., 13: 58–62, 1939.

Hemitragus hylocrius Ogilby. This tahr has no definite season and usually produces 2 kids at a birth (1).

1. Blanford, W. T. The Fauna of British India. Mammals. London, 1888–91.

Capra hircus L.

DOMESTIC GOAT

The goat is a seasonally polyestrous animal with a breeding season beginning about September, reaching its greatest intensity in October, and then gradually tapering off so that few goats, if held over, will come in heat in the New Year. The length of the breeding season is a matter of considerable practical importance, as it is difficult to arrange breedings so that a herd will give a uniform yield of milk all the year round. The lengths of the cycle and of heat periods are a little longer than those of the sheep. Ovulation is spontaneous.

THE ESTROUS CYCLE

The length of the estrous cycle is somewhat variable, and it is generally given as about 21 days. A series of accurate determinations which were made upon the Angora goat gives a mean cycle length of 19.4 ± .5 days, with a spread from 12 to 24 days. The same work reports a duration of heat of 39.2 ± 1.9 hours, with a spread from 1 to 4 days (1). For dairy goats a comparable series gave a mean cycle length of 17.8 ± .36 days, with a standard deviation of 5.76 days (2). This data, however, is peculiar in that there are a number of cycles of extremely short duration, from 6 to 12 days. There was a very strong mode at 20 days, and 72 per cent of the observations fell between 15 and 24 days; hence the normal duration may be regarded as 20 days. The number of aberrant cycles was greater in the kids than among mature does.

A study of the frequency of kiddings in the United States, month by month, has shown that the greatest number of conceptions occur in October and that fairly good breeding may be expected from August through January; in all other months very few conceptions occur (2). A similar, more extensive, study of American records gave the most intense breeding from mid-September to mid-November and the lowest in May (3). English records give the same results, but after a cool summer the peak of reproduction comes a month earlier, and after a hot summer, a month later, than usual (4). However, the shifts may be due to little and much sunlight rather than to the mean temperature, as it has been shown that the season may be modified by exposure of the goat to varied daylight periods. In India the Jumna Pari goat mates most frequently from July to October, but the Bar Bari goat, a dwarf type, will breed at any time of the year (5). Apparently the breeding season in that country is not so well defined as it is in temperate climates, since it is stated that 2 kiddings may occur in a year (6). The same remarks may apply to the Philippine Islands, as the mean interval between parturition and heat is 91.5 ± 8.0 days. Apparently goats in that country do not breed at so early an age as they do in temperate climates, since the mean age at first heat is given as 494 days (7). In the northern countries goats will breed as kids if they have been born early in the season. In Nigeria the native milk goat is ready for service 6 to 8 weeks after she has produced her kids (8). Wild goats in New Zealand breed all the year (9).

The usual number of young born at a kidding is 2, but 1 to 3 are common, and 4 and 5 are rare. Limited data, given in Table 11, indicate that some breed differences exist. The number of kids born to does below 18 months of age average 1.5, but to does above this age the average is 2.1 (10).

TABLE 11. Average Number of Kids per Birth

BREED	NUMBER OF KIDS
Toggenburg (15)	1.8
Saanen (15)	1.9
Anglo-Nubian (19)	2.1

The sex ratio in dairy goats is distinctly aberrant because of the large numbers of pseudohermaphrodites which are produced. One set of data gave 53.4 per cent males, but the incidence of pseudohermaphroditism was not recorded. In one Saanen herd in which the incidence was high the sex ratio was 49.3 per cent males, 39.6 per cent females, and 11.0 per cent pseudohermaphrodites. A Toggenburg herd gave 46.4 per cent males, 47.6 per cent females, and 6.0 per cent pseudohermaphrodites. A collection of data from Saanen breeders who had many pseudohermaphrodites in their herds gave 55.1 per cent males, 30.0 per cent females, and 14.9 per cent pseudohermaphrodites (11). The pseudohermaphrodites, despite the fact that they possess testes, are therefore reversed genetic females according to sex ratio data (12). In a recent study of 1,051 kids born, the sex proportion was 57.1 per cent males, 36.2 per cent females, and 6.9 per cent intersexes. The litter size in this sample was 2.02, and the mode 2 (13).

The pseudohermaphrodite has been often recorded in the literature. Testes are present in all cases, though the tubular and external genitalia present a wide variety of forms in which many gradations appear (14). The condition appears to be inherited as a simple recessive (15), and it is connected in some way with hornlessness, which is inherited as a dominant, since horned hermaphrodites are extremely rare if they occur at all (12, 16). A detailed study of its anatomy has been made (17).

True hermaphrodites are found among goats, and, while rare, they occur with greater frequency than in any other species except, perhaps, the pig.

The sex ratio in the Angora goat, in which pseudohermaphroditism seems to be absent, is 50.1 per cent males, but there is a very high incidence of ridglings (males with an undescended testis), about 5.5 per cent (18). These appear to be genetic males, as the sex ratio is normal if they are counted as such. In the majority of cases only one testis, almost always the right, has failed to descend. The condition is inherited (18).

The duration of gestation in the goat is similar to that in the sheep. A large series of goats of several breeds gave a mean of 150.807 ± .004 days, with a standard deviation of 3.26 days. The modal distribution was 151 days, and 86 per cent fell between 147 and 155 days (19). Some breed differences exist and are recorded in Table 12. The month of conception has some influence on the length of pregnancy; it averages 151.3 ± .1 days for August

TABLE 12. Breed Differences in the Duration of Gestation in Goats

BREED	DURATION OF GESTATION, DAYS
Bar Bari (5)	146
East African Dwarf Goat (21)	146.5
Angora (1)	148.08 ± .09
Philippine goats (7)	148.1 ± .07
Jumna Pari (5)	150
Anglo-Nubian (19)	150.0 ± .1
Schwartzwald (20)	150.8 ± .2

conceptions, and 149.8 ± .1 days for February conceptions. The difference is small, and it might be due to seasonal differences in the length of the heat period, if these exist, since the length is measured from the day of service and not from ovulation. The change between these months is fairly orderly. The age of the dam has some influence on the duration. It is least, 150.1 ± .1 days, when the goat conceives in her first year, as a kid; in the second year it is 150.61 ± .8 days; and it gradually rises to a maximum of 151.3 ± .1 days at 6 years. Differences with litter size are not statistically significant (19).

HISTOLOGY OF THE FEMALE TRACT

OVARY. Polynuclear ova are not found until 3 days before the onset of estrus; then they are 4 per cent of all oöcytes. The condition appears to be due to the fusion of oöcytes, as the size increases with the number of nuclei. Polyovular follicles are first encountered 4 days before the onset of estrus and they represent from 5 to 12 per cent of all follicles. They usually degenerate. In anestrum and metestrum a few medium-sized follicles, 2 to 5 mm. in diameter, and many small ones may be found. At 5 or 6 days before estrus many granulosa and thecal mitoses are present. The theca interna is very thick and is glandular in appearance. At the time of rupture the follicles are 8 mm. in diameter and the ovum, 95 μ. There is a slight hemorrhage into the ruptured follicle. The luteal cells are derived from the granulosa;

they reach a size of 30 to 40 μ. The corpus luteum reaches its maximum size at 12 days after ovulation when it is 11 mm. in diameter. In the nonpregnant doe retrogressive changes are in evidence at 14 to 15 days, but the changes are not as abrupt as they are in the pig (22).

VAGINA. At estrus the vaginal lining consists of stratified squamous epithelium. It is desquamating, and cornification is in evidence. At the second day after estrus it is thinner and it remains thin until 12 days, when it begins to thicken (23). During estrus a few leucocytes and many degenerating cells may be found in the smear. Towards the end of heat leucocytes become more abundant and reach their maximum at the sixth day (24).

UTERUS. At estrus the epithelium is tall-columnar (50–60 μ) and nonciliated. It is pseudostratified and the nuclei are fusiform or oval. The stroma is edematous and there is a superficial leucocytic invasion. From 6 to 18 hours after ovulation the epithelium is lower (36 μ); the nuclei are all oval. Edema has increased and there is increased vascularity also. The glands are more numerous but are inactive. In the first week the epithelium continues to reduce slightly. Glands are more numerous and secretion is impending. There is marked edema and increased vascularity. At about 12 days the epithelium is becoming columnar and the cells are vacuolated. The glands are large and dilated and they are secreting. The mucosa is edematous. In the third week the epithelial nuclei are basal. The stroma is less edematous and less glandular secretion is present. By 19 to 29 days the stroma is dense (23).

In the first week of the cycle cervical folds almost fill the lumen. This condition continues into the second week but in the third the folds diminish (23).

PHYSIOLOGY OF THE FEMALE TRACT

Adrenalin causes the muscles of the cornua uteri to relax and the cervix to contract during pregnancy. The cornua are sensitive to oxytocin, but the cervix is not (25).

The urine of the pregnant doe contains 500 to 1,000 M.U. of estrogen per liter from the seventy-fifth to the one-hundredth day (26). Estrone is present in the urine of pregnancy to the extent of 0.3 μg/l.; estradiol-17, 0.1 μg/l.; and equol, 10 to 15 μg/l. Pregnanediol, too, was detected in the urine (27). Urinary gonadotrophes have been detected both during the cycle and in pregnancy (28). Estrogens are present in the meconium of the newborn kid, and the level is high, especially of estradiol-17α (29). Serum prolactin is low in the first 4 months of pregnancy; then it increases abruptly and reaches its highest level by 10 days before parturition. There is a sudden drop at parturition. During early lactation the level is above that of pregnancy and it reaches 0.45 I.U./100 ml. of serum (30). Removal of the corpora lutea during mid and late pregnancy has invariably resulted in abortion (31). The daily amount of progesterone needed to continue the pregnancy is 15 mg.; 10 mg. was not quite sufficient (32). Daily injections of 5 mg. of progesterone inhibits estrous cycles (33).

The breeding season can be changed by modifying the amount of daylight to which the doe is exposed. By increasing the daily amount of light by 1 hour each 10 days after January 25, cycles are said to have ceased prematurely, though it must be remembered that reproductive intensity is normally at a low ebb at this time. By decreasing it by 1 hour a week from April to July, heat was induced in May and June in 3 of 6 does, though breeding was unsuccessful; but in July, after their return to normal light, 4 became pregnant in the first 10 days (34, 35).

A single injection of 400 R.U. of pregnant mares' serum given during anestrum causes the does to come in heat. Priming, which was found necessary in the sheep, is unnecessary. A dose of 200 R.U. is insufficient to produce this effect. Lactating does probably need more P.M.S. to bring them in heat than do dry ones. Several pregnancies resulted from services during the induced heat (2). An injection of 600 to 1,600 I.U. of P.M.S. was found to be a satisfactory dose for bringing the does into estrus but the fertility rate was low (36). This has been the experience with many attempts to breed out of the regular season. The probable reason is that the accessory sex organs are not in the proper condition owing to hormone imbalance. In ovariectomized goats the minimal amount of stilbestrol needed to induce estrous symptoms is 0.1 to 1.0 mg. (37).

THE MALE

The testes and epididymides develop rapidly from the age of 30 to 40 days until 140 to 150 days, when they are almost fully mature. The seminiferous tubules have reached mature diameter by 90 days and are fully mature at 150 days. Spermatozoa have appeared at 88 to 95 days (38, 39). The mature testes weigh from 90 to 135 g. the pair (39).

The average ejaculate of semen into an artificial vagina is lower for the goat than it is for the ram. For dairy goats it was found to be 0.57 cc., with a range of 0.10 to 1.25 cc. (40); and for Angoras it was 0.66 cc. (1). Another account gives 0.65 ml. with a sperm concentration of 2.7 billion per ml. The concentration is highest in spring and the volume in summer (41). For Angora bucks the concentration has been reported as 3 to 4 million per cu. mm. (1, 42). For two bucks the average pH of the semen was 6.5 (43). The fructose content is 465 mg./100 ml., and the total reducing value, 548 mg./100 ml. (44). The spermatozoa do not reach the upper part of the oviduct until at least 5 hours after mating (45).

There is some evidence that the beard is inherited in a sex-limited manner; the gene appears to be dominant in the male and recessive in the female (46). The excretion of 17-ketosteroids is higher during the mating season than at other times (47).

1. Poloceva, V. V., and M. V. Fomenko. Usp. Zooteh. Nauk, 3: 51–65, 1936.
2. Phillips, R. W., V. L. Simmons, and R. G. Schott. Am. J. Vet. Res., 4: 360–367, 1943.
3. Turner, C. W. J. Dairy Sci., 19: 619–622, 1936.
4. Asdell, S. A. J. Agric. Sci., 16: 632–639, 1926.
5. Slater, A. E., and S. S. Bhatia. Allahabad Farmer, 9: 156–159, 1935.
6. Goheen, J. L. Keeping Milk Goats in India. Sangli, India, 1933.
7. Arriola, G. C. Philippine Agric., 25: 11–29, 1936.
8. Vigo, A. H. S. Farm and Forest, 1: 119–123, 1946.
9. Wodzicki, K. A. New Zealand DSIR, Bull. 98, 1950.
10. Addington, L. H., and O. C. Cunningham. New Mexico Agric. Exp. Sta., Bull. 229, 1935.
11. Paget, R. F. British Goat Soc. Monthly J., 36: 57–59, 1943.
12. Asdell, S. A. Science, 99: 124, 1944.
13. Blokhuis, J. Tijdschr. Diergeneesk., 84: 347–351, 1959.
14. Eaton, O. N. Am. J. Vet. Res., 4: 333–343, 1943.
15. Eaton, O. N., and V. L. Simmons. J. Hered., 30: 261–266, 1939.
16. Eaton, O. N. Genetics, 30: 51–61, 1945.
17. Kondo, K. Japanese J. Zootech. Sci., 19: 55–62, 1949.
18. Lush, J. L., J. M. Jones, and W. H. Dameron. Texas Agric. Exp. Sta., Bull. 407, 1930.
19. Asdell, S. A. J. Agric. Sci., 19: 382–396, 1929.
20. Hinterthur, E. Zuchtungskunde, 8: 55–62, 1933.
21. Wilson, P. N. East African Agric. J., 23: 138–147, 1957.
22. Harrison, R. J. J. Anat., 82: 21–48, 1948.
23. Hamilton, W. J., and R. J. Harrison. J. Anat., 85: 316–324, 1951.
24. Barretto, J. F., and A. M. Filho. Bol. Ministr. Agric., Rio de Janeiro, 33(5): 1–11, 1944.
25. Newton, W. H. J. Physiol., 81: 277–282, 1934.
26. Kust, —, and — Vogt. Tierarztl. Rundschau, 40: 589–591, 1934.

27. Klyne, W., and A. A. Wright. Bioch. J., 66: 92–101, 1957.
28. Imai, K., Y. Hasegawa, and S. Nakajo. Japanese J. Zootech. Sci., 33: 59–64, 1962.
29. Velle, W. Undersökelser over Naturlig Forekommende Östrogener hos Drövtyggere og Gris. Oslo, 1958.
30. Goto, T., and I. Notsuki. Bull. Nat. Inst. Agric. Sci., Japan, G20: 199–209, 1961.
31. Drummond-Robinson, G., and S. A. Asdell. J. Physiol., 61: 608–614, 1926.
32. Meites, J., H. D. Webster, F. W. Young, F. Thorp, Jr., and R. N. Hatch. J. Animal Sci., 10: 411–416, 1951.
33. Dauzier, L., R. Ortevant, C. Thibault, and S. Wintenberger. Ann. d'Endocrinol., 14: 553–559, 1953.
34. Bissonnette, T. H. Physiol. Zool., 14: 379–383, 1941.
35. Eaton, O. N., and V. L. Simmons. United States Dept. Agric., Circ. 933, 1954.
36. Greenbaum, A. L. British Goat Soc., Year Book, pp. 47–48, 1947.
37. Nishikawa, Y., T. Horie, and H. Onuma. Bull. Nat. Inst. Agric. Sci., Japan, G10: 187–208, 1955.
38. Onuma, H., and Y. Nishikawa. Bull. Nat. Inst. Agric. Sci., Japan, G10: 365–375, 1955.
39. Yao, T. S., and O. N. Eaton. AJA., 95: 401–432, 1954.
40. Mockel, H. Diss., Leipzig, 1937.
41. Eaton, O. N., and V. L. Simmons. Am. J. Vet. Res., 13: 537–544, 1952.
42. Atabek, A. Türk Baytarlar Birligi Dergisi, No. 2: 275–285, 1936.
43. Dussardier, M., and P. Szumowski. Rec. Méd Vét., 128: 628–635, 1952.
44. Roy, A., Y. R. Karnik, S. N. Luktuke, S. Bhattacharya, and P. Bhattacharya. Indian J. Dairy Sci., 3: 42–45, 1950.
45. Ajello, P. Zooteh. e Vet., 13: 50–53, 1958.
46. Asdell, S. A., and A. D. B. Smith. J. Hered., 19: 425–430, 1928.
47. Leidl, W., and K. Bronsch. Zent. f. Vet. Med., 6: 28–36, 1959.

Capra falconeri Wagner

MARKHOR

In the wild this species is probably monestrous. It mates in December (1), and the gestation period is about 153 days (2). In the London zoo the single kids have been born in May and June (3).

1. Heape, W. QJMS., 44: 1–70, 1901.
2. Kenneth, J. H. Gestation Periods. Edinburgh, 1943.
3. Zuckerman, S. PZS., 122: 827–950, 1952–3.

Capra ibex L.

IBEX

In captivity the ibex experiences its first estrus at 1½ years old; in the wild at 2½ years. It has 1 kid every other year. Mating is between December 1 and January 15 and the births between May 25 and June 15 after a gestation of 165 to 170 days (1). The does are probably monestrous (2). The gestation period is between 150 and 180 days (3). In the London zoo births of single kids have been recorded in March, April, and May (4). The sperm head is 8 to 8.5 μ long, the midpiece 20 μ, and the tail 40 μ to 45 μ long (5).

1. Couturier, A. J. Terre et Vie, 108: 54–73, 1961.
2. Heape, W. QJMS., 44: 1–70, 1901.
3. Kenneth, J. H. Gestation Periods. Edinburgh, 1943.
4. Zuckerman, S. PZS., 122: 827–950, 1952–3.
5. Couturier, M. A. J. Mammalia. 20: 124–127, 1956.

Capra caucasica Güldenstaedt. In the London zoo the tur has had kids at the end of May or in June. One is usual; a single set of twins in 13 births has been recorded (1).

C. pyrenaica Schinz. This ibex gives birth to its kids in the latter half of April and early May after a gestation of 20 weeks, or perhaps more (2).

C. sibirica Meyer. This ibex ruts from November to December (3), and the kids are born in May or June (2). According to another account rut is in October, and the young, 2 usually, are born in June and July (4). In the London zoo single kids have been born in May and June (1).

1. Zuckerman, S. PZS., 122: 827–950, 1952–3.
2. Lydekker, R. The Game Animals of India, Burma, Malaya and Tibet. London, 1924.
3. Heape, W. QJMS., 44: 1–70, 1901.
4. Jerdon, T. C. The Mammals of India. Roorkee, India, 1867.

Pseudois nahoor Hodgson

BHARAL

In the wild (1) and in the London zoo (2) this goat is monestrous. Rut is usually in January and the kids are born in May (3) after a gestation of 160 days (4). In the London zoo single kids have been born from May to August, with a peak in June (5). Another account gives 2 as the usual number of kids, born in June or July (6).

1. Przewalsky, N. M. Mongolia, the Tangut Country and the Solitudes of Northern Tibet. London, 1876.
2. Heape, W. QJMS., 44: 1–70, 1901.
3. Wallace, H. F. The Big Game of Central and Western China. London, 1913.
4. Blanford, W. T. The Fauna of British India. Mammals. London, 1888–91.
5. Zuckerman, S. PZS., 122: 827–950, 1952–3.
6. Jerdon, T. C. The Mammals of India. Roorkee, India, 1867.

Ammotragus lervia Pallas

BARBARY SHEEP

This species usually lambs at the beginning of spring. In the region of Tigueddi this is early in March (1). In southern Morocco the period of birth is in December and the gestation period lasts 154 to 161 days (2). In the London zoo births have been spread throughout the year, with a marked peak in April. Twins have been born at about every sixth time (3).

1. Rode, P. Mammifères Ongulés de l'Afrique Noire. Paris, 1943.
2. Panouse, J. B. Trav. de l'Inst. Sci. Cérifien, Ser. Zool., 5, 1957.
3. Zuckerman, S. PZS., 122: 827–950, 1952–3.

Ovis aries L.

DOMESTIC SHEEP

The domestic sheep is polyestrous, with cycles recurring at about 16.5-day intervals. Fine-wooled breeds tend to be polyestrous all the year round, a result, perhaps, of the fact that they have been developed in warm conditions without extreme seasonal climatic changes. The coarse-wooled breeds are seasonally polyestrous, with their season beginning about September in the Northern Hemisphere and continuing, in the absence of pregnancy, well into winter. In the Southern Hemisphere the breeding season begins at the corresponding time of the year, i.e., about March. Sheep transferred from one hemisphere to the other rapidly adjust themselves to the changed seasons. Ovulation is spontaneous, occurring toward the end of heat. The corpus luteum is active for about 14 days, and proestrum is ill defined. Uterine changes during the cycle are not well marked, but the vaginal smear is a fair indication of the reproductive state. The pituitary of the sheep has a moderate F.S.H. content, and the L.H. level is fairly high. The period of gestation is about 150 days; the number of lambs varies from 1 to 3, depending largely upon the breed. The ram produces spermatozoa all the year round, but there is a tendency toward quiescence during the summer months in the coarse-wooled breeds. The chief practical problem in sheep breeding is to lengthen the season and to produce ovulation and pregnancy during the anestrum.

THE ESTROUS CYCLE

Merinos, Central European (1), Karakul (2), and Persian Blackhead (3) sheep tend to breed all the year round, but the presence or absence of an anestrous season during the summer depends largely upon climatic and feed conditions. Thus, in some regions of South Africa, Merinos have a definite breeding season from February to August (3). In the Tzurcana breed of Rumania there is a short anestrum in July (4), and in the Romanov breed of Russia ovulation often occurs in summer without external signs of heat (5). This is not the rule, however; usually during anestrum the ovaries are quiescent. The Dorset Horn also has a prolonged breeding season. In India the Bikaner ewe will breed at any time of the year, but two seasons predominate in lambings, spring and autumn (6). In the United States Karakuls tend not to conceive from April to June, and most ewes lambing in the spring do not conceive again until the fall (7).

The coarse-wool breeds generally begin to come in heat about September 1, and, if not bred, they continue to have cycles until about January 1, on the average, though many individuals continue even until May (8). There are breed differences in the incidence of the season: Rambouillets tend to come in heat earliest, followed by Hampshires, Southdowns, Shropshires, and Romneys in that order (9, 8). Lambs usually begin to breed during their first season, at a mean age of 213 days and a range of 187 to 250 days, but late-born lambs wait until the next season for their first heat period (8). Egyptian fat-tailed sheep have a late puberty, as their average age at first estrus is 300.8 ± 7.3 days and their body weight 30–35 kg. (10). The first heat period of lambs during the season is usually later than that of mature ewes (9). In Merino ewes with a season limited by climatic conditions the onset of the season is gradual, with several weak estrous periods without ovulation before the true breeding season sets in (11).

The duration of the cycles is remarkably constant, but the Merinos and Rambouillet cycles are about a day longer than those recorded for other breeds. A series of recorded means is given in Table 13. The mean length of succeeding cycles is variable and does not seem to follow any definite pattern. The whole question of age at puberty and the relation of breeding frequency and of cycle intervals to climatic conditions has been exhaustively reviewed (12).

No heat occurs in Merino sheep during lactation, but, if the lambs are weaned immediately after parturition, it usually follows in 60 to 150 days.

The mean length of the heat period for Scottish sheep is 36 hours, with a mode of 28 hours and a range from 3 to 84 hours (13). For Merinos it is slightly longer (8), with a mode from 36 to 48 hours (70 per cent of cases) (22); in South Africa the mean is 30 hours and the mode, 30 to 32 hours (3), but in Australia 50 per cent of Merinos are said to be in heat for less than 19 hours (18). Records for mixtures of breeds, mainly coarse-wool, give 30 to 40 hours, but rarely less than 24 hours or more than 48 hours (9), with a mean of 29.3 hours and mode of 21 to 27 hours (8). It may be concluded that the mean heat period is from 30 to 36 hours, with but little difference between breeds. Service has no effect upon the length, and long heat periods are usually more intense than short ones (13). However, a slight shortening after sterile service has been reported (8). Ovulation without heat occurs frequently at the beginning and the end of the breeding season, and there is no correlation between the length of heat and the length of the cycle (8).

Ovulation usually occurs from 18 to 24 hours from the beginning of heat, or from 12 to 24 hours before its end (13); at from 23 to 30 hours from its beginning (8); or, according to another report, in Merinos, at from 36 to 40 hours from the beginning, with heat finishing 6 hours after rupture of the follicle (22). The time of ovulation is more closely related to the end of heat than to the beginning, i.e., ovulation tends to cause its termination.

It may occur as early as 11 hours before its end, or as late as 6 hours afterwards, but, in general, it is before the end. In the case of twin ovulations, 1½ to 7½ hours may intervene between the two, with a mean of 1¾ hours (8). However, Merino and native Masai sheep in East Africa are said to ovulate shortly after the end of heat (23). The exact time of ovulation in relation to the end of heat is difficult to obtain as it can only be determined by laparotomy, which, in itself, may interfere somewhat with the length of the period or with its determination. The right ovary functions more frequently than the left, since ovulations have been found in that ovary in 57 per cent of cases (8), or 52 per cent (24), or 58.58 ± 1.85 per cent (25).

TABLE 13. Length of the Estrous Cycle in Sheep

BREED	LENGTH OF CYCLE, DAYS	RANGE, DAYS	REMARKS
Scottish (13)	16.4 ± .8	15–18.5	Mode, 16.5
Navaho (14)	16.44 ± .10		
Hampshire (15)	16.5		Flushed
Zackel (16)	16.5	11–19	Mode, 16
Shropshire and Hampshire (8)	16.7		Mode, 17; 68 per cent from 15.5–17.5
Romney (17)	16.7		Mode, 17: 89 per cent from 14–19
Rahmany (10)	16.75 ± .12		Lambs
Welsh Mountain (12)	16.8		
Border Leicester (12)	16.8		
Merino (3)	16.8	12.5–18.5	Mode, 16.5
Romney (12)	16.9		
Tzigai (18)	17	16–21	
Dorset (19)	17		
Merino (19)	17		
Romney (20)	17.0	14–39	91 per cent from 15–18
Dorset (12)	17.1		
Romney (21)	17.17 ± .17		93 per cent from 14–19
Hampshire (15)	17.2		Unflushed
Rahmany (10)	17.3		Adults
Suffolk (12)	17.4		
Merino (22)	17.4 ± .08	6–27	Mode, 17; S.D., 18; 85 per cent from 16–19
Rambouillet (15)	17.5	13–21	
Tzurcana (4)	17.5	15–20	

The onset of heat is usually abrupt, but the cessation is gradual. Heat does not begin more frequently during the day or night (8), though some have (19, 26) found a slight tendency for it to begin more often during the hours of daylight. There are breed differences in the intensity of heat; Dorsets are easier to detect than Merinos as there is more swelling and reddening of the vulva in the former. Moreover, the Dorset has a characteristic stance whereas the Merino tends to be more restless. Ewes attract the rams before they will accept coitus. Apparently the latter are attracted to the ewes by the olfactory sense as vaginal and perineal swabs from ewes in heat, when applied to pregnant ewes, make them attractive to the rams (19).

The time of mating or of artificial insemination within the heat period is not as important as it is in most other species. In one experiment with Kazac ewes service at 8 hours from the onset of heat resulted in 70 per cent conceptions, from 16 to 32 hours it varied from 82.5 to 86 per cent, from 40 to 48 hours it was 77 per cent, and at 56 hours it was 67 per cent. The onset of heat was detected by vaginal smears, a method which places this time rather earlier than does a test by the ram (28). Another experiment gave similar results (29). The optimum time of artificial insemination is said to be 18 to 26 hours after the onset of heat (30).

The rate of travel of ova through the oviduct is similar to that in other species. They are all in the middle third of the oviduct by 5 hours after heat has ceased. At 60 hours most are still in the middle third, but a few have reached the last third or even the uterus. By 72 hours most are in the uterus (19).

In those breeds which experience heat all the year round it is theoretically possible for the ewe to have two lambings a year. This is recorded, for instance, for the Wryosowka breed (31). The number of lambs produced at a lambing is very variable; 1 to 3 is the rule, but quadruplets are not infrequent in some breeds. Fecundity is a breed characteristic, and there are very wide differences between breeds. The hill breeds are the less fecund, while the lowland breeds, especially those which are kept for their milk, are the most fecund. It is generally agreed that fecundity is relatively low at first, rising to a peak at about the sixth year. The usual practice is to breed ewes at their second season, but most will breed as lambs in their first year, though it seems to affect their subsequent fecundity adversely. Table 14 illustrates this fact for Hampshire ewes and also shows the rise in fecundity with the age of the ewe (32). The percentage of multiple births tends to be highest in ewes bred early in the season. It may be stated as a general principle that ease of reproduction, fecundity, and the length of life during which reproduction

TABLE 14. Fecundity of Hampshire Ewes

	BRED AND CONCEIVED AS LAMBS	BRED AND CONCEIVED AS 1½-YEAR-OLDS
	Lamb crop, per cent	*Lamb crop, per cent*
First season	106	—
Second season	157	195
Third season	176	202
Fourth season	177	175
Fifth season	200	208

is possible are related, and the sheep is no exception to this rule. An animal's early performance is a good index of its lifetime capacity, provided that its readiness to breed is not abused. The variation of fecundity with the breed implies that it is an hereditary factor; not only is this so between breeds, but it holds true within a breed. For Romanov sheep this is well illustrated by Table 15 (33), which gives the numbers of lambs born to ewes which were,

TABLE 15. Effect of the Number Born on the Subsequent Fecundity of Romanov Sheep

EWES BORN AS	AVERAGE LAMBS PRODUCED BY THEM
Singles	2.17
Twins	2.36
Triplets	2.63
Quadruplets	3.01

themselves, singlets, twins, etc. In the Dorset breed there is a tendency for more multiple births in the spring than in the fall (34), which suggests that reproduction is more intense at the normal breeding season for the species than it is at other times.

Data on the sex proportion show great constancy. In general, analysis of farmers' records gives a proportion of 49 to 50 per cent males, but records from experiment station flocks have often given a slight preponderance of males. The fetal sex proportion is a little higher, averaging 50.9 ± 1.60 per cent males, with no significant difference with the age of the fetus (25). The proportion for stillbirths is 53.8 ± 2.9 per cent males (35). These figures suggest that the sex proportion at or near conception is not so markedly different from the secondary proportion as it is in the pig and in man. The proportion does not alter when singles, twins, and triplets are considered separately, but there is an excess of twins of opposite sex, which indicates that there is little or no monozygotic twinning in sheep (35, 36).

Mean gestation periods have varied from 144.1 days for the Dorset Horn to 151.8 days for the Karakul. In general, the early-maturing meat breeds have the shorter gestations. An analysis of periods from Rambouillet ewes gave a range from 143 to 159 days, with a standard deviation of 2.25 days, which agrees closely with that for other breeds (37). In this series there was a tendency for ewes bred early to have slightly longer periods than those bred late, and the period lengthened with age. Eight-year-old ewes had an average gestation period nearly 2 days longer than 2- and 3-year-olds. A gestation resulting in a single birth was about 0.6 of a day longer than that resulting in a twin birth. The sex of the lambs born had no effect on the duration (38). In Egyptian fat-tailed sheep a post-partum estrus follows lambing at an average interval of 40 to 41 days (39).

HISTOLOGY OF THE FEMALE TRACT

OVARY. While considerable work has been done upon the development of the follicles and corpora lutea during the breeding season, information on the development of the ovaries in lambs appears to be lacking. During anestrum the ovaries are quiescent. Toward the end of this period waves of follicles grow, but they become atretic before they mature, and no new corpora lutea are formed, while those from the previous season have disappeared. In proestrum it is easy to distinguish the follicles which are destined to rupture later unless large atretic follicles are present. Growth is rapid during heat, all the layers of the follicle become thin, vascularity increases, and the cumulus becomes dome-shaped and large. At the time of ovulation the follicle is about 1 cm. in diameter (9). A few hours before ovulation the follicle swells beyond the surface of the ovary, becoming conical. One hour before rupture a small round area becomes clear and transparent; a few minutes before rupture a cone, occasionally more than one, appears. This breaks, follicular fluid flows out, and the follicle gradually collapses. The fluid rarely spurts; at first it is thin, but it becomes viscous within 2 to 3 minutes (8). The follicle opens with ragged borders; a small blood clot closes the opening, though not unless a blood vessel is involved in the rupture area (8); but there is no hemorrhage into the cavity. The follicle walls grow in, and the cavity is completely filled by 30 hours after ovulation. The corpus luteum reaches its maximum size at about 6 to 8 days, and it is then about 9 mm. in diameter. It has lost its rupture point. At first it is reddish pink, becoming paler up to the next proestrum, when a yellowish tinge appears as fatty degeneration sets in (22). After this, degeneration is rapid, so that by the time it is 24 days old only vestiges remain. The corpus luteum of pregnancy remains until parturition (9), though atrophy begins about 2 to 3 weeks before (13).

The corpus luteum is developed from both granulosa and theca interna cells, but the lutein cells almost all have a granulosa origin as the theca lutein cells quickly degenerate. The lutein cells measure 25 to 30 μ at first, and they gradually increase in size until retrogression sets in at the fourteenth day (40). The size of the ovum is 147 μ, and the zona pellucida measures 15 μ (41). The ovarian tunic is a tough structure. No new ova are produced during sexual life from the germinal epithelium, but the neogenic layer below the tunica albuginea does produce them. The first polar body is usually extruded from the ovum just prior to or, in many cases, just after ovulation (9).

At the times of proestrum, estrus, and metestrum many polyovular follicles (1 per cent) and multinuclear oöcytes (12.5) may be found. These are all in small follicles. There are many atretic follicles in diestrum but few in proestrum (42).

VAGINA. The vagina does not undergo striking changes during the cycle. Except in anestrum there is continuous growth of the epithelium, slightly accelerated during heat; late in the heat period and in metestrum considerable desquamation occurs, often as many as 4 to 5 layers of cells being cast off. Partial keratinization is characteristic of the heat period, but it may occur at any time; it is always regional and never general. Lymphocytes and leucocytes may be found in the epithelium at any time, but the latter are most frequent early in diestrum (13). These cells are present in greatest numbers at about the seventh to eighth day after ovulation (8, 9). The vaginal stroma is most edematous in late proestrum and early in heat. During the first 10 days of diestrum most mitoses are found in the epithelium, and at this time the cells are most reactive to basic stains. Their greatest affinity for acid stains is shown at proestrum and on the first day of heat (43).

Some workers have found the smear to be more clear-cut than the changes in the vaginal epithelium would lead one to expect. During late proestrum and early heat the smear contains epithelial cells, leucocytes, and scant, thick mucus; the latter does not appear in diestrum. This stage lasts about 12 hours. Then the mucus becomes voluminous and transparent, with small epithelial cells, very often a few squamous cells, and usually leucocytes. This stage begins at about the fourth hour of heat and lasts for about 24 hours. On the second day of heat the leucocytes are reduced in numbers. This stage lasts through the first day of metestrum. On the second day desquamation increases, and the smear becomes voluminous, dry, and cheesy. It consists mostly of large squamous cells, with occasionally a few leucocytes, and it lasts for about 4 to 8 days. In late diestrum the number of small epithelial cells and leucocytes increases; the smear is still voluminous but is moist (9). Some workers, however, have found the smears much less distinct. The changes are said not be be characteristic of the stage of the cycle in the Karakul (44) and the Tzurcana (4) breeds, though in the latter cornified cells are absent during anestrum.

The ewe does not secrete the large quantities of mucin which are characteristic of the cow, probably because of the higher level of estrogen, but some is secreted from the glands of the cervix, not from the vaginal epithelium. Mucin builds up in the cells during diestrum and is greatest in amount late in this period. It is secreted in proestrum and early heat, becoming more fluid during heat (8). The glands are less complex in anestrum than at other times (9).

UTERUS. The chief characteristic of the uterine epithelium is the presence of black pigment in many ewes. This is present at all times and is melanoblastic in origin and not derived from blood pigment. The cell changes during the cycle are slight. There is considerable edema and increased vascularity in both the cotyledonary and intercotyledonary areas during heat and metestrum; the glands become more coiled, and their cells increase in height with the growth of the corpus luteum. The epithelium is folded most at the mid-luteal stage. Leucocytes are found at all stages, but mostly when the corpus luteum begins to retrogress (8). In anestrum the glands become scanty and small, with low-columnar cells (9). Alkaline phosphatase appears in the epithelium during the lutein phase (45).

In the cotyledonary areas the stroma is more densely cellular and more vascular than it is elsewhere. The cells are low-columnar just after heat, but they rapidly grow from 11 to 17 μ to 39 to 53 μ in about 5 to 6 days. At their maximum height they tend to be pseudostratified. They retrogress rapidly after this stage, which is of short duration (43). In the lutein phase the cotyledonary area is edematous, and fluid also infiltrates the glandular regions (46). Carbonic anhydrase is present in the mucosa of the nonpregnant uterus. The quantity does not change with the cycle nor after ovariectomy (47).

OVIDUCT. Changes in the oviduct are more definite than in any other part of the tract. The columnar ciliated epithelium of the branched folds in the mid-tubal area is highest in early diestrum, and the cilia are longest, while connective tissue is relatively abundant. Cytoplasmic extrusions are most apparent toward the end of diestrum. Changes at the fimbriated end precede those at mid-tube by a short time (43). During heat the ovaries are closely invested by the infundibulum (22). According to another account both the ciliated and the secretory cells are highest in estrus and metestrum. They are lowest in diestrum and anestrum. Probably the "early diestrum" of the earlier account and the "metestrum" of the later one coincide. Secretion is abundant during estrus and metestrum; this is an acid mucopolysaccharide. In the cytoplasm of the cells an alkaline phosphatase is present during proestrum and estrus (48).

The tubo-uterine junction contains no villi or sphincter, but the muscle is thickened at this point. Fluid passes easily from the uterus into the oviduct (52), but at estrus and for about 3 days afterwards this junction exerts a valvelike action that prevents fluid in the tube from entering the uterus. This action seems to be the result of an estrogen-induced edema and flexure of the wall of the utero-tubal junction (50).

PITUITARY. Changes in the histology of the anterior pituitary in relation to the cycle have been investigated. It was found to be difficult to employ the usual acidophil-basophil classification as the differences are not clear-cut. Several cell types have been described, and they seem to vary somewhat in relative numbers during the cycle, but the clearest changes are found in the degree of granulation. This increases gradually late is diestrum, is at a maximum at the beginning of heat, and rapidly decreases during heat (51).

An increase in the number of dark-staining, usually fuchsinophil, cells during heat has been described as taking place in the adrenal cortex (52).

THE PHYSIOLOGY OF THE FEMALE TRACT

The mean vaginal pH on the first day of heat is 6.65, with a range from 5.85 to 7.40. During mid-diestrum it is 6.69, with a range from 6.00 to 7.60. This is considerably more acid than the vagina of the cow, and the shift in the acid direction at the time of heat is much smaller (53). Perhaps it is correlated with the smaller amount of mucus secreted by the ewe.

The pH of the uterine secretions in a ewe in heat averages 7.50, and in diestrum it is 7.21 (54). The shift is in the reverse direction to that in the vagina. The motility of the uterine muscle has been investigated by the introduction of a balloon through a fistula. Spontaneous contractions of longitudinal muscle begin 1 to 2 days before heat. They reach their maximum amplitude during heat and cease after ovulation, to be replaced by segmental contractions of the circular muscle. By the fourth or fifth day of diestrum the uterus is quiescent. Contractions during heat occur at about 40 to 60 second intervals, and they become more widely spaced in early diestrum. The cervix uteri is open during heat and for a short time afterwards (55).

When Karakul sheep were subjected to reduced amounts of daylight in April, May, and June they bred earlier than usual. The percentage breeding rose from the control value of 44.7 per cent to 85.7 per cent (56). Increasing the light during the breeding season caused the sexual season to cease about 15 weeks earlier than usual (57). This question of light gradients is an important one since it is the basis for most seasonal breeding. The problem as it affects reproduction in sheep has been thoroughly reviewed (58).

Hysterectomy causes the corpus luteum to persist for at least 100 days (59). Distention of the uterus in the early lutein phase causes the cycle to be shortened, and the interval between heats is reduced to 11 or 12 days. This reaction depends upon the integrity of the nerve paths from the uterus (60).

The developing graafian follicle produces estrogenic activity equivalent to an injection of 20 μg of estradiol benzoate (61). There is no estrogen in the urine during anestrum. In the breeding season there may be up to 2.4 μg estrone equivalent per 24 hours, but the level is as high at mid-cycle as it is during estrus. Estrogens may be first detected in the urine of the pregnant ewe at 15 weeks; in late pregnancy, the Romney excretes 4.5 μg of estrone, 2.4 μg of estradiol, and 2.0 μg of estriol (62). With the corpus luteum of the cycle blood from the ovarian vein contains 0.5 μg/ml. of progesterone. During pregnancy the level may rise to 2.0 μg/ml. (63). Pregnanediol may be detected in pregnancy urine but it is absent from the urine of anestrous ewes (64). The placenta contains from 4 to 9 μg of progesterone per kg. of tissue and also 9 to 12 μg of 20α-hydroxypregn-4-en-3-one per kg. The amounts are highest in the placentas of ovariectomized ewes. These substances are also present in the blood, but not to a greater amount than in the nonpregnant ewe (65).

The level of F.S.H. in the anterior pituitary is high in proestrum; it falls throughout estrus and then gradually recovers until the proestrous level is again reached at 15 days. The level of luteotrophic hormone does not vary (66). L.H. is lowest at 36 hours after estrus begins, but by 10 days it is 3 times as high. The level is high during anestrum (67).

The pituitary of the sheep is about five times as rich in F.S.H. as is that of the cow, but even so the ewe stands low in the list of animals which have been tested (68). In L.H. content the pituitary is about ten times as potent as that of the cow (69).

In the ovariectomized ewe the median effective dose of estradiol benzoate required to induce psychic and vaginal estrous symptoms is 64 μg. If the treatment is preceded by progesterone injections the median effective dose is reduced to 22 to 25 μg (70, 71). During the cycle injection of 10 to 15 mg. of progesterone daily inhibited estrus (72).

Ovariectomy of pregnant ewes after the fifty-fifth day of gestation is not necessarily followed by resorption of the embryos or by abortion. The chances that the pregnancy will go on to term appear to increase the later the operation is performed (73). Fertilized ova were transferred to ewes that were ovariectomized at the time of transfer. Pregnancy was maintained at a dose level of 10 to 20 mg. of progesterone daily. It was found to be unnecessary to continue the injections beyond the sixtieth day. The length of gestation and the course of parturition were normal (74).

As the problem of extending the breeding season of the ewe is an important one in agricultural practice, a great many attempts have been made to achieve this object with gonad-stimulating hormones, but with varying degrees of success. It is very difficult to assess the value of most of this work as different breeds have been used, some with extended breeding seasons; and it has been done at different times of the year, often without any indication of the length of the normal breeding season in that particular year. Also, a great variety of extracts has been used, containing mixtures of hormones. The impression gained from the work as a whole is that it is more difficult to cause a ewe to ovulate early in anestrum than it is later, when the ovaries are undergoing some cyclical changes, not yet sufficient to bring about ovulation. If some attention were paid to the size of the follicles present when injections are made, a clue might be obtained as to the cause of the erratic results found.

The pituitary of the ewe appears to have the same potency in terms of rat units and rabbit ovulating units at all times. It is therefore reasonable to conclude that the cyclical changes in the ovary are determined by the release of hormones from the pituitary and not by the latter's hormone content. During anestrum 50 R.U. of pregnant mares' serum is the borderline dose for the induction of heat, but 100 R.U. produce much more consistent results. Two doses of this amount at an interval of 16 days are needed, and estrus occurs after the second injection (9). Usually 250 R.U. injected twice at an interval of 16 days is recommended; it is ineffective if the interval is 14 days (75). In this work no evidence of superovulation was found when the dose given was as high as 4,500 R.U. Such evidence has, however, been obtained by the use of an unspecified amount of F.S.H. in a pituitary extract digested with trypsin to destroy the L.H. The number of eggs recovered was much less than the number of corpora lutea found, and it was believed that most of them were entrapped in the developing corpora lutea (76). A clue to the erratic nature of the responses, particularly to the difficulty in obtaining conceptions, may be found in some work which shows that horse pituitary extract (very rich in F.S.H.) and P.M.S. evoke sexual receptivity only in animals having a regressing corpus luteum. In the absence of a corpus luteum ovulation occurs without heat; the presence of an active corpus luteum usually suppresses both ovulation and heat (77). In this work it was also found that the injection of 0.5 to 1.0 mg. of estradiol benzoate or of stilbestrol dipropionate induced ovulation, probably by stimulation of the animal's own pituitary. Superovulation was obtained with horse pituitary extract when it was given in mid-diestrum. However, the dosage may be as important as the condition of the ovary since the injection of 100 R.U. of P.M.S. into anestrous ewes induced ovulation without heat, while the injection of 1,000 R.U. induced heat (78, 79).

There is considerable evidence that the use of gonad-stimulating hormones about 2 days before the onset of a normal heat increases the number of multiple births, for instance, in Karakuls (80), and it may also increase the number of conceptions toward the beginning of the season (81).

Tables showing the prenatal growth of the embryo and fetus have been published (82, 83).

THE MALE

One characteristic of the reproductive tract of the ram is the presence of the filiform appendage, a process on the left side of the tip of the penis. This is perforated, and the tube is continuous with the urethra. Formerly it was believed that the appendage was instrumental in depositing the semen within the cervix uteri, but this is probably not so since its removal has no effect upon the ram's fertility (84).

Primary spermatocytes make their first appearance in the testis at about 63 days of age; at 126 days secondary spermatocytes appear, and at 147 days spermatozoa are present (85). Another account placed the appearance of primary spermatocytes at 4 weeks, of secondary spermatocytes at 10 weeks, and of spermatids and spermatozoa at 22 weeks (86). Testicular development is more closely associated with body weight than with age. In Merinos most growth took place between 23 and 27 kg. body weight. No spermatozoa could be found in the lumen of the testes of rams lighter than 27 kg. (87). The attainment of puberty is associated with the breakdown of the adhesion between the penis and the prepuce (88). In some breeds, e.g., the Shropshires, there is little seasonal variation in mating desire, but in others there is a strong seasonal tendency. In Hampshires it is highest from October to April. In Shropshires spermatogenesis is low in summer, but it is maintained in Hampshires (89). There is evidence that in summer spermatogenesis suffers in those breeds in which the scrotum is covered with wool, thus decreasing the heat loss. Usually the scrotal temperature is 5° C. below the rectal temperature. Insulation of the scrotum for as few as 4 days increases the percentage of abnormal spermatozoa (90). In Australia it has been found that exposure of rams to 4 weeks of sustained daily maximal temperatures of 90° F., and over, causes marked abnormalities of the spermatozoa, and spermatogenesis may cease completely (91). Much of the temperature control of the testis is produced by the dartos muscle of the scrotum, which relaxes in warm weather. Isolated strips of this muscle have little or no spontaneous rhythm, but temperature variations provide the stimulus. It relaxes with an increase of the temperature and shows rhythmic activity at low temperatures. It is also responsive to sympathetic and parasympathetic drugs (90).

Spermatogenesis takes 49 days to be performed (92). The length of the seminiferous tubules in a testis has been variously reported as 4,000 meters (93) and 10,737 meters (94). The sperm content of the testis varied between 21×10^8 and 413×10^8 (95). The cauda has been reported as containing from 40 to 60×10^9 spermatozoa (96). The epididymis is about 40 to 60 meters long when its coils are unraveled (97). It contains (average of 2 rams) 162×10^9 spermatozoa, of which 76 per cent are in the cauda. It took a mean number of 30.2 ejaculations to exhaust the rams, and by that time 84 per cent of the spermatozoa had been ejaculated from the cauda (98). Normally it takes an average of 8.8 days for the spermatozoa to travel through the epididymis (90).

During coitus the semen is deposited in the cranial end of the vagina. There has been a lively controversy concerning the rate at which the spermatozoa traverse the female tract and reach the upper end of the oviduct. Reports have varied from 6 minutes (99) to 14 hours (100). One report gives 3 hours as the average time and indicates that, once they are within the oviduct, they are transported by the flow of the tubal secretions towards the fimbriated end. Almost all the spermatozoa become immotile within 8 hours of their introduction into the oviduct (101). The duration of sperm fertility in the ewe's tract has been given as 2 days (102). Another account gives 30 hours as the upper limit of their fertility but they may remain motile in the cervix for up to 80 hours, but only for 20 hours in the vagina (103). The ovum survives for 15 hours, range 12 to 23 hours (103). The speed of ascent of the spermatozoa varies little whether the ewe is in heat or not (104), but it has also been stated that the oviduct becomes pervious to spermatozoa only about 15 to 17 hours before ovulation (105). The rate of travel of spermatozoa is said to average 1.26 mm. per minute in the female tract (19) or 4.6 mm. per minute *in vitro* (106).

Semen may be collected in an artificial vagina, or the ram may be caused to ejaculate with an electric shock through the lumbar portion of the spinal cord. The average volume, etc., of the ejaculate collected by the two methods are given in Table 16 (107). The buffering capacity of the semen is fairly

TABLE 16. Rams' Semen Collected by Different Methods

	SEMEN COLLECTED BY	
	*artificial vagina**	*electrical ejaculation*
Mean volume, cc.	1.18 ± .14	1.46 ± .06
Spermatozoa, thousands per cu. mm.	1,537 ± 41	937 ± 35
Spermatozoa per ejaculate, millions	1,875 ± 110	1,354 ± 102
Mean pH	6.3	6.8

* These figures are fairly typical for rams' semen, though the sperm count is a little low.

high, and it is comparable to that of the bull (108). Limited analyses of certain secretions of the male tract have been made and are reported in Table 17

TABLE 17. Composition of Rams' Semen

	K mg. per cent	Na mg. per cent	Ca mg. per cent
Epididymal secretion	165.9	87.5	18.3
Sperm serum	87.3	143.6	18.1

(109). The figures are considerably lower than corresponding ones for the bull.

The total reducing substance in the semen is 623 mg./100 ml., and 529 mg. of this is fructose (110). Inositol is present in the semen plasma to the extent of 41 mg./100 ml., range 37 to 45 (111). Plasmalogen, the bulk of which is in the spermatozoon, is 1.2 g. per 100 g. of sperm weight (112).

The inheritance and growth of horns is complicated and is, to a certain extent, apparently under the influence of sex hormones. In some breeds, for example, the Dorset Horn, both sexes are horned, though the ewe has smaller horns than the ram. In Merinos and Rambouillets the ram is usually horned and the ewe not, though she has prominences often capped by scurs at the usual points of growth. In the Suffolks, Southdowns, and other breeds both ram and ewe are hornless, and the usual growth points are represented by depressions in the skull. Crosses between horned and hornless breeds and studies of inheritance in the Rambouillet breed have shown that horns develop in the ram if he is either homozygous or heterozygous for horns, but in the ewe only if she is homozygous, though in the latter the picture is frequently obscured by the growth of loose scurs (113). Another interpretation is that inhibiting factors with a sex-limited expression may account for the results obtained on crossing (114).

In breeds which are horned in both sexes, removal of the gonads does not interrupt horn growth, except that in the wether they resemble those of the ewe and do not become as heavy or as coiled as they do in the ram. In a breed such as the Herdwick, horned in the ram and not in the ewe, horn growth ceases when the ram lamb is castrated, but ovariectomy has no effect; the ewes continue to be hornless (115, 116). A similar result has been recorded for the Merino (117). It is evident that there is an interplay between hormones and genes the exact nature of which is still obscure. It would be worth while to observe the effects of the implantation of tablets of gonadic hormones in gonadectomized sheep of known genetic make-up.

1. Kupfer, M. Union of South Africa, Dept. of Agric., Dept. of Vet. Education and Res., 13th and 14th Rpt., 1211–1270, 1928.
2. Pomanskic, E. A., and V. I. Stojanovaskaja. Probl. Zivotn., No. 3: 77–82, 1936.
3. Roux, L. L. Onderstepoort J. Vet. Sci. and Animal Ind., 6: 465–717, 1936.
4. Mihaila, S. Ann. Inst. Nat. Zootech. Rouman., 5: 190–209, 1936.
5. Panyseva, L. V. Dokl. Akad. Seljskohoz Nauk, 1: 37–39, 1939.
6. Smith, L. W., and G. Singh. Agric. Livestock, India, 8: 683–688, 1938.
7. Phillips, R. W., R. G. Schott, and V. L. Simmons. J. Animal Sci., 6: 123–132, 1947.
8. McKenzie, F. F., and C. E. Terrill. Missouri Agric. Exp. Sta., Res. Bull. 264, 1937.
9. Cole, H. H., and R. F. Miller. AJA., 57: 39–97, 1935.
10. Mounib, M. S., I. A. Ahmed, and M. K. O. Hamada. Alexandria J. Agric. Res., 4(2): 85–108, 1956.
11. Quinlan, J., H. P. Steyn, and D. de Vos. Onderstepoort J. Vet. Sci. and Animal Ind., 16: 243–262, 1941.
12. Hafez, E. S. E. J. Agric. Sci., 42: 189–265, 1952.
13. Grant, R. Trans. Roy. Soc., Edinburgh, 58: 1–47, 1936.
14. Blum, C. T. J. Hered., 34: 141–152, 1943.
15. Briggs, H. M., A. E. Darlow, L. E. Hawkins, O. S. Wilham, and E. R. Hauser. Oklahoma Agric. Exp. Sta., Bull. 255, 1942.
16. Olbrychtowa, F. M. Zesz. Probl. Postep. Nauk, Polsk Akad. Nauk, 11: 35–41, 1958.

17. Dry, F. W. New Zealand J. Agric., 47: 386–387, 1933.
18. Bonfert, A. Bulet. Asoc. Gen. Med. Vet. Romania, 45: 215–233, 1933.
19. Kelley, R. B. CSIRO Australia, Bull. 112, 1937.
20. Gill, D. A. New Zealand J. Agric., 47: 305–307, 1933.
21. Goot, H. New Zealand J. Sci. Tech., Sect. A, 30: 330–344, 1949.
22. Quinlan, J., and G. S. Maré. Union of South Africa, Director of Vet. Services and Animal Ind., 17th Rpt., 663–703, 1931.
23. Anderson, J. J. Agric. Sci., 28: 64–72, 1938.
24. Clark, R. T. AR., 60: 125–134, 1934.
25. Henning, W. L. J. Agric. Res., 58: 565–580, 1939.
26. Eastoe, R. D., P. B. Sutton, and J. McDonald. Agr. Gaz., New South Wales, 59: 93–98, 1948.
27. Sinclair, A. N. Australian Vet., 33: 88–91, 1957.
28. Zajac, T. Ovcevodstvo, No. 8: 15–19, 1935.
29. Polovtzeva, V., and M. Fomenko. Ovcevodstvo, No. 4: 11–13, 1933.
30. Kardymovic, M., A. Marsakova, and V. Pavljucek. Probl. Zivotn., No. 5: 105–110, 1934.
31. Ozaja, M. Studia nad Wryosowka, Warsaw, 1947.
32. Spencer, D. A., R. G. Schott, R. W. Phillips, and B. Aune. J. Animal Sci., 1: 27–33, 1942.
33. Belogradskii, A. P. Sovetsk. Zooteh., No. 7: 88–90, 1940.
34. Roberts, E. J. Agric. Res., 22: 231–234, 1921.
35. Johansson, I. Proc. Am. Soc. Animal Prod., 285–291, 1932.
36. Barton, R. A. New Zealand J. Sci. Tech., 31: 24, 1950.
37. Asdell, S. A. J. Agric. Sci., 19: 382–396, 1929.
38. Terrill, C. E. J. Animal Sci., 3: 434–435, 1944.
39. Hafez, E. S. E. Bull. Fac. Agric. Cairo Univ., No. 34, 1953.
40. Warbritton, V. J. Morphol., 56: 181–202, 1934.
41. Clark, R. T. AR., 60: 135–159, 1934.
42. Hadek, R. Am. J. Vet. Res., 19: 873–881, 1958.
43. Casida, L. E., and F. F. McKenzie. Missouri Agric. Exp. Sta., Res. Bull. 170, 1932.
44. Höcker, U. Kuhn Arch., 47: 27–77, 1938.
45. Hadek, R. Am. J. Vet. Res., 19: 882–886, 1958.
46. Carles, J. Compt. Rend. Assn. Anat., 36: 91–101, 1949.
47. Lutwak-Mann, C. JE., 13: 26–38, 1955–6.
48. Hadek, R. AR., 121: 187–206, 1955.
49. Andersen, D. H. AJA., 42: 255–305, 1928.
50. Edgar, D. G., and S. A. Asdell. JE., 21: 315–320, 1960.
51. Warbritton, V., and F. F. McKenzie. Missouri Agric. Exp. Sta., Res. Bull. 257, 1937.
52. Nahm, L. J., and F. F. McKenzie. Missouri Agric. Exp. Sta., Res. Bull. 251, 1937.
53. Quinlan, J., S. J. Myburgh, and D. de Vos. Onderstepoort J. Vet. Sci. and Animal Ind., 17: 105–114, 1941.
54. Polovcova, V. V., and S. S. Judovic. Trud. Inst. Ovcevod., Kozovod, No. 10: 125–154, 1939.
55. Polovcova, V. V. Vestn. Sljskohoz. Nauki Zivotn., No. 1: 127–138, 1940.
56. Eaton, O. N., and V. L. Simmons. United States Dept. Agric., Circ. 933, 1953.

57. Hafez, E. S. E. Nature, 168: 336, 1951.
58. Yeates, N. T. M. J. Agric. Sci., 39: 1–43, 1949.
59. Wiltbank, J. N., and L. E. Casida. J. Animal Sci., 15: 134–140, 1956.
60. Moore, W. W., and A. V. Nalbandov. E., 53: 1–11, 1953.
61. Robinson, T. J. J. Agric. Sci., 46: 37–43, 1955.
62. Bassett, E. G., O. K. Sewell, and E. P. White. New Zealand J. Sci. Tech., Ser. A, 36: 437–449, 1955.
63. Edgar, D. G. Nature, 170: 543–544, 1952.
64. Robertson, H., and W. F. Coulson. Nature, 182: 1512–1513, 1958.
65. Short, R. V., and N. W. Moore. JE., 19: 288–293, 1959.
66. Santolucito, J. A., M. T. Clegg, and H. H. Cole. E., 66: 273–279, 1960.
67. Hutchinson, J. S. M., and H. Robertson. Nature, 188: 585–586, 1960.
68. Witschi, E. E., 27: 437–446, 1940.
69. West, E., and H. L. Fevold. PSEBM., 44: 446–449, 1940.
70. McDonald, M. F., and J. L. Raeside. JE., 18: 359–365, 1959.
71. Robinson, T. J. JE., 12: 163–173, 1955.
72. Dauzier, L., R. Ortavant, C. Thibault, and S. Wintenberger. Ann. d'Endocrinol., 14: 553–559, 1953.
73. Casida, L. E., and E. J. Warwick. J. Animal Sci., 4: 34–36, 1945.
74. Moore, N. W., and L. E. Rowson. Nature, 184: 1410, 1959.
75. Cole, H. H., and R. F. Miller. AJP., 104: 165–171, 1933.
76. Murphree, R. L., E. J. Warwick, L. E. Casida, and W. H. McShan. J. Animal Sci., 3: 12–21, 1944.
77. Hammond, Jr., J., J. Hammond, and A. S. Parkes. J. Agric. Sci., 32: 308–323, 1942.
78. Bell, T. D., L. E. Casida, G. Bohstedt, and A. E. Darlow. J. Agric. Res., 62: 619–625, 1941.
79. Robinson, T. J. J. Agric. Sci., 40: 275–307, 1950.
80. Lopyrin, A. I., N. V. Loginova, and I. G. Babicev. Sovetsk. Zooteh., No. 1: 82–88, 1940.
81. O'Neal, F. L. North Am. Vet., 19(10): 25–27, 1938.
82. Winters, L. M., and G. Feuffel. Minnesota Agric. Exp. Sta., Tech. Bull. 118, 1936.
83. Malan, A. P., and H. H. Curson. Ondestepoort J. Vet. Sci. and Animal Ind., 7: 239–249, 1936.
84. Iwanow, E. I. Vet. J., 85: 351–355, 1929.
85. Phillips, R. W., and F. N. Andrews. Massachusetts Agric. Exp. Sta., Bull. 331, 1936.
86. Carmon, J. L., and W. W. Green. J. Animal Sci., 11: 674–687, 1952.
87. Watson, R. H., C. S. Sapsford, and I. McCance. Australian J. Agric. Sci., 7: 574–590, 1956.
88. Dun, R. B. Australian Vet. J., 31: 104–106, 1955.
89. McKenzie, F. F., and V. Berliner. Missouri Agric. Exp. Sta., Res. Bull. 265, 1937.
90. Phillips, R. W., and F. F. McKenzie. Missouri Agric. Exp. Sta., Res. Bull. 217, 1934.
91. Gunn, R. M. C., R. N. Sanders, and W. Granger. CSIRO Australia, Bull. 148, 1942.
92. Ortavant, R. Ann. Zootech., 8: 271–321, 1959.

93. Bascom, K. F., and H. L. Osterud. AR., 31: 159–169, 1925.
94. Knepp, T. H. Proc. Pennsylvania Acad. Sci., 10: 39–42, 1936.
95. Ortavant, R. Rpt. 2nd Internat. Congr. Animal Reprod., Copenhagen, 1: 105–112, 1952.
96. Ortavant, R. Pap. 3rd Internat. Congr. Animal Reprod., Cambridge, Sect. 1: 44, 1956.
97. Buchman, E. G. Dokl. Akad. Seljskohoz. Nauk, No. 5–6: 31–36, 1931.
98. Polovceva, V. V. Dokl. Seljskohoz. Nauk, No. 15–16: 43–52, 1938.
99. Starke, N. C. Onderstepoort J. Vet. Sci., 22: 415–425, 1949.
100. Dausier, L., and S. Wintenberger. CRSB., 146: 67–70, 1952.
101. Edgar, D. G., and S. A. Asdell. JE., 21: 321–326, 1960.
102. Green, W. W. Am. J. Vet. Res., 8: 299–300, 1947.
103. Dausier, L., and S. Wintenberger. CRSB., 146: 660–663, 663–665, 1952.
104. Green, W. W., and L. M. Winters. AR., 61: 457–469, 1935.
105. Lopyrin, A. I., and N. V. Loginova. Sovetsk. Zooteh., No. 2–3: 144–149, 1939.
106. Phillips, R. W. Proc. Am. Soc. Animal Prod., 222–235, 1935.
107. Brady, D. E., and E. M. Gildow. Proc. Am. Soc. Animal Prod., 250–254, 1939.
108. Sergin, N. P. Probl. Zivotn., No. 12: 100–122, 1935.
109. Nesmejanova, T. N. Usp. Zooteh. Nauk, 5: 65–72, 1938.
110. Roy, A., Y. R. Karnik, S. N. Lukyuke, S. Bhattacharya, and P. Bhattacharya. Indian J. Dairy Sci., 3: 42–45, 1950.
111. Hartree, E. F. Bioch. J., 66: 131–137, 1957.
112. Hartree, E. F., and T. Mann. Bioch. J., 71: 423–433, 1959.
113. Warwick, B. L., and P. B. Dunkle. J. Hered., 30: 325–329, 1939.
114. Ibsen, H. L., and R. F. Cox. J. Hered., 31: 327–336, 1940.
115. Marshall, F. H. A. PRS., 85B: 27–32, 1912.
116. Marshall, F. H. A., and J. Hammond. J. Physiol., 48: 171–176, 1914.
117. Zavadovskii, M. M. Trudy Lab. Eksp. Biol. Moskov. Zooparka, 1: 49–66, 1926.

Ovis ammon L.

(= *O. poli* Blyth)

ARGALI, ARKAR

The arkar has one breeding season a year, from the end of October to December in Siberia, and the period of gestation lasts 5 months (1). Lambing is usually in early April, and the young ewes have 1 lamb while the older ones usually bear 2 (2). Older reports suggest that the arkar is monestrous and that the time of lambing varies considerably with the climate of the habitat, being later in the severer parts of the Himalayas (3).

1. Rumjancev, B. F., N. S. Butarin, and V. F. Denisov. Trud. Kirgiz Kompl. Exsp., 4: 15–58, 1933–34.
2. Sokolov, S. S. ZZ., 18: 444–450, 1939.
3. Blanford, W. T. Fauna of British India. Mammals. London, 1888–91.

Ovis canadensis Shaw

BIGHORN

Puberty is usually reached at 2½ to 3 years of age. The breeding season is from mid-November to mid-December. One lamb is usually born, and the period of gestation is about 180 days (1). In the southern part of its range, e.g., in Texas, rut appears to be a little earlier as the lambs are dropped from March to April (2). Most does experience their first estrus at 2½ years old (3). In Siberia puberty in both sexes is at 2 years old. They mate in December and the kids are born in June (4).

1. Spencer, C. C. JM., 24: 1–11, 1943.
2. Davis, W. B., and W. P. Taylor. JM., 20: 440–455, 1939.
3. Smith, D. R. Idaho Dept. Fish and Game Wildlife Bull. 1: 1–154, 1944.
4. Tchernyavsky, F. B. ZZ., 41: 1556–1566, 1962.

Ovis musimon Schreber

MOUFLON

The mouflon is seasonally polyestrous (1), but in the London zoo births have been spread throughout the year, though with a very marked peak in April. Twins are produced once in about 10 births (2). The males are fertile all the year (3). The gestation period is about 150 days. The testis weight has been recorded as 41 and 65 g. In the first of these the seminiferous tubules measured 2,040 meters, and in the second 6,816 meters. The spermatozoon length is given as 31.1 μ (4, 5).

1. Heape, W. QJMS., 44: 1–70, 1901.
2. Zuckerman, S. PZS., 122: 827–950, 1952–3.
3. Mottl, S. Zool. List., 21: 343–352, 1958.
4. Knepp, T. H. Zoologica, 24: 329–332, 1939.
5. Knepp, T. H. Proc. Pennsylvania Acad. Sci., 13: 58–62, 1939.

Ovis vignei Blyth

URIAL

This species comes in heat and breeds from September to December, but mating is earlier in the Punjab than in Astor (1). In the London zoo births have been from March to July, with most of them at the end of May and in June (2). The period of gestation is about 6 months and the number of kids is usually 1 or 2 (3).

One testis weighed 70.2 g., and the length of seminiferous tubules was 2,461 meters (4).

1. Blanford, W. T. Fauna of British India. Mammals. London, 1888–91.
2. Zuckerman, S. PZS., 122: 827–950, 1952–3.
3. Lydekker, R. Game Animals of India, Burma, Malaya and Tibet. London, 1924.
4. Knepp, T. H. Zoologica, 24: 329–332, 1939.

INDEX OF LATIN NAMES

Abrocoma cinerea, 407
Abrocomidae, 407
Acinonyx jubatus, 493
Acomys caharinus, 372; *dimidiatus*, 372; *hunteri*, 372; *ignitis*, 372; *percivali*, 372; *wilsoni*, 372
Aconaemus fuscus, 406
Acrobates pygmaeus, 21
Addax nasomaculatus, 612
Adenota kob, 609; *vardoni*, 609
Aepyceros melampus, 617
Aepyprymnus rufescens, 31
Aethechinus frontalis, 36
Aethomys chrysophilus, 322; *kaiseri*, 322; *thomasi*, 322
Ailurinae, 449
Ailuropoda melanoleuca, 449
Ailurus fulgens, 449
Akodon amoenus, 271; *andinus*, 271; *berlepschii*, 271; *bolivensis*, 271; *jelskii*, 271
Alcelaphini, 536
Alcelaphus caama, 613; *cokei*, 613; *lelwel*, 613; *lichtensteinii*, 612; *major*, 613
Alces alces, 573
Allactaga elater, 383; *major* 383; *sibirica*, 383
Allactagalus pumilio, 384
Allenopithecus nigroviridis, 162
Alopex lagopus, 435
Alouatta caraya, 136; *palliata*, 135, 138; *seniculus*, 136
Alouattinae, 135
Amblonyx cinerea, 470
Ammotragus lervia, 632
Anoa depressicornis, 584
Anomaluridae, 257
Anomalurinae, 257
Anomalurus beecrofti, 257; *fraseri*, 257; *pelii*, 257
Antechinomys laniger, 13
Antechinus, 11
Anthropoidea, 134
Antidorcas marsupialis, 619
Antilocapra americana, 578
Antilocapridae, 614
Antilope cervicapra, 617
Antilopinae, 614
Antrozous pallidus, 118
Aonyx capensis, 471
Aotes trivirgatus, 134
Aotinae, 134
Aplodontia rufa, 214
Aplodontidae, 214
Apodemus agrarius, 318; *flavicollis*, 318; *mystacinus*, 318; *speciosus*, 318; *sylvaticus*, 317
Arctictis binturong, 475
Arctocephalus australis, 494; *forsteri*, 494; *pusillus*, 493
Arctogalidia trivirgata, 474
Artibeus cinereus, 91; *hirsutus*, 91; *jamaicensis*, 90; *lituratus*, 90; *planirostris*, 91
Artiodactyla, 536
Arvicanthus abyssinicus, 320; *niloticus*, 320
Arvicola amphibius, 293; *terrestris*, 293
Atelerix albiventris, 39
Ateles geoffroyi, 137; *paniscus*, 138
Atelinae, 137
Atherurinae, 385
Atherurus africanus, 385; *macrourus*, 385
Atilax paludinosus, 479
Avahi, 128
Axis axis, 460; *porcinus*, 561

Babirussa babyrussa, 553
Baiomys musculus, 278; *taylori*, 270

Balaena mysticetus, 424
Balaenidae, 423
Balaenoptera acutirostrata, 418; *borealis*, 419; *physalus*, 420
Balaenopteridae, 418
Balaenopteryx, 73
Bandicota bengalensis, 372; *gracilis*, 373; *kok*, 372; *malabarica*, 373
Barbastella barbastellus, 113
Bassariscus astutus, 446; *sumichraasti*, 446
Bathyergidae, 408
Beamys major, 373
Berardius bairdi, 411
Bettongia cuniculus, 30; *lesueuri*, 31
Bibos banteng, 605; *frontalis*, 603; *gaurus*, 603; *indicus*, 603
Bison bison, 605; *bonasus*, 606
Blarina brevicauda, 51; *telmalestes*, 52
Blastocerus besoarticus, 572; *dichotoma*, 572
Bos grunniens, 602; *taurus*, 584
Boselaphus tragocamelus, 580
Bovidae, 536, 579
Bovinae, 579
Brachylagus idahoensis, 212
Brachyphylla cavernarum, 89
Bradypodidae, 178
Bradypus cuculliger, 178; *griseus*, 178
Bubalus bubalis, 581
Budorcas bedfordi, 622

Callithricidae, 139
Callithrix argentata, 139; *jacchus*, 139
Callorhinus ursinus, 494
Callosciurus caniceps, 224; *erythraeus*, 224; *hippurus*, 224; *lokroides*, 224; *lowi*, 224; *nigrovattus*, 224; *notatus*, 224; *prevosti*, 225; *tenuis*, 225; *vittatus*, 225
Calomyscus bailwardi, 269
Caloprymnus campestris, 31
Camelidae, 555
Camelus bactrianus, 556; *dromedarius*, 557
Canidae, 425
Caninae, 426
Canis adustus, 435; *aureus*, 435; *azarae*, 435; *familiaris*, 426; *latrans*, 434; *lupus*, 434; *mesomelas*, 435; *mexicanus*, 435
Capra caucasica, 631; *falconeri*, 630; *hircus*, 623; *ibex*, 631; *pyrenaica*, 631; *sibirica*, 631
Capreolus capreolus, 575
Capricornis sumatrensis, 621
Caprinae, 620
Caprini, 536
Capromyidae, 404
Capromys pilorides, 404
Cardioderma cor, 78
Carnivora, 425
Carollia castanea, 88; *perspicillata*, 88
Carolliinae, 88
Casinycteris orgynnis, 71
Castor canadensis, 255; *fiber*, 256
Castoridae, 255
Cavia porcellus, 386
Caviidae, 214, 386
Cebidae, 134
Cebinae, 136
Cebus albifrons, 136; *apella*, 136; *capucinus*, 137
Centurio senex, 91
Cephalophinae, 607
Cephalophus dorsalis, 607; *harveyi*, 607; *maxwelli*, 607; *monticola*, 607; *natalensis*, 607; *sylvicultrix*, 607
Ceratotherium simum, 534; *unicornis*, 534
Cercaertus concinnus, 21
Cercocebus albigena, 155; *torquatus*, 155
Corcopithecidae, 140
Cercopithecinae, 140
Cercopithecus aethiops, 161; *diana*, 161; *l'hoesti*, 161; *mitis*, 161; *mona*, 161; *nictitans*, 161
Cervidae, 536, 558
Cervinae, 559
Cervus alfredi, 566; *canadensis*, 561; *cashmiriensis*, 561; *duvauceli*, 564; *elaphus*, 563; *eldi*, 564; *nippon*, 566; *timoriensis*, 565; *unicolor*, 565
Cetacea, 410
Chaerophon, 120
Chaeropus castanotis, 17
Chalinolobus argentatus, 112; *gouldi*, 112; *humeralis*, 112
Cheirogaleinae, 127
Cheirogaleus major, 127; *medius*, 127
Chilonycterinae, 82
Chilonycteris, personata, 82; *rubiginosa*, 82
Chinchilla laniger, 403
Chinchillidae, 402
Chiroderma trinitatum, 89; *villosum*, 89
Chironectes minimus, 11; *panamensis*, 11

Chiropodomys gliroides, 377
Chiroptera, 64
Choeronycteris mexicana, 87
Choeronyscus inca, 88; *intermedius*, 88
Choeropsis liberiensis, 555
Choloepus didactylus, 178; *hoffmanni*, 178
Chrysochloridae, 35
Chrysochloris damarensis, 35
Chrysocyon brachyurus, 442
Chrysospalax villosus, 36
Cingulata, 179
Citellus armatus, 237; *beecheyi*, 230; *beldingi*, 237; *citellus*, 237; *columbianus*, 231; *eversmanni*, 237; *franklini*, 237; *fulvus*, 237; *harrisii*, 238; *interpres*, 238; *lateralis*, 238; *leucurus*, 231; *major*, 238; *mexicanus*, 238; *mohavensis*, 238; *nelsoni*, 238; *oregonus*, 237; *pygmaeus*, 238; *relictus*, 238; *richardsoni*, 231; *spilosoma*, 238; *suslicus*, 232; *tereticaudus*, 238; *townsendii*, 232; *tridecimlineatus*, 233; *undulatus*, 237; *varigatus*, 238; *washingtoni*, 238
Civettictis civetta, 473
Clethrionomys gapperi, 290; *glareolus*, 290; *occidentalis*, 292; *rufocanus*, 291; *rutilis*, 292
Coelomys bicolor, 356
Coendu mexicanum, 386; *villosus*, 386
Coleura afra, 72
Colobinae, 162
Colobus badius, 164; *polykomus*, 164
Colomys goslingi, 371
Comopithecus hamadryas, 156, 158
Condylura cristata, 62
Condylurinae, 62
Conepatus chinga, 469; *mapurito*, 469; *mesoleucas*, 469
Connochoetes gnu, 613; *taurinus*, 614
Corynorhinus, 114
Cratogeomys castanops, 248; *fulvescens*, 248; *merriami*, 248; *perotensis*, 248
Cricetidae, 213, 214, 258
Cricetinae, 258
Cricetomys gambianus, 374
Cricetulus barabensis, 280; *eversmanni*, 281; *migratorius*, 282; *triton*, 281
Cricetus cricetus, 280
Crocidura attila, 54; *deserti*, 53; *dsi-nezumi*, 54; *flavescens*, 54; *hildegardeae*, 54; *hirta*, 54; *jacksoni*, 54; *leucodon*, 54; *luna*, 54; *occidentalis*, 54; *poensis*, 54; *russula*, 54; *turba*, 54
Crocidurinae, 53
Crocuta crocuta, 481
Cryptomys damarensis, 408; *hottentotus*, 409; *mechowii*, 409; *mellandi*, 409
Cryptoprocta ferox, 480
Cryptoproctinae, 480
Cryptotis parva, 53
Ctenodactylidae, 409
Ctenodactylus gundi, 409
Ctenomyidae, 407
Ctenomysopimus, 407; *peruanus*, 407
Cuniculinae, 401
Cuniculus paca, 401
Cuon alpinus, 442
Cyclopes didactylis, 177
Cynictis penicillata, 480; *selousi*, 480
Cynocephalidae, 63
Cynocephalus variegatus, 63
Cynogale bennetti, 476
Cynomys gunnisoni, 230; *leucurus*, 228; *ludovicianus*, 229; *mexicanus*, 230
Cynopterus brachyotis, 66; *sphinx*, 65
Cystophora cristata, 505
Cystophorinae, 505

Dama dama, 559; *mesopotamica*, 560
Damaliscus albifrons, 612; *korrigum*, 612; *lunatus*, 612; *pygargus*, 612
Dasycercus cristicauda, 12
Dasymys incomtus, 319
Dasypodidae, 179
Dasypodinae, 179
Dasyprocta agouti, 402; *azarae*, 402; *cristata*, 402; *prymnolopha*, 402; *punctata*, 402; *variegata*, 402
Dasyproctidae, 401
Dasypterus, 113
Dasypus hydridus, 181; *novemcinctus*, 180, 181
Dasyuridae, 11
Dasyurinae, 13
Dasyurinus, 15
Dasyurops, 15
Dasyurus geoffroyi, 15; *hallucatus*, 15; *maculatus*, 15; *quoll*, 15; *viverrinus*, 13
Daubentonia madagascariensis, 129
Daubentoniidae, 129
Delphinapterus leucas, 414

Delphinidae, 415
Delphinus delphis, 415
Dendrohyrax arboreus, 513; *dorsalis*, 513
Dendrolagus matschiei, 30; *ursinus*, 30
Dendromus acraeus, 374; *arenarius*, 374; *insignis*, 374; *melanotis*, 374; *messorius*, 374; *ochropus*, 374; *whytei*, 375
Deomys ferrugineus, 376
Dermoptera, 63
Desmana moschata, 56
Desmaninae, 56
Desmodillus auricularis, 313
Desmodontidae, 64, 92
Desmodus rotundus, 92; *youngi*, 92
Dicerorhinus sumatrensis, 534
Diceros bicornis, 535
Dicrostonyx groenlandicus, 287; *rubricatus*, 288; *torquatus*, 288
Didelphidae, 4
Didelphis marsupialis, 6; *paraguayensis*, 11; *virginianus*, 6
Diphylla ecaudata, 92
Diplomesodon pulchellum, 56
Dipodidae, 382
Dipodinae, 382
Dipodomyinae, 251
Dipodomys deserti, 253; *heermannii*, 251; *ingens*, 254; *merriami*, 252; *microps*, 254; *nitratoides*, 254; *ordii*, 252; *spectabilis*, 253
Dipus sagitta, 382
Dolichotinae, 400
Dolichotis patagona, 400; *salincola*, 400
Dromiciops australis, 5
Dryomys nitedula, 379
Dugong dugon, 515
Dugongidae, 515

Echimyidae, 407
Echinosorex gymnurus, 36
Echinosoricinae, 36
Edentata, 177
Eidolon helvum, 71
Elaphurus davidianus, 567
Elephantidae, 510
Elephantulus, 33, 41; *capensis*, 44; *intufi*, 44; *jamesoni*, 41; *myurus*, 41; *renatus*, 44; *rozeti*, 44; *rupestris*, 41, 44
Elephas maximus, 511
Eliomys quercinus, 379
Ellobius talpinus, 308
Emballuronidae, 64, 72
Emballuroninae, 72
Enhydris lutris, 471
Eothenomys fidelis, 292; *kageus*, 293; *melanogaster*, 293; *simithii*, 293
Epomophorus angolensis, 70; *anurus*, 69; *crypturus*, 70; *dobsoni*, 70; *franqueti*, 70; *gambianus*, 70; *labiatus*, 70; *monstrosus*, 70; *pusillus*, 70; *wahlbergi*, 70
Epomops, 70
Eptesicus capensis, 108; *fuscus*, 106; *garambae*, 108; *rendelli*, 109; *serotinus*, 108; *tenuipinnis*, 109; *tickelli*, 109
Equidae, 516
Equus asinus, 530; *burchellii*, 532; *caballus*, 516; *grevyi*, 533; *hemionus*, 532; *quagga*, 533; *zebra*, 533
Erethizon dorsatum, 385; *epixanthum*, 386
Erethizontidae, 385
Erethizontinae, 385
Ericulus, 34
Erignathus barbatus, 502
Erinaceidae, 36
Erinaceinae, 36
Erinaceus europaeus, 37
Erophylla sezekorni, 91
Erythrocebus patas, 162
Eubalaena australis, 423; *glacialis*, 423
Euderma maculatum, 116
Eumetopias jubata, 496
Eumops, 121, 122
Euoticus elegantulus, 133
Euphractus sexcinctus, 179; *villosus*, 179
Eupleres goudotii, 476
Eutamias amoenus, 240; *bulleri*, 241; *cinereicollis*, 241; *dorsalis*, 241; *merriami*, 241; *minimus*, 240; *quadrimaculatus*, 241; *quadrivittatus*, 241; *sibiricus*, 241; *speciosus*, 241; *townsendii*, 242; *umbrinus*, 242

Felidae, 425, 482
Felinae, 482
Felis bengalensis, 490; *canadensis*, 490; *caracal*, 490; *catus*, 482; *chaus*, 490; *concolor*, 489; *geoffroyi*, 490; *libyca*, 490; *lynx*, 490; *margarita*, 490; *nigripes*, 490; *pardalis*, 490; *planiceps*, 490; *rubiginosa*, 490; *serval*, 490; *silvestris*, 488; *tigrina*, 490; *tufus*, 488; *yaguarondi*, 490
Fennecus zerda, 440

Funambulus palmarum, 221; *pennanti*, 222
Funisciurus auriculatus, 223; *congicus*, 223; *layardi*, 222; *pyrrhopus*, 223

Galaginae, 131
Galago alleni, 132; *crassicaudatus*, 131; *demidovii*, 132; *senegalensis*, 131
Galea musteloides, 400
Galemys pyrenaicus, 57
Galera barbara, 463
Galidia elegans, 476
Galidiinae, 476
Gazella arabica, 618; *bennettii*, 618; *cuvieri*, 618; *dama*, 618; *dorcas*, 618; *granti*, 619; *marica*, 619; *muscatensis*, 619; *subgutturosa*, 619; *thomsoni*, 618
Genetta genetta, 472; *tigrina*, 472
Geomyidae, 213, 244
Geomyinae, 244
Geomys arenarius, 245; *bursarius*, 244; *pinetis*, 245; *tuza*, 245
Gerbillinae, 214, 309
Gerbillus campestris, 309; *cheesmani*, 309; *dasyurus*, 309; *gerbillus*, 309; *pyramidum*, 309; *quadrimaculatus*, 309; *simoni*, 309; *swalius*, 309
Giraffa camelopardalis, 577
Giraffidae, 576
Giraffinae, 577
Glaucomys sabrinus, 242; *volans*, 243
Gliridae, 378
Glirinae, 378
Glis glis, 378
Globicephala macrorhyncha, 417; *melaena*, 416; *scammoni*, 417
Glossophaga longirostris, 86; *soricina*, 86
Glossophaginae, 86
Golunda ellioti, 320
Gorilla gorilla, 168
Grampus rectipinna, 415
Graphiurinae, 379
Graphiurus christyi, 379; *kelleni*, 380; *lorraneus*, 380; *murinus*, 380; *nanus*, 380; *vulcanicus*, 380
Grison canaster, 464
Gulo gulo, 465

Hadromys humei, 320
Halichoerus grypus, 501
Hapale, 139
Hapalemur griseus, 125
Helarctos malayana, 446
Helictis personata, 468
Heliophobus argentocinereus, 409
Heliosciurus gambianus, 224; *rufobrachium*, 224
Helogale dybowskii, 478; *ivori*, 479; *parvula*, 479; *vetula*, 479
Hemicentetes semispinosus, 35
Hemiechinus auritus, 40
Hemigalinae, 476
Hemigalus derbyianus, 476
Hemitragus hylocrius, 623; *jemlahicus*, 623
Herpailurus, 490
Herpestes auropunctatus, 477; *cauui*, 478; *edwardsi*, 478; *fuscus*, 478; *ichneumon*, 478; *javanicus*, 478; *sanguineus*, 478; *smithi*, 478; *urva*, 478; *viticollis*, 478
Herpestinae, 477
Hesperomys lauchia, 272; *sorella*, 272
Hesperotinus, 109
Heterogeomys hispidis, 249
Heterohyrax chapini, 513; *syriacus*, 513
Heteromyidae, 213, 249
Heteromyinae, 254
Heteromys desmarestianus, 254; *gaumeri*, 255; *oresterus*, 255
Hippocamelus antesensis, 572
Hippopotamidae, 554
Hippopotamus amphibius, 554
Hipposideridae, 64, 65, 80
Hipposideros abae, 80; *armiger*, 81; *beatus*, 608; *bicolor*, 81; *caffer*, 81; *cyclops*, 81; *galeritus*, 81; *lankadiva*, 81; *nanus*, 81; *speoris*, 81
Hippotraginae, 608
Hippotragus equinus, 611; *niger*, 611
Holochilus sciureus, 273
Hominidae, 168
Homo sapiens, 168
Hyaena brunnea, 482; *hyaena*, 482
Hyaenidae, 481
Hyaeninae, 481
Hybomys univattatus, 322
Hydrochoeridae, 401
Hydrochoerus hydrochaeris, 401; *isthmus*, 401
Hydromyinae, 377
Hydromys chrysogaster, 377
Hydropotes inermis, 575

Hydrurga leptonyx, 503
Hyelaphus, 561
Hyemoschus aquaticus, 557
Hylobates concolor, 165; *hoolock*, 164; *lar*, 164
Hylobatinae, 164
Hylochoerus meinertzhageni, 552
Hylomyscus, 355
Hyperoodon ampullatus, 412
Hypsignathus, 70
Hypsiprymnodon moschatus, 32
Hyracoidea, 513
Hystricidae, 384
Hystricinae, 384
Hystrix africae-australis, 384; *brachyurus*, 384; *cristata*, 384; *hodgsoni*, 384; *leucura*, 384; *subcristatus*, 384

Ichneumia albicauda, 480
Ictailurus, 490
Idionycteris phyllotis, 115
Indri indri, 129
Indriidae, 128
Inia geoffroyensis, 411
Iniinae, 411
Insectivora, 33
Iomys horsfieldi, 243
Isoodon, 18

Jaculus jaculus, 383; *orientalis*, 383
Jaguarius, 492

Kerivoula cuprosa, 118; *lanosa*, 118; *picta*, 118
Kerivoulinae, 118
Kobus defassa, 608; *ellipsiprymnus*, 608
Kogia breviceps, 414
Kogiinae, 414

Lagenorhynchus acutus, 416; *obliquidens*, 416
Lagidium peruanum, 403
Lagomorpha, 183
Lagorchestes fasciatus, 25
Lagostomus maximus, 402
Lagothrix cana, 138; *lagotricha*, 138
Lagurus curtatus, 308; *lagurus*, 308
Lama glama, 555; *vicugna*, 555
Lariscus insignis, 225
Lascionycteris noctivagans, 102
Lasiorhinus latifrons, 25
Lasiurus borealis, 112; *cinereus*, 112; *ega*, 113; *floridanus*, 113; *seminolus*, 113; *semotus*, 113
Lavia frons, 78
Lemmus lemmus, 289; *trimucronatus*, 289
Lemniscomys barbara, 321; *griselda*, 321; *striatus*, 321
Lemur catta, 125; *fulvus*, 126; *macaco*, 125; *mongoz*, 126; *rubriventer*, 126
Lemuridae, 125
Lemurinae, 125
Lemuroidea, 123
Leo, 492
Leontocebus midas, 140; *nigricollis*, 140; *rosalia*, 139; *sphinx*, 140
Leopardus, 490
Lepilemus mustilenus, 126
Leporidae, 184.
Leporinae, 185
Leptailurus, 490
Leptonychotes weddellii, 503
Lepus allenie, 190; *americanus*, 186; *arcticus*, 185; *californicus*, 187; *capensis*, 190; *europaeus*, 188; *marjorita*, 190; *mexicanus*, 190; *nigricollis*, 190; *saxatilis*, 189; *timidus*, 189; *townsendii*, 190
Lichanotus laniger, 128
Limnotragus, 579
Liomys pictus, 254; *salvini*, 254
Lobodon carcinophagus, 502
Lobodontinae, 502
Lonchoglossa ecaudata, 87; *geoffroyi*, 87
Lonchorhino aurita, 84
Lophuromys aquilus, 371; *sikapusi*, 371
Loris tardigradus, 130
Lorisidae, 130
Lorisinae, 130
Loxondonta africanus, 510
Lutra annectens, 470; *canadensis*, 469; *lutra*, 470; *maculicollis*, 470
Lutrinae, 469
Lycaon pictus, 442
Lynx, 488, 490
Lyroderma, 77

Macaca cyclopis, 140; *cynomolgus*, 141; *fascicularis*, 141; *fuscata*, 140; *irus*, 141, 156; *maurus*, 154; *mordax*, 141; *mulatta*, 143; *nemestrina*, 152; *pileatas*, 153; *radiata*,

152; *rhesus*, 143; *silenus*, 153; *sinica*, 153; *sylvanus*, 154
Macrochiroptera, 65
Macroglossinae, 71
Macroglossus lagochilus, 71
Macrophyllum macrophyllum, 84
Macropodidae, 4, 25
Macropodinae, 25
Macropus fuliginosus, 29; *major*, 27; *robustus*, 28; *rufus*, 28
Macroscelides proboscideus, 40
Macroscelididae, 40
Macrotarsomys bastardi, 287
Macrotus californicus, 84; *mexicanus*, 84
Malacomys longipes, 357
Malacothrix typicus, 375
Mandrillus leucophaeus, 160; *sphinx*, 160
Manidae, 182
Manis crassicaudata, 182; *gigantea*, 182; *javanica*, 182; *pentadactyla*, 182; *temmincki*, 182; *tricuspis*, 182
Marmosa cinerea, 5; *elegans*, 5; *mexicana*, 5; *murina*, 5
Marmota bobak, 227; *caligata*, 227; *camschatica*, 227; *flaviventris*, 227; *marmota*, 226, 227; *monax*, 226; *sibirica*, 228
Marsupialia, 4
Martes americana, 461; *martes*, 462; *pennanti*, 462; *zibellina*, 460
Mastomys, 352
Mazama americana, 571; *simplicicornis*, 571; *tema*, 571
Megaderma cor, 78; *lyra*, 77; *spasma*, 78
Megadermatidae, 64, 77
Megaptera novae-angliae, 421
Meles meles, 466
Melinae, 466
Mellivora capensis, 465
Mellivorinae, 465
Melomys cervinipes, 356
Melursus ursinus, 446
Mephitinae, 468
Mephitis macroura, 468; *mephitis*, 468
Meriones hurrianae, 314; *libycus*, 314; *longifrons*, 313; *meridianus*, 314; *persicus* 314; *tamaricinus*, 314; *tristrami*, 314; *vinogradovi*, 314
Mesembriomys gouldi, 319
Mesocricetus auratus, 282; *raddei*, 286
Mesoplodon gervaisi, 411; *mirus*, 411
Metachirus nudicaudatus, 6
Mico, 139
Microcavia australis, 400
Microcebus coquereli, 128; *murinus*, 127
Microchiroptera, 72
Microdipodops megacephalus, 251; *pallidus*, 251
Microfelis, 490
Micromys minutus, 317
Micronycteris brachyotis, 83; *hirsuta*, 83; *megalotis*, 83; *minuta*, 83; *nicefori*, 83
Micropteropus, 70
Microsciurus alfori, 220
Microsorex hoyi, 50
Microtinae, 287
Microtus agrestis, 298; *arvalis*, 299; *blythi*, 307; *californicus*, 300; *chrotorrhinus*, 306; *gregalis*, 306; *guentheri*, 301; *incertus*, 306; *longicaudus*, 306; *ludovicianus*, 307; *mexicanus*, 306; *miurus*, 306; *montanus*, 302; *montebelloi*, 306; *mordax*, 306; *nanus*, 306; *nivalis*, 306; *ochrogaster*, 302; *oeconomus*, 303; *operarius*, 306; *orcadensis*, 306; *oregoni*, 304; *pennsylvanicus*, 304; *richardsoni*, 307; *sikkimensis*, 307; *socialis*, 307; *townsendii*, 307; *xanthognathus*, 307
Millardia meltada, 322
Mimon crenulatum, 85
Miniopterinae, 116
Miniopterus, 64; *australis*, 116; *macrocneme*, 117; *minor*, 116; *natalensis*, 116; *rufus*, 117; *schreibersii*, 117
Mirounga angustirostris, 508; *leonina*, 506
Mogera latouchei, 59
Molossidae, 65, 118
Molossops greenhalli, 118
Molossus ater, 121; *centralis*, 122; *major*, 121; *nigricans*, 121; *perotis*, 121; *underwoodi*, 122
Monachinae, 505
Monachus monachus, 505; *schauinslandi*, 505; *tropicalis*, 505
Monodon monoceros, 414
Monodontidae, 414
Monotremata, 1
Mops, 120
Mormoops megalophylla, 83
Moschinae, 558
Moschus moschiferus, 558

Mungos mungo, 479
Muntiacinae, 559
Muntiacus muntjak, 559; *reevesi*, 559
Muridae, 213, 214, 316
Murinae, 316
Mus bactrianus, 357; *bellus*, 370; *booduga*, 370; *calliwaerti*, 370; *gratus*, 370; *minutoides*, 370; *musculoides*, 370; *musculus*, 357; *triton*, 370
Muscardinus avallanarius, 378
Mustela altaica, 460; *arctica*, 450; *cicognanii*, 450; *erminea*, 449; *eversmanni*, 452; *flavigula*, 460; *foina*, 461; *frenata*, 451; *furo*, 452; *kathiac*, 460; *lutreola*, 460; *nivalis*, 451; *putorius*, 452; *rixosa*, 460; *sibirica*, 460; *vison*, 457
Mustelidae, 425, 449
Mustelinae, 449
Myocaster coypus, 405
Myonax, 478
Myonycteris, 66
Myosorex varius, 56
Myotis adversus, 100; *austroriparius*, 93; *bechsteini*, 100; *bocagii*, 100; *californicus*, 100; *capaccinii*, 94; *chiloensis*, 100; *daubentoni*, 100; *emarginatus*, 94; *evotis*, 100; *fortidens*, 100; *grisescens*, 94, 97; *keeni*, 95; *longicaudatus*, 100; *lucifugus*, 96; *myotis*, 97, 109; *mystacinus*, 99; *nattereri*, 100; *nigricans*, 100; *occulutus*, 100; *peytoni*, 100; *sodalis*, 99; *thysanodes*, 101; *tricolor*, 101; *velifer*, 99; *vivesi*, 101; *volans*, 101; *yumaensis*, 101
Myrmecobiinae, 16
Myrmecobius fasciatus, 16
Myrmecophaga tridactyla, 177
Myrmecophagidae, 177
Mystomys albicaudatus, 286

Nandinia binotata, 474
Napaeozapus insignis, 382
Nasalis larvatus, 163
Nasilio brachyrhynchus, 40
Nasua nasua, 448
Natalidae, 93
Natalus mexicanus, 93; *tumidirostris*, 93
Nemorhaedus goral, 621
Neofiber alleni, 296
Neomys fodiens, 50
Neophoca hookeri, 497
Neophocaena phocaenoides, 418
Neotoma albigula, 275; *cinerea*, 276; *floridana*, 276; *fuscipes*, 278; *goldmani*, 279; *lepida*, 278; *mexicana*, 279; *micropus*, 279; *phenax*, 279
Neotomodon alstoni, 275
Neotomys ebriosus, 273
Neotragini, 536
Nesokia indica, 373
Nesomyinae, 287
Nesotragus livingstonianus, 615; *moschatus*, 616
Neurotrichus gibbsii, 60
Noctifelis, 490
Noctilio labialis, 75; *leporinus*, 75
Noctilionidae, 75
Notiosorex crawfordi, 53
Notomys gouldii, 371; *mitchelli*, 372
Notopteris macdonaldi, 71
Nyctalus, 105, 106
Nyctereutes procyonides, 441
Nycteridae, 64, 75
Nycteris aethiopica, 76; *arge*, 76; *capensis*, 76; *damarensis*, 76; *grandis*, 76; *hispida*, 75; *luteola*, 75, 76; *nana*, 76; *pallida*, 77; *thebaica*, 77
Nycterophilinae, 118
Nycticebus coucang, 130; *intermedius*, 130
Nycticeius humeralis, 109
Nyctinomis, 120
Nyctomys sumichrasti, 261

Ochotona alpina, 183; *collaris*, 184; *daurica*, 184; *hyperborea*, 184; *princeps*, 183; *pusilla*, 184; *roylei*, 184
Ochotonidae, 183
Ochromys woosnami, 355
Octodon degus, 406
Octodontidae, 406
Odobenidae, 497
Odobenus rosmarus, 497
Odocoileinae, 567
Odocoileus hemionus, 567; *virginianus*, 569; *v. acapulcensis*, 571
Oecomys, 259
Oedipomidas, 140
Oenomys hypoxanthus, 319
Okapia johnstoni, 576
Ondatra zibethica, 294
Onotragus leche, 609

Onychogalea frenata, 26
Onychomys leucogaster, 270; *torridus*, 271
Oreamnos americanus, 621
Oreotragus oreotragus, 614
Ornithorhynchidae, 2
Ornithorhynchus anatinus, 2
Orycteropodidae, 509
Orycteropus afer, 509
Oryctolagus cuniculus, 195
Oryx algazel, 611; *beisa*, 611; *gazella*, 611
Oryzomys alfaroi, 258; *bicolor*, 259; *caliginosus*, 259; *couesi*, 259; *devius*, 269; *endersi*. 259; *fulvascens*, 259; *palustris*, 258; *pyrrorhinus*, 259; *talamancae*, 259
Osphranter robustus, 28
Otaria byroni, 496
Otariidae, 493
Otocyon megalotis, 443
Otocyoninae, 443
Otomops, 120
Otomyinae, 376
Otomys anchieta, 376; *denti*, 376; *irroratus*, 376; *tropicalis*, 376
Ototylomys phyllotis, 261
Ourebia ourebi, 615
Ovibos moschatus, 622
Ovis ammon, 650; *aries*, 632; *canadensis*, 651; *musimon*, 651; *poli*, 650; *vignei*, 652

Pachyuromys duprasi, 313
Paguma lanigera, 475; *larvata*, 475
Palaeotraginae, 576
Paleolaginae, 184
Pan satyrus, 166
Panthera leo, 492; *onca*, 492; *pardus*, 491; *tigris*, 492; *uncia*, 493
Pantholops hodgsoni, 620
Papio anubis, 158; *comatus*, 155, 159; *cynocephalus*, 158; *dogurea*, 158; *hamadryas*, 158
Paradoxurinae, 474
Paradoxurus hermaphroditus, 474; *jerdoni*, 474; *zeylonensis*, 475
Paraechinus micropus, 40
Parascalops breweri, 60
Paratomys brantsii, 377; *littledalei*, 377
Paraxerus cepapi, 223; *emeni*, 223; *flavivittis*, 224; *ochraceus*, 224
Pedetes cafer, 257; *surdaster*, 257
Pedetidae, 257
Pedomys, 307
Pelea capreolus, 610
Pelomys fallax, 321; *frater*, 321
Perameles eremiana, 17; *fasciata*, 17; *myosura*, 17; *nasuta*, 17
Peramelidae, 17
Perissodactyla, 516
Perodicticus potto, 131
Perognathinae, 249
Perognathus baileyi, 249; *fasciatus*, 259; *flavescens*, 249; *flavus*, 250; *formosus*, 250; *hispidus*, 250; *longimembris*, 250; *lordi*, 249; *merriami*, 250; *nelsoni*, 250; *parvus*, 250; *pennicillatus*, 250
Peromyscus, 213; *bairdii*, 266; *banderanus*, 268; *boylii*, 261; *californicus*, 262; *crinitus*, 268; *difficilis*, 268; *eremicus*, 262; *floridanus*, 263; *gossypinus*, 263; *gracilis*, 266; *leucopus*, 264; *maniculatus*, 265; *megalops*, 268; *melanophrys*, 268; *melanotis*, 268; *mexicanus*, 268; *nasutus*, 268; *nudipes*, 269; *nuttalli*, 269; *pectoralis*, 269; *polionotus*, 267; *truei*, 267; *yucatenicus*, 269
Peropteryx, 73
Petaurista magnificus, 242; *petaurista*, 242; *volans*, 242
Petauristinae, 242
Petaurus breviceps, 22; *sciureus*, 22; *s. norfolcensis*, 22
Petrodromus robustus, 45; *rovunae*, 45; *tetradactylus*, 45
Petrogale inornata, 25; *pearsoni*, 25; *penicillata*, 25; *xanthopus*, 25
Petromyidae, 408
Petromys typicus, 408
Petromyscus collinus, 375; *shortridgei*, 375
Phacochoerus aethiopicus, 552
Phaiomys, 307
Phalanger orientalis, 18
Phalangeridae, 18
Phalangerinae, 18
Phaner furcifer, 128
Phascogale apicalis, 12; *flavipes*, 11; *ingrami*, 12; *maculatus*, 12; *tapoatafa*, 12
Phascogalinae, 11
Phascolarctinae, 23
Phascolarctos cinereus, 23
Phascolomidae, 24

Phascolomis hirsutus, 24; *mitchelli*, 24; *ursinus*, 24
Phataginus, 182
Phenacomys intermedius, 296; *longicaudus*, 297; *sylvicola*, 297
Philander opossum, 4
Phloeomyinae, 377
Phoca caspica, 500; *fasciata*, 500; *groenlandica*, 498; *hispida*, 498; *vitulina*, 499
Phocaena phocoena, 417
Phocaenidae, 417
Phocidae, 498
Phocinae, 498
Pholidota, 182
Phyllonycterinae, 91
Phyllonycteris poeyi, 91
Phyllostomatidae, 64, 82
Phyllostomatinae, 83
Phyllostomus discolor, 85; *hastatus*, 85
Phyllotis boliviensis, 272; *darwinii*, 272; *pictus*, 272; *sublimis*, 272
Physeter catodon, 412
Physeteridae, 412
Physeterinae, 412
Pinnipedia, 425
Pipistrellus abramus, 102; *ceylonicus*, 105; *coromandra*, 105; *culex*, 105; *foureri*, 105; *hesperus*, 105; *kuhli*, 105; *leisleri*, 106; *mimus*, 105; *nanus*, 106; *nathusii*, 106; *noctula*, 105; *pipistrellus*, 103; *savii*, 106; *subflavus*, 104
Pitymys duodecimcostatus, 298; *pinetorum*, 297; *quasiter*, 298; *subterraneus*, 298
Pizonyx, 101
Planigale, 12
Platanista gangetica, 410
Platanistidae, 410
Platanistinae, 410
Plecotus auritus, 113; *rafinesquii*, 114
Poecilictis libyca, 464
Poecilogale albinucha, 464
Poelagus, 190
Poephagus, 602
Pogonomys macrourus, 377
Pongidae, 164
Ponginae, 165
Pongo pygmaeus, 165
Potamochoerus larvatus, 537; *porcus*, 537
Potamogale velox, 35
Potamogalidae, 35
Potororoinae, 30
Potorous tridactylus, 32
Potos flavus, 448
Praomys morio, 355; *taitae*, 355; *tulbergi*, 355
Presbytis aygula, 163; *cristatus*, 163; *entellus*, 162; *johni*, 163; *senex*, 163
Primates, 33, 123
Priodontes gigantes, 179
Prionailurus, 490
Prionodon pardicolor, 473
Proboscidea, 510
Procapra gutturosa, 619; *picticaudata*, 620
Procavia capensis, 514; *habessinica*, 514; *johnstoni*, 514; *ruficeps*, 514
Procaviidae, 513
Procyon cancrivorus, 448; *lotor*, 447
Procyonidae, 425, 446
Procyoninae, 446
Proechimys cayennensis, 407
Prometheomys schaposchnikovi, 295
Promops, 122
Pronolagus crassicaudatus, 184; *randensis*, 184
Propithecus verreauxi, 129
Proteles cristatus, 481
Protelinae, 481
Protemnodon agilis, 27; *bicolor*, 27; *cangura*, 27; *dorsalis*, 27; *elegans*, 27; *rufogrisea*, 27
Protoxerus stangeri, 223
Psammomys obesus, 315
Pseudocheirus convolutor, 23; *cooki*, 24; *laniginosus*, 24; *peregrinus*, 24
Pseudois nahoor, 632
Pseudorca crassidens, 416
Pteronotus davyi, 82; *suapurensis*, 82
Pteronura braziliensis, 470
Pteropus alecto, 68; *ariel*, 68; *conspicillatus*, 68; *eotinus*, 68; *geddiei*, 67; *giganteus*, 67; *gouldi*, 68; *melanotus*, 68; *natalis*, 68; *niger*, 69; *ornatus*, 69; *poliocephalus*, 69; *rufus*, 69; *scapulatus*, 69; *subniger*, 69
Pudu, *pudu*, 572
Puma, 489
Punomys lemminus, 296

Rangifer tarandus, 573; *t. arcticus*, 574; *t. caribou*, 574; *t. sibiricus*, 574
Raphiceros campestris, 615; *sharpei*, 615

Rattus alleni, 355; *assimilis*, 323; *bowersi*, 354; *canus*, 354; *carrillus*, 355; *conatus*, 324; *coucha*, 352; *cremoriventer*, 354; *erythroleucus*, 351; *exulans*, 354; *hawaiiensis*, 354; *longicaudatus*, 355; *morio*, 355; *mulleri*, 354; *natalensis*, 353; *norvegicus*, 324; *pellex*, 355; *rajah*, 355; *rattus*, 351; *r. jalorensis*, 352; *r. kandianus*, 352; *sabanus*, 355; *surifer*, 355; *taitae*, 355; *tullbergi*, 355; *whiteheadi*, 355; *woosnami*, 355
Ratufa macroura, 222
Redunca arundinum, 610; *fulvorufula*, 610; *redunca*, 610
Reithrodon typicus, 273
Reithrodontomys creper, 260; *fulvescens*, 260; *gracilis*, 260; *humilis*, 259; *megalotis*, 260; *mexicanus*, 260; *montanus*, 260
Rhabdomys pumilia, 321
Rhachianectes gibbosus, 418
Rhachianectidae, 418
Rhemoys thomasi, 280
Rhinocerotidae, 534
Rhinolophidae, 64, 65, 78
Rhinolophus ferrum-equinum, 78; *geoffroyi*, 80; *hipposideros*, 79; *lepidus*, 80; *luctus*, 80; *megaphyllus*, 80; *rouxi*, 80; *tragatus*, 80
Rhinopoma hardwickei, 72; *kinneari*, 72; *microphyllum*, 72
Rhinopomatidae, 64, 72
Rhinosciurus laticaudatus, 225
Rhinzomyidae, 316
Rhizomys pruinosus, 316; *vestitus*, 316
Rhogeessa parvula, 110; *tumida*, 110
Rhombomys opimus, 315
Rhynchocyon cirnei, 45
Rhynchonycteris naso, 73
Rhynchotragus guentheri, 616; *kirkii*, 616
Rodentia, 213
Romerolagus diazi, 185
Rousettus aegyptiacus, 66; *angolensis*, 66; *leachi*, 66; *leschenaulti*, 66; *seminudus*, 66; *wroughtoni*, 66
Rucervus, 564
Rupicapra rupicapra, 622
Rusa, 565, 566

Saccolaimus, 74
Saccopteryx bilineata, 73; *canina*, 73; *io*, 73; *leptura*, 4; *macrotis*, 73; *plicata*, 73
Saccostomus campestris, 373
Saiga tartarica, 620
Saimiri oerstedi, 137; *sciurea*, 137
Sarcophilus harrisi, 15
Satanellus, 15
Scalopinae, 60
Scalopus aquaticus, 61
Scapanus orarius, 60; *townsendii*, 61
Schoinobates volans, 24
Scirtopoda telum, 383
Sciuridae, 213, 215
Sciurillus pusillus, 220
Sciurinae, 215
Sciurus aberti, 218; *alleni*, 218; *apache*, 219; *aureogaster*, 219; *carolinensis*, 215; *deppei*, 219; *gerrardi*, 219; *griseus*, 219; *hoffmanni*, 219; *negligens*, 219; *niger*, 217; *poliopus*, 219; *variegatoides*, 219; *vulgaris*, 218
Scotinomys teguina, 272
Scotophilus heathi, 111; *murino-flavus*, 111; *nigrita*, 111; *temmincki*, 110
Selenarctos thibetanus, 443
Semnopithecus, 162
Setifer setosus, 34
Setonix brachyurus, 29
Sibbaldus musculus, 422
Sicista betulina, 380; *subtilis*, 380
Sicistinae, 380
Sigmondon bogotensis, 275; *hispidus*, 273; *minimus*, 275; *Ochrognathus*, 275
Sika, 566
Simocyoninae, 442
Sirenia, 515
Sminthopsis crassicaudata, 13
Smutsia, 182
Sorex araneus, 45; *arcticus*, 49; *cinereus*, 49; *dispar*, 49; *fumeus*, 49; *longirostris*, 49; *merriami*, 49; *milleri*, 49; *minutus*, 47; *palustris*, 48; *trowbridgei*, 49; *vagrans*, 48
Soricidae, 33, 45
Soricinae, 45
Spalacidae, 315
Spalax kirgisorum, 315; *microphthalamus*, 316
Spilocuscus nudicaudatus, 22
Spilogale gracilis, 468; *putorius*, 469
Steatomys opimus, 375; *pratensis*, 376

Stenoderminae, 89
Stochomys, 355
Strepsiceros angasi, 579; *imberbis*, 579; *scriptus*, 579; *spekei*, 579; *strepsiceros*, 579
Sturnira lilium, 88; *tildae*, 89
Sturnirinae, 88
Suidae, 537
Suncus lixus, 55; *murinus*, 55; *varilla*, 55
Suricata suricatta, 477
Sus barbatus, 551; *salvanius*, 552; *scrofa*, 537; *vittatus*, 552
Sylvicapra grimmia, 607
Sylvilagus aquaticus, 191; *auduboni*, 191; *bachmani*, 192; *braziliensis*, 192; *cunicularis*, 194; *floridanus*, 193; *nuttalli*, 194; *palustris*, 194; *transitionalis*, 194
Sylvisorex granti, 55; *megalura*, 55; *sorella*, 55
Symphalangus syndactylus, 165
Synaptomys borealis, 289; *cooperi*, 288
Syncerus caffer, 605; *nana*, 605

Tachyglossidae, 1
Tachyglossus aculeata, 1
Tadarida aegyptiaca, 119; *ansorgei*, 120; *braziliensis*, 119; *condylura*, 120; *congica*, 120; *faradjius*, 120; *femorosacca*, 119; *limbata*, 120; *midas*, 120; *molossa*, 120; *occipitalis*, 120; *ochracea*, 120; *pumila*, 120; *russatus*, 120; *teniotis*, 120; *thersites*, 120; *trevori*, 120; *wroughtoni*, 120
Talpa europaea, 57
Talpidae, 33, 56
Talpinae, 57
Tamandua tetradactyla, 177
Tamarin, 140
Tamias striatus, 239
Tamiasciurus douglasii, 221; *fremonti*, 221; *hudsonicus*, 220
Taphozous kachhensis, 74; *longimanus*, 74; *mauritanus*, 74; *melanopogon*, 74; *peli*, 74; *perforatus*, 74; *saccolaimus*, 74
Tapiridae, 533
Tapirus bairdii, 533; *indicus*, 533; *terrestris*, 533
Tarsiidae, 133
Tarsipedinae, 23
Tarsipes spenserae, 23
Tarsius spectrum, 133; *syrichta*, 134
Tatera afra, 310; *bransti*, 311; *dichrura*, 312; *indica*, 312; *lobengulae*, 312; *nigricauda*, 312; *nyasae*, 312; *schinzi*, 312; *valida*, 312; *vicina*, 312
Taterillus emini, 313
Taurotragus derbianus, 580; *oryx*, 580
Taxidea taxus, 467
Tayassu angulatus, 553; *pecari*, 554; *tajacu*, 554
Tayassuidae, 553
Teanopus, 279
Tenrec ecaudatus, 33
Tenrecidae, 33
Tetracerus quadricornis, 581
Thalarctos maritimus, 445
Thallomys damarensis, 323; *namaquensis*, 323; *nigricauda*, 323
Thamnomys surdaster, 319
Theropithecus gelada, 160
Thomomys bottae, 246; *bulbivorus*, 248; *monticola*, 248; *talpoides*, 247; *townsendii*, 248; *umbrinus*, 248
Thos, 435
Thryonomyidae, 408
Thryonomys harrisoni, 408; *swindlerianus*, 408
Thylacininae, 16
Thylacinus cynocephalus, 16
Thylacis macrourus, 18; *obesulus*, 18
Thylacomys leucurus, 17
Thylogale billardierii, 26; *eugenii*, 26; *stigmatica*, 26; *thetis*, 26
Tigris, 492
Tolypeutes conurus, 180; *tricinctus*, 180
Tonatia bidens, 85
Trachops cirrhosus, 86
Tragelaphus, 579
Tragulidae, 557
Tragulus javanicus, 558; *meminna*, 558; *pretullius*, 558; *russeus*, 558; *stanleyanus*, 558
Triaenops afer, 81
Trichechidae, 515
Trichechus manatus, 515
Trichosurus caninus, 21; *vulpecula*, 18
Tubulidentata, 509
Tupaia castanea, 124; *glis*, 123; *javanica*, 124; *minor*, 124; *tana*, 124
Tupaiidae, 123
Tupaiinae, 123

Tursicops truncatus, 415
Tylomys watsoni, 261

Uncia, 493
Urocyon cineroargenteus, 441
Uroderma bilobatum, 89
Urogale everetti, 124
Uromys caudimaculatus, 356
Ursidae, 425, 443
Ursus americanus, 443; *arctos*, 444; *gyas*, 444; *horribilis*, 444; *middendorfi*, 445; *ornatus*, 445

Vampyrum spectrum, 86
Vandeleuria oleracea, 316
Vespertilio murinus, 109
Vespertilionidae, 64, 65, 93
Vespertilioninae, 93
Viverra megaspila, 473; *zibetha*, 473
Viverricula indica, 472; *rasse*, 472
Viverridae, 425, 472
Viverrinae, 472
Vombatus, 24
Vormela peregusna, 460
Vulpes bengalensis, 440; *chama*, 440; *corsac*, 440; *fulva*, 436; *macroura*, 436; *regalis*, 436; *velox*, 440; *vulpes*, 439

Wallabia, 27

Xenarthra, 177
Xerus erythropus, 225; *inaurus*, 225; *rutilis*, 226

Zalophus californianus, 495
Zapodidae, 380
Zapodinae, 381
Zapus hudsonius, 381; *princeps*, 381; *trinotatus*, 381
Ziphiidae, 411
Zorilla striatus, 464
Zygodontomys cherriei, 271

INDEX OF VERNACULAR NAMES

Aard-vark, 509
Addax, 612
Agouti, 402
Anoa, 584
Anteater, giant, 177; two-toed, 177
Antelope, four-horned, 581; roan, 611; sable, 611
Ape, Barbary, 154
Argali, 650
Arkhar, 650
Armadillo, hairy, 179; nine-banded, 181; six-banded, 179
Ass, mulita, 181; domestic, 530
Aye-aye, 129

Babirusa, 553
Baboon, chacma, 155; dogurea, 158; gelada, 160; sacred, 158; yellow, 158
Badger, American, 467; European, 466, ferret, 468; honey, 465
Bandicoot, 17
Banteng, 605
Barasingha, 564
Bat, American fruit, 90; Asiatic pipistrelle, 102; barbastelle, 113; big brown, 106; bulldog, 75; cinnamon, 83; common European, 97; common pipistrelle, 103; epaulette, 70; evening, 109; fish-eating, 101; fruit, 66; gray, 94; greater horseshoe, 78; greater yellow, 111; hoary, 112; Indian fruit, 67; Indian sheath-tailed, 74; Indian vampire, 77; javelin, 85; leaf-nosed, 81, 84; leaf-tipped, 82; lesser horseshoe, 79; lesser yellow, 110; little brown, 96; little canyon, 107; long-eared, 84, 113; long-tongued, 87, long-tongued vampire, 86; lump-nosed, 114; mastiff, 121; mouse-tailed, 72; noctule, 105; pale, 118, particolored, 109; red, 112; red-necked fruit, 69; Rhoads', 93; serotine, 108; sharp-nosed, 73; sheath-tailed, 72; silvery, 102; spotted, 116; vampire, 92; white-eared fruit, 66; white-lined, 73; yellow, 110
Bear, Asiatic black, 443; black, 443; European brown, 444; grizzly, 444; koala, 23; polar, 445; sloth, 446; sun, 446
Beaver, 255, 256; mountain, 214
Bharal, 632
Bighorn, 651
Bilby, 17
Binturong, 475
Bison, American, 605; European, 606
Blackbuck, 617
Blackfish, 416
Blesbok, 612
Bobcat, 488
Bontebok, 612
Boubieda, 313
Brocket, red, 571; wood, 571
Buffalo, African, 665; American (bison), 605; Eurasian, 581; European (wisent), 606
Bushbaby, 131
Bushbok, harnessed, 579

Cacomistle, 446
Camel, bactrian, 556; dromedary, 557; two-humped, 556
Capuchin, weeping, 136
Capybara, 401
Caracal, 490
Caribou, 573
Cat, domestic, 482; European wild, 488; jungle, 490; leopard, 490; ring-tailed, 446
Cattle, domestic, 584; humped, 603
Cavy, domestic, 386; Patagonian hare, 400; spotted, 401

Chamois, 622
Cheetah, African, 493
Chevrotain, 558; water, 557
Chickaree, 221
Chimpanzee, 166
Chinchilla, 403
Chipmunk, 246; eastern, 239; little northern, 240
Chiru, 620
Chital, 560
Civet, 473; African, 473; African palm, 474; golden palm, 475; palm, 474, 475
Coati, 448
Coyote, 434
Coypu, 405
Cuscus, 22; gray, 18

Dassie, rock, 514
Deer, Chinese water, 575; fallow, 559; hog, 561; Japanese, 566; mule, 567; musk, 558; pampas, 572; Pere David's, 567; red, 563; roe, 575; spotted, 560; Virginia, 569; white-tailed, 569
Dhole, 442
Dik-dik, 616
Dingo, 426
Dog, Cape hunting, 442; domestic, 426
Dolphin, Amazon, 411; bottle-nose, 415; common ocean, 415; Ganges, 410; striped, 416
Donkey, 530
Dormouse, 379; African, 379; edible, 378; tree, 379
Dourocouli, 134
Drill, 160
Dromedary, 557
Dugong, 515
Duiker, 607; blue, 607; red, 607

Echnidna, 1
Eland, 580
Elephant, African, 510; Indian, 511
Elephant shrew, 41
Elk, moose, 573; wapiti, 561
Ermine, 449

Fennec, 440
Ferret, 452
Fisher, 462
Flying fox, 67
Fossa, 480
Fox, Arctic, 435; big-eared, 443; English red, 439; gray, 441; Indian, 440; red, 436; silver, 436

Gaur, 603
Gayal, 603
Gazelle, 618; Indian, 618; Persian, 619
Gemsbok, 611
Genet, 472
Gerbil, 287, 309, 310, 311; mouse, 375; pampas, 273; short-eared, 313
Gibbon, 165; black, 165; white-browed, 164
Giraffe, 577
Glider, 22; pigmy, 21; sugar, 22
Glider-possum, 24
Glutton, 465
Gnu, 613
Goat, domestic, 623; Rocky Mountain, 621
Gopher, pocket, 244, 246, 247, 248, 249
Goral, 621
Gorilla, 168
Grison, 464
Ground hog, 226
Guinea pig, 386
Gundi, 409

Hamster, 280; Chinese, 280; golden, 282
Hanglu, 562
Hare, Arctic, 185; black-naped, 190; common, 188; mountain, 189; rock, 184; snowshoe, 186; spring, 257; varying, 189
Hartebeest, 612
Hedgehog, European, 37; long-eared, 40; South African, 36
Hippopotamus, 554; pygmy, 555
Hog, pygmy, 552; river, 537; wart, 552
Horse, 516
Huemul, 572
Hutia, 404
Hyaena, 482; brown, 482; spotted, 481
Hyrax, rock, 513; tree, 513

Ibex, 631
Impala, 617
Indri, 129

Jackal, black-backed, 435; common, 435; side-striped, 435
Jaguar, 490, 492
Jerboa, 382, 383; African, 383
Jird, 313, 314

Kangaroo, gray, 27; muskrat, 32; rat, 30, 31; red, 28; sooty, 29; tree, 30
Kinkajou, 448
Klipspringer, 614
Koala, 23
Kob, 609
Kongoni, 613
Kudu, greater, 579; lesser, 579

Langur, 162
Lechwe, 609
Lelwel, 613
Lemming, 289; bog, 288; collared, 287; mole, 308
Lemur, black, 125; brown, 126; flying, 63; gentle, 126; mongoose, 126; mouse, 127; ring-tailed, 125; woolly, 128
Leopard, 491; snow, 492
Linsang, 473
Lion, 492; mountain, 489
Llama, 555
Loris, slender, 130; slow, 130
Lynx, 490

Macaque, bonnet, 152; crab-eating, 141; Formosan, 140; Japanese, 140; lion-tailed, 153; pig-tailed, 152; rhesus, 143; toque, 153
Man, 168
Manatee, 515
Mandrill, 160
Mangabey, 155; sooty, 155
Margay, 490
Markhor, 630
Marl, 17
Marmoset, 139; silky, 139
Marmot, 226
Marsupial cat, 13; mouse, 11; rat, 12
Marten, beech, 461; European pine, 462; pine, 461; stone, 461
Mierkat, 477, 480
Mink, 457; European, 460
Mole, common American, 61; European, 57; golden, 35, 36; hairy-tailed, 60; rough-haired golden, 36; star-nosed, 62; Townsend's, 61; Western, 60
Mongoose, 476; crab-eating, 478; dwarf, 478; gray, 478; Madagascar, 476; striped, 479; water, 479
Monkey, black howler, 136; black spider, 138; grivet, 161; howler, 135; macaque, 140; pluto, 161; proboscis, 163; red, 162; spider, 137; squirrel, 137; woolly, 138
Moonrat, 36
Moose, 573
Mouflon, 651
Mouse, back-striped, 322; birch, 380; blind, 316; bush, 320; California meadow, 300; cotton, 263; deer, 263, 265; desert, 262; dwarf meadow, 306; fat, 375; field, 298, 299, 304, 317; grasshopper, 270; harvest, 259, 260; house, 357; jerboa pouched, 13; jumping, 381, 383; laboratory, 359; leaf-eared, 272; lemming, 296; little harvest, 317; night, 261; Oregon creeping, 304; Persian house, 357; pine, 297; pinon, 267; pocket, 249; pouched, 373; pygmy, 270; pygmy kangaroo, 251; pygmy rock, 375; Rocky Mountain meadow, 306; spiny, 372; spiny pocket, 254; striped, 321; tree, 297, 316, 374, 377; volcano, 275; white-footed, 261; wild, 357; wood, 264; woodland jumping, 382
Mundarda, 21
Muntjac, 559
Musk ox, 622
Muskrat, 294; kangaroo, 32; round-tailed, 296

Narwhal, 414
Nilgai, 580
Noolbenger, 23
Numbat, 16
Nutria, 405
Nyala, 579

Ocelot, 490
Okapi, 576
Onager, 532
Opossum, 6; mountain, 5; mouse, 5; rat-tailed, 6; Virginia, 6; water, 11; woolly, 4
Orang-utan, 165
Oribi, 615
Oryx, beisa, 611; white, 611
Otter, Canadian, 469; clawless, 470, 471; European, 470; giant, 470; sea, 471
Ounce, 492

Pademelon, 26
Panda, 449; giant, 449

Pangolin, 182
Peccary, 553; collared, 554
Peromyscus, desert, 262
Pig, domestic, 538; wild, 537
Pika, 183
Platypus, 2
Polecat, 452; African, 464; marbled, 460
Porcupine, 385; hairy tree, 386; Old World, 384
Porpoise, 417; Indian, 418
Possum, brush-tail, 18; glider, 24; honey, 23; ring-tailed, 23
Potoroo, 32
Potto, 131
Prairie dog, 228, 229
Pronghorn, 578
Pudu, 572
Puku, 609
Puma, 489

Quagga, 533
Quokka, 29

Rabbit, brush, 192; domestic, 195; Eastern cottontail, 193; highland, 194; jack, 187, 190; marsh, 194; New England cottontail, 194; swamp, 191; volcano, 185
Raccoon, 447; dog, 441
Rasse, 472
Rat, bamboo, 316; black, 351; brown, 324; bush, 322, 406; cane, 271, 408; chinchilla, 407; coffee, 320; cotton, 273; creek, 321; dassie, 408; giant, 374; Indian antelope, 312; Indian mole, 372; jerboa, 319; kangaroo, 251, 252, 253; laboratory, 324; mole, 315, 408; multimammate, 352; Norway, 324; rice, 258; spiny, 407; tree, 323, 356; water, 319, 377; wood, 275, 276, 278, 279, 371
Ratel, 465
Reedbuck, 610
Reindeer, 573
Rhebok, Vaal, 610
Rhinoceros, Asiatic two-horned, 534; black, 535; white, 534
Rodents, 213
Rusa, 565

Sable, Siberian, 463
Saiga, 620
Sambar, 565
Sassaby, 612
Sea elephant, northern, 508; southern, 506
Sea lion, New Zealand, 497; northern, 496; southern, 496
Seal, bearded, 502; Californian fur, 495; Cape fur, 493; Caspian, 500; crab-eating, 502; gray, 501; Greenland, 498; harbor, 499; harp, 498; Hawaiian monk, 505; hooded, 505; leopard, 503; monk, 505; New Zealand fur, 494; Northern fur, 494; ribbon, 500; ringed, 498; Weddell's, 503
Seren, 619
Serow, 621
Serval, 490
Sheep, Barbary, 632; bighorn, 651; domestic, 632; mouflon, 651
Shrew, common American, 49; common European, 45; Dismal Swamp, 52; elephant, 41; forest, 55; greater tree, 123; house, 53; lesser, 45; mole, 51; mouse, 55, 56; Philippine tree, 124; pygmy, 50; short-tailed, 51, 53; smoky, 49; spotted, 56; tree, 124; water, 48, 50
Sifaca, 129
Simang, 165
Skunk, 468; spotted, 468
Sloth, three-toed, 178; two-toed, 178
Spring hare, 257
Springbok, 619
Squirrel, African ground, 225; American red, 220; antelope, 231; bush, 223; Californian gray, 219; Californian ground, 230; Columbian ground, 230; European red, 218; flying, 242, 243; fox, 217; giant, 223; gray, 215; mantled ground, 238; palm, 221; pygmy, 220; rock, 238; thirteen-lined ground, 233
Steinbok, 615
Stoat, 449, 452
Suni, 615
Suslik, 232; Arctic, 237

Tahr, 623
Takin, 622
Tamarin, 140
Tammar, 26
Tapir, Asiatic, 533; Brazilian, 533
Tarsier, 133
Tasmanian devil, 15; wolf, 16

Tayra, 463
Tenrec, 32
Thameng, 564
Tiger, 492
Titi, 137
Tuco-tuco, 407
Tur, 631

Urial, 652

Vicugna, 555
Viscacha, 402; mountain, 403
Vole, 269, 302; Asiatic, 301; bank, 290; Orkney, 306; red-backed, 290; rock, 306; singing, 306; water, 293, 307

Wallaby, black, 27; brush, 27; hare, 25; nail-tailed, 26; rock, 25
Wallaroo, 28
Walrus, 497
Wapiti, 561
Wart hog, 552
Waterbuck, common, 608; defassa, 608
Weasel, Alpine, 460; European, 451; large brown, 451; least, 460; yellow-bellied, 460
Whale, Atlantic right, 423; beaked, 411; blue, 422; bottle-nose, 412; false killer, 416; finner, 420; gray, 418; Greenland, 424; humpback, 421; little piked, 418; Pacific killer, 415; pygmy sperm, 414; sei, 419; southern right, 423; sperm, 412; white, 414
Wildebeest, black, 613; blue, 614
Wisent, 606
Wolf, 434; aard, 481; maned, 442
Wolverine, 465
Wombat, 24
Woodchuck, 226

Yak, 602

Zebra, Burchell's, 532; Grevy's, 533; mountain, 533
Zebu, 603